Instructor's Resource Manual

KATHERINE M. WHATLEY • JUDITH A. BECK

University of North Carolina - Asheville

AND

Instructor Notes on ConcepTest Questions

CORNELIUS BENNHOLD • GERALD FELDMAN

The George Washington University

PHYSICS

PRINCIPLES WITH APPLICATIONS

GIANCOLI

Associate Editor: Christian Botting
Senior Editor: Erik Fahlgren
Editor in Chief, Science: John Challice
Vice President of Production & Manufacturing: David W. Riccardi
Executive Managing Editor: Kathleen Schiaparelli
Assistant Managing Editor: Becca Richter
Production Editor: Rhonda Aversa
Supplement Cover Designer: Elizabeth Wright
Manufacturing Buyer: Ilene Kahn

PEARSON
Prentice Hall

Pearson Prentice Hall
Pearson Education, Inc.
Upper Saddle River, NJ 07458

Printed in the United States of America

10 9

ISBN 0-13-035251-9

Pearson Education Ltd., *London*
Pearson Education Australia Pty. Ltd., *Sydney*
Pearson Education Singapore, Pte. Ltd.
Pearson Education North Asia Ltd., *Hong Kong*
Pearson Education Canada, Inc., *Toronto*
Pearson Educación de Mexico, S.A. de C.V.
Pearson Education—Japan, *Tokyo*
Pearson Education Malaysia, Pte. Ltd.

Instructor's Resource Manual

to accompany *Physics, sixth edition,* by Douglas C. Giancoli

Table of Contents

Introduction to the Instructor's Resource Manual

This Instructor's Resource Manual for *Physics, sixth edition,* by Douglas C. Giancoli, is aimed primarily at instructors who are relatively new to teaching or new to this course. We hope it will be helpful for more experienced instructors, as well. It is based on many years of experience with algebra- and trigonometry-based physics courses. These are ideas and suggestions, not a blueprint for the "right way" to teach. Teaching a course is a cooperative enterprise that actively involves the students and the instructor. We hope you find this manual helpful as you develop your own unique approach to teaching and learning.

Physics for the Non-Major

Most of the students who enroll in an algebra- or trigonometry-based physics course are not planning to be physics majors. Some will be majoring in other sciences (biology, environmental studies, etc.) and some will be non-science majors. These students often bring fear of physics to the course. It is important to address these fears early and to reassure the students that physics is something they can learn. It is also important to discuss why physics is important to them. Physics provides ways of modeling and understanding the behavior of the universe; if we can model behavior, then we can predict it and determine how to influence it. How well we can predict behavior depends on the quality of the model and on how well information can be known. Most of these students will have a Newtonian worldview. Discussions of how well information can be known lead into discussions of the quantum mechanical worldview, which many students find confusing but interesting.

Viewing physics as a way of modeling the universe also gives justification for some of the simplifying assumptions we as instructors take for granted. Why do we neglect air resistance when talking about objects falling near the surface of the Earth? Because it makes the model simpler and easier to understand. We know that in later courses air resistance will be added into the equations to build a more complex model. Did the first automobiles have all the extras that modern cars have? No—at first it was enough to build a car that ran and didn't explode. Later, after the simple model worked in a predictable way, engineers could add additional details (windscreens, windows, streamlining, heaters, fuel injectors), and the car evolved into a very complex machine.

Examples and demonstrations are critical for these students. The more concrete, everyday, visual examples you can provide the better. It is also useful to tailor the examples to the class. If most of the students in your class are interested in the health professions, biological examples that involve the human body are appealing. If the class is full of literature or history majors, bring in examples of the writings of famous physicists or examples of when physics was important in history.

Course Organization and Structure

Physics often will not be the most important course non-majors will take in a given semester, at least from their point of view. It is advisable to provide sufficient structure so that the students know where they are in the course, know what will happen and when, and are encouraged to stay up to date. It is very helpful for students to have a complete syllabus, including homework assignments and test dates, at the beginning of the semester. This helps them plan out their semester and enables them to inform the instructor in advance if there are scheduling conflicts.

Each class has its own rhythm. As the instructor, you can set the rhythm by providing a variety of formats throughout the class to keep students alert and involved. For example, the class could begin with a short quiz over the previous class material, then proceed with a question-and-answer session on the homework, a demonstration illustrating the concepts to be covered that day, a discussion of the concepts, and a series of examples. Conclude with an active learning experience (doing a short experiment or working an example in teams). Students generally like structure and being able to anticipate what will happen next. Do remember to be flexible and look for "teachable moments," such as questions from the class that lead into discussions pertaining to the concepts of the day.

Problem solving can be very difficult for beginning physics students. Many are not familiar with the approaches and techniques of physics, and others feel their math skills are a little rusty. Consider designating one of your office hours per week as a problem session. Schedule it one or two days before homework is due each week, and hold it in a classroom or seminar room instead of your office so that you can circulate among students as they work. This approach allows you to help many students at once, including some who might be reluctant to come individually to your office hours. An added benefit is that students will help each other, and as teachers, we recognize that one of the best ways to learn something is to try to teach it to someone else!

There are a variety of ways to encourage students to keep up with the class. Collecting **homework** every week is quite effective; it gives the instructor immediate feedback about which concepts were not clear to the students, and it provides the students with incentive to practice problem solving. Homework is most effective when at least a portion of it is graded and returned at the next class, or when solutions are discussed as soon as the homework is collected. It is generally not effective to assign massive amounts of homework, collect it, and never give the students feedback on it. The students tend to get discouraged and stop doing homework at all. Posting solutions on the Web, on reserve in the campus library, or on a bulletin board will allow you to provide complete explanations without spending a lot of class time on solutions.

Quizzes are another way to encourage students to stay engaged and caught up. A relatively simple and short quiz given at the beginning of class once or twice a week can ask the students questions about the last class (which encourages them to look over their notes) or questions about the material to be covered that day (which encourages them to read the material before coming to class). Alternatively, use technology and via computer give a quiz that must be completed before class. The in-class quiz can also be used as an attendance check and as an incentive to get to class on time; the Web quiz allows you more class time for discussion and examples.

It is a good idea to give a longer quiz that requires problem solving every few weeks and at least once before the first test. No matter how much you tell students to practice, those who have not taken physics before may not understand the critical difference between understanding a problem solution when the instructor writes it on the board and being able to do it themselves. Bombing the first quiz is a great learning experience and won't affect the overall grade too much in the long run. Bombing the first test is a much more costly learning experience.

Tests are an important component of a physics class. Think about giving three per semester or two per quarter (instead of a single midterm), and a comprehensive final exam. Although it's hard to give up class time for tests, students benefit from frequent evaluation, especially the first semester or quarter when they are learning *how* to study physics in addition to learning the content. Include some conceptual questions on the tests as well as problems, particularly if you have been emphasizing conceptual understanding in class. These can include ranking exercises, qualitative sketches of diagrams or graphs, short explanations, and multiple-choice questions. A test format with about 25 percent conceptual questions and 75 percent problem solving is one that works well. Beginning physics students often worry about memorizing all the formulas they need to solve problems. Consider allowing them to bring an index card of formulas for use on the test. This will enable them to concentrate more on *how* to solve problems and less on memorization. Be very specific about what is allowed on the card, and point out to them that after all the practice they are doing, they should be quite familiar with the formulas and will probably know them anyway.

Consider using a variety of **evaluation methods**. It is best for student learning to have a variety of ways to earn points and not to have a majority of the course grade dependent on the final exam. How the instructor structures the grading for a course depends on the instructor, the number of students, the presence of a lab component, and the institutional culture. It is critical to be consistent, fair, and prompt with feedback. Students need to know at the *beginning* of the course how they will be graded. Several possible grading schemes follow.

Scheme 1: Semester course with lab

Labs 5%; Homework 10%; Quizzes 5%; Tests (3 @ 20%) 60%; Final Exam 20%

Grading scale: A: 92–100, B: 82–90, C: 72–80, D: 62–70, F: 0–60

In this scheme, a gap is left between the letter-grade ranges for the discretion of the instructor. Often, borderline cases are decided primarily by the student's performance on the final exam.

Scheme 2: Semester course with lab

Labs 6%; Homework and Quizzes 15%; Tests (3 @ 18%) 54%; Final Exam 25%

Grading scale: A: 90–100, B: 80–89, C: 70–79, D: 60–69, F: 0–59

Scheme 3: Quarter course with lab

Labs 5%; Homework 10%; Quizzes 5%; Tests (2 @ 25%) 50%; Final Exam 30%

Grading scale: A: 92–100, B: 82–90, C: 72–80, D: 62–70, F: 0–60

In this scheme, a gap is left between the letter-grade ranges for the discretion of the instructor. Often, borderline cases are decided primarily by the student's performance on the final exam.

Scheme 4: Semester or quarter course without lab

Homework 15%; Quizzes 10%; Tests 50% (3 per semester or 2 per quarter); Final Exam 25%

Grading scale: A: 90–100, B: 80–89, C: 70–79, D: 60–69, F: 0–59

Sample schedules

Suggested times to spend on the material in each chapter for both a two-semester and a three-quarter course are given in the following tables. We present schedules for two formats: three 50-minute class meetings per week and two 75-minute class meetings per week. The schedules assume 15 weeks per semester or 10 weeks per quarter, for a total of 30 weeks for the entire course in either case. The order of topics is different from the textbook for the three-quarter course to avoid splitting the electricity and magnetism section between quarters. Note that the schedule includes time for three tests per semester (or two tests per quarter) instead of a single midterm exam. The schedules below assume a separate final exam period at the end of the semester or quarter, during which you will give a comprehensive final.

SCHEDULE FOR A TWO-SEMESTER COURSE: FIRST SEMESTER Mechanics, Thermal Physics		
Topic	**# of 75-min. lectures (2/week)**	**# of 50-min. lectures (3/week)**
Ch. 1	1	1
Ch. 2	1.5	2
Ch. 3	2.5	4
Ch. 4	4	6
Test #1	1	1
Ch. 5	1	2
Ch. 6	3	5
Ch. 7	2	3
Ch. 8	1.5	2
Ch. 9	1.5	2
Test #2	1	1
Ch. 10	1.5	2
Ch. 11	1.5	3
Ch. 12	2	3
Test #3	1	1
Ch. 13	1	2
Ch. 14	1.5	2
Ch. 15	1.5	3
TOTAL	**30**	**45**

SCHEDULE FOR A TWO-SEMESTER COURSE: SECOND SEMESTER Electromagnetism, Light and Optics, Modern Physics		
Topic	**# of 75-min. lectures (2/week)**	**# of 50-min. lectures (3/week)**
Ch. 16	2	3
Ch. 17	2	3
Ch. 18	1.5	2.5
Ch. 19	1.5	2.5
Test #1	1	1
Ch. 20	2	3
Ch. 21	4	6
Test #2	1	1
Ch. 22	1.5	2
Ch. 23	2.5	4
Ch. 24	2	3
Ch. 25	1	2
Test #3	1	1
Ch. 26	1.5	2
Ch. 27	2	3
Ch. 28	1	1
Ch. 29	.5	1
Ch. 30	.5	1
Ch. 31	.5	1
Ch. 29	.5	1
Ch. 33	.5	1
TOTAL	**30**	**45**

SCHEDULE FOR A THREE-QUARTER COURSE: FIRST QUARTER Mechanics		
Topic	**# of 75-min. lect. (2/wk)**	**# of 50-min. lect. (3/wk)**
Ch. 1	1	1
Ch. 2	1.5	2
Ch. 3	2.5	5
Test #1	1	1
Ch. 4	4	6
Ch. 5	1	2
Ch. 6	3	5
Test #2	1	1
Ch. 7	2	3
Ch. 8	1.5	2
Ch. 9	1.5	2
TOTAL	**20**	**30**

SCHEDULE FOR A THREE-QUARTER COURSE: SECOND QUARTER Waves, Electromagnetism		
Topic	**# of 75-min. lect. (2/wk)**	**# of 50-min. lect. (3/wk)**
Ch. 11	2	3
Ch. 12	2	3
Test #1	1	1
Ch. 16	2	3
Ch. 17	2	3
Ch. 18	1.5	3
Ch. 19	1.5	2
Test #2	1	1
Ch. 20	2	3
Ch. 21	3	5
Ch. 22	2	3
TOTAL	**20**	**30**

SCHEDULE FOR A THREE-QUARTER COURSE: THIRD QUARTER Fluids, Thermal Physics, Light and Optics, Modern Physics		
Topic	**# of 75-min. lect. (2/wk)**	**# of 50-min. lect. (3/wk)**
Ch. 10	2	3
Ch. 13	1	2
Ch. 14	1.5	2
Ch. 15	1.5	3
Test #1	1	1
Ch. 23	2	3
Ch. 24	2.5	3
Ch. 25	1	2
Test #2	1	1
Ch. 26	1.5	2
Ch. 27	2	2
Ch. 28	.5	1
Ch. 29	.5	1
Ch. 30	.5	1
Ch. 31	.5	1
Ch. 32	.5	1
Ch. 33	.5	1
TOTAL	**20**	**30**

Sample Syllabus

The following is a syllabus from a first-semester physics class. It contains information about class policies, test dates, grading, a list of topics covered by date and the associated reading assignments, and student strategies for success. Homework assignments with due dates can be included in the syllabus or on a separate handout. If your course has a laboratory component, you will also need to provide students with a lab schedule and guidelines for laboratory write-ups.

PHYSICS 131 SYLLABUS

Instructor: Dr. Katherine Whatley
Office: PH 222, Office Phone: 250-3880, E-mail: whatley@unca.edu
Office hours: MWF: 1:00-2:00 pm, TR 3:00-4:00 pm, and by appointment, or feel free to drop in if my door is open.

Welcome to Introductory Physics I! In this semester, we will discuss motion, energy, momentum, fluids, waves, and thermodynamics. We will cover most of the first 15 chapters (volume 1) of *Physics* by Douglas C. Giancoli. Please read over the policies and guidelines below, and see me if you have any questions or concerns.

Class Attendance: Attendance and participation in classroom activities are extremely important. Bring your calculator to class every day. It is also important to be on time for class. Handouts are distributed and announcements are made at the beginning of class. You are responsible for all information and instructions discussed in class whether or not you were present.

Prerequisite: MATH 164 or equivalent background in algebra, trigonometry, and pre-calculus is required. Do not take PHYS 131 at this time if you do not have this minimum mathematics prerequisite.

Laboratory: Laboratory attendance is required. One absence may be excused, but a second absence will lower your final grade for the course by one full level. Lab manuals should be purchased from the bookstore before the first lab meeting. You must also supply graph paper. Lab reports are to be completed and turned in by the end of the lab period unless otherwise noted.

Homework and Quizzes: Homework is due at the beginning of class on the dates indicated on the assignment sheet. Late homework will not be accepted. Solutions to all homework problems, practice problems, quizzes, and tests will be placed on reserve in the library. Discussion of homework problems is allowed and encouraged; however, copying of homework is not. All work submitted should represent your own best effort. Announced quizzes consisting of problems and conceptual questions will occasionally be given. Reading quizzes, taken on the Web, are due prior to class each Thursday.

Tests: Three tests will be given during the semester on the following dates: Tuesday, February 19; Thursday, March 28; Tuesday, April 25. Each test will include material covered in class up to that point, with emphasis on material covered since the last test. The final exam will be comprehensive and is scheduled for Tuesday, May 14, at 9:25 a.m. Please be sure to bring your calculator to all tests. Calculators may not be shared.

Grade Composition: The final grade will be determined by the following weighting:

Tests (3 @ 18% each): 54%; Final Exam: 25%; Homework & Quizzes: 15%; Lab reports: 6%

Grading Scale: The following scale will be used for letter grades:

A: 90–100 B: 80–89 C: 70–79 D: 60–69 F: less than 60

Academic Dishonesty: Students are expected to perform their own work on all assignments in this course. Dishonesty on an exam, quiz, homework, or lab report will result in a grade of zero for that assignment. Severe cases will result in a failing grade for the course.

Schedule of topics and text assignments: In order to prepare for class, please complete each reading assignment before the class during which the topic is discussed, and take the WebCT reading quiz for each chapter.

Week	Dates	Topics	Reading Assignment
1	Jan. 17	introduction; math review	Ch 1: all
2	Jan. 22, 24	kinematics: speed, velocity, acceleration, free fall	Ch 2: all
3	Jan. 29, 31	2-D motion: vectors, relative velocity, projectile motion	Ch 3: all
4	Feb. 5, 7	more on projectiles; force, inertia, Newton's 1st law	Ch 3: sections 5–7 Ch 4: sections 1–3
5	Feb. 12, 14	Newton's 2nd and 3rd laws; friction	Ch 4: sections 4–9
6	Feb. 19	**Test #1: Chapters 1–4**	
	Feb. 21	gravitation, circular motion	Ch 5: all
7	Feb. 26, 28	energy, work, kinetic energy, potential energy, conservation of energy, power	Ch 6: all
8	Mar. 5, 7	linear momentum and impulse, conservation of momentum, collisions, center of mass;	Ch 7: sections 1–8
9	Mar. 12, 14	**Spring Break!**	
10	Mar. 19, 21	rotational motion, torque	Ch 8: all
11	Mar. 26	rotational motion, continued, equilibrium	Ch 9: sections 1–4
	Mar. 28	**Test #2: Chapters 5–8**	
12	Apr. 2, 4	solids and fluids	Ch 10: sections 1–10
13	Apr. 9, 11	vibrations and waves	Ch 11: all
14	Apr. 16, 18	sound	Ch 12: all
15	Apr. 23	temperature; kinetic theory	Ch 13: sections 1–10
	Apr. 25	**Test #3: Chapters 9–12**	
16	Apr. 30, May 2	heat, specific heat, phase changes, heat transfer	Ch 14: all
17	May 7, 9	thermodynamics	Ch 15: sections 1–9
18	May 14	**Final Exam**	

<u>**Hints for best performance**</u>:

- Prepare for class, read material in the text *before* the lecture, then read the material again after class discussions of the topics.
- Use your resources, including the information and hints on WebCT.
- Don't miss class; get notes from someone if you have an unavoidable absence.
- Study.
- Review and practice math as necessary.
- Participate in class. Bring your calculator every day.
- Practice, practice, practice. Work lots of problems
- Let me know how you're doing!

Specifically related to problem solving and homework:

- Make sure you understand what the problem is asking before you begin plugging numbers into equations.
- Try the worked examples in the book and from your notes—without peeking!
- Try problems that have solutions in the back of the book.
- Draw a picture whenever possible.
- Pay attention to units.
- Don't expect to always work straight through a problem. Wrong turns and dead ends are often instructive.
- Check results to make sure they are reasonable.
- Get help—but only after trying the problem yourself.

<u>**And finally**</u>:

Keep in mind that learning physics involves both understanding concepts and solving problems. The two processes are inseparable and support each other. For instance, you will probably find that you understand a concept somewhat, then work your way through a related problem, which illuminates the concept even more, making the next problem easier to solve, etc. Finally it will all come together and you will understand the concepts and can whiz through the problems!

ENJOY!

Teaching Resources

Supplementary materials

In addition to this *Instructor's Resource Manual*, there are a number of other supplements to *Physics, sixth edition,* by Douglas C. Giancoli available from the publisher. The *Instructor's Solutions Manual* (Vol. I: 0-13-035237-3; Vol. II: 0-13-141545-X), by Bob Davis (Taylor University) and J. Erik Hendrickson (University of Wisconsin-Eau Claire), contains detailed, worked solutions to every problem in the text. The *Test Item File* (0-13-047311-1) contains approximately 2800 multiple-choice, short answer, and true/false questions, many conceptual in nature, referenced to the corresponding text sections. The *Transparency Pack* (0-13-035245-4) has 400 full-color transparencies of images from the text, including many of those mentioned in this manual. Finally, the *Instructor's Resource Center on CD-ROM* (0-13-035246-2) contains all text illustrations and tables, in various formats. The CD set also contains

TestGenerator, an easy to use networkable program for creating tests ranging from short quizzes to long exams, plus additional PowerPoint Lectures, electronic versions of the Instructor's Resource Manual, Instructor's Solutions Manual, and End-of-Chapter Questions and Problems.

Be sure to visit the *Companion Website* (http://physics.prenhall.com/giancolippa), which contains excellent resources for both instructors and students. Students can view simulations of concepts from the text, practice problem-solving skills in self-assessment modules, prepare for the MCAT exam, explore and refine their understanding of physics, and learn to connect physics to the world around them. All the materials on the Website are in addition to and not duplicative of material in the text. The site grades and scores all objective questions, the results of which can be automatically e-mailed directly to a professor or TA if so desired. See the text Preface for more details.

Demonstration resources

What would physics be without demonstrations? They provide excellent illustrations of important concepts, and besides that, they're fun. Even when they don't work (and indeed, sometimes they won't!), something can be learned from them. Demonstrations are most effective when they involve students as much as possible. In addition to having students actually help perform the demonstrations, engage the whole class by asking for predictions or explanations. A few demonstrations are included in each of the chapter sections that follow, but you are also encouraged to check the resources listed below, ask your colleagues for favorites, and experiment with ideas of your own.

Edge, R. D., *String and Sticky Tape Experiments*. College Park, MD: American Association of Physics Teachers, 1981.

Ehrlich, R., *Turning the World Inside Out and 174 Other Simple Physics Demonstrations*. Princeton, NJ: Princeton University Press, 1990.

Ehrlich, R., *Why Toast Lands Jelly-Side Down: Zen and the Art of Physics Demonstrations*. Princeton, NJ: Princeton University Press, 1997.

Freier, G. D. and F. J. Anderson, *A Demonstration Handbook for Physics*. College Park, MD: American Association of Physics Teachers, 1981.

Meiners, H., editor, *Physics Demonstration Experiments*. New York: The Ronald Press Company, 1970.

Interactive teaching resources

There are a number of excellent books available to help instructors create a more active and interactive learning environment for students. A few are mentioned here that include valuable suggestions about teaching practices as well as many specific activities, quizzes, warm-up questions, and exercises. (For more information on these resources, see http://www.prenhall.com/tiponline/.)

Just-In-Time Teaching: Blending Active Learning with Web Technology, by G. Novak, E. Patterson, A. Gavrin and W. Christian (Prentice Hall, 1999), contains a description of the *Just-in-Time Teaching* strategy designed to enhance student learning. The book also contains many content-specific Web assignments students perform outside of class. Chapters 8 and 9 consist of warm-ups and puzzles from mechanics, thermodynamics, electricity and magnetism, and optics. Warm-ups are designed to help pique students' interest and prepare them for class. Puzzles are more complex and are attempted by students after the topic has been covered in the course.

Peer Instruction: A User's Manual, by Eric Mazur (Prentice Hall, 1997), describes an interactive teaching style developed by the author. It includes two nationally recognized evaluation tools, the Force Concept Inventory and the Mechanics Baseline Test, which can be used as pretests and post-tests to evaluate teaching effectiveness and student learning. In addition, it contains reading quizzes, "Conceptests," and conceptual exam questions for all main topics covered in a one-year introductory physics course. All questions are also in ready-to-print format on the enclosed diskette.

Physlets: Teaching Physics with Interactive Curricular Material, by W. Christian and M. Belloni (Prentice Hall, 2001), introduces Java physics applets (physlets) that are designed to deal with individual concepts in physics. Physlets are interactive Web animations and can be used in a variety of applications. This book teaches readers how to author physlets and includes a CD with physlets from all major topics in introductory physics.

Ranking Task Exercises in Physics, by T. O'Kuma, D. Maloney, and C. Hieggelke (Prentice Hall, 2000), is a thorough collection of exercises that require students to rank variations of a particular physical situation on a specified basis. These exercises are quite useful in developing and testing conceptual understanding.

Tutorials in Introductory Physics, by L. McDermott, P. Shaffer, and the Physics Education Group (Prentice Hall, 2002), contains supplementary instructional materials for a standard introductory physics course. The emphasis in the tutorials is on the development of important physical concepts and scientific reasoning skills rather than quantitative problem solving. The tutorials include pretests, worksheets, and homework assignments for topics from mechanics, electricity and magnetism, waves, and optics.

Physics education resources

The area of physics education research has produced lively discussions and resulted in numerous contributions to the literature in recent years. The following articles about teaching physics represent a sample that should prove interesting to both new and veteran instructors.

Appelquist, T. and Shapero, D., "Physics in a New Era," *Physics Today* (November 2001), pp. 34–39. Contains recommendations for strengthening physics in light of the needs of a rapidly changing society.

Bianchini, J., Whitney, D., Breton, T. and Hilton-Brown, B., "Toward Inclusive Science Education: University Scientists' Views of Students, Instructional Practices, and the Nature of Science," *Science Education* (January 2002), pp. 42–78. An excellent article with recommendations on ways to improve undergraduate science education.

Crouch, C., Fagan, J., Callan, J., and Mazur, E., "Classroom Demonstrations: Learning Tools or Entertainment?" *American Journal of Physics* (June 2004), pp. 835–838. A discussion of the benefits of engaging students in classroom demonstrations.

Crouch, C. and Mazur, E., "Peer Instruction: Ten Years of Experience and Results," *American Journal of Physics* (September 2001), pp. 970–977. A description of the Peer Instruction teaching technique and an analysis of its success as compared to traditional lecture methods of teaching introductory physics.

Ehrlich, R., "How Do We Know if We Are Doing a Good Job in Physics Teaching?" *American Journal of Physics* (January 2002), pp. 24–29. An article based on a talk delivered by the author upon receiving the 2001 AAPT Award for Excellence in Undergraduate Teaching.

Fagen, A., Crouch, C. and Mazur, E., "Peer Instruction: Results from a Range of Classrooms," *The Physics Teacher* (April 2002), pp. 206–209. Reports student gains in the Force Concept Inventory score after participation in classes using the Peer Instruction pedagogical method.

Greca, I. and Moreira, M., "Mental, Physical, and Mathematical Models in the Teaching and Learning of Physics," *Science Education* (January 2002), pp. 106–121. An excellent article on the value of mental modeling in the teaching of physics.

Kim, E. and Pak, S.-J., "Students Do Not Overcome Conceptual Difficulties After Solving 1000 Traditional Problems," *American Journal of Physics* (July 2002), pp. 759–765. Investigates the relationship between traditional problem solving and conceptual understanding.

Lindenfeld, P., "Format and Content in Introductory Physics," *American Journal of Physics* (January 2002), pp. 12–13. A thoughtful guest comment on how little introductory physics courses have changed compared to the great changes in the field of physics.

Lopez, R. and Schultz, T., "Two Revolutions in K-8 Science Education," *Physics Today* (September 2001), pp. 44–49. Discusses two major changes taking place in precollege science education and the role of the scientific community in affecting change. Particularly interesting for students who are planning on teaching.

May, D., and Etkina, E., "College Physics Students' Epistemological Self-Reflection and Its Relationship to Conceptual Learning," *American Journal of Physics* (December 2002), pp. 1249–1258. An interesting article on students' self-reflection and its use to improve conceptual learning in physics classes.

Meltzer, D., "The Relationship Between Mathematical Preparation and Conceptual Learning Gains in Physics: A Possible 'Hidden Variable' in Diagnostic Pretest Scores," *American Journal of Physics* (December 2002), pp. 1259–1268. Suggests that students' preinstructional mathematical skills may be a significant factor in learning gains during physics instruction.

Meltzer, D. and Manivannan, K., "Transforming the Lecture-Hall Environment: The Fully Interactive Physics Lecture," *American Journal of Physics* (June 2002), pp. 639–654. Description of the development, application, and evaluation of active-learning techniques to large lecture classes.

Seymour, E., "Tracking the Processes of Change in U.S. Undergraduate Education in Science, Mathematics, Engineering, and Technology," *Science Education* (January 2002), pp. 79–105. A history of efforts to improve science, math, engineering, and technology education.

Steinberg, R. and Donnelly, K., "PER-Based Reform at a Multicultural Institution," *The Physics Teacher* (February 2002), pp. 108–114. Describes reforms made in physics education at the City College of New York, based on physics education research.

Van Domelen, D. and Van Heuvelen, A., "The Effects of a Concept-Construction Lab Course on FCI Performance," *American Journal of Physics* (July 2002), pp. 779–780. An interesting report on the effects of two different lab curricula on student learning in introductory physics.

See also *Physics Education Research: A Supplement to the American Journal of Physics*, Supplement 1 to vol. 68, No. 7 (July 2000). Contains many wonderful articles on physics education.

Physics organizations

The following organizations are concerned about issues in the teaching of physics and have helpful resources available.

American Association of Physics Teachers (AAPT). Publishes the *American Journal of Physics* and *The Physics Teacher*. Sponsors workshops and meetings on teaching innovations. (http://www.aapt.org)

American Institute of Physics (AIP). Publishes *Physics Today.* Excellent educational materials. (http://www.aip.org)

American Physical Society (APS). Publishes the *Physical Review* journals. (http://www.aps.org)

Astronomical Society of the Pacific (ASP). Excellent educational materials related to astronomy. (http://www.aspsky.org)

Materials and equipment

The following companies are good sources of materials and equipment for demonstrations and labs.

Edmund Scientific (http://www.edsci.com). Educational science products can be found at http://www.scientificonline.com.

Fisher Scientific (http://www.fishersci.com/main.jsp). Fisher Science Education (http://www.fisheredu.com).

Frey Scientific (http://www.freyscientific.com).

Klinger Educational Products Corporation (http://www.KlingerEducational.com).

Ohaus Corporation (http://www.ohaus.com).

PASCO (http://www.pasco.com).

Sargent-Welch (http://www.sargentwelch.com). Also contains Cenco Physics product listings.

Organization of the IRM

The rest of this manual is organized by textbook chapters. In each section you will find a chapter outline, summary, list of major concepts, and teaching suggestions and demonstrations. At the end of each section the transparencies available for the chapter are listed. In addition, there are readings pertaining to the chapter content and materials that may be useful for demonstrations. Each section concludes with space for your notes.

A Final Word

We find the teaching of introductory physics extremely stimulating and rewarding and hope you do, as well. One of our goals is to change—and indeed enrich—the way our students look at the world. After taking physics, one has a new perspective on the world, whether admiring a rainbow or riding a bicycle. Although the specifics of formulas may eventually be forgotten, we hope that our students retain a sense of the relationships in physics and an awe for the workings of the universe based on knowledge and understanding. Students can always look up equations, but the skills to think critically, solve problems, and evaluate issues from an educated perspective will serve them well no matter what their future endeavors may be.

We would like to thank Christian Botting of Prentice Hall for his guidance and support during this project. A special thank you goes to Jac, Katie, and Michael Whatley and to Jerome, Caitlin, and Duncan Hay for their encouragement and understanding throughout the writing process. We are indebted to our mentors, colleagues, and students who inspire our love of physics and our teaching pursuits.

Katherine M. Whatley
Judith A. Beck
Department of Physics
The University of North Carolina at Asheville
whatley@unca.edu
jbeck@unca.edu
May 2004

Chapter 1: Introduction, Measurement, Estimating

Outline

1-1 The Nature of Science
1-2 Physics and its Relation to Other Fields
1-3 Models, Theories, and Laws
1-4 Measurement and Uncertainty; Significant Figures
1-5 Units, Standards, and the SI System
1-6 Converting Units
1-7 Order of Magnitude: Rapid Estimating
*1-8 Dimensions and Dimensional Analysis

Summary

Chapter 1 presents the general definitions of *science* and *physics* and some of the conceptual tools needed to begin their study. The chapter includes discussion of units and conversion factors, dimensional analysis, significant figures, and scientific notation. In addition, the usefulness of order-of-magnitude calculations in physics is addressed.

Major Concepts

By the end of the chapter, students should understand each of the following and be able to demonstrate their understanding in problem applications as well as in conceptual situations.

- General definitions of science and physics
- Significant figures
    - Addition and multiplication
    - Scientific notation
- Systems of units
    - Length
    - Mass
    - Time
- Unit conversions
- Estimating (order of magnitude calculations)
- Dimensional analysis

Teaching Suggestions and Demonstrations

Many of the students who take a college physics course will have seen most of the material presented in this chapter in previous courses. The general tendency is to skip over this chapter or have the students read it and not discuss it in class. We don't recommend this. While many students may have seen the material, most will not have internalized it. It is a good idea to spend a little time introducing these concepts. Talk about them again each time you work an example in class. Even good students seem to have trouble remembering to use significant figures correctly.

Sections 1-1 through 1-3

Begin by **justifying the study of science in general and physics in particular**. Many of the students will be in this course because it satisfies a requirement. You need to convince them that it will also be interesting (and fun!). Physics provides us with ways to model the universe. Models are useful in the understanding of phenomena and also in the prediction of future behavior. (You might point out how physics is different from what psychics claim to do!)

Section 1-4

Most students will also have heard of the concept of **significant figures** but will not have any idea of its importance or know the rules on how to treat them. It is a good idea to go over the rules and to discuss how they follow common sense. For example, consider a highway engineer who has a 100-ft tape measure. The tape is cut accidentally. When the engineer splices it back together, she loses about half an inch near the beginning of the tape. Does that half inch matter? It depends. If she is using the tape to measure flower beds for a highway beautification program, then no, the lost half inch doesn't matter. If she is using the tape to measure replacement glass for a garage window, the half inch matters a great deal!

You will also need to talk to students about significant figures and calculators. Just because nine digits appear on the calculator display does *not* mean there are nine significant figures!

➲ **DEMO 1-1** The 2000 U.S. Presidential election is a good source for discussion material on **significant figures**, especially if you have political science or history majors in your class. Choose any state where the vote count was close (Florida, New Mexico, and Wisconsin are examples) and discuss the vote difference in terms of significant figures. (You can begin talking about error analysis with this example, too.)

It is sometimes easier to discuss significant figures in the context of **scientific notation**. The number of significant figures is often more clear when a number is written in scientific notation. Some students may not be comfortable with exponents; you will need to make sure that the class understands that $10 \cdot 10 = 10^2$, etc.

Section 1-5

Most students are familiar with **definitions of units** and **systems of units**. They will be comfortable with British and metric units of length and time. Plan to spend a little time on the idea and unit of mass; most students will not have heard of the British unit (slug). This is a good time to emphasize the importance of units. (See Table 1-4 for a list of common prefixes.)

➲ **DEMO 1-2** Measure a convenient object (a table is often available and works well). Announce to the class that its length is "two" (or something close to its length in meters). Ask the class if this information conveys anything useful. Usually someone will say "Two what?," which gives you the perfect opening to talk about the importance of **units**.

In September 1999, a probe launched by NASA was lost in the atmosphere of Mars because the engineers who built the engines were working in British units and the scientists who were controlling the engines were working in SI units. (See Resource Information for reference.) This makes a good story and (again) lets you emphasize that units are part of every result.

Sections 1-6 and 1-7

Students generally catch on to basic **unit conversion** rather quickly, but they have trouble with converting compound units (miles/hour to m/s, for example) and powers of units (cm^2 to m^2). Go over these conversions now, and be prepared to go over them often.

➲ **DEMO 1-3** Make a cube measuring 10 cm on a side. Cardboard or Styrofoam work well and are light. Mark off the centimeter lines. You can use this to illustrate the relationship between 1 cm and 1 m, between 1 cm^2 and 1 m^2, and between 1 cm^3 and 1 m^3. This is particularly effective if you also have a cube that is 1 m on a side. (A 1-m cube is much harder to store.)

Emphasize the usefulness of **order-of-magnitude calculations**. You can estimate the amount of paint you need to paint a room, the amount of water needed to wash a car, the amount of food needed for a cookout – anything that catches the students' attention. Then discuss how order-of-magnitude calculations can help check for "reasonableness" of problem solutions.

➲ **DEMO 1-4** You can do a quick estimate of the number of times your heart beats in a year. Take your pulse for 6 seconds. Multiply by 10 and round off to a reasonable number. That gives the number of times your heart beats per minute. Then use rounded-off unit conversions to estimate the number of beats in a year. (Number of beats per minute times 60 minutes/hour times 20 hours/day times 400 days/year.)

Section 1-8

Dimensional analysis will likely be a new concept for most students. Its usefulness as a problem-solving strategy will not be immediately obvious. Even though this section is optional, we recommend that you take time to discuss this section in class and continue to point out the consistency of units every time a new formula is introduced.

Resource Information

Transparencies

T1.	Table 1-1	Some Typical Lengths or Distances (order of magnitude)
T2.	Table 1-2	Some Typical Time Intervals
	Table 1-3	Some Masses
T3.	Table 1-4	Metric (SI) Prefixes
T4.	Figure 1-12	Example 1-9 (Height by triangulation)

Suggested Readings

Allie, S., Buffler, A., Campbell, B., Lubben, F., Evangelinos, D., Psillos, D., and Valassiades, O., "Teaching Measurement in the Introductory Physics Laboratory," *The Physics Teacher* (October 2003), pp. 394–401. An interesting presentation of a first measurement lab.

Bergquist, J., Jefferts, S. and Wineland, D., "Time Measurement at the Millennium," *Physics Today* (March 2001), pp. 37–42. One of two articles celebrating the centennial of the National Institute of Standards and Technology (NIST).

Frisch, H., "No Kid Should Know That Much About . . .," *Physics Today* (October 1999), pp. 71–72. An opinion piece on science education. Mentions several great demonstrations.

Goodwin, I., "Washington Briefings: One Too Many Mishaps on Voyages to Mars," *Physics Today* (January 2000), p. 47. A short summary of the problems with recent NASA Mars missions.

Haseltine, E., "The Greatest Unanswered Questions of Physics," *Discover* (February 2002), pp. 36–42. An interesting general article on current issues in physics.

Hillger, D., "Metric Units and Postage Stamps," *The Physics Teacher* (November 1999), pp. 507–510. Great article on SI units.

Keeports, D., "Addressing Physical Intuition—A First-Day Event," *The Physics Teacher* (May 2000), pp. 318–319. An example of what to do on the first day of class.

Mohr, P. and Taylor, B., "Adjusting the Values of the Fundamental Constants," *Physics Today* (March 2001), pp. 29–34. One of two articles celebrating the centennial of the National Institute of Standards and Technology (NIST).

Romer, R., "Units—SI-Only, or Multicultural Diversity?," *American Journal of Physics* (January 1999), pp. 13–16. An editorial discussing choice of units in physics. See also numerous letters to the editor in response in the June 1999 issue.

Wheeler, D. and Mazur, E., "The Great Thermometer Challenge," *The Physics Teacher* (April 2000), p. 235. An interesting activity that emphasizes the importance of critical thinking.

Notes and Ideas

*Class time spent on material: Estimated:*______ *Actual:*_____________

Related laboratory activities:

Demonstration materials:

Notes for next time:

Chapter 2: Describing Motion: Kinematics in One Dimension

Outline

2-1 Reference Frames and Displacement
2-2 Average Velocity
2-3 Instantaneous Velocity
2-4 Acceleration
2-5 Motion at Constant Acceleration
2-6 Solving Problems
2-7 Falling Objects
*2-8 Graphical Analysis of Linear Motion

Summary

After the groundwork is laid in Chapter 1, Chapter 2 begins the real study of physics. One-dimensional kinematics is the study of motion in a straight line without regard to its causes. Many of the concepts in this chapter, such as velocity and acceleration, are familiar to students from everyday experiences, like driving a car. However, most students will not know the physics definitions of these same terms or how they relate to one another. In addition, common misconceptions due to the presence of air resistance and friction in the real world need to be addressed. Chapter 2 lays the foundation for the treatment of two-dimensional motion addressed in later chapters.

Major Concepts

By the end of the chapter, students should understand each of the following and be able to demonstrate their understanding in problem applications as well as in conceptual situations.

- Reference frames
- Position, distance, and displacement
- Speed and velocity
  - Average
  - Instantaneous
  - Constant
- Acceleration
  - Average
  - Instantaneous
  - Constant
- Equations of motion with constant acceleration
- Free fall
- Graphs of position versus time, velocity versus time, and acceleration versus time

Teaching Suggestions and Demonstrations

Since problem solving is a large component of most physics courses, it is a good idea to emphasize it from the beginning. A short quiz at the end of this chapter can help give students practice solving problems on their own.

Sections 2-1 through 2-5

Begin by discussing the concept of a reference frame and typical assumptions used when describing motion. For example, we describe everyday motion as relative to the Earth. Beginning physics students already know the difference between distance and speed. However, you will need to help them differentiate between **distance** and **displacement** and between **speed** and **velocity**. Also discuss constant, average, and instantaneous speed and velocity.

The **units of acceleration** are initially puzzling. Have students calculate a common acceleration, such as the acceleration of a car on the entrance ramp to the highway. They can initially use units that are intuitively sensible, such as miles per hour per second, and then convert to the more conventional m/s^2 to get a feel for the units they will be using in physics class. Conceptual Example 2-4 is an excellent question for students to think through.

Students often have trouble understanding the important differences between **constant velocity** and **constant acceleration**. A helpful approach is to directly compare these two cases. Imagine, for instance, a cart with a spark timer that leaves marks on a straight track at one-second intervals. Students can determine distance traveled by the cart during each second when it is moving with a constant velocity and then again when it is moving with a constant acceleration. When the distances are plotted on a one-dimensional line, it is apparent that the cart covers equal distances in each one-second interval for the constant velocity case but increasingly larger distances in each successive second for the constant acceleration case.

Throughout this chapter, the significance of **positive and negative signs** needs to be emphasized. The meanings of positive and negative displacement and velocity are fairly straightforward. Acceleration is more confusing. Some numerical examples can help convince students that a negative acceleration does not necessarily mean an object is slowing down. (An object slowing down while traveling in the positive direction and an object speeding up while traveling in the negative direction both have negative accelerations.)

A solid understanding of the **equations of motion for constant acceleration** (Equations 2-11a, b, c, and d) provides students with an excellent base upon which to build in Chapter 3, when two-dimensional motion is considered. Initially, students may need guidance in using the equations and in recognizing which terms are constants and which are variables. The importance of consistency with sign conventions also needs to be emphasized, as can be demonstrated by comparing a problem in which an object is thrown straight up from a bridge to one in which an object is thrown straight down at the same initial speed from the same bridge. Although the speeds are the same, the velocities have different signs and so are not the same, resulting in very different outcomes. Explicitly choosing a positive direction at the outset of each problem reminds students of this important distinction.

Section 2-6

This section contains a very helpful discussion of problem solving in physics. You can talk about this in general, but students understand problem-solving techniques best in the context of solving a problem. Students generally like a little routine and a little predictability in class. Write down a few key rules (keep them short) and repeat them mantra-like every time you solve a problem. (What is given? Draw a picture. Think. What are the equations? Solve. Check your answer. UNITS!) Soon the students will repeat them as well, both in class as you are working examples, and out of class when they are working on their own. Go through Example 2-8, Estimate Air Bags, with the class; it is a good example of physics in the real world, and it might encourage the students to wear their seat belts!

Section 2-7

Free fall provides a wonderful opportunity to emphasize major points regarding position, velocity, and acceleration for the case of constant acceleration. Although most students understand intuitively that the velocity of a ball thrown up in the air is zero at the top of its trajectory, many assume incorrectly that the acceleration there is zero as well.

➲ **DEMO 2-1** The simple demonstration of dropping objects simultaneously from the same height helps convince students that **acceleration due to gravity** really is the same for all objects. Choosing a wide variety of objects can also help them get a feel for the conditions under which air resistance can and cannot be ignored.

➲ **DEMO 2-2** One very simple but effective demonstration is to drop a quarter or a ball and a piece of notebook paper simultaneously from the same height, as illustrated in Figure 2-19. First give the class a chance to guess what they think will happen. They have just learned that the acceleration due to gravity is the same for all objects, but they will usually revert to their experience and guess that the quarter will fall first. Drop the quarter and paper. (Be dramatic – climb up on a stool or table before dropping the objects to get a greater distance of fall.) Discuss with the class why the quarter fell first. Someone will mention air resistance. Discuss ways in which air resistance could be taken out of the problem. Crush the paper into a ball and again poll the class about which object will hit the ground first. This time about half the class will guess that the two objects will fall at the same rate. Drop the quarter and paper and discuss the results. Students respond well to this demonstration; it gives them dramatic evidence of the role of air resistance in free fall and information about when it can be ignored.

➲ **DEMO 2-3** Another very simple demonstration involves filling a Styrofoam (or paper) cup with water and punching a hole with a pencil in the side near the bottom. (It is best to do this over a sink or a trash can, since it can be messy.) When the cup is held stationary, water shoots out of the hole. (You can hold a nice discussion about the shape of the curve the water makes if the cup is large enough to hold sufficient water.) When the cup is dropped, water will not come out of the hole, since the cup and the water are falling at the same rate. Be sure to ask the class what they think will happen before dropping the cup; often no one will guess that no water will emerge from the hole.

➲ **DEMO 2-4** Students get a kick out of measuring their own reaction time and comparing it to others. Have students work in pairs with one holding a ruler vertically so that the zero mark is level with the second student's fingertips. The first student lets go, and the second catches the ruler. The time it took to catch the ruler is calculated using g and the distance the ruler fell read directly from where the catcher's fingers snagged the ruler. As an added bonus, make a bar graph of the reaction times from the whole class and use the graph to introduce bell curves and standard deviations.

➲ **DEMO 2-5** A free-fall apparatus with spark timer can demonstrate change in position for objects undergoing acceleration due to gravity. The sparks mark the paper at equal time intervals, therefore the distance between adjacent sparks increases as the plummet accelerates. This demonstration is useful for helping students visualize constant acceleration and can also be used to calculate the acceleration due to gravity. (See Resource Information.)

Graphs of y versus t, v versus t, and a versus t are very useful in finding and correcting student misconceptions about acceleration for objects in free fall. Carefully go through Example 2-12, Conceptual Example 2-13, Example 2-14, and Example 2-15 and the graphs in Figure 2-23 for a ball projected straight upward. Emphasize the fact that acceleration is a rate of change. Even though the object moves first up and then down, the rate of change of its velocity is a constant $-9.81\ \text{m/s}^2$.

Section 2-8

Graphs are useful, but can often be difficult for beginning physics students to interpret. The material included in this optional section is valuable for all students; we recommend covering the concepts. Relating the three graphs of position, velocity, and acceleration versus time to each other and to the description of the motion requires practice and is a very valuable exercise, whether used quantitatively or conceptually.

➲ **DEMO 2-6** A motion sensor with computer interface and software to plot the graphs of actual moving objects is an excellent demonstration tool for graphical understanding and interpretation. (See Resource Information.) One approach is to have students sketch their predictions for a certain motion and then receive immediate feedback from the computer. Alternatively, students can be shown a graph on the computer and try to move an object or themselves in front of the motion sensor to duplicate it.

Resource Information

Transparencies

T5.	Figure 2-9	Velocity of a car as a function of time
T6.	Figure 2-10	Example 2-3 (Average acceleration.)
	Figure 2-11	Example 2-5 (Car slowing down.)
T7.	Equations 2-11a-d	Kinematic equations for constant acceleration
T8.	Figure 2-15	Example 2-9 (Braking distances)
T9.	Figure 2-21	Example 2-10 (Falling from a tower)
T10.	Figure 2-22	An object thrown upward into the air
T11.	Figure 2-23	Graphs of (a) y vs. t, (b) v vs. t for a ball thrown upward
T12.	Figure 2-24	Graph of position vs. time for an object moving at a uniform velocity
T13.	Figure 2-25	(a) Velocity vs. time and (b) displacement vs. time
T14.	Figure 2-26	Determining the displacement from the graph of v vs. t

Suggested Readings

Andereck, B., "Measurement of Air Resistance on an Air Track," *American Journal of Physics* (June 1999), pp. 528–533. Excellent experiment for a more in-depth examination of air resistance.

Conderle, L., "Extending the Analysis of One-Dimensional Motion," *The Physics Teacher* (November 1999), pp. 486–489. Thorough treatment of displacement, velocity, and acceleration-versus-time graphs for one-dimensional motion.

Levi, B., "Atom Interferometer Measures *g* with Same Accuracy as Optical Devices," *Physics Today* (November 1999), p. 20. A description of an extremely accurate measurement of *g*.

McClelland, J., "g-whizz," *The Physics Teacher* (March 2000), p. 150. A discussion of the constant *g*.

Moreland, P., "Improving Precision and Accuracy in the *g* Lab," *The Physics Teacher* (September 2000), pp. 367–369. Discussion of a lab to determine acceleration due to gravity.

Patterson, J., "Physical Principles versus Mathematical Rigor," *The Physics Teacher* (April 2000), p. 214. Using a 1-D kinematics problem to illustrate how physics principles can simplify calculations.

Rist, C., "The Physics of . . . Baseballs: Foul Ball?" *Discover* (May 2001), pp. 26–27. An interesting article on how the construction of a baseball affects its performance.

Singh, K., "The Flight of the Bagel," *The Physics Teacher* (October 2000), pp. 432–433. Using a bagel and camcorder to determine the value of *g*.

Wick, K. and Ruddick, K., "An Accurate Measurement of *g* Using Falling Balls," *American Journal of Physics* (November 1999), pp. 962–965. A description of an experiment to determine the acceleration due to gravity, *g*, with an accuracy of about 1 part in 10^4.

Materials

The Pasco *Science Workshop* is a good system for use in computer-based labs. Many bundles that include the Motion Sensor II (CI-6742) are available.

One free-fall apparatus is the Sargent-Welch/Cenco/Physics Behr Free-Fall Apparatus, model number CP00749-05. There are several required accessories: a six-volt ac/dc power supply (CP33031-00), a spark timer (CP31755-01) and wax-coated recording tape (WLS-65250-C).

Notes and Ideas

*Class time spent on material: Estimated:*______ *Actual:*____________

Related laboratory activities:

Demonstration materials:

Notes for next time:

Chapter 3: Kinematics in Two Dimensions; Vectors

Outline

Summary

This chapter moves the concepts of position, displacement, velocity, and acceleration into the two-dimensional world. In two dimensions, direction can no longer be indicated simply by positive and negative signs. Vectors and vector manipulations, important for the remainder of the course, are introduced. Projectile motion, or motion of an object under the influence of gravity only, is treated thoroughly.

Major Concepts

By the end of the chapter, students should understand each of the following and be able to demonstrate their understanding in problem applications as well as in conceptual situations.

- Scalars (magnitude only)
- Vectors (magnitude and direction)
  - Components
  - Addition and subtraction
- Unit vectors
- Vector position, displacement, velocity, and acceleration
- Motion in two dimensions
  - Components of velocity and acceleration
  - Equations of motion for constant acceleration and constant velocity
- Projectile Motion
  - Acceleration due to gravity: g
  - Independence of horizontal and vertical motions
  - Air resistance
  - Basic equations
    - Special case: zero launch angle
    - General case
  - Characteristics of projectile motion
- Relative motion

Teaching Suggestions and Demonstrations

Understanding the difference between scalar and vector quantities is key to understanding many of the concepts in an introductory physics course. Vector manipulation is also an important skill for the students to master. Many of the students will not have seen vectors before. Spending class time going over scalars, vectors, and vector manipulations *now* will pay off later in the course.

The two-dimensional equations of motion may at first appear intimidating to students. Point out that they are really nothing more than applications of equations already introduced in the previous chapters. Encourage students to work many problems in this chapter to become comfortable with the variables and the initial conditions. Also remind them to use sketches to illustrate the problems for clarity.

Sections 3-1 through 3-4

Students usually understand the concept of **scalar** quantities. Ask the students to give examples of scalars, but be prepared to supply a few yourself (e.g., mass, time, speed).

The concept of **vector** quantities is more problematic. Begin with the definition of a vector quantity (magnitude and direction) and decide how you will indicate these quantities for class. Books usually use boldface type, but it is difficult to do boldface on the board or on an overhead. An arrow over the vector quantity works well.

Position and displacement vectors are the simplest to describe. It is relatively easy for students to see that a position vector has both a magnitude and a direction.

➲ **DEMO 3-1** If you are in a room with a tile floor, you can use the tile as a grid and have students stand at the "head" and "tail" of a position vector. (If there is no tile pattern in your room, you can copy a piece of graph paper onto a transparency and use an overhead projector to show a grid.) Have a student walk through several displacements and see how the position vector, measured from the origin, changes. This is also a good time to define the negative of a vector, a concept needed for vector subtraction.

Before tackling vector **components**, you will need to review the definitions of the sine, cosine, and tangent of an angle, the relationship between degrees and radians, and the Pythagorean theorem. There is usually a wide range of understanding of these concepts among students enrolled in introductory physics classes.

As you define the **scalar components of a vector**, be sure the students understand that the components are scalars and can be positive or negative. Demonstrate that the trigonometric function associated with a particular component of a vector depends on the *angle* chosen. (The x component is not always associated with the cosine function, and the y component is not always associated with the sine function. Many students have difficulty with this idea.)

Begin treatment of the rules for **vector addition and subtraction** with the graphical method. You can use the tile floor or the graph paper on an overhead projector mentioned previously. When the students understand that vector addition is different from scalar addition, move on to the component method of vector addition and subtraction. (See Figures 3-3, 3-4, 3-5, and 3-14.)

It's nice to remind students that they can use the graphical method to check their work with addition and subtraction of vectors by the component method. They can do a quick sketch of a problem and estimate the answer then check to make sure their careful calculation makes sense. The Problem Solving box on Adding Vectors in Section 3-4 is a great summary.

Sections 3-5 and 3-6

Now that the students are familiar with vectors in general and with position and displacement vectors, introduce **velocity and acceleration vectors**. Point out the vector nature of both quantities.

➲ **DEMO 3-2** Have the students think about velocity and acceleration in a car. If they travel around a curve at constant speed, does the velocity change? (Yes, because the direction of the velocity vector changes.) How many "accelerators" are there in a car? (Three—the accelerator pedal and the brake, which both change the speed and therefore change the velocity, and the steering wheel, which changes the direction of the velocity vector.)

To derive the **equations of motion in two dimensions,** listed in Table 3-1, students must first understand that the horizontal and vertical motions are independent of each other. Acceleration in the x direction will not affect velocity in the y direction and vice versa. Once this concept is addressed, begin with the equations for one dimension and replace v, v_o, and a with v_x, v_{ox}, and a_x, pointing out that only the x components of the velocity and acceleration are relevant to equations describing motion in the x direction. The equations for the y direction are identical, except that all x's are replaced with v's. After some examples, students should notice the importance of **time** in the equations. It is the only variable that does not have components and is also the only one shared by both x and y directions. Work several examples to illustrate both constant velocity and constant acceleration in two dimensions.

A **projectile** is any object that, after being launched into motion, is under the influence of gravity only. This is a broader definition than the common usage of the word, which conjures up images of cannonballs or food fights, so plenty of examples are helpful. A thrown baseball and a dropped rock are projectiles; a rocket firing its engines is not. As with free-fall motion (see Chapter 2), we assume that air resistance is negligible. In addition, the rotation and curvature of the Earth are ignored. Discuss the conditions under which these assumptions are valid.

The **basic equations of projectile motion,** listed in Table 3-2, are obtained directly from the two-dimensional equations of motion for constant acceleration with $a_x = 0$ and $a_y = -g$. The equations describing motion in the x direction simplify to their constant velocity form. Also, the motion in the y direction is independent of the x velocity. For example, with the assumptions given above, a ball dropped and a ball thrown horizontally from the same height over level ground will hit the ground at the same moment. Go over the Problem Solving box in Section 3-6 and work several examples to illustrate the method.

➲ **DEMO 3-3** To demonstrate the **independence of the x and y components of the velocity of a projectile**, have one student slide a quarter off a level table so that it leaves the table with an initial velocity only in the x direction. A second student holds a second quarter at the edge of the table and, with it at eye level, attempts to drop it as the first quarter leaves the table. Students can predict which will hit the ground first and listen for the results. It should be apparent that they hit the ground at the same time. (If not, discuss reaction time and the difficulty of dropping the second quarter at exactly the right moment.) Since each has the same y component of initial velocity, namely zero, they travel the vertical distance to the floor in the same amount of time. However, they do not land at the same place. Since their initial velocities have different x components, their x displacements are not identical.

➲ **DEMO 3-4** If you have or can construct the apparatus, the classic *Monkey and Hunter* (or *Shoot the Target*) demonstration is an excellent way to show the independence of the horizontal and vertical motions for projectiles as well as the fact that g is constant. (See Resource Information.) A hunter wants to shoot a monkey and isn't sure where to aim. He knows that as soon as the gun is fired, the monkey will be startled by the noise and let go of his branch and drop. So should the hunter fire directly at, above, or below the monkey? Most students will want to fire below, since they know that the monkey will fall while the bullet is on the way. However, the bullet is accelerated downward at the same rate, so in fact the hunter needs to aim directly at the monkey. (See Conceptual Example 3-7.)

Time is again important in the projectile motion equations. If a problem gives information about one direction, for example the initial height of a projectile, and asks for information about the other direction, for example the horizontal range, students must use the time variable to connect the two sets of equations.

Example 3-8 involves the **range** equation for the specific case of a projectile starting and stopping at the same vertical height. Before working this example, have students predict the angle of the initial velocity that will maximize the horizontal range of a projectile. Most will correctly predict 45°. See Figure 3-25 to compare the paths of projectiles with different launch angles.

➲ **DEMO 3-5** To test that the **maximum range of a projectile** is obtained with an initial velocity at an angle of 45°, try projecting a ball at various angles with a compressed spring. It is important to compress the spring the same amount each time, so that the initial speed is constant and the only quantity varying is the angle. (See Resource Information or build your own launcher!) If air resistance is indeed negligible, students will also discover that angles less than 45° and angles more than 45° by the same amount (such as 30° and 60°) will result in equal ranges.

The concept of **symmetry in projectile motion** is extremely useful, and students can be encouraged to use it to their advantage. For example, a projectile launched over level ground will spend as much time on the trip up as on the trip down. If the projectile is launched from a cliff, the time it takes to reach its maximum height will be the same as the time it takes to fall from that height back to the level of the top of the cliff. Also, the projectile will be traveling the same speed when it is back to the same level, although the y component of its velocity will now be down instead of up.

Sections 3-7 and 3-8

These last two sections are optional, and both can be skipped without loss of continuity. Section 3-7 derives y as a function of x, demonstrating that the motion of the projectile is indeed parabolic. Section 3-8 is a treatment of **relative motion,** involving the addition of velocity vectors. If you choose to cover this section, it will be the first time the students have tried to add and subtract vectors other than position vectors. Assure them that the rules for vector addition and subtraction are the same no matter what kind of vector is under consideration.

Resource Information

Transparencies

T15.	Figure 3-2	Combining vectors in one dimension
	Figure 3-3	A person walks 10.0 km east and then 5.0 km north
T16.	Figure 3-5	The resultant of three vectors
T17.	Figure 3-6	Vector addition by two different methods
T18.	Figure 3-8	Subtracting two vectors

T19.	Figure 3-12	Finding the components of a vector using trigonometric functions
T20.	Figure 3-13	*x* and *y* components of a vector
T21.	Figure 3-14	Components of a resultant vector
T22.	Figure 3-18	Projectile motion
T23.	Figure 3-20	Path of a projectile
T24.	Figure 3-25	Example 3-8 (Level horizontal range)
T25.	Figure 3-28	Crossing a river

Suggested Readings

Black, H., "Vector Toy," *The Physics Teacher* (September 1998), p. 375. Describes the use of "walking toys" to demonstrate vector forces.

Chow, J., Carlton, L., Ekkekakis, P. and Hay, J., "A Web-Based Video Digitizing System for the Study of Projectile Motion," *The Physics Teacher* (January 2000), pp. 37–40. A two-dimensional projectile motion lab using a camcorder for data collection.

Deakin, M. and Troup, G., "Approximate Trajectories for Projectile Motion with Air Resistance," *The American Journal of Physics* (January 1998), pp. 34–37. Describes approximations for the trajectories of projectiles under various laws of resistance.

Durkin, T. and Graf, E., "Quibbles, Misunderstandings, and Egregious Mistakes," *The Physics Teacher* (May 1999), pp. 297–306. Discusses common mistakes in high school textbooks, with a nice section on vectors.

Larson, R., "Measuring Displacement Vectors with the GPS," *The Physics Teacher* (March 1998), p. 161. A nice use of GPS to talk about vectors.

Molina, M., "More on Projectile Motion," *The Physics Teacher* (February 2000), pp. 90-91. Discussion of the range of a projectile.

Price, R. and Romano, J., "Aim High and Go Far—Optimal Projectile Launch Angles Greater Than 45°," *The American Journal of Physics* (February 1998), pp. 109–113. An investigation of optimal projectile launch angles when air resistance is taken into account, with simple physical arguments that help explain the results.

Sheets, H., "Communicating with Vectors," *The Physics Teacher* (December 1998), pp. 520–521. A good exercise for teaching vectors.

Taylor, D., "Vector Video," *The Physics Teacher* (January 2001), p. 14. A great illustration of vector addition and relative motion.

Van den Berg, W. and Burbank, A., "Sliding Off a Roof: How Does the Landing Point Depend on the Steepness?" *The Physics Teacher* (February 2002), pp. 84–85. Examination of an object sliding off an elevated ramp, with a plot of horizontal projection versus angle of inclination.

Wetherhold, J., "A Toy Airplane for Projectile Motion Experiments," *The Physics Teacher* (February 2001), pp. 116–119. An inexpensive toy airplane modified to demonstrate various characteristics of projectile motion.

Wheeler, D. and Charoenkul, N., "Whole Vectors," *The Physics Teacher* (May 1998), p. 274. A nice treatment of vectors and vector diagrams.

Widmark, S., "Vector Treasure Hunt," *The Physics Teacher* (May 1998), p. 319. A fun activity for teaching vectors.

Materials

A "Projectile Launcher" can be obtained from Frey Scientific, catalog number S1900503.

Two variations of the "Shoot the Target" apparatus are available from Pasco, catalog numbers ME-6805 and ME-6826. (Projectile launchers from Pasco for use with the apparatus include ME-6801 and ME-6825.)

Notes and Ideas

Class time spent on material: Estimated:______ Actual:_____________

Related laboratory activities:

Demonstration materials:

Notes for next time:

Chapter 4: Dynamics: Newton's Laws of Motion

Outline

4-1 Force
4-2 Newton's First Law of Motion
4-3 Mass
4-4 Newton's Second Law of Motion
4-5 Newton's Third Law of Motion
4-6 Weight—the Force of Gravity; and the Normal Force
4-7 Solving Problems with Newton’s Laws: Free-Body Diagrams
4-8 Applications Involving Friction, Inclines
4-9 Problem Solving—A General Approach

Summary

Until now we have been studying the *effects* of motion (kinematics). Chapter 4 begins the study of the *causes* of motion (dynamics). An unbalanced force is one cause of motion. We will consider *normal-sized* objects moving at *normal* speeds, keeping us in the realm of Newtonian physics. Newton's three laws are quite powerful and elegant and explain how an object moves when acted on by one or more forces. Chapter 4 introduces force as a *push* or *pull*. The vector nature of force is discussed and contrasted with the scalar nature of mass. Weight and the normal force are presented as examples of forces. Frictional forces, both static and kinetic, are explained in terms of the microscopic interactions of surfaces. Strings transmit force along their length and can only pull on an object. (You can't push with a string!) Pulleys and springs are considered massless for now. Pulleys simply change the direction of the tension in a string. Finally, general methods of problem solving are discussed.

Major Concepts

By the end of the chapter, students should understand each of the following and be able to demonstrate their understanding in problem applications as well as in conceptual situations.

- Force
  - Vector nature of force
  - Weight
  - Normal force
- Mass
- Reference frames
  - Inertial
  - Noninertial
- Newton's laws
  - First law (law of inertia)
  - Second law ($\vec{\mathbf{F}} = m\vec{\mathbf{a}}$)
  - Third law (action-reaction force pairs)
- Free-body diagrams
- Friction
  - Static friction
  - Kinetic friction
- Strings and pulleys

- Assumptions
- Transmission of force

Teaching Suggestions and Demonstrations

The challenge of this chapter is not learning Newton's laws; it is learning how to apply them to problems. Students often have difficulty with the vector nature of forces and with the difference between the concepts of weight and mass. This should not come as a surprise. Vectors are still new to these students. The rules for vector addition will have to be reinforced continually. The students will need to see many examples. Be sure to work several that are not worked in the text.

Sections 4-1 and 4-2

Force is a vector, with magnitude and direction. Units of force are the newton (SI) or the pound (British units). The forces that act on an object to cause motion are usually external to the object. (One exception is the case of explosions; another is the movement of the human body.)

As stated in the text **Newton's first law** is "every object continues in its state of rest, or of uniform velocity, in a straight line, as long as no net force acts on it." This first law is sometimes called the law of inertia. It is important to note that in both cases above, whether the object is at rest or moving with constant velocity, the net force acting on the object is zero.

➲ **DEMO 4-1** If you have access to an air track or an air table, set it up and let students suggest experiments to demonstrate the first law. (See Resource Information.)

If you don't have access to an air track, you can demonstrate Newton's first law with a series of small model cars. Start with one that has lots of friction in its axles. If you roll it across the floor, it will not go very far. Next find one that has very little friction in the axles. (Hot Wheels work well and are inexpensive.) If you roll it across a smooth floor, it will go much farther than the first car. You can then talk about (theoretically) removing the friction entirely, and discuss what would happen.

You can also do a nice demonstration with an ice cube on one of the shiny black tabletops that are in many science classrooms. The ice surface melts slightly, so that the ice travels on a thin layer of water, a nearly frictionless situation.

Section 4-2 contains a quick discussion of **inertial and non-inertial reference frames**. If you would like to add a little historical perspective, you can discuss Newton's idea of the absolute reference frame of the fixed stars. Since we now know that the stars are not really "fixed" in space, we know there are no absolute reference frames. We use the definition given in the text: If Newton's laws are true, then the reference frame is an inertial frame.

Encourage students to notice the many common examples of inertial and non-inertial reference frames, such as a car traveling at constant speed in a straight line, a car accelerating in a straight line, and a car traveling around a curve. What happens to a ball thrown up in the air in each case?

Sections 4-3 and 4-4

Walk down any supermarket aisle and you will see that common usage confuses **mass** and **weight** (a **force**). Most food packages list the weight of the contents in pounds or ounces and the mass in kilograms or grams as if weight and mass were the same concept. (Why? The fundamental quantities in SI units are mass, time, and length; in British units they are force, time, and length. Besides, who wants to buy food by the slug, the unit for mass in the British system?) As the concepts are introduced, emphasize the physics definitions. Be sure to remind the students of the different definitions of mass and weight each time you work an example problem.

Mass is an intrinsic property of an object, the amount of matter contained in the object. Mass is a scalar. Units of mass are the kilogram (SI) or the slug (British units). The mass of an object stays constant (unless a piece falls off!) as the object moves from the Earth to another planet. Table 4-1 in the text gives a useful summary of the units for mass and force.

Newton's second law can be stated as "a net force causes accelerations," or $\vec{\mathbf{a}} = \vec{\mathbf{F}} / m$. In its more familiar form, this is: $\Sigma \vec{\mathbf{F}} = m\vec{\mathbf{a}}$. Note that the sum of the forces is the sum of the *external* forces acting on the object and that mass (NOT WEIGHT!) is the constant of proportionality between the net force and the acceleration. Point out that the acceleration is a *result* of the application of a force to a mass.

➲ **DEMO 4-2** With a spring scale and several different masses, measure the amount of force required to pull each of the masses across a tabletop at approximately constant velocity. (See Figure 4-2.) Does it take more force to pull a larger mass? (Friction is at work here!) (See Resource Information.)

Be sure to make the point that the equation given above is a vector equation that can be written as three independent scalar equations. Example 4-3 gives a nice illustration of how Newton's laws can be applied to real-world situations.

Sections 4-5 and 4-6

Newton's third law states "whenever one object exerts a force on a second object, the second exerts an equal and opposite force on the first." Students often have trouble with this concept. Remind them that the two forces act on *different* objects, so they appear in *different* free-body diagrams and therefore cannot cancel. Also, since the two forces act on different objects, they usually produce different accelerations. Discuss Conceptual Examples 4-4 and 4-5 with the class.

A discussion of **weight** presents the opportunity to demonstrate again the difference between weight (or force) and mass. Remind students that we sometimes use the terms mass and weight interchangeably in everyday language. This works because we are all confined to the surface of the Earth (or at least close to it!) and g is a constant in this region. If g changed from place to place, weight would also change, but mass would remain constant. (Remember, astronauts in space are "weightless" but not massless.)

➲ **DEMO 4-3** If your class or lab section is small enough, send the students out in pairs with a bathroom scale to ride an elevator and measure **apparent weight** (the normal force read by the scale). (See Example 4-8.) If the students can travel between the same floors, you can use the measurements of the apparent and actual weights to calculate the acceleration of the elevator. If your class is too large, recommend this as an out-of-class assignment.

As you talk about apparent weight, you can lead right into a discussion of the **normal force**. Work through Example 4-6 in class. Students usually catch on quickly to the concept of the normal force on a flat, horizontal surface; they have a little more trouble with an inclined or curved surface. Remind the students that the normal force is always perpendicular to the surface, though not always along the same axis as the weight.

Section 4-7

Free-body diagrams are essential for solving problems using Newton's second law. Show your students lots of examples, such as those in Figure 4-18 and Example 4-9, and provide plenty of opportunities for practice. The problem-solving steps outlined in the Problem Solving box in this section can be turned into another mantra:

- Draw a picture.
- Isolate the object of interest.
- Draw in the external forces acting on the object of interest.
- Choose a coordinate system.
- Resolve the forces into components.
- Apply Newton's second law in each coordinate direction.
- Solve the equations.

It is important to work many examples so that students can see how to apply these techniques in a variety of situations. Go through some examples that are in the text (Conceptual Example 4-10 is a good one), but be sure to do some that are not worked in the text as well.

Understanding the idea of **forces in two dimensions** is difficult for students who are not comfortable working with vectors. It is important to emphasize that motion in each coordinate direction is independent and that problems can be made easier by a careful choice of axis. You will have to remind students how to take the components of a vector and point out that the x and y components of a force can be associated with the sine *or* cosine of an angle, depending on which angle is chosen.

You may want to begin the discussion of **strings** by justifying the assumption that most strings can be considered massless. Remind the students that you can only pull (not push!) with a string.

➲ **DEMO 4-4** Hang a block from a **string**. Find the weight of the block and the weight of the string. Calculate the tension at the top of the string, where the string is supporting its own weight as well as the weight of the block, and at the bottom of the string, where just the weight of the block is supported. How different are these tensions? Are we justified in making the assumption that an ideal string is massless?

Ideal **pulleys** are also considered massless and serve only to change the direction of the tension in the string. Example 4-13 is a good one to work in class.

When two **connected objects** are in motion, how do they move? We begin by assuming that the string that connects the objects is inextensible. We can draw a free-body diagram for each object individually and then write separate equations of motion for each object. Since the objects are tied together, the magnitudes of the velocities and accelerations of the objects must be the same. It is possible for the directions of the velocity and acceleration to be different for the two objects, especially if pulleys are in the problem. (It is sometimes necessary to draw a free-body diagram for the pulley in order to solve a problem.) Objects that are connected along a straight line can be treated as a single system, as shown in Figure 4-22.

Sections 4-8 and 4-9

Most students do not have much trouble with the concept of **kinetic friction**. They experience it every day, so it seems familiar, and they have an intuitive understanding of the direction (opposing motion). **Static friction** is not as simple as kinetic friction. The inequality in the equation tends to confuse the students.

➲ **DEMO 4-5** Do the demonstration shown in Figure 4-27 with a block and a spring scale. Use several different blocks with different masses to show that the **force of static friction** can vary. Find the weight of each block and calculate the coefficient of static friction in each case. Then let the block move and determine the coefficient of kinetic friction. Which coefficient is larger?

Note: Accident reconstructionists will drag a tire across the roadway at the scene of an accident to determine the coefficient of kinetic friction between the tire and the road. This helps with determining how cars moved during the accident.

Table 4-2 lists typical coefficients of friction between common surfaces. Be sure to work problems in which the force of static friction is *not* at its maximum value to show students that the value of the force can vary. Go through Example 4-16 in class.

Be prepared to discuss the "rule of thumb" that the frictional force is independent of the area of contact. This often comes up in the context of car tires, particularly race car tires. You can point out that for a car traveling normally, the friction between the tires and the road is static friction. The tread expands and contracts as the tire surface contacts the road. This provides extra interactions between the tire and the road and so increases the coefficient of static friction.

The chapter ends with an excellent discussion of general methods of problem solving. Be sure to go over this in class, and talk through the procedure every time you work an example in class. This is a good time to schedule a problem session. If you have an extra class, lab time or a recitation time available, you can use that. Otherwise, schedule a problem session outside of class. The students need to see lots of examples, preferably before and after they have tried to work problems on their own.

Resource Information

Transparencies

T26.	Figure 4-2	A spring scale used to measure a force
T27.	Table 4-1	Units for Mass and Force
T28.	Figure 4-11	We can walk forward
T29.	Figure 4-12	Example 4-5 (Third law clarification)
T30.	Figure 4-14	The net force on a object at rest is zero
T31.	Figure 4-15	Example 4-6 (Weight, normal force, and a box)
T32.	Figure 4-17	Example 4-8 (Apparent weight loss)
T33.	Figure 4-19	Example 4-9 (Adding force vectors)
T34.	Figure 4-20	Example 4-10 (The hockey puck.)
T35.	Figure 4-22	Example 4-12 (Two boxes connected by a cord)
T36.	Figure 4-23	Example 4-13 (Atwood's machine)
T37.	Figure 4-26	An object moving to the right on a table or floor
T38.	Table 4-2	Coefficients of Friction

T39.	Figure 4-28	Example 4-16 (Friction: static and kinetic)
T40.	Figure 4-32	Example 4-20 (Two boxes and a pulley.)
T41.	Figure 4-33	Forces on an object sliding down an incline

Suggested Readings

Bao, L., Hogg, K., and Zollman, D., "Model Analysis of Fine Structures of Student Models: An Example With Newton's Third Law," *American Journal of Physics* (July 2002), pp. 766–778. A study of how the contextual features of problems affect student reasoning.

Bernhard, K. and Bernhard, J., "Mechanics in a Wheelchair," *The Physics Teacher* (December 1999), pp. 555–556. A description of a kinesthetic experience of Newton's laws.

Brand, H., "Action-Reaction at a Distance," *The Physics Teacher* (March 2002), pp. 136–137. Describes a procedure for demonstrating action-reaction in a case when the two bodies are not in contact.

Chandler, D., "Newton's Second Law for Systems with Variable Mass," *The Physics Teacher* (October 2000), p. 396. This example does involve calculus but is a good illustration of the variable mass problem.

Court, J.E., "Free-Body Diagrams Revisited—I," *The Physics Teacher* (October 1999), pp. 427–433. Free- body exercises with solutions in linear and circular motion.

Cross, R., "Standing, Walking, Running, and Jumping on a Force Plate," *American Journal of Physics* (April 1999), pp. 304–309. Details of an inexpensive force plate designed to measure ground reaction forces involved in human movement.

Dalton, R., "Caught on Camera," *Nature* (15 August 2002), pp. 721–722. An interesting note on the use of high-speed cameras in biomechanics.

Gettrust, E., "An Extraordinary Demonstration of Newton's Third Law," *The Physics Teacher* (October 2001), pp. 392–393. Description of an apparatus using magnets and force probes to demonstrate that the action and reaction forces are equal in magnitude.

Haugland, O., "Physics Measurements for Sports," *The Physics Teacher* (September 2001), pp. 350–353. Contains instructions for making a simple force platform and for using ultrasonic motion detectors in helping athletes gain insight into the physics of their sports.

Kunzig, R., "The Physics of Walking: Falling Forward," *Discover* (July 2001), pp. 24–25. A fascinating general article on the physics involved in walking.

Larabee, D., "Car Collisions, Physics, and the State Highway Patrol," *The Physics Teacher* (September 2000), pp. 334–336. A real-world application of the laws of motion and friction.

Leonard, W., "Dragging a Box: The Representation of Constraints and the Constraint of Representations," *The Physics Teacher* (October 2001), pp. 412–414. A non-calculus derivation of the best angle to drag a box, along with a discussion of presentations of normal, friction, and tension forces.

Linthorne, N., "Analysis of Standing Vertical Jumps Using a Force Platform," *American Journal of Physics* (November 2001), pp. 1198–1204. Describes the use of a force platform to demonstrate the kinematics and dynamics of vertical jumping.

Liphardt, J., Bibiana, O., Smith, S., Tinoco, I. and Bustamante, C., "Reversible Unfolding of a Single RNA Molecule by Mechanical Force," *Science* (27 April 2001), v. 292, pp. 733–737. A nice application of forces in biology.

Mainardi, R., "Demonstration Experiments with Platform Scales," *The Physics Teacher* (November 2001), pp. 488–489. Describes three experiments, including one on Newton's third law.

Morrow, R., Grant, A. and Jackson, D., "A Strange Behavior of Friction," *The Physics Teacher* (October 1999), pp. 412–415. A more in-depth look at the subtleties of kinetic friction.

Reichert, J., "How Did Friction Get So 'Smart'?," *The Physics Teacher* (January 2001), pp. 29–31. An interesting discussion of the frictional force.

Styer, D., "The Word 'Force'," *American Journal of Physics* (June 2001), pp. 631–632. A letter to the editor containing many examples to help students differentiate the physics meaning of "force" from its everyday meaning.

Van den Berg, W., "The Best Angle for Dragging a Box," *The Physics Teacher* (November 2000), pp. 506–508. A great exercise in calculation of vector components and friction. Requires a graphing calculator or computer.

Wilczek, F., "Reference Frame: Mass without Mass I: Most of Matter," *Physics Today* (November 1999), pp. 11–13. Discusses where mass comes from.

Wilczek, F., "Reference Frame: Mass without Mass II: Most of Matter," *Physics Today* (January 2000), pp. 13–14. Deducing mass as a secondary property of matter.

Williams, K., "Inexpensive Demonstrator of Newton's First Law," *The Physics Teacher* (February 2000), p. 80. Uses a Downy® Ball fabric-softener dispenser!

Materials

Two air-track systems are available from Fisher Scientific: model number S52229, for use with a computer interface, or model number S52227A, for use as a stand-alone.

Spring scales are available from Ohaus in a variety of measurement ranges.

A metal block set is also available from Fisher Scientific, model number S41245.

Notes and Ideas

*Class time spent on material: Estimated:*______ *Actual:*______________

Related laboratory activities:

Demonstration materials:

Notes for next time:

Chapter 5: Circular Motion; Gravitation

Outline

Summary

Chapter 5 begins with the study of uniform circular motion. This chapter introduces Newton's law of universal gravitation and applies it to spherical objects. Kepler's laws are stated and discussed, as is the general equation for gravitational potential energy and its role in the conservation of energy. The chapter ends with an optional section on tides.

Major Concepts

By the end of the chapter, students should understand each of the following and be able to demonstrate their understanding in problem applications as well as in conceptual situations.

- Uniform circular motion
  - Centripetal acceleration
  - Centripetal force
  - Banked and unbanked highway curves
- Newton's law of universal gravitation
  - Universal gravitation constant *G*
  - Inverse square dependence on the distance
  - Point and spherical objects
  - Cavendish experiment
- Kepler's laws of orbital motion
  - Law of orbits
  - Law of areas
  - Law of periods

Teaching Suggestions and Demonstrations

Many students have difficulty with **circular motion** and **centripetal acceleration**. Carefully work out the vectors for centripetal acceleration. All students have experience with the **gravitational force**, but not all of them will connect the force we feel on Earth with the one that holds the solar system (and the galaxy and the universe) together. Some of their intuitive ideas about gravity will be correct; others will not be correct. You will have to make careful assessment of student progress for this chapter.

Sections 5-1 through 5-3

Uniform circular motion is another concept that is not intuitive for students. The discussion at the beginning of Section 5-1 is excellent. Remind the students that if an object is not moving in a straight line, it must have a force on it, and use that idea to justify the direction of the force. (See Figure 5-4.) You then can define the **centripetal acceleration**. You will need to go through the derivation of Equation 5-1 carefully. There are several steps that are tricky for the students. Point out that for uniform circular motion, the acceleration vector and the velocity vector are perpendicular to one another, as in Figure 5-3.

You will need to spend some time defining the concepts of **frequency** and **period**. These concepts are introduced here and will come up again in later chapters. Go through Example 5-1 to illustrate how the period and frequency relate to centripetal acceleration.

Emphasize that the **centripetal force** is not a mysterious force imposed on a situation from the outside. It must be provided by some force in the problem (e.g., gravity, the tension in a string, the normal force).

Any force that is serving as a centripetal force can be written in the form: $f_{cp} = ma_{cp} = m\dfrac{v^2}{r}$.

There is a very good discussion of the centripetal force and the **centrifugal force** in Section 5-2, illustrated in Figures 5-5 and 5-6.

➲ **DEMO 5-1** The traditional ball on a string swung in a horizontal circle works well for demonstrating **circular motion**. Variations include a (small) bucket of water or sand swung in a vertical circle (very dramatic!). See Example 5-4.

Students like to see the mechanics of **amusement park rides** worked out. You can do the case of the loop-the-loop roller coaster (where the cars go upside down – see Problem 5-13) and the spinning cylinder ride (see Problem 5-18). Have the students work together in pairs or groups of three to figure out why the riders don't fall out of the ride in either case!

Section 5-3 shows the forces on a car going around a flat curve and a banked curve. (See Figures 5-13 and 5-14.) Work through Examples 5-6 and 5-7 and lead students in a discussion of the advantages of banked roadways.

Sections 5-4 and 5-5

Sections 5-4 and 5-5 are optional discussions of **nonuniform circular motion** and **centrifugation**. If you have pre-health professions students in your class, it is a good idea to go over the acceleration vectors for nonuniform circular motion and the principles behind the centrifuge. (These topics may appear on the MCAT.) These two sections may be skipped without loss of continuity.

Sections 5-6 through 5-8

Newton's law of universal gravitation is elegant and simple, especially for spherical or point objects. Emphasize that the force acts between centers of the objects, is always attractive, is proportional to both masses, and is inversely proportional to the distance between the centers squared. You will need to discuss the implications of the inverse square nature of the force. The **infinite range of the gravitational force** and the fact that it is the weakest of the **four fundamental forces** are often surprising to students. It is a good idea to talk about why we don't have to worry about the gravitational force from everyday objects. (See Example 5-10.) For most cases, the only time we have to take the gravitational force into account is when at least one of the two objects is the size of a planet.

There is a nice treatment of the **Cavendish experiment** in Section 5-6 and Figure 5-20. Point out that ***G*** is a universal constant; it is the same for all pairs of objects. Students often find it amazing that the mass of the Earth wasn't known until 1798, long after Newton's death, since the radius of the Earth was known to the ancient world.

The students will find the discussion of **satellite motion**, particularly **geosynchronous satellites**, interesting. Be sure to go over Figures 5-24 and 5-25 and Example 5-14. The following section on **apparent weightlessness** will be very useful in later discussion of relativity; plan to spend a few minutes of class time on it here.

Section 5-9

This section is marked as optional and can be skipped if you are short on time however, we recommend that you include it if possible. The students find **Kepler's laws** interesting and understandable. Kepler's laws can be derived from Newton's laws, but Kepler deduced them from astronomical data collected by visual observation. The story of **Brahe** and Kepler makes interesting reading; students find it fascinating. (See Resource Information.)

Kepler's first law is also called the **law of orbits**. Giving up the idea of circular orbits for planets involved a major shift of worldview for Kepler. You will probably need to spend some time talking about the properties of ellipses before continuing to the second law.

➲ **DEMO 5-2** Using suction cups attached to the blackboard or whiteboard and a piece of string tied in a loop, draw an **ellipse**. Show the students what happens if the foci are moved farther apart or closer together. Show them how a circle is a special case of an ellipse.

Kepler's second law is called the **law of areas**. Point out that it is based in the conservation of angular momentum.

Kepler's third law is called the **law of periods**. It is important to emphasize to the students that the "constant" that appears in the third law is *not* the universal constant *G*. It is a "constant" that is the same for all objects orbiting the same mass. Plan to go through Example 5-16, the calculation of the Sun's mass, in class.

Section 5-10

Section 5-10 is a concise treatment of the **fundamental forces of nature**. You will need to allow some time for general discussion of these force and their relative strengths.

Resource Information

Transparencies

T42.	Figure 5-2	Determining the change in velocity for a particle moving in a circle
T43.	Figure 5-3	For uniform circular motion, ***a*** is always perpendicular to ***v***
	Figure 5-4	A force is required to keep an object moving in a circle
T44.	Figure 5-6	If centrifugal force existed

T45.	Figure 5-9	Exercise C
T46.	Figure 5-13	Example 5-6 (Skidding on a curve)
	Figure 5-14	Normal force on a car rounding a banked curve
T47.	Figure 5-15	The speed of an object moving in a circle changes
T48.	Figure 5-17	Two positions of a rotating test tube in a centrifuge
T49.	Figure 5-20	Schematic diagram of Cavendish's apparatus
T50.	Figure 5-24	Artificial satellites launched at different speeds
	Figure 5-25	A moving satellite "falls" out of a straight-line path toward the Earth
T51.	Figure 5-26	An object in an elevator
T52.	Figure 5-28	Kepler's first law
	Figure 5-29	Kepler's second law
T53.	Table 5-2	Planaetary Data Applied to Kepler's Third Law
T54.	Figure 5-30	Our solar system system, compared to recently discovered planets

Suggested Readings

Adam, D., "Amazing Grace," *Nature* (7 March 2002), v. 416, pp. 10–11. A very accessible article describing the precise mapping of Earth's gravitational field by two satellites.

Bouffard, K., "Physics Olympics: The Inertia Ball," *The Physics Teacher* (January 2001), pp. 46–47. A nice demonstration for centripetal force.

Charoenkul, N., Wheeler, D. and Dejasvanong, C., "The Wall of Death: Newtons, Nerves, and Nausea," *The Physics Teacher* (December 1999), pp. 533–535. A dramatic example of circular motion, centripetal force, and friction.

Court, J.E., "Free-Body Diagrams Revisited—I," *The Physics Teacher* (October 1999), pp. 427–433. Free body exercises with solutions in linear and circular motion.

Larson, R., "Centrifugal Force and Friction," *The Physics Teacher* (October 1999), pp. 426–427. Uses a rotary motion probe to investigate circular motion.

Metz, J., "Finding Kepler's Third Law with a Graphing Calculator," *The Physics Teacher* (April 2000), p. 242. Instructions for using a TI-83 to find the relationship between orbital radius and period for planets.

Morris, R., *Dismantling the Universe: The Nature of Scientific Discovery*, Simon and Schuster, New York, 1983. A great book to read before teaching a physics class. Chapter 4, pp. 81–101, covers the story of Brahe and Kepler (and Galileo).

Newton, I. and Henry, R., "Circular Motion," *American Journal of Physics* (July 2000), pp. 637–639. Presentation of a simple derivation of the formula for the acceleration that occurs in uniform circular motion.

Ronhovde, P. and Sirochman, R., "Center of Mass Correction to an Error-Prone Undergraduate Centripetal Force Lab," *American Journal of Physics* (February 2003), pp. 185–188. Describes a simpler correction to a centripetal force laboratory experiment.

Sawicki, M., "Myths About Gravity and Tides," *The Physics Teacher* (October 1999), pp. 438–441. Explores popular misconceptions about gravity and tides.

Schwarzschild, B., "Theorists and Experimenters Seek To Learn Why Gravity Is So Weak," *Physics Today* (September 2000), pp. 22–24. A description of the search for departures from the inverse-square law at millimeter separations.

Toepker, T., "Babies and the Moon," *The Physics Teacher* (April 2000), p. 242. A graph of birth data to dispel the popular myth that more babies are born under a full moon.

Vogt, E., "Elementary Derivation of Kepler's Laws," *American Journal of Physics* (April 1996), pp. 392–396. A proof of Kepler's laws that follows from conservation of energy and angular momentum, with further discussion.

Wright, K., "Very Dark Energy," *Discover* (March 2001), pp. 70–76. A discussion of the new ideas on the accelerating expansion of the universe.

Notes and Ideas

*Class time spent on material: Estimated:*______ *Actual:*_____________

Related laboratory activities:

Demonstration materials:

Notes for next time:

Chapter 6: Work and Energy

Outline

Summary

Chapter 6 introduces students to the important concepts of work and energy. Kinetic energy, potential energy, and the law of conservation of energy are all covered. The important distinction between conservative and nonconservative forces is made. Conservation of mechanical energy is discussed in detail. Methods of solving problems using energy considerations are demonstrated; these methods often lead to easier solutions than methods involving kinematic equations. Power, the time rate of change of doing work, is also defined.

Major Concepts

By the end of the chapter, students should understand each of the following and be able to demonstrate their understanding in problem applications as well as in conceptual situations.

- Work
  - Force in the direction of displacement
  - Force at an angle to displacement
  - Positive, negative, and zero work
  - Constant force and variable force
- Kinetic energy
- Work-energy theorem
- Potential energy
  - Gravitational
  - Spring (Hooke's law)
- Conservative and nonconservative forces
  - Work and stored energy
  - Path dependence or independence of work
- Conservation of mechanical energy
- Work done by nonconservative forces; changing mechanical energy
- Law of conservation of energy
- Power

Teaching Suggestions and Demonstrations

Throughout this chapter, students encounter terms that have meanings in English that are different from their meanings in physics. Emphasize to students that the words *work*, *energy*, and *power* have specific definitions in physics that are somewhat related to but definitely not the same as their meanings in everyday conversations. Point out that when they say they've *worked* really hard on their homework, they are speaking English and not physics! Chapter 6 is your first chance to show the power of conservation laws in physics. Plan to work several problems using methods from earlier chapters, then work them again using energy conservation methods. The (usually) dramatic difference between the complexity of the first solution and the simplicity of the second is enough to convince most students to learn the new method.

Sections 6-1 through 6-3

Students may at first be confused to find that they do no **work** in trying to push a stalled car that won't move or in holding a heavy box up in the air. Reassure them that on the microscopic level, the cells in their muscles are contracting and expanding and therefore doing work, however, no net work is done on the car or the box. The person shown holding a bag of groceries in Figure 6-2 is a good example. How much work does he do *holding* the bag of groceries? None.

➲ **DEMO 6-1** To demonstrate the relationships among **work, force,** and **displacement**, attach a block to a spring scale and use the scale to drag the block across a table. The work performed by the scale on the block is equal to the component of force parallel to the tabletop times the distance the block moves along the table. (In order to actually do the calculations, you will need to pull the block so that the force stays constant.) Students can try this exercise in small groups or you can have a few demonstrate to the whole class and then involve the others in the calculations. Try pulling the block with a horizontal force and then with a force applied at an angle.

This section also provides a good opportunity to point out that numbers don't always tell the whole story when it comes to **work** and **force**. The person pulling the box along the floor in Figure 6-1 could do the same amount of work with less force if she pulled the box horizontally. Why doesn't she? Well, one reason could be that bending over to pull the box is uncomfortable! You can also discuss what the vertical component of the pulling force is accomplishing. It reduces the normal force on the box, which doesn't contribute to the work done but does reduce friction if any is present. Go over Example 6-1 and Exercise A to illustrate these concepts. Example 6-1 shows how work can be positive, negative, or zero. Give some examples of **negative work**, such as the work done in catching a baseball or braking a car.

If no friction is present, the same amount of work is done in lifting a box straight up or in sliding it up a ramp. If friction is present, more work is needed to slide it up the ramp. Even though real-life ramps have friction, most of us would choose to do more work and use the ramp if we are moving heavy boxes. Why? Because we can apply less force. Although power has not yet been introduced, you can also point out that less power is needed to push the box up the ramp.

The SI unit of work and energy, the **joule**, will most likely be unfamiliar to students and will require explanation. The problem-solving hints in Section 6-1 are quite helpful. Go over them and work through Example 6-2 to demonstrate how to use them.

Section 6-2 is optional and can be skipped, but it may be interesting for those students who have had calculus. In the case of a **variable force**, work done is the **area under the curve** on a force-position graph. Students may be confused by the fact that an area has units of joules instead of square meters. Remind them of another instance when the area under a curve has important physical significance; the area under the curve on a velocity-time graph is displacement. For those students who have had some calculus, point out that Figure 6-6 is a graphical representation of an integral.

When introducing the **work-energy principle**, emphasize that it is the **total work** done on an object that is equal to its change in kinetic energy.

➲ **DEMO 6-2** Demonstration 6-1 can be extended to illustrate the **work-energy principle**. Try pulling the block with just the constant force necessary to keep it moving at a constant velocity. Then apply a larger constant force and note that it speeds up. In the first case, the change in kinetic energy is zero, so the total work done is zero. Friction did an amount of **negative work** on the block just equal to the positive work done by the spring scale. (If you have time, you can even use the exercise to calculate the coefficient of friction between the block and the table.) In the second case, the change in kinetic energy is positive as is the net work done on the block.

Students may at this point be tempted to think of work as a vector. It isn't. Work is a **scalar** that can have positive, negative, or zero values depending on the angle between the force and the displacement.

➲ **DEMO 6-3** If you feel particularly adventuresome, an egg-catching demonstration is a great way to investigate both the **work-energy principle** and the relationships among **work**, **force**, and **distance**. Toss an egg to a student and have the class discuss why she moves her hand back in catching it. An important point for them to understand is that the catcher has no control over the amount of work she has to do. (The work-energy theorem says that her work equals the change in kinetic energy; the egg has a certain initial speed and the catcher wants to bring it to rest, so the change in kinetic energy is fixed.) However, she can control force and displacement. The same work will be accomplished by applying a small force over a large distance (i.e., pulling her hand back) or a large force over a small distance (splat!). The same relationships can be demonstrated with less mess by catching baseballs. It hurts a lot less to move your hand back with the ball because your force on the ball is less and therefore, according to Newton's third law, the ball's force on your hand is less as well.

Work several examples using the work-energy principle before moving on to the discussion of potential energy.

Section 6-4

This section covers the definition of **potential energy** (energy of position or configuration) and also the definition of the spring force. The potential energy of the system in a certain state is equal to the work done in order to put it in that state. Begin by discussing potential energy in terms of the gravitational force near the surface of the earth. Students are already familiar with the formula for work needed to raise an object in a uniform gravitational field. Point out that the equation for **gravitational potential energy near the surface of the earth** is the same. Be sure to emphasize that it is the ***change* in potential energy** that is important, not its absolute value, and that the zero level for the gravitational potential energy can be set at any convenient level.

You will need to spend significant class time on **springs** and the **elastic force**. **Hooke's Law** ($F_x = -kx$) seems very simple, but it is the first non-constant force law that the students have seen. Point out that ideal springs are considered massless. Show the students some examples of springs that can be compressed as well as stretched. These can be found in dart guns, retractable ballpoint pens, or pinball machines.

➲ **DEMO 6-4** Hang a spring vertically and note its hanging length. (See Resource Information.) Hang several different masses from it, one at a time, and note the change in the length for each mass. Calculate the **spring constant** of the spring.

Once the students are comfortable with the concept of the spring force, you can introduce the **elastic potential energy**. Students are very likely to assume that it will take twice as much work to stretch a **spring** twice as far. Use Figure 6-15 and the discussion of work done by a variable force in Section 6-2 to address this misconception and to continue the discussion of force and work for the specific case of a spring.

Sections 6-5 through 6-9

The distinction between **conservative and nonconservative forces** is made clearly in Section 6-5. Table 6-1 categorizes some common forces as either conservative or nonconservative. Energy can be stored and recovered by conservative forces. For instance, a pile driver is simply a large mass raised in a gravitational field. Energy put into the system raises it in a gravitational field; energy is released in the form of work done on the pile when the mass falls.

Conservation of mechanical energy is the first conservation law encountered in physics, so it is important to ensure that students understand what *conservation* means. If a quantity is conserved, then the total value of that quantity remains constant. Conservation of energy in physics refers to the fact that energy isn't created or destroyed, although it does change in form. Kinetic energy is converted to potential energy when a book is tossed up into the air. Kinetic energy is converted into heat when a book slides across a table and comes to rest due to friction. Mechanical energy (potential plus kinetic) is conserved only in the first case (when only conservative forces are acting), although total energy is conserved in both. The difference is that the heat energy cannot be easily converted back to organized, macroscopic motion; there is no way to gather the energy back out of the particles and organize it to spontaneously make the book move again. (Note that "energy conservation" in environmental science means something entirely different!)

➲ **DEMO 6-5** A pendulum and a spring are both simple demonstrations of **conservation of mechanical energy**. Set a mass on a spring oscillating and a pendulum swinging and point out when the energy is all kinetic or all potential. By measuring the vertical displacement from equilibrium of either the mass on the end of the pendulum or the mass on the end of the spring, students can calculate the velocity the mass has as it passes through equilibrium. In both cases, the mechanical energy is not perfectly conserved, as is made clear by the fact that the mass eventually stops swinging or stops bobbing up and down. Ask students what nonconservative forces (friction and air resistance) may be acting on the system.

Example 6-9 uses energy considerations to solve a motion problem and shows that the total mechanical energy at different points in the motion is the same. This example gives students direct evidence for the **conservation of mechanical energy,** and they are subsequently encouraged to use energy considerations to solve problems. Go over the Problem Solving suggestions on Conservation of Energy in Section 6-9.

➲ **DEMO 6-6** **Roller coasters** provide excellent illustrations of kinetic and potential energy and the conservation of mechanical energy. Demonstration setups complete with photogates for determining velocities are available for purchase. (See Resource Information.) Alternatively, you could use a toy car, train, or marble and an appropriate track. Comparison of initial height and final height will allow students to determine the total mechanical energy "lost" due to nonconservative forces and the coefficient of friction between the track and the moving object. In addition, there are numerous Websites devoted to roller coaster design and other applications of physics in amusement parks. (See Resource Information.)

Be sure to work several examples that illustrate how energy considerations can be useful even if mechanical energy is not conserved, as is the case when **nonconservative forces** are acting.

Section 6-10

Power is work (or energy) per unit time, or the rate of doing work. Introduce the SI unit of power, the **watt**, and contrast it with a kilowatt-hour, the unit that typically appears on a "power" bill. Point out that people actually pay the power company for energy consumed, not power, since a kilowatt-hour (or power times time) is a unit of energy. As an example, convert a typical monthly energy usage in kilowatt-hours to joules so that students will begin to get a feel for the relationship between these two energy units. As more and more new units are introduced in the course, it is helpful to occasionally do a "units check" of an equation. Two equations are given for power in this section, work per time and force times speed. Have students show that the units from these two expressions are the same.

➲ **DEMO 6-7** Example 6-14 involves calculating the **power output** of a jogger running up a flight of stairs. Your students can run up bleachers or stairs and calculate their own power outputs.

➲ **DEMO 6-8** If you go refrigerator shopping, you will probably find **energy efficiency ratings** for various models prominently displayed. Bring in prices and ratings of a few different refrigerators, as well as a local power bill, so you will know the cost of electricity per kilowatt-hour. Students can estimate the monthly cost of running a refrigerator, and if they find that the more-efficient models are more expensive to buy than the less-efficient models, they can also estimate the payback time for investing in the more-efficient model.

Example 6-15 discusses the power required of a car going up a hill. Extend this example by discussing why trucks are so slow going up hills. Since the mass of the truck is much greater, the force needed to keep it moving up a hill is also much greater.

Resource Information

Transparencies

T55.	Figure 6-1	A person pulling a crate along the floor
T56.	Figure 6-2	A person does no work on the bag of groceries
T57.	Figure 6-4	Example 6-2 (Work on a backpack)
T58.	Figure 6-6	The work done by a force *F*
T59.	Figure 6-11	A person exerts an upward force to lift a brick
T60.	Figure 6-12	Example 6-7 (Potential energy changes for a roller coaster)
T61.	Figure 6-13	A spring can store energy
	Figure 6-14	Spring in unstretched, stretched, and compressed positions
T62.	Figure 6-15	As a spring is stretched, the force needed increases linearly

T63.	Table 6-1	Conservative and Nonconservative Forces
T64.	Figure 6-18	Energy buckets
T65.	Figure 6-20	Example 6-10 (Speeds on two water slides)
T66.	Figure 6-24	Example 6-12 (Two kinds of PE)
T67.	Figure 6-29	Example 6-15 (Power needs of a car)

Suggested Readings

Arons, A., "Development of Energy Concepts in Introductory Physics Courses," *American Journal of Physics*, (December 1999), pp. 1063–1067. Discussion of the use of work in conservation-of-energy calculations.

Ehrlich, R., "Using a Retractable Ball Point Pen to Test the Law of Conservation of Energy," *American Journal of Physics* (February 1996), p. 176. A clever demonstration/experiment that uses the elastic potential energy of the compressed spring from a pen to launch a projectile.

Harrison, E., "Mining Energy in an Expanding Universe," *The Astrophysical Journal* (10 June 1995) v. 446, pp. 63–66. An intriguing look at whether or not energy is conserved on a cosmic scale in an expanding universe.

Hilborn, R., "Let's Ban *Work* from Physics," *The Physics Teacher* (October 2000), p. 447. A commentary on problematic terminology.

Ingham, W., "A Consistent Sign Convention for Work," *The Physics Teacher* (March 2000), p. 160. Discussion of positive and negative sign conventions for work.

Jolly, P., Zollman, D., Rebello, N. and Dimitrova, A., "Visualizing Motion in Potential Wells," *American Journal of Physics*, (January 1998), pp. 57–63. Description of a unit designed to aid learning of potential-energy diagrams that incorporates a sequence of computer-interfaced experiments using dynamics or air-track systems.

Keeports, D., "How Does the Potential Energy of a Rising Helium Balloon Change?" *The Physics Teacher* (March 2002), pp. 164–165. Considers gravity and the buoyant force, two conservative forces acting on a rising helium balloon.

Malone, J. and Holzwarth, D., "A Real Look at Speeds and Stopping Distances," *The Physics Teacher* (February 1998), pp. 95–96. A real world demonstration of the relationship of speed and stopping distance.

Mendelson, K., "Physical and Colloquial Meanings of the Term 'Work'," *American Journal of Physics* (March 2003), pp. 279–281. An interesting article on the history of the term work.

Piatek, S. and Gautreau, R., "Constant Acceleration and Kinetic Friction," *The Physics Teacher* (May 1998), p. 316. A very simple experiment connecting work, acceleration, and kinetic friction.

Sawicki, M., "Static Elongation of a Suspended Slinky™," *The Physics Teacher* (May 2002), pp. 276–278. A precalculus treatment of the statics of a loosely wound spring, using both mathemathical summation techniques and physical considerations.

Tanner, R., "Roller-Coaster Design Project," *The Physics Teacher* (March 1997), pp. 148–149. Description of a project requiring conservation of energy and other physics concepts, including a sample handout of the assignment.

Taylor, J., Carpenter, A. and Bunton, P., "Conservation of Energy with a Rubber Ramp," *The Physics Teacher* (March 1997), pp. 146–147. Instructions for construction of a conservation of energy lab and demonstration apparatus.

Velázquez-Avilés, A., "Using the Work–Energy Theorem with *Car and Driver* Website Data," *The Physics Teacher* (April 2002), pp. 235–237. A great exercise to illustrate the work-energy theorem using real data.

http://www.learner.org/exhibits/parkphysics/ is a Website devoted to amusement park physics and includes roller coasters and pendulum rides. It is maintained by The Annenberg/CPB Project. A Web search for "roller-coaster design" will lead you to numerous other related sites.

Materials

A Hooke's law spring set is available from Sargent-Welch, model number CP33855-00.

AAPT's Official Guide: Amusement Park Physics, available from www.aapt.org/catalog or by calling 301-209-3333.

One roller-coaster apparatus is available from Frey Scientific, model number S618337.

Drilled balls for use as pendulum bobs are also available from Frey Scientific.

Notes and Ideas

Class time spent on material: Estimated:______ *Actual:_____________*

Related laboratory activities:

Demonstration materials:

Notes for next time:

Chapter 7: Linear Momentum

Outline

Summary

Chapter 7 introduces the concepts of linear momentum and impulse. Momentum is a vector quantity. The conservation of linear momentum is extremely important in the treatment of collisions, both elastic and inelastic. The center of mass is defined. The cases of collisions in two or three dimensions are covered as optional topics.

Major Concepts

By the end of the chapter, students should understand each of the following and be able to demonstrate their understanding in problem applications as well as in conceptual situations.

- Linear momentum
  - $\vec{\mathbf{p}} = m\vec{\mathbf{v}}$
  - General form of Newton's second law $\sum \vec{\mathbf{F}} = \Delta\vec{\mathbf{p}} / \Delta t$
- Impulse $\vec{\mathbf{I}} = \vec{\mathbf{F}}_{\mathbf{av}} \Delta t = \Delta\vec{\mathbf{p}}$
- Conservation of momentum $\sum \vec{\mathbf{F}} = \Delta\vec{\mathbf{p}} / \Delta t = 0$
  - Internal and external forces
  - Recoil
- Collisions
  - Inelastic
  - Elastic
- Center of mass

Teaching Suggestions and Demonstrations

Most students do not have an intrinsic feel for the concept of linear momentum. You will need to work many examples comparing the momenta of everyday objects. The conservation of momentum is the second major conservation law that students encounter. Be sure to discuss the usefulness of collision theory in many areas of physics.

Sections 7-1 through 7-3

Linear momentum is defined as $\vec{\mathbf{p}} = m\vec{\mathbf{v}}$, with units kg·m/s. The term linear momentum is often shortened to "momentum." Point out that the combination $m\vec{\mathbf{v}}$ appears in Newton's second law in its original form $\sum \vec{\mathbf{F}} = \Delta\vec{\mathbf{p}} / \Delta t$ and that it comes up often enough to earn a special name. Go through Example 7-1 to show how a force can cause a change in momentum. Work several examples to illustrate the trade-off between mass and velocity in the calculation of momentum. It is very important to remind students that momentum is a **vector quantity**. Point out that the **net momentum of a system of objects** is just the vector sum of the momenta of all the objects in the system.

Conservation of momentum ranks as one of the most powerful ideas in physics. The full statement of the law of conservation of momentum is if the net force acting on an object is zero, its momentum is conserved. It is important at this point to distinguish between **external** and **internal forces** acting on an object or a system of objects. The internal forces acting on a system will always sum to zero (Newton's third law). For momentum to be conserved, the external forces must also sum to zero.

Go through Example 7-3. Point out that in this case the *system* is made up of both railroad cars. Be sure to go through Example 7-4, or another problem like it, and talk about recoil. Recoil is evident when throwing a ball on ice or when turning on a hose.

Work through the calculation in the text at the beginning of Section 7-3 to illustrate the vector nature of the **change in momentum. Impulse** is defined as $\vec{\mathbf{I}} = \vec{\mathbf{F}}_{\mathbf{av}}\Delta t = \Delta\vec{\mathbf{p}}$. Students find this concept nonintuitive. Why do we need a separate quantity for the change in momentum? Why does it have the same units as momentum? Again, the justification can come from Newton's second law. Be sure to go over the concept of average force (see Figures 7-9 and 7-10) and work several examples. Talk about the picture of the tennis ball and racket in Figure 7-8.

It is interesting to talk about impulse and momentum in relation to **traffic laws**. Why are pedestrians required to walk facing car traffic, but bicyclists required to ride in the same direction as car traffic? (It is in the best interest of a pedestrian to see traffic as it approaches; the momentum of a pedestrian is negligible compared to that of a moving car. It is in the best interest of a bicyclist to move in the same direction as the cars; if the bicyclist is hit, the impulse delivered by the car is less if the car and bicycle are moving in the same direction than if they meet head-on.)

You can also have a good class discussion about impulse and the **safety features for cars** (or bicycle helmets). Note that LARGE FORCE · short time = smaller force · LONGER TIME. Crumple zones in cars and the padding in helmets are all there to increase the time over which a collision will take place, thus reducing the force necessary for the same impulse. Refer back to the egg-catching Demo suggested for Chapter 6 as well. In that case we discussed the work done to stop an egg. If we increase the distance over which the egg is stopped, we can decrease the force required for stopping and yet do the same amount of work. In the current discussion, impulse stays constant and increasing the stopping time decreases the force required.

Sections 7-4 through 7-7

Collisions are an important tool for finding out about the atomic and subatomic world. They also have many macroscopic applications. Collisions are divided into two general types: **inelastic collisions**, which conserve only momentum, and **elastic collisions**, which conserve momentum and kinetic energy. The special case of a one-dimensional, completely inelastic collision, in which the objects stick together after the collision, is the simplest and should be treated first. If the masses of the two colliding objects and their initial velocities are known, the only unknown is the final velocity of the combined objects. Analyses of simple (rear-end) traffic collisions are interesting to the students and can give object lessons on the importance of wearing a seat belt.

If you choose to cover **two-dimensional collisions**, we recommend that you start with the simple case of a completely inelastic collision between objects with perpendicular initial velocities.

Analysis of one-dimensional elastic collisions involves the simultaneous solution of two equations, one of which is second order in velocity. Take the time to work out the equations for the final velocities (Section 7-5) and go over either Example 7-7 or 7-8. These can help the students develop intuition about collisions. The Problem Solving box on Momentum Conservation and Collisions is quite clear.

➲ **DEMO 7-1** An air track (or an air table) is an excellent demonstration device for collisions. Set up a couple of photogates for timing purposes and use carts with different masses to explore a variety of cases. Use metal spring bumpers on the carts to simulate elastic collisions and Velcro to simulate inelastic collisions. Demonstrate a variety of elastic and inelastic collisions.

➲ **DEMO 7-2** Collect three different sizes of superballs. Stack them on top of each other, with the smallest on top, and drop them together. (This takes some practice, but the result is spectacular!) Careful analysis of the collisions, assuming that the approximate ratio of the masses is 1:2:4, determines that the smallest ball bounces to nine times the original height. (The largest ball impacts with the floor and hits the middle ball, which then collides with the smallest ball. Since the balls fall about the same distance, they all have the same speed just before the initial impact.) (Caution – DO NOT DO THIS in a room with a low ceiling. In that case, use only two of the balls.)

➲ **DEMO 7-3** A Newtonian collision ball apparatus is a fairly easy and inexpensive way to demonstrate pendulum motion as well as conservation of energy and conservation of linear momentum. (See Resource Information.)

Section 7-8

The **center of mass** of a system of masses can be defined as the point where the system can be balanced in a uniform gravitational field.

➲ **DEMO 7-4** Cut a random shape out of cardboard. Punch two small holes in it near the edge at positions about 120° from each other. Suspend the cardboard from one hole, hang a plumb bob from the hole, and mark a vertical line. Repeat the process for the second hole. The point where the two lines intersect is the center of mass of the shape. The shape can be balanced horizontally at that point. (See Figure 7-25.)

For two or more objects, the position of the center of mass can be calculated as the weighted average of the positions of the objects:

$$x_{CM} = \frac{\Sigma m_i x_i}{M} \qquad y_{CM} = \frac{\Sigma m_i y_i}{M}$$

You can explain briefly (for any students who have had calculus) that these sums become integrals for the case of continuous objects. Note that the velocity and acceleration equations follow this same form, as does Newton's second law.

➲ **DEMO 7-5** Throw several extended objects (carefully) across the front of the room. Many common objects (erasers, pens, etc.) work well. A baton (or a meter stick) is particularly easy to see. Be sure to give each object some rotation about the center of mass as you throw it. Have the students watch the center of mass. They should be able to see it follow a parabolic path in each case, even as the rest of the object rotates around it.

Sections 7-9 and 7-10

These sections are both optional but interesting. If you have pre-health professions students in your class, you should plan to go through Section 7-9 and Example 7-13.

Resource Information

Transparencies

T68.	Figure 7-3	Momentum is conserved in a collision of two balls
	Figure 7-4	Forces on the balls during the collision
T69.	Figure 7-5	Example 7-3 (Railroad cars collide: momentum conserved)
T70.	Figure 7-6	A rocket at rest and in motion
T71.	Figure 7-9	Force as a function of time during a typical collision
	Figure 7-10	The average force *F* acting over an interval of time
T72.	Figure 7-14	Two small objects collide
T73.	Figure 7-17	Ballistic pendulum
T74.	Figure 7-19	Object A, the projectile, collides with object B, the target
T75.	Figure 7-21	The motion of a diver
T76.	Figure 7-23	The center of mass of a two-particle system lies on the line joining the masses
T77.	Figure 7-27	Example 7-13 (A leg's CM)
T78.	Figure 7-29	Example 7-14 (A two-stage rocket)

Suggested Readings

Cross, R., "The Bounce of a Ball," *American Journal of Physics* (March 1999), pp. 222–227. A study of the dynamics of bouncing for several common ball types, including a discussion of energy loss during each collision and differing coefficients of restitution.

DeYoung, P. and Mulder, B., "Studying Collisions in the General Physics Laboratory with Quadrature Light Emitting Diode Sensors," *American Journal of Physics* (December 2002), pp. 1226–1230. Describes a technique to measure position as a function of time in the general physics lab.

Edge, R., "The Physics of Crème Brulée," *The Physics Teacher* (October 2000), p. 441. An interesting application of an impulsive force.

Ha, M., Kim, Y. and Lee, S., "Development of an Apparatus for Two-Dimensional Collision Experiments Using a Cycloidal Slide," *American Journal of Physics* (November 2001), pp. 1187–1190. Describes an apparatus for performing high-precision experiments of conservation of linear momentum in two-dimensional collisions.

Hu, H., "More on One-Dimensional Collisions," *The Physics Teacher* (February 2002), p. 72. A rearrangement of the equations for one-dimensional collisions that results in an alternate definition of the coefficient of restitution.

Loveland, K., "Simple Equations for Linear Partially Elastic Collisions," *The Physics Teacher* (September 2000), pp. 380–381. A nice treatment of inelastic collisions that are not completely inelastic.

Millet, L., "The One-Dimensional Elastic Collision Equation: $v_f = 2v_c - v_i$," *The Physics Teacher* (March 1998), p. 186. An interesting proposal for a new treatment of one-dimensional collisions.

Nathan, A., "Dynamics of the Baseball-Bat Collision," *American Journal of Physics* (November 2000) pp. 979–990. A complete description of the collision process including bending vibrations of the bat.

Penner, A. R., "The Physics of Golf: The Optimum Loft of a Driver," *American Journal of Physics* (May 2001), pp. 563–568. Detailed consideration of the motion of a golf ball, from the impact with the club to the run after landing.

Rist, C., "Breaking Boards," *Discover* (May 2000), pp. 28–31. A very accessible article on the application of physics principles to karate.

Turner, W. and Ellis, G., "The Energetics of a Bouncing Ball," *The Physics Teacher* (November 1999), pp. 496–498. An experiment involving kinetic and potential energy of a ball during collisions with the ground.

Van den Berg, E., Nuñez, J., Guirit, A. and Van Huis, C., "Cotton Buds, Momentum, and Impulse," *The Physics Teacher* (January 2000), pp. 52–53. An interesting experiment demonstrating impulse and momentum.

Materials

See air-track listing in Chapter 4.

A Newtonian collision ball apparatus is available from Fisher Scientific, model number S40976.

Notes and Ideas

*Class time spent on material: Estimated:*______ *Actual:*_____________

Related laboratory activities:

Demonstration materials:

Notes for next time:

Chapter 8: Rotational Motion

Outline

Summary

Chapter 8 begins the study of rotational motion. The variables of rotational motion are defined, used in kinematic equations, and connected to their linear counterparts. A description of rolling is presented. The *causes* of rotational motions are also considered. Just as forces cause translational motion, torques cause rotational motion. The moment of inertia and rotational kinetic energy are defined, and energy considerations are discussed. Angular momentum and rotational work are defined, and the conservation of angular momentum in the absence of external torques is discussed. The chapter ends with an optional section on vectors in rotational motion.

Major Concepts

By the end of the chapter, students should understand each of the following and be able to demonstrate their understanding in problem applications as well as in conceptual situations.

- Angular variables
  - Angular position θ
  - Angular velocity ω
  - Angular acceleration α
- Equations for rotational kinematics
  - Connections with linear variables
  - Rolling
- Rotational kinetic energy
- Moment of inertia
- Conservation of mechanical energy
- Torque
  - Definitions
  - Dynamic applications
- Angular momentum
  - Definitions
  - Conservation of angular momentum
- Rotational work
- Vectors in rotational motion

Teaching Suggestions and Demonstrations

The mathematical description of **angular motion** is sometimes difficult for students to understand. Use lots of demonstrations, pictures, and diagrams. Point out the parallels and connections between linear and rotational motion. Many students find the concepts of **torque** and **angular momentum** confusing at best. It is important to give many examples (as always!) and to check on the students' progress at regular intervals.

Sections 8-1 and 8-2

➲ **DEMO 8-1** It is a good idea to have a bicycle wheel handy for this entire chapter. (See Resource Information.) Since the demise of the turntable, students often have trouble visualizing angular motion. You can demonstrate the angular variables easily and dramatically with a bicycle wheel. Use chalk to mark on the rim, then move the wheel through an angle θ. Tie a piece of red ribbon or yarn to a spot on the rim, and another to a spot on one of the spokes to demonstrate the difference in tangential velocity at different distances from the center of rotation.

As you discuss the **angular position**, you will need to go over the concepts of **radians**, the **sign conventions for angular position**, **conversions** (**radians**, **revolutions**, and **degrees**), and **arc length**. Most students have covered these topics before, but most will not be familiar enough with them to use them easily.

There are many definitions and small equations in this chapter. Have students begin an equation card to help keep them organized. (A 3" by 5" card is fine.) Include the definitions for ω, α, the period T, tangential speed v, centripetal acceleration, and tangential acceleration. Remind the students of the value of dimensional analysis to help deduce the form of these equations.

Refer to Table 8-1, which gives the **correspondences between linear variables and angular variables**, and to Equations 8-9 a–d, which give the linear kinematic equations and the angular kinematic equations. These will aid in making the connections between the variables and in recognizing that these "new" equations aren't really new; they are just restatements of the familiar kinematic equations. It is important to point out that the angular quantities θ, ω, and α are the same for all parts of a rotating rigid object. The linear quantities x, v, and a will be different for different parts of the object.

Take time to work several kinematic examples. The students will need to see how various quantities fit into the solutions of problems, especially when the angular variables are given in different units (revolutions, radians, degrees).

Be sure to discuss the two **components (centripetal** and **tangential) of acceleration**. Students who have become used to one-dimensional acceleration will need extra help with this concept.

Section 8-3

Rolling is a combination of rotational motion and translational motion.

➲ **DEMO 8-2** Use the bicycle wheel to demonstrate **rolling**. Mark a point on the rim of the wheel, and roll the wheel along a tabletop. Ask a series of questions. What kind of motion is the center of mass of the wheel experiencing? What about the point on the rim? How does the point on the rim move in the vertical direction? In the horizontal direction? When is it moving fastest? Slowest? This will help the students develop an intuitive feel for rolling and make the mathematical treatment easier to understand.

Figure 8-8 gives a good illustration of the **horizontal velocity vectors** for different points on the wheel considered in different reference frames.

Sections 8-4 through 8-6

All students have experienced **torque**. Talk about the many everyday examples. Go over Figure 8-10 and discuss the torque involved in opening a standard door, turning a doorknob, and using a lug wrench or screwdriver. For the health-profession majors, you can point out that special doorknobs and water faucets made for people with arthritis are designed to increase r, decrease F, and allow straight-line (rather than turning) motions. You will need to spend some time on the general definition of torque and on the **sign conventions** in Section 8-4. Work Example 8-8 in class.

➲ **DEMO 8-3** Bring in several tools, especially various kinds of screwdrivers and wrenches. Ask the students to decide which would provide more torque for the same force. (Ask them to justify their answers.)

Torques cause **angular accelerations**. The equation that connects torque and angular acceleration is $\tau = I\alpha$, where I is the **moment of inertia**. Remind the students that in rotational motion I plays the role that m plays in translational motion. (See Figure 8-21 for a collection of moments of inertia for common solids.)

➲ **DEMO 8-4** It is useful to bring a meter stick to class for these sections. You can let it rotate about the center or about an end, and easily measure the distance from the axis of rotation to the line of action of the force.

Justifying the **moment of inertia** without calculus is difficult. The discussion at the beginning of Section 8-5 is quite good. Go through Example 8-20 in class and talk about the idea that the moment of inertia gives information about the distribution of mass about an axis of rotation. The farther the mass is from the axis, the larger the moment of inertia. The larger the moment of inertia, the greater the resistance to change in rotational motion. For instance, tightrope walkers hold long poles to increase their moments of inertia. Gymnasts on balance beams hold out their arms for the same reason. Try balancing a tennis racket (or hammer) on your fingertip. It is easier to balance with the heavy end up, away from your fingertip.

The Problem Solving box in Section 8-6 is very useful. Go over it in class and work Examples 8-11 and 8-12.

Section 8-7

You can use the correspondence of the angular quantity I to the translational quantity m to deduce the form of the **rotational kinetic energy**.

The inclusion of the rotational kinetic energy in the **total mechanical energy** of a rolling object seems reasonable but is often not obvious to students. Use Example 8-13 to help illustrate this concept.

➲ **DEMO 8-5** Set up the apparatus for Conceptual Example 8-14 and do the experiment with several different solids. You can make this into a real competition that the students will enjoy and remember. Use the disk, hoop, and sphere, but also use small cars. Have the students calculate the moment of inertia for a toy car (remember, only the wheels rotate!) and predict which car or other object will get to the bottom of the incline the fastest.

The correspondence between the equation for translational work and the equation for **rotational work** makes this concept clear for most students. You will need to discuss the work-energy theorem again and do several examples. Emphasize the simplification of the calculation if problems are treated with the work-energy theorem rather than with rotational kinematics.

Section 8-8

There are several points about **angular momentum** that students will not understand immediately. One is that the angular momentum, like torque, depends on the **choice of axis**. Another is that a particle can have angular momentum about an axis while moving in a straight line. Still another is that a particle moving along a radial path has no angular momentum about an axis through the center of the circle. Go through examples that illustrate these points.

Remind the students that the angular momentum plays the role in rotational motion that the linear momentum plays in translational motion. Just like the change in linear momentum with time is equal to the force, the change in angular momentum with time is equal to the torque. This implies that if the net torque is zero, angular momentum is constant in time. There are many good examples of **conservation of angular momentum**. Talk about divers, skaters, and gymnasts. Be sure to point out that in all of these cases it is possible to change the moment of inertia, and thus change the angular velocity, without changing the angular momentum.

➲ **DEMO 8-6** **Conservation of angular momentum** can be a fun concept to illustrate. Sit on a rotating stool (or better yet, have a student sit on the stool) while holding dumbbells or hand weights at arm's length. (See Resource Information.) Have a student help you start spinning (slowly!). Once you are in motion, bring the weights in close to your torso. As you reduce your moment of inertia, you will increase your angular velocity.

Try this while holding a bicycle wheel instead of the weights. Hold the wheel by the axle and have a student spin it. As you sit on the stool, you can make yourself rotate by trying to change the position of the axis of the wheel. See if the students can explain why!

Section 8-9

This optional section covers the **right-hand rules for angular velocity, angular momentum, and torque**. If you have access to a bicycle wheel (or other disk), you can demonstrate these rules quite easily. Many students find the right-hand rules helpful in visualizing rotational motions. It is a good idea to spend a little time on the qualitative aspects of these concepts.

➲ **DEMO 8-7** Demonstrate the **precession of the axis of a gyroscope**. While you have the gyroscope out, you can also do other fun tricks, such as having it balance on a tightrope. Use this demo to lead into a discussion of why it is easy to stay upright on a moving bicycle (with turning wheels) but nearly impossible to keep from tipping over if the bicycle is stopped.

Resource Information

Transparencies

T79.	Figure 8-1	A wheel rotating about an axis through its center
T80.	Figure 8-2	Example 8-2 (Birds of prey—in radians)
T81.	Figure 8-5	A wheel rotating uniformly counterclockwise
T82.	Figure 8-6	A rotating wheel whose angular speed is increasing
	Figure 8-7b	Example 8-4 (Angular and linear velocities and accelerations)
T83.	Table 8-1	Linear and Rotational Quantities
T84.	Figure 8-10	Applying the same force with different lever arms
T85.	Figure 8-13	Torque = $r_{\perp}F = rF_{\perp}$
T86.	Figure 8-14	Example 8-8 (Biceps torque)
T87.	Figure 8-21	Moments of inertia for various objects of uniform composition
T88.	Figure 8-26	A sphere rolling to the right on a plane surface
T89.	Figure 8-27	Torque $\tau = rF$ does work when rotating a wheel
T90.	Figure 8-28	A skater doing a spin on ice
T91.	Figure 8-31	Rotating wheel

Suggested Readings

Altshuler, K. and Pollock, P., "Inexpensive Rotating-Arm Device for Angular Motion Labs," *The Physics Teacher* (October 1998), pp. 424–425. Describes a very inexpensive device that can be constructed and used for demonstration of angular motion.

Bracikowski, C., "Feeling the Forces That Produce Torques," *The Physics Teacher* (January 1998), p. 15. A simple modification to the classic introductory experiment that allows students to experience the forces that produce torques.

Cox, A., "Angular Momentum and Motorcycle Counter-Steering: A Discussion and Demonstration," *American Journal of Physics* (November 1998), pp. 1018–1020. A demonstration of torque and angular momentum.

Cross, R., "The Trajectory of a Ball in Lawn Bowls," *American Journal of Physics* (August 1998), pp. 735–738. A discussion of the trajectory and precession of a weighted ball rolling on a smooth horizontal surface.

Denardo, B., "Demonstration of the Parallel-Axis Theorem," *The Physics Teacher* (January 1998), pp. 56–57. A simple demonstration for the parallel-axis theorem.

Gluck, P., "MBL Experiment in Angular Momentum," *The Physics Teacher* (April 2002), pp. 230–234. Studies the loss and conservation of angular momentum using a small direct current motor as generator.

Henderson, C., "Measuring the Forces Required for Circular Motion," *The Physics Teacher* (February 1998), pp. 118–121. A description of an apparatus that allows students to measure the forces exerted in circular motion.

Ho, A., Contardi, L., Dion, P. and Griffioen, E., "Rotating Wheels as Seen on Television," *The Physics Teacher* (September 1998), pp. 367–369. An analysis of the motion of rotating wheels as seen under stroboscopic lighting.

Johns, R., "Physics on a Rotating Reference Frame," *The Physics Teacher* (March 1998), pp. 178–180. A description of a simple apparatus useful in illustrating motion in rotating reference frames.

McDonald, K., "Physics in the Laundromat," *American Journal of Physics* (March 1998), pp. 209–211. Analysis of the spin cycle of a washing machine drum.

Mei, W. and Wilkins, D., "Making a Pitch for the Center of Mass and the Moment of Inertia," *The American Journal of Physics* (September 1997), pp. 903–907. Description of some simple experiments to measure moments of inertia of certain physical pendula.

Mohazzabi, P., "Free Fall and Angular Momentum," *American Journal of Physics* (November 1999), pp. 1017–1020. A discussion of the influence of the Earth's rotation on freely falling objects.

Pechan, M., O'Brien, A. and Burgei, W., "Conservation of Angular Momentum Apparatus Using Magnetic Bearings," *The Physics Teacher* (January 2001), pp. 26–28. A simple laboratory exercise on angular momentum.

Pritchett, T., Nelson, R., Creamer, T. and Oldaker, B., "Does an Ideal Wheel Really Rotate About Its Instantaneous Point of Contact?," *The Physics Teacher* (March 1998), pp. 167–170. An interesting article on rolling.

Schönhammer, K., "Elementary Theoretical Description of the Heavy Symmetric Top," *American Journal of Physics* (November 1998), pp. 1003–1007. A theoretical description of a symmetric spinning top employing conservation of energy and presented without the use of Lagrangian mechanics.

Materials

A bicycle wheel gyroscope is available from Fisher Scientific, model number S40965.

A ring and disk suitable for rolling down an inclined plane are also available from Fisher Scientific, model number S41013.

A rotating stool is available from Fisher Scientific, model number S51543.

Notes and Ideas

*Class time spent on material: Estimated:*______ *Actual:*______________

Related laboratory activities:

Demonstration materials:

Notes for next time:

Chapter 9: Static Equilibrium; Elasticity and Fracture

Outline

Summary

This chapter treats static equilibrium, the case in mechanics when both the sum of the forces and the sum of the torques on an object are zero. The principles of static equilibrium are applied to muscles and joints and to balance. Stress, strain, and fracture, and arch and dome construction are covered in optional sections.

Major Concepts

By the end of the chapter, students should understand each of the following and be able to demonstrate their understanding in problem applications as well as in conceptual situations.

- Conditions for static equilibrium
  - Sum of the forces = 0
  - Sum of the torques = 0
- Types of equilibrium
  - Stable
  - Unstable
  - Neutral
- Mechanical properties of solids
  - Young's modulus
  - Stress and strain
  - Bulk modulus
  - Fracture

Teaching Suggestions and Demonstrations

The conditions for static equilibrium seem reasonable to most students. Students often expect the problems for this case to be much easier than previous problems and are quite surprised when they are not. The key here (as always!) is to work as many examples in class as possible.

Sections 9-1 through 9-3

Translational equilibrium combines ideas from Chapter 4 with new concepts from Chapter 8. Remind students that $\sum \vec{\mathbf{F}} = 0$ is a special case of $\sum \vec{\mathbf{F}} = m\vec{\mathbf{a}}$. Students will need to see many examples. Emphasize the importance of resolving vectors into their components. Be sure to give several examples of objects in **static equilibrium**, such as bridges, buildings, playground structures, or sawhorses. Point out in each case that just having the **sum of the forces** equal to zero is not enough. The **sum of the torques** must also be zero. You may want to mention that the text restricts consideration to planar forces (*x* and *y* directions), which restricts the torques to a single axis.

Emphasize that the **torque is always taken about an axis of rotation**. The axis can be placed at any location, but once placed, it must stay put for the rest of the problem. All torques in the problem must be computed about the same axis. Computations in problems can be made harder or easier by the choice of axis. It is important to choose wisely. Be sure to work several examples. Example 9-1 is a good one to work, since many of the students will have worn braces and be familiar with this situation. The Problem Solving box in Section 9-2 is especially helpful. Examples 9-6 and 9-7 will allow you to talk about the role of the normal force and friction in static equilibrium.

If you have health professions majors in your class, plan to spend class time on Section 9-3. Examples 9-8 and 9-9 and problems like them will be of interest.

Section 9-4

The idea that an object **balances** if it is suspended from its center of mass is easy to understand when the object is symmetric, like a meter stick or a barbell. It is much more difficult to understand when the object has a nonuniform mass distribution.

➲ **DEMO 9-1** Bring in a mobile. You can make one out of coat hanger and cardboard and have the students help balance it. Ask them how they know how to balance the system. Once they have figured out what they are doing intuitively, you can quantify their reasoning with the torque equation for static equilibrium and the center-of-mass equation.

Remind the students that the center of mass is a weighted average of the positions of the masses and not necessarily the geometric center of the mass distribution. Figures 9-16 and 9-17 are good for beginning discussion.

Sections 9-5 through 9-7

These sections contain some really fascinating material, but if time is limited, you can skip them entirely. If you do have time, **stretching and Young's modulus** and **shear deformation and the shear modulus** are explained simply and quickly in the text. The discussion of the change in the volume of a solid with increasing pressure includes the definition of the **bulk modulus**. Point out the very different behavior of gases, liquids, and solids when external pressures are applied. **Stress**, **strain**, and **elastic deformation** are also presented clearly; be sure to talk about the **elastic limit**. Make sure students understand that strain is a ratio and therefore has no units. Conceptual Example 9-12, on the Kansas City hotel lobby collapse, is interesting and can be covered quickly if you have limited time for these sections.

➲ **DEMO 9-2** Use a foam cube or a Matter Model (see Resource Information) to demonstrate the three types of **stress**.

Resource Information

Transparencies

T92. Figure 9-3 Forces on a tooth

T93.	Figure 9-4	Example 9-2 (Chandelier cord tension)
T94.	Figure 9-8	A 1500-kg beam supports 15,000-kg machine
T95.	Figure 9-9	A cantilever
T96.	Figure 9-10	Example 9-6 (Hinged beam and cable)
T97.	Figure 9-11	A ladder leaning against a wall
T98.	Figure 9-12	The biceps (flexor) and triceps (extensor) muscles in the human arm
T99.	Figure 9-14	A person bending over
T100.	Figure 9-15	(a) Stable equilibrium, and (b) unstable equilibrium
T101.	Figure 9-19	Applied force versus elongation for a typical metal under tension
T102.	Figure 9-22	The three types of stress for rigid objects
T103.	Figure 9-31	Forces in (a) a round arch, compared with those in (b) a pointed arch

Suggested Reading

Gallant, J., "The Shape of the Eiffel Tower," *American Journal of Physics* (February 2002), pp. 160–162. An interesting discussion of the design of the Eiffel Tower and the balance of the maximum torque created by wind and the torque due to the tower's weight.

King, R., *Brunelleschi's Dome*, New York, NY, Penguin Group (USA) Inc., 2000. An excellent short biography of the architect of the dome of Santa Maria del Fiore in Florence.

Materials

A Matter Model is available from Pasco (ME-9825).

A Stress/Strain Apparatus is also available from Pasco (AP-8216 or AP-8214)

Notes and Ideas

*Class time spent on material: Estimated:*______ *Actual:*____________

Related laboratory activities:

Demonstration materials:

Notes for next time:

Chapter 10: Fluids

Outline

Summary

Chapter 10 begins with general definitions of density and pressure and then moves to a study of fluids, substances that readily flow. Both liquids and gases are fluids. Fluid statics, the study of fluids at rest, and fluid dynamics, the study of moving fluids, are both examined in this chapter.

Major Concepts

By the end of the chapter, students should understand each of the following and be able to demonstrate their understanding in problem applications as well as in conceptual situations.

- Density
- Pressure
  - General definition
  - Pressure exerted by a fluid
  - Atmospheric, gauge, and actual pressure
- Fluid statics
  - Equilibrium
  - Pascal's principle
  - Archimedes' principle and buoyancy
- Fluid dynamics
  - Equation of continuity
  - Bernoulli's equation

Teaching Suggestions and Demonstrations

There are many everyday applications of the concepts in Chapter 10. Students will know that airplanes can fly and boats can float. They've put air in their car tires and heard that the barometer is falling. And if their dance partner is going to step on their toes, they would rather it be with flat-bottomed shoes than with high heels! In this chapter, you will help them relate all these experiences to physics.

Sections 10-1 through 10-4

Before beginning a study of **fluids**, students need to understand **phases of matter**, **density**, and **pressure**. Make sure some of your examples involve solids so that students realize that these general definitions apply to substances that aren't fluids as well as to liquids and gases. Emphasize the difference between **gauge pressure** and **actual pressure**. A completely flat tire has a gauge pressure of zero but an actual pressure of about 14.7 psi, since the pressure in the tire equals the atmospheric pressure.

➲ **DEMO 10-1** An assortment of different blocks, some the same size but made out of different materials, others of the same material but different sizes, can help you check student understanding of **density**. Hold up pairs of blocks and ask students which has the greater density. The key is to pay attention to the material, not the size. You can also preview Archimedes' principle by asking them to guess which blocks would float in water and which would sink.

➲ **DEMO 10-2** An assortment of shoes makes a nice demonstration of **pressure**. A person with a given weight will exert different pressures on the floor depending on the surface area of the shoes in contact with the floor. You can do rough calculations by measuring the bottoms of the shoes. On a hot summer day, high heels can actually sink into asphalt parking lots! Winter backpackers wear snowshoes or cross-country skis to keep from sinking into the snow. Ice skates work because the thin blade creates so much pressure that the ice right under the blade melts.

Students will intuitively understand that **pressure in a liquid** increases with depth, but you may need to work harder to convince them that only the depth matters. If a funnel-shaped container and a cylindrical container are filled to the same height with water, the pressure at the bottom of each will be the same. Students will be tempted to think that the funnel will somehow "concentrate" the pressure. Figure 10-4 and the discussion with it will also help clarify the relationship between depth and pressure. Point out that the formula doesn't work for atmospheric pressure since the density of the atmosphere is not constant.

Sections 10-5 and 10-6

The hydraulic lift shown in Figure 10-7 is a great example of **Pascal's principle**. Remind students that energy is still conserved, so the magnification of the force has to be "paid for" somehow. The distance through which the force is applied is the key. A small force applied over a large distance results in the same work as a large force applied over a small distance.

There are many ways to measure pressure. Figure 10-8 shows several kinds of pressure gauges. A barometer is a clever device that balances atmospheric pressure with pressure due to an incompressible fluid so that the formula can be applied. The common unit of atmospheric pressure, **millimeters of mercury**, is a direct result of this measuring device (see Figure 10-9). Have students calculate what atmospheric pressure in millimeters of water would be and they will understand why using a denser liquid such as mercury is more practical. Table 10-2 lists other common units of pressure and conversion factors.

While on the topic of barometers, you can take a few minutes to discuss other **weather topics** related to pressure. Point out that fair weather is associated with high-pressure systems and stormy or rainy weather with low. Hurricanes and tornadoes have very low pressure at their centers. When a hurricane passes over a house, the outside pressure drops suddenly, and if the inside pressure doesn't drop also, the roof may get blown off. Students living in hurricane-prone areas will have heard that it is a good idea to open windows if a hurricane is on the way.

Section 10-7

➲ **DEMO 10-3** Introduce **Archimedes' principle and buoyancy** through a demonstration. Suspend a metal block from a spring scale to find its weight. Then lower the block into a beaker of water and notice that the force registered by the spring scale is less. Draw a force diagram to illustrate that another force is acting upward on the block when it is submerged in the water. This is the **buoyant force**, equal to the weight of the displaced water.

Use Figures 10-11 and 10-12 and derive Archimedes' principle, then do lots of examples with objects that float or sink in water. Students will quickly notice that objects with densities less than the density of water will float and objects with densities greater than the density of water will sink. They will tend to think that the buoyant force on an object depends on its density, which isn't quite true. If blocks with different densities but equal volumes are submerged, they will all have the same buoyant force since they all displace the same volume of water. Also do some examples for objects that float in liquids other than water. Examples 10-9 and 10-10 provide a good check of student understanding.

Sections 10-8 through 10-10

These sections deal with fluids in motion. The **equation of continuity** will probably be intuitive for students. They will be familiar with the fact that when the cross-sectional area of a pipe decreases, the fluid flowing through it speeds up. That's why you put your thumb over the end of a hose when you want the water to spray. Go through Example 10-11 in class.

Bernoulli's equation may actually be counterintuitive for many students. In the case where potential energy remains constant, the pressure in a flowing fluid will be less where the speed is greatest.

➲ **DEMO 10-4** Begin your demonstrations of **Bernoulli's equation** by making a paper arch out of a piece of paper and placing it on a tabletop. Blow through the arch. Instead of blowing the arch off the table, your breath flattens it! For another dramatic demonstration, place a Ping Pong ball in a funnel and challenge a student to blow it out. When air is blown through the funnel, the air-pressure difference between the top and bottom of the Ping-Pong ball will hold the ball in place, even when the funnel is inverted (see Figure 10-25).

Two more examples to discuss are airplane wings (see Figure 10-26) and curve balls in baseball (see Figure 10-28). The velocity of the air on either side of a spinning ball is different because of the drag caused by the ball itself. The side with higher-speed air will have lower pressure, so the ball will curve unexpectedly in that direction. **Torricelli's theorem** is a good example to include in your discussion as well because it illustrates an application involving change in potential energy. Health professions majors will be interested in the discussion of transient ischemic attacks (TIAs) near the end of Section 10-10.

Sections 10-11 through 10-14

These optional sections discuss viscosity, surface tension, and a number of other interesting topics. They can be skipped without loss of continuity.

Resource Information

Transparencies

T104.	Table 10-1	Densities of Substances
T105.	Figure 10-4	Forces on a thin slab of fluid
T106.	Figure 10-5	Example 10-3. (Pressure at a faucet)
T107.	Figure 10-7	Applications of Pascal's principle
T108.	Figure 10-8	Pressure gauges
T109.	Table 10-2	Conversion Factors Between Different Units of Pressure
T110.	Figure 10-11	Determnation of the buoyant force
T111.	Figure 10-14	Buoyant force
T112.	Figure 10-15	The fully submerged log accelerates upward
T113.	Figure 10-19	(a) Streamline, or laminar, flow; (b) turbulent flow
T114.	Figure 10-20	Fluid flow through a pipe of varying diameter
T115.	Figure 10-21	Human circulatory system
T116.	Figure 10-23	Fluid flow: for derivation of Bernoulli's equation
T117.	Figure 10-25	Examples of Bernoulli's principle
	Figure 10-26	Lift on an airplane wing
T118.	Figure 10-27	Sailboat sailing against the wind
T119.	Figure 10-29	Rear of the head and shoulders showing arteries leading to the brain
T120.	Figure 10-35	U-shaped wire apparatus holding a film of liquid
	Figure 10-37	Surface tension acting on (a) a sphere, and (b) an insect leg
T121.	Figure 10-40	One kind of pump
	Figure 10-41	Additional pumps
T122.	Figure 10-42	The human heart

Suggested Readings

Denardo, B., Pringle, L., DeGrace, C. and McGuire, M., "When Do Bubbles Cause a Floating Body to Sink?" *American Journal of Physics* (October 2001), pp. 1064–1072. Describes qualitative lecture demonstrations that show that bubbles can sink a floating body, as well as a related quantitative experiment.

Eastlake, C., "An Aerodynamicist's View of Lift, Bernoulli, and Newton," *The Physics Teacher* (March 2002), pp. 166–173. Outlines alternative, but not conflicting, models to explain lift on an airfoil.

Gaffney, C., "The Hydrostatics of Trapped Bubbles in Fluids," *The Physics Teacher* (November 2000), pp. 458–460. A clever look at bubbles trapped in fluids and their effect on fluid level height.

Greene, N. and Dworsak, M., "Bernoulli at the Gas Pump," *The Physics Teacher* (September 2001), pp. 346–348. Discusses the basic physics of the shutoff sensing mechanism in a gas pump.

Heron, P., Loverude, M., Shaffer, P., and McDermott, L., "Helping Students Develop an Understanding of Archimedes' Principle. II. Development of Research-based Instructional Material," *American Journal of Physics* (November 2003), pp. 1188–1195. Describes materials developed after research on student understanding of Archimedes' principle.

Loverude, M., Kautz, C. and Heron, P., "Helping Students Develop an Understanding of Archimedes' Principle. I. Research on Student Understanding," *American Journal of Physics* (November 2003), pp. 1178–1187. Describes research on student understanding of Archimedes' principle.

Noll, E., "Confronting the Buoyant Force," *The Physics Teacher* (January 2002), pp. 8–10. A description of a challenge problem involving the buoyant force, and student-designed methods for its solution.

Ouellette, J., "The Physics of . . . Foam," *Discover* (June 2002), pp. 26–27. A short article on the physics of soft matter.

Smith, D., "Flexural Strength in Windows During Hurricanes," *The Physics Teacher* (October 2000), pp. 400–402. Nice discussion of pressure changes and bulging windows during hurricanes.

Velasco, S., González, A., Román, F. and White, J., "A Simple Method for Measuring Atmospheric Pressure," *American Journal of Physics* (December 2002), pp. 1236–1237. Describes a simple, low-cost method for measuring atmospheric pressure.

Waltham, C., "The Flight of a Balsa Glider," *American Journal of Physics* (July 1999), pp. 620–623. Analysis of the flight of a small toy glider, with comparison to full-sized aircraft.

Whitesides, G. and Stroock, A., "Flexible Methods for Microfluidics," *Physics Today* (June 2001), pp. 42–48. Applications for handling nanoliter quantities of fluids.

Notes and Ideas

Class time spent on material: Estimated:______ *Actual:_____________*

Related laboratory activities:

Demonstration materials:

Notes for next time:

Chapter 11: Vibrations and Waves

Outline

Summary

General periodic motion and the special case of simple harmonic motion are treated thoroughly in Chapter 11. Students use their knowledge of circular motion to derive equations for the position, velocity and acceleration of an object undergoing simple harmonic motion. The specific cases of a mass on a spring and a pendulum are emphasized. Conservation of energy is applied to oscillating systems. Real-world situations involving physical pendulums, damped oscillations, and resonance are discussed. The second half of this chapter introduces the student to basic wave properties and characteristics: superposition and interference, standing waves, and refraction and diffraction.

Major Concepts

By the end of the chapter, students should understand each of the following and be able to demonstrate their understanding in problem applications as well as in conceptual situations.

- Periodic motion
  - Frequency
  - Period
- Simple harmonic motion
  - Sine and cosine curves
  - Connection to uniform circular motion
  - Position, velocity, acceleration
  - Angular frequency
- Mass on a spring
- Simple pendulum
- Conservation of energy applied to oscillating systems
- Damped and driven oscillations and resonance
- Waves
  - Transverse and longitudinal

  - Wavelength and frequency
  - Speed of a wave
- Superposition and interference
  - Constructive and destructive
  - Phase
  - Standing waves

Teaching Suggestions and Demonstrations

Until this point in the course, most of the trigonometry that students have been using has involved the basic sine, cosine, and tangent function definitions applied to triangles. Now is the time to remind students to review the **graphs of sine and cosine functions** or even review briefly in class if you have time. In order to tackle oscillatory motion, students need to be comfortable expressing angles in radians and sketching graphs of sine and cosine functions.

Sections 11-1 through 11-3

Begin by asking students to name as many examples of **periodic motion** as they can. The human heartbeat is a common one. You can also bring in a graph of tides to illustrate the wide variety in motions that are periodic. Next, students can derive Equation 11-2, the relationship between **frequency** and **period**. They will understand the inverse relationship better if they think about it before they actually see the equation. For instance, ask them, "If you run around a track six times in one hour, how long did it take you to complete just one lap?" Six cycles per hour is the frequency and one sixth of an hour (or ten minutes) is the period. Try other examples with the more common units of **hertz** (cycles per second) for the frequency and seconds for the period.

➲ **DEMO 11-1** Throughout your discussion of this chapter, it's a good idea to have a **spring** and a **simple pendulum** on hand. Set up an oscillation with either and have students determine frequency and period and verify the relationship between them. Some students will probably have stopwatches built into their wristwatches. Others can try Galileo's technique and use their pulse rate as a timing device!

The spring and pendulum lead the discussion nicely from general periodic motion to **simple harmonic motion**. Derive the equations for the position, velocity and acceleration of an object in simple harmonic motion by using the connection to **uniform circular motion**, as shown in Section 11-3. Students will need to spend some time with these equations after they have been derived to fully understand them. Superimposing all three graphs can help emphasize the phase relationships. To tie it in to real-life motion, set the spring oscillating again and ask them to describe the relationship between velocity and position. Conceptual questions are also helpful. If the position of the mass is positive and the velocity is negative, where might the mass be? Which direction is it moving? Is it moving toward or away from its equilibrium position? Is it speeding up or slowing down? Is its acceleration positive or negative? What point on the graph might represent this point in the oscillation? Use Figure 11-9 to illustrate answers to these questions.

Often, students have the most trouble conceptualizing the **acceleration graph**. Remind them of Newton's second law: The acceleration is proportional to the force exerted by the spring. From Hooke's law, we also know that the force is a restoring force, proportional to the displacement but in the opposite direction. Relating both position and acceleration directly to force in this way clarifies the relationship between them. Go through Example 11-8 to check for understanding.

➲ **DEMO 11-2** Before deriving the equation for the **period of a mass on a spring**, bring in a number of different springs and masses and lead students in a determination of which factors influence period and which do not. Does increasing the mass increase, decrease, or have no effect on the period? Does the same mass on a stiffer spring have a period that is longer, shorter, or the same? Does increasing the amplitude (that is, how far the mass is pulled from equilibrium before it is released) have any effect?

The demonstration will lead students to qualitative results, such as the fact that when mass increases, so does period. The exact dependence—period increases as the square root of mass—will become clear in the derivation. Equating Hooke's law and Newton's second law leads to an expression for the **angular frequency for a mass on a spring** in terms of the mass and the spring constant. The period is defined from the angular frequency.

The concept of **potential energy of a spring** was introduced in Chapter 6. Use Figure 11-5 to discuss how the total energy of a mass on a spring remains constant (neglecting friction) but the proportions of kinetic and potential vary. Work Conceptual Example 11-3 or an example like it. Adding friction between the block and the table is a nice extension and provides students with a review of friction and of the usefulness of energy considerations, even when mechanical energy is not conserved.

Section 11-4

➲ **DEMO 11-3** The **period of a simple pendulum** can be introduced the same way as the period of a mass on a spring, by letting students adjust different variables and determine which ones affect the period and how. It will probably come as a surprise to them that changing the mass does not change the period. (Remind them that if air resistance can be neglected, all objects near the surface of the Earth accelerate at the same rate.)

In deriving the equation for the **period of a simple pendulum**, refer to Figure 11-12 and Table 11-1 to help students understand the small angle approximation and that the equation is valid only for oscillations with small amplitudes. You may not have enough time to discuss the case of the physical pendulum in detail, but it is worth pointing out that legs are essentially pendulums, though not simple ones. Ask students if they have seen Dachshunds walk. The period of oscillation for a Dachshund leg is very different from that of a Great Dane!

Another interesting application is the **Foucault pendulum**, invented in 1851 and used to demonstrate the Earth's rotation. The pendulum appears to change its path throughout the course of the day, but in fact the floor under it is moving, carried by the rotation of the Earth.

➲ **DEMO 11-4** A Newtonian collision ball apparatus is a fairly easy and inexpensive way to demonstrate pendulum motion as well as conservation of energy and conservation of linear momentum. (See Resource Information.)

Sections 11-5 and 11-6

Resonance and natural frequency and **damping** are important concepts to touch on here because both appear later in the course with waves and again with ac circuits. Most students will be familiar with the "feel" of pushing someone on a swing; they just won't realize that what they are doing is driving a system at its resonant frequency! The video of the collapse of the Tacoma Narrows bridge is quite impressive and well worth showing to your class. (See Resource Information.) Go over Figure 11-15 and discuss the three general categories of damping an oscillating system.

Sections 11-7 through 11-11

Basic definitions and **wave characteristics** are covered in these sections. Emphasize Equation 11-12, relating **speed, wavelength, and frequency**, which will reappear when light is discussed later in the course.

➲ **DEMO 11-5** A Slinky is another handy prop to bring to class throughout this chapter. (See Resource Information.) As your first demonstration, use it to illustrate the difference between **transverse and longitudinal waves**. Transverse waves are set up by shaking one end of the stretched Slinky up and down, longitudinal waves by gathering a number of coils in your hand and releasing them. (See Figure 11-24.) Tie a piece of red yarn somewhere on the Slinky and repeat the demonstrations a few times so that students can watch the whole Slinky and then concentrate on just the motion of the yarn. The yarn not only emphasizes the difference between the types of waves, since it moves perpendicular to the direction of propagation of the wave pulse in one case and parallel in the other, it also illustrates the basic fact that a wave is the **propagation of energy**, not of matter. Even though the pulse makes it to the opposite end of the Slinky, the yarn never moves far from its original position.

➲ **DEMO 11-6** A climbing rope or long jump rope works well for investigating the **speed of a wave on a string**. Fix one end of the rope and send a pulse along its length; repeat for different tensions. It should be apparent that speed increases if tension increases. The square root dependence will not be obvious from this qualitative demonstration, but dimensional analysis, while not a derivation, makes a convincing argument. The rope can also be used to demonstrate **reflections**. The behavior of a pulse when it reaches the end of the rope depends on whether or not the end is fixed. A similar phenomenon for light waves reflecting off boundaries between different media will be very important to the thin film interference discussion later in the course.

The typical *picture* of a **harmonic wave** is a sine or cosine curve. It is worth spending a little time ensuring that students know what they are looking at. Point out that you can't really have a complete **representation of a wave with a graph** because a wave is moving. What you can have is either a snapshot of the whole wave at one instant in time (y versus x) or a graph of the displacement of one bit of medium over time (y versus t). In the Slinky demonstration, the *snapshot* would be a picture of the whole Slinky at one moment, and the *graph* would be the position of the red yarn as a function of time.

Section 11-9 gives a nice treatment of the energy transported by waves. Go through Example 11-13 in class. Section 11-10 is optional, but plan to mention that the intensity of a wave is proportional to the square of the amplitude and the square of the frequency.

Sections 11-12 through 11-16

To introduce students to **superposition and interference** of sound waves, have them consider interfering water waves first. They will probably be familiar with the patterns that arise from interfering circular waves created when two pebbles are dropped in water. Figure 11-37 is a good pictorial representation. Figures 11-36 and 11-38 are also clear and useful.

Introduce **standing waves** as an interference of a wave with its reflection under the right conditions. Drawing the two interfering waves at quarter-period intervals helps show students that the interference will alternate between constructive and destructive. Even when the interference is constructive, the nodes remain still.

➲ **DEMO 11-7** A Slinky or a long spring can be used to demonstrate **standing waves**. If students are doing the demonstration, challenge them to create standing waves representing the different harmonics in order, with an increasing number of nodes and antinodes. In addition to seeing the standing waves, they will recognize that they need to increase the frequency of vibration to create more antinodes. Relate this observation to the relevant formulas. In this case, the length and the tension are kept constant.

➲ **DEMO 11-8** You can also demonstrate **standing waves** with a string attached to a vibrator. (See Resource Information.) Orient the string horizontally with the free end over a pulley and attached to a weight hanger. As different amounts of weight are added to the vibrating string, different standing wave patterns will be produced. In this case, frequency and length are kept constant, but the tension in the string, provided by the weights, is a variable. Notice that a higher tension is required to produce fewer antinodes. Again, have students relate the observations to the relevant equations. This demonstration can also be done quantitatively. Measure the mass and length of the string as well as the force on the end, and the frequency can be calculated and compared to the known vibrator frequency.

Sections 11-14 and 11-15 are optional here and are treated in greater depth in the chapters on optics. If you have time, go over the concepts briefly, and plan to spend more time on them in the second half of the course. We do, however, recommend covering Section 11-16, which is optional and can be skipped without loss of continuity. We think it is important for the students to see the equations for traveling waves.

Resource Information

Transparencies

T123.	Figure 11-2	Force on, and velocity of, a mass at different positions
T124.	Figure 11-3	A spring, hung vertically
T125.	Figure 11-5	Energy changes from potential energy to kinetic energy and back again
T126.	Figure 11-6	Circular motion of a small object
T127.	Figure 11-8	Position as a function of time
T128.	Figure 11-10	Graphs of displacement, velocity, and acceleration as a function of time
T129.	Figure 11-12	Simple pendulum, and a free-body diagram
	Table 11-1	Sin θ at Small Angles
T130.	Figure 11-14	Damped harmonic motion
	Figure 11-15	Graphs that represent other damped motion
T131.	Figure 11-16	Automobile spring and shock absorber
T132.	Figure 11-22	Motion of a wave pulse to the right
T133.	Figure 11-23	Characteristics of a single-frequency continuous wave
T134.	Figure 11-24	(a) Transverse wave; (b) longitudinal wave
T135.	Figure 11-28	A water wave is an example of a *surface wave*
	Figure 11-29	How a wave breaks
T136.	Figure 11-30	A wave traveling outward in three dimensions
T137.	Figure 11-32	Reflection of a wave pulse on a rope

	Figure 11-33	A wave pulse traveling to the right along a thin cord
T138.	Figure 11-35	Law of reflection
T139.	Figure 11-36	Two wave pulses pass each other
T140.	Figure 11-37b	Interference of water waves
T141.	Figure 11-38	Graphs showing two waves, and their sum, as a function of time
T142.	Figure 11-39	Standing waves corresponding to three resonant frequencies
	Figure 11-40	A string is plucked
T143.	Figure 11-43	(a) Soldier analogy to derive (b) law of refraction for waves
T144.	Figure 11-47	A traveling wave

Suggested Readings

Bensky, T., "Measuring *g* with a Joystick Pendulum," *The Physics Teacher* (February 2001), pp. 88–89. Instructions for making a physical pendulum to measure *g*.

Carnevali, A. and Newton, C., "Coupled Harmonic Oscillators Made Easy," *The Physics Teacher* (November 2000), pp. 503–505. An experiment for analyzing the frequencies of coupled harmonic oscillators.

Erlichson. H., "Galileo's Pendulum," *The Physics Teacher* (November 1999), pp. 478–479. Description of a lab experiment investigating the small angle approximation and discussion of Galileo's assertion that the period of a pendulum is independent of the amplitude.

Greene, N. and Dunn, R., "A Conical Spring—Which End Up?," *The Physics Teacher* (April 2000), pp. 228–231. Useful information about the conical spring commonly used in physics labs.

Greenslade, T., "Models of Traveling Waves," *The Physics Teacher* (November 2001), p. 466. Suggestion for how to make a three-dimensional model of a traveling wave, in order to help students understand that *y* is a function of both *x* and *t*.

Greenslade, T., "Waves in the Movies," *The Physics Teacher* (February 2000), p. 78. Nice demonstration of transverse waves.

Holzwarth, D. and Malone, J., "Pendulum Period versus Hanging-Spring Period," *The Physics Teacher* (January 2000), p. 47. A student exercise in matching periods of oscillating masses.

Kanamori, H. and Brodsky, E., "The Physics of Earthquakes," *Physics Today* (June 2001), pp. 34–40. A discussion of information gleaned from seismic waves as well as many other applications of physics to earthquakes.

Kidd, R. and Fogg, S., "A Simple Formula for the Large-Angle Pendulum Period," *The Physics Teacher* (February 2002), pp. 81–83. Extends the theory and experiments typical of introductory physics discussions to include large-amplitude pendulum oscillations.

Peters, R., "Student-Friendly Precision Pendulum," *The Physics Teacher* (October 1999), pp. 390–393. Instructions for constructing a Kater pendulum to accurately measure *g*.

Rowland, D. and Pask, C., "The Missing Wave Momentum Mystery," *American Journal of Physics* (May 1999), pp. 378–388. Discussion of paradoxes involved in the study of momentum carried by transverse waves on a string.

Szapiro, B., "Simple-Pendulum Lab with a Twist," *The Physics Teacher* (March 2002), pp. 158–163. Uses a force sensor connected to a computer to overcome some of the difficulties with a standard pendulum lab.

Slogoff, H. and Berner, B., "A Simulation of the Tacoma Narrows Bridge Oscillations," *The Physics Teacher* (October 2000), pp. 442–443. A demonstration of bridge oscillations using a fan and a box with detailed instructions.

Materials

AAPT, Twin Views of the Tacoma Narrows Bridge Collapse, video, user guide, and student activities. Order online at: www.aapt.org/catalog.

A Newtonian collision ball apparatus is available from Fisher Scientific, model number S40976.

Drilled balls for use as pendulum bobs are available from Frey Scientific.

A Hooke's law spring set is available from Sargent-Welch, model number CP33855-00.

A super Slinky is available from Fisher Scientific, model number S42162.

A string vibrator is available from Sargent-Welch, model number CP85102-00.

Notes and Ideas

Class time spent on material: Estimated:______ *Actual:_____________*

Related laboratory activities:

Demonstration materials:

Notes for next time:

Chapter 12: Sound

Outline

Summary

Chapter 12 covers the production, transmission, and detection of sound. Sound intensity, beats, and the Doppler effect are also treated here. This chapter contains many interesting applications of sound in modern technologies.

Major Concepts

By the end of the chapter, students should understand each of the following and be able to demonstrate their understanding in problem applications as well as in conceptual situations.

- Sound waves
  - Speed of sound
  - Frequency and pitch
- Intensity and intensity level
- The Doppler effect
- Beats

Teaching Suggestions and Demonstrations

Now is a good time to find out if you have any musicians or music majors in your class. If so, this is the chapter for them! Elicit their help whenever possible in this chapter. Health professions majors will be interested in the applications to medical imaging.

Sections 12-1 through 12-3

Most students will probably be familiar with the rule of thumb that relates the time elapsed between seeing a lightning flash and hearing a thunder clap to the distance to the lightning strike. Use Conceptual Example 12-1 to illustrate the **speed of sound** and explain why the rule of thumb works. You can point out that light travels at a finite speed too, but that the speed is so great (3×10^8 m/s or 186,000 miles/s) that the light from a source as close as a lightning bolt arrives practically instantaneously.

You won't need to spend much class time on **ultrasonic and infrasonic sound**, although students should be aware that these are nothing more than sounds with frequencies beyond the human range of perception. Example 12-2 is a nice application.

The difference between **sound intensity and intensity level** tends to be a hard concept for students because of the logarithmic nature of the decibel scale. Convince them that the use of the **logarithmic scale** makes sense for two reasons: first, the range of sound intensities we can hear is very large (refer to Table 12-2), and second, our ears perceive sound logarithmically. That is, if we hear one sound to be twice as loud as another, it has an intensity (energy per time per area) that is 10 times as great. Discuss Table 12-2 in class to help students become familiar with the intensity levels for familiar sounds. Do Example 12-3 and Exercise B to further illustrate the nature of a logarithmic scale.

Health professions majors will be very interested in Section 12-3 on the ear and its response to sound. Take time to go over it quickly.

Sections 12-4 and 12-5

The strings on **instruments** such as guitars and violins oscillate in standing wave patterns to produce notes. Considering such instruments helps students review many of the concepts and formulas from Chapter 11. For instance, most students will know that when you tighten a string, the pitch increases. Tightening corresponds to increasing the tension, which will increase the speed of waves on the string. Increasing the speed while keeping the wavelength (or the length of the string) constant increases the frequency, which is related to pitch. You can also examine what happens to the pitch when a thicker string is used, which changes the mass per unit length, or when the musician uses a capo, which effectively changes the length of the string. Work Example 12-8 with the class.

Plan to spend a little more time on sound produced by air columns. Point out the similarities in the fundamental and the harmonics for the cases of strings and air columns.

➲ **DEMO 12-1** **Standing waves in a column of air** can be demonstrated by blowing across the top of a bottle. Better yet, use a resonance tube so that the water level, and therefore the length of the air column, can be adjusted. Strike a tuning fork at the top of the tube and quickly raise or lower the water level until resonance is heard. (See Resource Information.) You might also have students who play instruments bring them to class or lab one day to demonstrate how they produce sound.

Work Example 12-9 with the class to illustrate these concepts. Section 12-5 is optional. If you have time, plan to go over it briefly.

Section 12-6

Use the information from Chapter 11 on interference to discuss the interference of sound waves in space. Then move on to the discussion of the interference of sound waves in time.

➲ **DEMO 12-2** To illustrate the interference of sound waves in space, use two speakers in a demonstration of Example 12-12. You will need to hook up both speakers to the same signal for the interference to be heard. (A frequency generator works best as a source of sound.)

➲ **DEMO 12-3** Use two tuning forks with frequencies close to each other or try a demonstration of Example 12-13 to help students hear **beats**. (You can also use two identical tuning forks and wrap a rubber band around one tine of one fork. This reduces the frequency slightly, producing beats when both forks are struck.) Strike both of the tuning forks, and hold them close to each other. You will be able to hear the sound intensity vary with time. This works best if the tuning forks are mounted over a resonance cavity (like a wooden box). Use Figure 12-19 to illustrate how the sound waves interfere in time.

Section 12-7

To introduce the **Doppler effect**, find a student to imitate the sound of a siren coming toward, passing and then speeding away from a stationary car. As an analogy, have them imagine a person in a boat in the middle of a pond dropping pebbles into the water at a constant frequency. The frequency and wavelength of the ripples that reach an observer on the shore will depend on whether the boat is stationary, moving toward the shore, or moving away from the shore.

➲ **DEMO 12-4** A Nerf ball with a beeper or bell embedded inside it makes a nice **Doppler ball.** Students throw the ball back and forth while the class listens to the sound and compares it to the sound they hear when the ball is stationary. (See Resource Information.)

The paragraph on the Doppler effect for light at the end of this section is a good introduction to a topic to be covered in more depth in the second half of the course.

Sections 12-8 and 12-9

Both of these sections are optional and can be skipped if you are short on time. However, the students will find this material very interesting, and it is worth a little class time to discuss them briefly.

Resource Information

Transparencies

T145.	Table 12-1	Speed of Sound in Various Materials
T146.	Figure 12-2	The membrane of a drum
	Figure 12-3	Representation of a sound wave in space
T147.	Table 12-2	Intensity of Various Sounds
T148.	Figure 12-5	Diagram of the human ear
T149.	Figure 12-6	Sensitivity of the human ear as a function of frequency
T150.	Table 12-3	Equally Tempered Chromatic Scale
T151.	Figure 12-7	Standing waves on a string
T152.	Figure 12-11	Modes of vibration (standing waves) for a uniform tube open at both ends
T153.	Figure 12-12	Modes of vibration (standing waves) for a uniform tube closed at one end
T154.	Figure 12-14	The amplitudes of the fundamental and first two overtones
	Figure 12-15	Sound spectra for different instruments
T155.	Figure 12-16	Sound waves from two loudspeakers interfere
T156.	Figure 12-18	Beats occur as a result of the superposition of two sound waves
T157.	Figure 12-20	Determination of the frequency shift in the Doppler effect
T158.	Figure 12-23	Sound waves emitted by an object (a) at rest or (b, c, and d) moving
T159.	Figure 12-26	Ultrasound pulse passes through the abdomen

Suggested Readings

Bedard, A. and Georges, T., "Atmospheric Infrasound," *Physics Today* (March 2000), pp. 32–37. Discussion of detection of sounds with frequencies too low for humans to hear and various applications including monitoring compliance with the Comprehensive Test Ban Treaty.

Dibble, W., "A Pedagogical Note on the Doppler-Effect Formulas," *The Physics Teacher* (September 2000), pp. 362–363. A quick and simple derivation for the Doppler effect formulas useful at the introductory level.

Gibson, G. and Johnston, I., "New Themes and Audiences for the Physics of Music," *Physics Today* (January 2002), pp. 42–48. Discussion of the treatment of music in physics class.

Haar, G., "Acoustic Surgery," *Physics Today* (December 2001), pp. 29–34. Describes the focusing of high intensity ultrasound fields for medical procedures, including the treatment of cancer.

Hartmann, W., "How We Localize Sound," *Physics Today* (November 1999), pp. 24–29. Describes a variety of cues our brain uses to construct a three-dimensional image of the acoustic landscape.

Hoyt, D., "A Different Viewpoint on Doppler-Effect Calculations," *The Physics Teacher* (January 2002), pp. 14–16. Analysis of the classical Doppler-effect formula, including a method for determining signs and a useful approximation.

Krider, J., Cook, T., Easterling, W. and Rzanek, I., "Hear and 'See' Sound Interference," *The Physics Teacher* (April 2001), pp. 220–222. A visual demonstration of wave interference.

Martin, B., "Measuring the Speed of Sound—Variation on a Familiar Theme," *The Physics Teacher* (October 2001), pp. 424–426. A new twist on the classic resonance tube lab measurement of the speed of sound.

Pettersen, I., "Speed of Sound in Gases Using an Ultrasound Motion Detector," *The Physics Teacher* (May 2002), pp. 284–286. Experiments to determine the speed of sound in air and other gases, with suggestions for possible areas of further investigation.

Saba, M. and Rosa, R., "A Quantitative Demonstration of the Doppler Effect," *The Physics Teacher* (October 2001), pp. 431–433. Uses a recording of the sound of a passing car horn and spectrogram software to quantitatively investigate the Doppler effect.

Sanchez-Lavega, A., "Acoustic Waves in Planetary Atmospheres," *The Physics Teacher* (April 2002), pp. 239–242. An interesting discussion of sound on other planets and moons in our solar system.

Materials

One example of a "Doppler ball" is the Vortex Mega Howler, available from Toys R Us or other toy stores.

A resonance apparatus and tuning forks are available from Fisher Scientific, model numbers S42105 and S42036.

Notes and Ideas

Class time spent on material: Estimated:______ Actual:____________

Related laboratory activities:

Demonstration materials:

Notes for next time:

Chapter 13: Temperature and Kinetic Theory

Outline

Summary

Chapter 13 is the first of three chapters on thermal physics. The concepts of temperature and thermal energy are introduced. Ideal gases are defined and used to present the kinetic theory of gases, which connects macroscopic variables (temperature and pressure) to microscopic variables (molecular speed and mass).

Major Concepts

By the end of the chapter, students should understand each of the following and be able to demonstrate their understanding in problem applications as well as in conceptual situations.

- Temperature
  - The zeroth law of thermodynamics
  - Temperature scales
  - Absolute zero
- Thermal expansion
  - Linear
  - Area
  - Volume
- Ideal gases
  - Definition
  - Equation of state
- Mole (Avogadro's number)
- Kinetic theory of gases
  - Molecular speed distribution
  - Kinetic energy, pressure, temperature

Teaching Suggestions and Demonstrations

This chapter lays the foundation for Chapters 14 and 15. Although the material can be covered fairly quickly, it is important to ensure that students understand the basics before moving on.

Sections 13-1 through 13-3

Most students will have covered the **atomic theory of matter** in high school or in general chemistry. Be sure students are familiar with units and definitions. Figure 13-2 gives a nice picture of the atomic arrangements in a solid, liquid, and gas.

Students will most likely be familiar with all three **temperature scales** discussed in this chapter. Derive one of the conversion formulas with them to point out that both zero point and degree size must be taken into account. The **zeroth law of thermodynamics** is essentially common sense, but don't be tempted to skip over it. The concept is critical for an understanding of calorimetry, and students need to become comfortable with terminology such as "thermal contact" and "thermal equilibrium."

➲ **DEMO 13-1** Make your own **thermometer**. Use a blank alcohol thermometer tube (or just put tape over the scale on a thermometer). (See Resource Information.) Put the tube in an ice-water bath and mark the level of the alcohol. Put the tube in a boiling-water bath and mark the level again. Define your own scale! Derive a conversion formula between your scale and the Celsius scale.

Sections 13-4 and 13-5

Thermal expansion is the principle that causes a thermometer to work. As the liquid in the tube is heated, its volume expands, causing the liquid level to rise. Bridges and railroad tracks must be built with expansion joints to allow for varying temperatures, as shown in the photographs in Figure 13-3. Conceptual Example 13-4 makes the very important point that when a container expands, the volume it encloses expands as well. Try Example 13-7 with your students to remind them to take into account both the expansion of the liquid and the expansion of the container holding it. As another example, ask students if they have ever held a jar with a tight lid under hot water in order to loosen the lid. Both the jar and the lid expand, but the lid expands more, making it easier to unscrew. (See Conceptual Example 13-6.) Compare typical coefficients of thermal expansion for glass and metal in Table 13-1.

Students will find the **anomalous behavior of water close to its freezing point** interesting. Take time to go through this material. Go through Example 13-8 and briefly discuss thermal stress in materials.

Sections 13-6 through 13-10

Plan to spend a little time on the definition of an **ideal gas** even if all the students have seen the equation of state before. Point out how an ideal gas differs from a real gas, and why the concept of an ideal gas is useful. Go over the **constant-volume gas thermometer** (Figure 13-8) and connect it to the concepts from the previous sections. The concept of absolute zero is often difficult for students. The graphs in Figure 13-13 are good illustrations of the origin of this idea and of the **Kelvin scale**.

➲ **DEMO 13-2** The form of the **equation of state for an ideal gas** is deduced quite nicely from observations. Bring in a beach ball (or any kind of inflatable ball) and an air pump. Use the air pump to increase the number of air molecules in the beach ball. This will raise the pressure of the air in the beach ball and will cause it to inflate. You can also sit on the ball, which will decrease its volume, and also increase the pressure of the gas inside.

Even if most of the students have had chemistry, you will need to discuss the concepts of the **mole** and **Avogadro's number**. Be sure to go over the relationship between the **Boltzmann constant (k)** and the **universal gas constant (R)**. Example 13-10 will help students develop an intuitive feel for the "size" of a mole of air.

If you are pressed for time, you can skip the individual discussions of **Boyle's law** and **Charles's law**. However, you do need to go over the definitions of **isotherms** and **isobars**, since they will be needed again in later chapters.

The **kinetic theory of gases** connects the macroscopic quantities that describe the gas as a whole to the microscopic quantities that describe the individual molecules of the gas. Students sometimes have trouble believing that this can actually be done; be prepared to present evidence and argue the case. Go over Examples 13-16 and 13-17 to help students relate temperature and the kinetic energy of molecules.

➲ **DEMO 13-3** You can use the evidence of the scarcity of low atomic weight gases (hydrogen and helium) in the atmosphere as an argument in support of the **kinetic theory of gases**. Since the masses of these gas molecules are small compared to the molecular masses of nitrogen or oxygen, the root mean square (rms) velocity of hydrogen or helium molecules will be larger than that of other molecules at the same pressure. In fact, the rms velocity of hydrogen molecules at atmospheric pressure and normal temperature is on the same order of magnitude as escape velocity! Do the calculation and discuss its implications with the class.

Sections 13-11 through 13-14

These sections contain lots of interesting material and good information for students, but they are optional and can be skipped without loss of continuity. If you have time, consider covering the section on vapor pressure and humidity.

Resource Information

Transparencies

T160.	Figure 13-2	Atomic arrangements in (a) a crystalline solid, (b) a liquid, and (c) a gas
T161.	Figure 13-7	Celsius and Fahrenheit scales compared
T162.	Figure 13-8	A constant-volume gas thermometer
T163.	Table 13-1	Coefficients of Expansion, near 20°C
T164.	Figure 13-11	Behavior of water as a function of temperature near 4°C
T165.	Figure 13-13	Volume of a fixed amount of gas as a function of temperature
T166.	Figure 13-16	Molecules of a gas moving about in a rectangular container
T167.	Figure 13-17	Distribution of speeds of molecules in an ideal gas
T168.	Figure 13-20	Phase diagram for water
T169.	Figure 13-26	Diffusion

Suggested Readings

Denardo, B., "Temperature of a Lightbulb Filament," *The Physics Teacher* (February 2002), pp. 101–105. Investigates the error introduced by the assumption that the variation of resistivity with temperature is linear.

Graham, M., "Investigating Gases' Masses in Impecunious Classes," *The Physics Teacher* (March 2002), pp. 144–147. Directions for building and using an inexpensive apparatus to measure molecular mass of a gas to an accuracy of about 10 percent.

Gutiérrez, G. and Yáñez, J., "Can an Ideal Gas Feel the Shape of its Container?," *The American Journal of Physics* (August 1997), pp. 739–743. Examination of thermodynamic quantities of an ideal gas enclosed in a finite container.

Lee, W., Gilley, H. and Caris, J., "Finding the Surface Temperature of the Sun Using a Parked Car," *American Journal of Physics* (November 1997), pp. 1105–1109. An activity involving both quantitative observations and theoretical analysis for estimating the surface temperature of the Sun.

Reif, F., "Thermal Physics in The Introductory Physics Course: Why and How to Teach It From a Unified Atomic Perspective," *American Journal of Physics* (December 1999), pp. 1051–1062. An argument for and description of using an atomic approach for teaching thermal physics.

Thompson, D., "Derivation of the Ideal Gas Law from Kinetic Theory," *The Physics Teacher* (April 1997), pp. 238–239. A short derivation of the ideal gas law using collisions of particles with container walls.

Velasco, S., Faro, J. and Roman, F., "An Experiment for Measuring the Low Temperature Vapor Line of Water," *American Journal of Physics* (December 2000), pp. 1154–1157. A description of a method to measure the vapor pressure curve of water below the normal boiling temperature.

Materials

Thermometers are available from a variety of sources, including Fisher Scientific, model number S41577A or S41580 (ungraduated).

Notes and Ideas

*Class time spent on material: Estimated:*______ *Actual:*______________

Related laboratory activities:

Demonstration materials:

Notes for next time:

Chapter 14: Heat

Outline

14-1 Heat as Energy Transfer
14-2 Internal Energy
14-3 Specific Heat
14-4 Calorimetry—Solving Problems
14-5 Latent Heat
14-6 Heat Transfer: Conduction
14-7 Heat Transfer: Convection
14-8 Heat Transfer: Radiation

Summary

Chapter 14 continues the study of thermodynamics begun in Chapter 13. Applications of conservation of energy in this chapter include the connection between mechanical work and heat. Phase changes, the relationship between heat given off or absorbed by a material and the corresponding temperature or phase change, and calorimetry are discussed. The last three sections of this chapter deal with the processes of heat exchange.

Major Concepts

By the end of the chapter, students should understand each of the following and be able to demonstrate their understanding in problem applications as well as in conceptual situations.

- Heat
  - Energy transfer
  - Mechanical equivalent
  - Specific heat
  - Calorimetry
- Internal energy of a gas
- Phase equilibrium
- Latent heat
- Phase changes and energy conservation
- Mechanisms of heat exchange
  - Conduction
  - Convection
  - Radiation

Teaching Suggestions and Demonstrations

Students who have had chemistry will find some familiar material in this chapter. It is useful to find out how many students have seen this material before. You may be able to skim through the sections on latent heat and heat transfer and give more time to the less-familiar parts of the chapter.

Sections 14-1 and 14-2

Conservation of energy is back again! **Heat** is energy transferred from one object to another because of a difference in temperature. Mechanical energy can also be converted into heat. Remind students of the problems in Chapter 6 in which mechanical energy was not conserved in the presence of friction. Now you can explain that the energy wasn't really lost; it was just converted to thermal energy. Vigorously rubbing your hands together provides a quick demonstration of the conversion of mechanical to thermal energy. Discuss the units of heat, the **calorie** and the **kilocalorie**, and their relationship to the common energy unit, the joule. Work through Example 14-1 in class.

It is important for students to distinguish between temperature, heat, and internal energy. Section 14-2 is quite clear about these concepts. Plan to go through it carefully.

Sections 14-3 through 14-5

Different substances require different amounts of heat to raise a unit mass through the same temperature difference. Water has a fairly high **specific heat**, and, as a result, it takes addition or removal of larger amounts of thermal energy to change its temperature. Thus, oceans have moderating effects on the climate of coastal areas. Work through Example 14-2 with the class to demonstrate this concept.

Calorimetry is an application of the conservation of energy when substances with different temperatures are brought into thermal contact. If no heat escapes to the external environment, then the heat lost by the warmer substance(s) equals the heat gained by the cooler.

➲ **DEMO 14-1** Heat metal shot to a known temperature and pour it into a **calorimeter** can filled with water at room temperature. (See Resource Information.) The shot releases heat and cools, while the can and water, which were initially in thermal equilibrium, warm up until the entire mixture reaches a stable temperature. To quantify the demonstration, measure the increase in temperature of the can and water and calculate the specific heat of the metal.

Be sure to work several examples in class to reinforce these ideas. Calorimetry problems in general are not difficult, but students can easily become confused by the details in them.

Heat entering or leaving a substance can result in a temperature change. Phase changes also involve loss or gain of heat, but have no accompanying change in temperature. Students usually have trouble with the concept that during a (quasi-static) **phase change** a system does not change temperature. Go over the three latent heats: **latent heat of fusion**, **latent heat of vaporization**, and **latent heat of sublimation**, and the phase changes to which they apply. Table 14-3 lists latent heats for common substances. Discuss Figure 14-5 in detail and go over Example 14-7.

➲ **DEMO 14-2** For this demo you will need a saucepan or Pyrex beaker, crushed ice (as cold as you can get it), a hot plate, and a thermometer. Fill the saucepan or Pyrex beaker with the ice, insert the thermometer and have a student record the temperature. Place the pan or beaker on the hot plate. Preheat the burner on low in advance, so that it has reached a stable temperature before you place the pan on it.) Have a student check the temperature of the ice (which will eventually melt) every two minutes (or one minute, or five minutes – as long as the time interval is constant). The hot plate should deliver heat to the pan at (approximately!) a constant rate. If you plot **temperature vs. time**, you should be able to reproduce Figure 14-5. This demonstration is a little crude, but it will give students a concrete idea of what Figure 14-5 represents. You can take the temperature of the water all the way up to the boiling point.

➲ **DEMO 14-3** Make ice cream in an electric freezer. Connect this exercise to energy transfer and latent heat. Point out that the cranking, whether done by machine or by human power, serves to break up the ice crystals and keep them from getting too large as they freeze. This is a good exercise to do on a Friday – you can set it up at the beginning of class, and it should be ready to eat by the end!

The ideas of specific heat and the latent heats combine to give a general **conservation of energy** equation for isolated systems. You can treat this concept as ***total heat is conserved***, or ***heat lost by part of the system equals heat gained by the rest of the system***. Either way, you will need to work many problems, such as Example 14-8, that involve both temperature and phase changes. The discussion on **evaporative cooling** at the end of Section 14-5 is interesting and informative.

Sections 14-6 through 14-8

The three mechanisms of **heat exchange** should be introduced at least conceptually, although you can go through these sections rather quickly. As examples, give several instances of heat exchange and have students name the mechanism(s) responsible in each case. Figures 14-8 and 14-9 are very good illustrations of convection heating

➲ **DEMO 14-4** **Conduction, convection, and radiation** can all be demonstrated using a pot filled with water on a hot plate. The bottom of the pot heats up due to conduction. Glitter sprinkled into the water as it is heating will show the convection currents. If you hold your hand close to the glowing hot plate, you can feel the radiation.

Resource Information

Transparencies

T170.	Figure 14-1	Joule's experiment on the mechnical equivalent of heat
T171.	Table 14-1	Specific Heats
	Table 14-2	Specific Heats of Gases
T172.	Figure 14-4	Simple water calorimeter
T173.	Figure 14-5	Temperature as a function of heat added
T174.	Table 14-3	Latent Heats
T175.	Figure 14-6	Heat conduction between areas at temperatures T_1 and T_2
T176.	Table 14-4	Thermal Conductivities
	Table 14-5	*R*-values
T177.	Figure 14-8	Convection currents in a pot of water being heated on a stove
	Figure 14-9	Convection plays a role in heating a house
T178.	Figure 14-13	Earth's seasons

Suggested Readings

Capitolo, M., "Phase-Change Demonstration—Instant Gratification," *The Physics Teacher* (September 1998), p. 349. A simple but dramatic phase-change demonstration.

Gash, P., "So You Thought a Glass Thermometer Measured Temperature," *The Physics Teacher* (February 2002), pp. 74–76. An application of calorimetry, which includes the heat absorbed or lost by the glass of the thermometer.

Hart, F., "A Microcomputer-Based Phase-Change Experiment," *The Physics Teacher* (February 1998), pp. 98–99. A laboratory exercise for examining a phase change.

McGee, H., McInerney, J. and Harrus, A., "The Virtual Cook: Modeling Heat Transfer in the Kitchen," *Physics Today* (November 1999), pp. 30–36. An interesting, though advanced, article on what happens in the cooking process.

Schmidhuber, C., "On Water, Steam, and String Theory," *The American Journal of Physics* (November 1997), pp. 1042–1050. A detailed description of a second-order phase transition between water and steam, with applications in particle physics and string theory.

Silverman, M. and Silverman, C., "Cool in the Kitchen: Radiation, Conduction, and the Newton 'Hot Block' Experiment," *The Physics Teacher* (February 2000), pp. 82–88. Valuable information about heat loss and thermal properties.

Wolbarst, A., "Effective Thermal Conduction Model for Estimating Global Warming," *American Journal of Physics* (October 1999), pp. 885–890. Illustrates thermal conduction and radiation using the important example of greenhouse warming of the Earth.

Materials

A student calorimeter is available from Fisher Scientific, model number S41746.

Notes and Ideas

*Class time spent on material: Estimated:*______ *Actual:*____________

Related laboratory activities:

Demonstration materials:

Notes for next time:

Chapter 15: The Laws of Thermodynamics

Outline

Summary

Chapter 15 presents the two great laws of thermodynamics. The first law applies conservation of energy to thermodynamic systems. The second law defines the concept of entropy and the limits on the ability to use the energy of the system to do useful work. The chapter concludes with a discussion of practical devices and optional sections on the application of thermodynamics to current topics.

Major Concepts

By the end of the chapter, students should understand each of the following and be able to demonstrate their understanding in problem applications as well as in conceptual situations.

- The first law of thermodynamics
  - Definitions of Q, W and ΔU
  - Sign conventions
  - Specific heat at constant pressure
  - Specific heat at constant volume
- The second law of thermodynamics
  - Carnot's theorem
  - Heat engines
  - Refrigerators and heat pumps
- Entropy

Teaching Suggestions and Demonstrations

Thermodynamics remains a great mystery to many students in introductory physics. Parts of it (such as the zeroth law) seem obvious, but other parts (such as entropy) seem very difficult. Use lots of examples and try to tie the concepts to practical applications.

Sections 15-1 through 15-3

Go over the **zeroth law** again for completeness. (It was covered in Chapter 13.) Then start on the **first law of thermodynamics**. Be sure to go over the definitions and sign conventions for $\boldsymbol{Q}$, $\boldsymbol{W}$, and $\boldsymbol{\Delta U}$. It is important for students to understand the differences between Q, which represents the **flow of thermal energy**; W, which represents **physical work involving a force**; and U, which is a **state function**, determined by the state variables of pressure, temperature, and volume. Go through Example 15-1 to illustrate these concepts.

The idea of a **quasi-static or reversible process** will be new to the students. Discuss the definitions of reversible and **irreversible processes** thoroughly.

As you present **constant pressure**, **constant volume**, **constant temperature,** and **adiabatic processes,** use the illustrations in Figures 15-1 through 15-6. Figures 15-2, 15-3, 15-4, and 15-6 introduce ***PV* plots**, which can illustrate many thermodynamic processes. Make sure students are comfortable interpreting and sketching these plots. Work many examples, including Conceptual Examples 15-3 and 15-4.

In an **adiabatic process** no heat is exchanged with the environment, but the temperature of the system can change. This concept is difficult for students, who often equate heat exchange with a change in temperature. Example 15-6 is a good discussion starter. Table 15-1 gives a compact summary of the definitions presented in these sections and how the first law applies to each process. As a review, have students apply the first law to each different type of process and see if they can correctly generate Table 15-1.

Section 15-3 is optional, but the students will find it interesting and informative. Go over Table 15-2 if you have time.

Sections 15-4 through 15-6

The **second law of thermodynamics** gives a direction to the flow of energy in a thermodynamic process.

➲ **DEMO 15-1** For a very simple demonstration, bring several (around 10) beakers to class. Fill each partially with water and put some ice in each one. Ask the class to predict the number of beakers in which the ice will melt and the number of beakers in which the water will freeze. Most of the class will realize that, of course, the ice will melt in all of the beakers. You can use this as a discussion starter on the **second law**.

A **heat engine** converts energy into work, with waste heat (ALWAYS) ejected to the environment. Go over the steam engine (see Figure 15-12) and the schematic diagram of a heat engine in Figure 15-11. The concepts of a **hot reservoir** and a **cold reservoir** may be difficult for some students. The discussion of the internal combustion engine in Section 15-5 is quite good. Go through Figure 15-13 with the class. Define the **efficiency** of a heat engine and point out that you can never have a heat engine that is 100 percent efficient. Work Example 15-9 and several other examples from the homework problems.

Carnot's theorem gives the upper limit on the efficiency of a heat pump operating between two constant-temperature reservoirs. If the students have understood the difference between reversible and irreversible processes, they should follow the theorem without too much trouble. Be sure to explain the equation for the **maximum efficiency of a heat engine** and work several examples. The diagram of the Carnot cycle in Figure 15-14 is very clear.

Refrigerators, air conditioners, and **heat pumps** all use work to move energy from a cold reservoir to a hot reservoir. Use Figures 15-15, 15-16, and 15-18 to illustrate the similarities and differences between these devices and the heat engine. Define the **coefficient of performance for a refrigerator** and the **coefficient of performance for a heat pump**. Be sure to work several examples from these sections and from the homework problems.

Sections 15-7 through 15-12

Entropy is a difficult concept for students to grasp. They can learn to do calculations, and they can develop intuition about which processes create more **disorder**. Try to be as concrete as possible when describing entropy and its applications in physics. If you are short on time, you can just discuss the concept of entropy and skip the equations.

➲ **DEMO 15-2** Bring in a new deck of cards. Show the class that the cards are sorted in order by suit. The new deck is an **ordered system**. Shuffle the cards and in the process spray them all over the room. (This usually takes the students by surprise.) Ask if the new configuration of cards (all over the room) has more or less order than the original configuration. Can the original configuration be restored? Yes, but (relatively easy) work must be done!

➲ **DEMO 15-3** Bring in a clear beaker of water and a bottle of food coloring. Place the beaker where it can be seen by the whole class and where it can stay for the entire class period. Point out that the system of the beaker of water and the bottle of food coloring is an ordered system. Place one or two drops of food coloring in the beaker, being careful not to bump the beaker or stir the mixture. The food coloring will disperse throughout the beaker by the end of class. Is the mixture more or less disordered than the original system? Can the original configuration be restored? No, not without using a tremendous amount of energy!

Discuss the definition of ΔS and point out that it is the **change in entropy** in a process that is important, not the total entropy of a system. Work Examples 15-14 and 15-15 in class. The discussions of **heat death** and **entropy and living systems** in Sections 15-9 and 15-10 are well worth some class time, as is the material in Section 15-12 on **global warming**. Section 15-11 covers the statistical interpretation of entropy quite well, but may be skipped if you are out of time.

Resource Information

Transparencies

T179.	Figure 15-1	An ideal gas in a cylinder fitted with a movable piston
T180.	Figure 15-3	PV diagram for adiabatic (AC) and isothermal (AB) processes on an ideal gas
	Figure 15-4	(a) Isobaric process (b) Isovolumetric process
T181.	Figure 15-7	Work done by a gas is equal to the area under the PV curve
T182.	Table 15-1	Simple Thermodynamic Processes and the First Law
	Figure 15-8	Example 15-5 (First law in isobaric and isovolumetric processes)
T183.	Table 15-2	Metabolic Rates
T184.	Figure 15-11	Schematic diagram of energy transfers for a heat engine
T185.	Figure 15-12	Steam engines
T186.	Figure 15-13	Four-cycle internal combustion engine
T187.	Figure 15-14	The Carnot cycle
T188.	Figure 15-16	Schematic diagram of energy transfers for a refrigerator or air conditioner

	Figure 15-17	Typical refrigerator system
T189.	Figure 15-18	A heat pump
T190.	Table 15-3	Probablilities of Various Microstates for 100 Coin Tosses
T191.	Table 15-4	Electric Energy Resources

Suggested Readings

Graham, D. and Schacht, D., "Simple Estimate of the Human Metabolic Rate," *American Journal of Physics* (June 2001), pp. 723–724. An experiment involving energy and entropy that will be of particular interest to students in the pre-health professions.

Leff, H., "What If Entropy Were Dimensionless?," *American Journal of Physics* (December 1999), pp. 1114–1122. Conclusions from an examination of the puzzling aspects of the dimensions of entropy.

Lieb, E. and Yngvason, J., "A Fresh Look at Entropy and the Second Law of Thermodynamics," *Physics Today* (April 2000), pp. 32–37. An accessible and different approach to understanding entropy without applying statistical mechanics.

Loverude, M., Kautz, C. and Heron P., "Student Understanding of the First Law of Thermodynamics," *American Journal of Physics* (February 2002), pp. 137–148. Report on an investigation indicating that students often fail to differentiate among the concepts of heat, temperature, work, and internal energy. Includes implications for instruction.

Saslow, W., "An Economic Analogy to Thermodynamics," *American Journal of Physics* (December 1999), pp. 1239–1247. An interesting analogy, especially if there are students from a variety of majors in your class.

Schoepf, D., "A Statistical Development of Entropy for the Introductory Physics Course," *American Journal of Physics* (February 2002), pp. 128–136. A model for introducing the statistical basis for the definition of entropy, including a correspondence between the statistical and the classical Clausius definitions.

Styer, D., "Insight into Entropy," *American Journal of Physics* (December 2000), pp. 1090–1096. Several examples from statistical mechanics demonstrating the qualitative character of entropy.

Velasco, S., Román, F. and Faro, J., "A Simple Experiment for Measuring the Adiabatic Coefficient of Air," *American Journal of Physics* (July 1998), pp. 642–645. Instructions for performing an experiment to measure compressibility and the adiabatic coefficient of air.

Yeo, S. and Zadnik, M., "Introductory Thermal Concept Evaluation: Assessing Students' Understanding," *The Physics Teacher* (November 2001), pp. 496–504. Presentation of a multiple-choice evaluation instrument designed to test student understanding of thermal concepts.

Notes and Ideas

Class time spent on material: Estimated:______ *Actual:_____________*

Related laboratory activities:

Demonstration materials:

Notes for next time:

Chapter 16: Electric Charge and Electric Field

Outline

16-1 Static Electricity; Electric Charge and Its Conservation
16-2 Electric Charge in the Atom
16-3 Insulators and Conductors
16-4 Induced Charge; the Electroscope
16-5 Coulomb's Law
16-6 Solving Problems Involving Coulomb's Law and Vectors
16-7 The Electric Field
16-8 Field Lines
16-9 Electric Fields and Conductors
*16-10 Gauss's Law
*16-11 Electric Forces in Molecular Biology: DNA Structure and Replication
*16-12 Photocopy Machines and Computer Printers Use Electrostatics

Summary

The electric force affects our daily lives more than any other force. Chapter 16 introduces the concepts of charge, Coulomb's law, and the electric field. The electric properties of insulators and conductors are discussed. Charging by induction, electric flux, and Gauss's law are also presented. The chapter ends with applications of electrostatics in biology and technology.

Major Concepts

By the end of the chapter, students should understand each of the following and be able to demonstrate their understanding in problem applications as well as in conceptual situations.

- Electric charge
  - Positive and negative
  - Quantization of charge
  - Conservation of charge
- Electric properties of materials
  - Insulators
  - Conductors
  - Semiconductors
- Coulomb's law
- The electric field
  - Definition
  - Field lines
- Induction
- Gauss's law
  - Electric flux

Teaching Suggestions and Demonstrations

Electric charges and the basic rules that govern their behavior (opposite charges attract; like charges repel) are familiar to most students. The challenge for this chapter is to build on that intuition so more complex situations can be treated.

Sections 16-1 and 16-2

Emphasize to the students that the labels of **positive and negative charge** are arbitrary (assigned by Benjamin Franklin!) and that they work well for most situations. Charge is **quantized**, occurring in "bits" of ***e***, the magnitude of the fundamental charge on the electron or proton.

➲ **DEMO 16-1** Rub a glass or Lucite rod with animal fur (or wool or silk). (See Resource Information.) You will charge the rod. Use the charged rod to pick up small pieces of paper. If it is winter, you can also do this demo with a plastic comb and clean hair, or a balloon and clean hair. If you are teaching this class in the summer you are probably out of luck – very few electrostatic demonstrations "work" during the summer due to higher levels of humidity!

➲ **DEMO 16-2** Charge a balloon by rubbing it on your head, and place it near a thin stream of water from a faucet. The movement of the stream of water is very dramatic!

Be sure to go over the **conservation of charge**. Remind the students that when we charge objects as in the above demonstration, we are transferring charge from one object to another, not creating charge. Discuss how **polarization of charge** within the atom enables these demonstrations to work. (See Figures 16-4 and 16-9.)

Sections 16-3 and 16-4

Go over the definitions of **insulators** and **conductors**. Students should understand that in insulators any excess charge is not free to move, while in conductors any excess charge is free to move about the surface. **Semiconductors** are intermediate between conductors and insulators; their properties can be controlled by adding small amounts of other materials.

Take the students step-by-step through the **charging by induction process**; the logical progression of steps will make sense to most students. Go over parts of the electroscope and how it works.

➲ **DEMO 16-3** Bring in a demonstration electroscope and use it to demonstrate the process of charging by induction. (See Resource Information.)

Sections 16-5 and 16-6

Present the definition of **Coulomb's law**, which is true for point charges and spherically symmetric charge distributions. Be sure to point out the similarities between Coulomb's law and the law of universal gravitation. Note that the gravitational force is always attractive, but the **electrostatic force** can be attractive or repulsive. (See Figure 16-15.) Conceptual Example 16-2 should generate some interesting class discussion about the electrostatic force. The calculation of the **relative size of the electrostatic and gravitational forces** is startling to the students and is good to go through in class.

In general, students will need to be reminded about **vectors**, **vector addition**, and the **superposition of forces**. Plan to work many examples of force calculations. Conceptual Example 16-5 is a good test of student understanding of the vector addition of forces.

Sections 16-7 through 16-9

The idea of a **force field** will be alien to most students. Be as concrete as possible in describing this concept. Figure 16-23 will help. Go over the definition of the field (force per unit charge) and compare it to the definition of *g* (force per unit mass).

➲ **DEMO 16-4** On the board or on an overhead, use the idea of a small **positive test charge** to map out the **force field** from a positive charge *q* located at the origin. Use arrows to indicate the direction and magnitude of the field at many points. (Figure 16-23 will give you a start.) Remind the students that the field is three-dimensional, not flat like the board. Once you have the field set up, be sure to discuss the direction of the force on a **negative test charge**.

Note that the **superposition of electric fields** from several charges involves vector addition. Plan to spend some time on this concept in class and work several examples. Figure 16-28 will help. Go over Examples 16-8 and 16-9 to assess student understanding.

The above demonstration leads right into a discussion of **electric field lines**. The rules for drawing electric field lines are given at the beginning of Section 16-8. Figures 16-29 through 16-31 illustrate several of the key concepts. Figure 16-31d is a picture of the ideal **parallel-plate capacitor**, which will be discussed in later chapters in more detail.

Section 16-9 begins with an excellent description of how **excess charge** is distributed on a conductor. This leads logically into a discussion of electrostatic **shielding**. You will need to spend some time on this concept, students find it somewhat confusing. Go over Conceptual Example 16-10. Use Figures 16-33 and 16-34 to talk about **electric fields at conductor surfaces**. Be sure to point out how they differ from electric fields at insulator surfaces, which are not required to be perpendicular to the surface.

Section 16-10

This section is labeled optional, but we recommend you cover it in detail. **Electric flux**, like the electric field, is a difficult concept for many students because they can't see it. The more visual you can make your presentation of this section, the better.

➲ **DEMO 16-5** Bend some wire (coat hangers do nicely) into a square loop. Draw parallel electric field lines on the board. Hold up the loop so that its face is perpendicular to the field lines. Discuss the **normal to the surface**, and point out that in this position the normal is parallel to the field lines and the most field lines pass through the loop. Rotate the loop until the normal is perpendicular to the field lines. Point out that in this position no field lines pass through the loop. Once you have discussed how the electric field lines pass through (or just pass by) the loop, you can define the electric flux. Be sure to talk about **electric flux through closed surfaces** and give several examples.

You can present **Gauss's law** by talking about "sources" (faucets) and "sinks" (drains) of water. If you place a closed surface made of screen wire in a flowing stream, the amount of water entering the surface is equal to the amount of water leaving the surface. The "net flux" of water through the surface is zero. If you place the closed surface over the end of a faucet, there is a **net positive flux** through the surface. If you place the closed surface over a drain, there is a **net negative flux** through the surface. Gauss's law works just this way for electric fields. Positive charges are "sources" of electric field lines; negative charges are "sinks" of electric field lines. If a closed surface (usually imaginary) encloses no charge, then the net flux of electric field lines through the surface is zero. If it contains a net positive or negative charge, then the net flux of electric field lines is positive or negative, respectively. Go over Examples 16-11 and 16-12, which cover the two primary types of Gaussian surfaces (spheres and cylinders). (See Figures 16-39 and 16-40.)

Sections 16-11 and 16-12

These two sections are also optional, but contain material the students will find of great interest. If you have time, plan to go through them quickly.

Resource Information

Transparencies

T192.	Figure 16-6	Charging by conduction
	Figure 16-7	Charging by induction
	Figure 16-8	Inducing a charge on an object connected to ground
T193.	Figure 16-10	Electroscope
T194.	Figure 16-11	Electroscope charged by (a) induction, (b) by conduction
	Figure 16-12	Determining the sign of a charged object
T195.	Figure 16-15	Direction of the force
T196.	Figure 16-18	Review of vector addition
T197.	Figure 16-20	Determining the forces for Example 16-4
T198.	Figure 16-24	Electric field at a given point in space
T199.	Figure 16-25	Example 16-6 (Photocopy machine)
T200.	Figure 16-28	Calculation of the electric field at points A and B for Example 16-9
T201.	Figure 16-31	Electric field lines for four arrangements of charges
T202.	Figure 16-33	A charge inside a neutral spherical metal shell
	Figure 16-34	Electric field $\vec{E}$ at the surface of a conductor
T203.	Figure 16-37	A uniform electric field $\vec{E}$ passing through a flat area A
T204.	Figure 16-38	Electric field lines passing through a closed surface
T205.	Figure 16-39	Cross-sectional drawing of a thin spherical shell
	Figure 16-40	Calculation of $\vec{E}$ due to a very long line of charge
T206.	Figure 16-44	Section of a DNA double helix
T207.	Figure 16-45	Replication of DNA
T208.	Figure 16-46	Inside a photocopy machine
	Figure 16-47	Inside a laser printer

Suggested Readings

Avants, B., Soodak, D. and Ruppeiner, G., "Measuring the Electrical Conductivity of the Earth," *American Journal of Physics* (July 1999), pp. 593–598. An experiment for measuring the electrical conductivity of the Earth to yield information about what is underground.

Bracikowski, C., "Graphical Analysis of Electric Fields of Dipoles and Bipoles," *The Physics Teacher* (January 2000), pp. 20–21. A qualitative explanation of the dependence of field strength on distance.

Cortel, A., "Demonstrations of Coulomb's Law with an Electronic Balance," *The Physics Teacher* (October 1999), pp. 447–448. A method of measuring weak repulsive forces between electric charges.

Gelbart, W., Bruinsma, R., Pincus, P. and Parsegian, V., "DNA-Inspired Electrostatics," *Physics Today* (September 2000), pp. 38–44. Research concerning electrostatic phenomena and electrostatic interactions between biomolecules that will be of particular interest to biology majors and students in the pre-health professions.

Harrington, R., "Getting a Charge Out of Transparent Tape," *The Physics Teacher* (January 2000), pp. 23–25. Two experiments using charging by adhesion-separation.

Jefimenko, O., "Dynamic Electric Field Maps of Point Charge Moving with Constant Velocity," *The Physics Teacher* (March 2000), pp. 154–157. An alternative way to graphically represent the electric fields of moving charges.

Materials

One electrostatics demonstration kit is available from Fisher Science Education, model number S57200-1. Other electrostatic demonstration equipment is available from Fisher and other science equipment companies.

A demonstration electroscope is available from Pasco, model number SF-9069.

Notes and Ideas

*Class time spent on material: Estimated:*______ *Actual:*_____________

Related laboratory activities:

Demonstration materials:

Notes for next time:

Chapter 17: Electric Potential

Outline

Summary

Chapter 17 continues the study of electricity begun in Chapter 16. Electric potential energy and electric potential are defined. Conservation of energy is applied to electrical systems. Equipotential surfaces and their relationships to the electric field and electric field lines are covered. The chapter ends with a discussion of capacitors, with and without dielectrics, how they store electrical energy, and with two optional sections on applications of these ideas.

Major Concepts

By the end of the chapter, students should understand each of the following and be able to demonstrate their understanding in problem applications as well as in conceptual situations.

- Electric potential energy
- Electric potential
  - Definition
  - Equipotential surfaces
- Capacitors
  - Definition
  - Dielectrics
  - Electrical energy storage

Teaching Suggestions and Demonstrations

Students generally understand the concept of electrical potential energy. Electrical potential is more difficult for them. You may need to review the definition of work from Chapter 6. Capacitors also seem to give students trouble. Work as many examples as possible.

Sections 17-1 and 17-2

Use the correspondence between the electric and gravitational forces (both forces are conservative, and both have the same inverse-square dependence) to argue for an **electric potential energy**. Be very careful to remind students (again!) that while the gravitational force is always attractive, the electric force can be attractive or repulsive. This affects the potential energy. Go over the calculation of the work done on a test charge (see Section 17-1). Point out that (just as in the gravitational case) it is the **change in potential energy** that is significant, not the absolute value.

Go over the definition of the **electric potential**, the change in potential energy per unit charge, and its units. Work Conceptual Example 17-1. Plan to work several examples before moving on to the connection between the **change in the electric potential with position** and the **electric field**. Students who have had some calculus will catch on to this concept relatively quickly; it will take more time for the students who have had no calculus. Go through Example 17-2. Work several other examples in class to give the students practice with this concept.

Students should remember the **work-energy theorem** and the **conservation of mechanical energy** from Chapter 6. (You may need to review these briefly.) The application of these ideas to the electric force and the electric potential energy is straightforward. Again, remind students that they must be very careful with signs, since the electric force can be attractive or repulsive. Example 17-3 will provide good structure for a class discussion.

Section 17-3

The analogy of the contour map at the beginning of this section works well in the introduction of **equipotential lines and surfaces**.

➲ **DEMO 17-1** If possible, obtain a contour map of your campus (or your area). Make a transparency of part of it, and go over it with the class. Point out that the curves represent different altitudes and that there is a constant altitude difference between contour lines. Closely spaced curves represent hills or places where the altitude changes rapidly. Widely spaced curves represent more gentle changes in altitude.

Also go over transparencies of Figures 17-6 and 17-7 in class. Point out that the lines represent **lines of constant potential** (in three dimensions these would be **surfaces of constant potential**). There is a constant potential difference between contour lines, just as in the contour map above. In the region where the lines are close together, the potential is changing rapidly. In the region where the circles are farther apart, the potential is not changing as fast.

Be sure to tell the students that the **electric field points in the direction of decreasing electric potential** and that the **electric field lines are perpendicular to the equipotential surfaces**. Remind them that when the electric field lines are close together, the electric field is large.

The concept that the **surface of a conductor is an equipotential surface** is relatively easy for beginning physics students to understand. The implications of that concept for charge density of surfaces with different radii of curvature are much more difficult to grasp, but worth going over with the students. Discuss the **concentration of charge on surfaces with smaller radii** thoroughly and apply this concept to the development of lightning rods.

Sections 17-4 through 17-6

The **electric potential of a point charge** is more difficult to describe without calculus because the field decreases as $1/r^2$. The qualitative argument at the beginning of Section 17-5 is useful. Be sure to point out to the students that setting the potential equal to zero for a particle located at infinity is a *choice*, and that (again) it is the change in the potential that is significant. Also note that the sign of the potential depends on the sign of the charge.

If more than one charge is present, the net electric field is the *vector* sum of the electric fields from all the charges. Electric potential is a scalar; if more than one charge is present, the potential is the *algebraic* sum of the potential created by each charge. This is the **principle of superposition for potentials**. The same principle works for the electric potential energy, since energy is also a scalar. Go over Example 17-6 and Conceptual Example 17-7 in class.

Optional Section 17-6 discusses the potential due to an electric dipole and the electric dipole moment. If you have time, it is a good idea to define a dipole and the dipole moment and talk about the dipole moments of molecules, particularly those listed in Table 17-2.

Sections 17-7 through 17-9

The definition of **capacitance** is deceptively simple. Many students have a difficult time with this concept. Go over the definition of a **capacitor** and the definition of capacitance, pointing out that the capacitance serves as the constant of proportionality between the voltage applied to a capacitor and the amount of charge it can hold. Go through the derivation of the capacitance of a **parallel-plate capacitor**. Discuss how the capacitance depends on the area of the capacitor and the distance between the plates. Example 17-8 is good for review.

It helps to remind students that **dielectrics** are just insulators placed between the plates of a capacitor. Figure 17-16 is a great diagram of what happens in the dielectric as it is moved between the capacitor plates. Since the field inside the dielectric is less than the field outside (due to the polarization of charge in the dielectric), it makes sense to divide the outside field strength by the **dielectric constant** to get the inside field strength. If you can convince the students that this is true, the rest of the derivation follows. Go through Conceptual Examples 17-9 and 17-10 in class. The discussion on computer keyboards in Section 17-7 is quite interesting. If your students have heard the spooky music played for early science-fiction movies, they have heard a theremin, which is based on the principles of capacitance. (See Resource Information.)

➲ **DEMO 17-2** Bring in several capacitors of various sizes and shapes. Discuss why they are made the way they are. Point out that normal-sized capacitors have capacitances measured in the picofarad or microfarad range.

Because capacitors store charge, they also store electrical energy. Discuss the derivation of the **electrical energy stored in a parallel-plate capacitor** in Section 17-9. Work Example 17-11 in class. Be sure to cover the hazards of capacitors and review the idea that the energy stored in the capacitor is stored in the electric field between the plates.

Sections 17-10 and 17-11

These two optional sections discuss applications of the ideas from this chapter to CRTs and the electrocardiogram. The material is quite interesting and can be covered quickly.

Resource Information

Transparencies

T209.	Figure 17-1	Work is done by the electric field in moving the positive charge
T210.	Table 17-1	Some Typical Potential Differences
T211.	Figure 17-6	Equipotential lines between two charged parallel plates
T212.	Figure 17-7	Equipotential lines are always perpendicular to the electric field lines
T213.	Table 17-2	Dipole Moments of Selected Molecules
T214.	Figure 17-13	Capacitors
T215.	Figure 17-15	Key of a computer keyboard
T216.	Table 17-3	Dielectric Constants
T217.	Figure 17-16	Molecular view of the effects of a dielectric
T218.	Figure 17-19	Evacuated glass tube
T219.	Figure 17-20	A cathode-ray tube
T220.	Figure 17-23	Typical ECG
	Figure 17-24	Heart muscle cell

Suggested Readings

Beichner, R., "Visualizing Potential Surfaces with a Spreadsheet," *The Physics Teacher* (February 1997), pp. 95–97. Presents a technique for developing three-dimensional plots of potential surfaces.

Moelter, M., Evans, J. and Elliott, G., "Electric Potential in the Classical Hall Effect: An Unusual Boundary-Value Problem," *American Journal of Physics* (August 1998), pp. 668–677. Detailed presentation of a solution to a nonstandard boundary value problem in a Hall effect experiment.

Parker, G., "Electric Field Outside a Parallel Plate Capacitor," *American Journal of Physics* (May 2002), pp. 502–507. An interesting topic to mention, although it is beyond the scope of most introductory-level physics courses.

Saslow, W., "Voltaic Cells for Physicists: Two Surface Pumps and an Internal Resistance," *American Journal of Physics* (July 1999), pp. 574–583. A discussion of the basic properties of voltaic cells considering the chemical reactions that take place at the electrode–electrolyte interfaces.

Skeldon, K., Reid, L., McInally, V., Dougan, B. and Fulton, C., "Physics of the Theremin," *American Journal of Physics* (November 1998), pp. 945–955. A discussion of some of the interesting physics behind the design and operation of the theremin electronic musical instrument.

Zimmerman, N., "A Primer on Electrical Units in the Système International," *American Journal of Physics* (April 1998), pp. 324–331. A detailed discussion of electrical units, from basic physical laws to commercial calibrations.

Notes and Ideas

Class time spent on material: Estimated:______ *Actual:_____________*

Related laboratory activities:

Demonstration materials:

Notes for next time:

Chapter 18: Electric Currents

Outline

Summary

Basic concepts of electric currents are covered in Chapter 18. Resistance, Ohm's law, and power are introduced and applied first to a simple circuit consisting of a battery and a single resistor. Household circuits are also discussed.

Major Concepts

By the end of the chapter, students should understand each of the following and be able to demonstrate their understanding in problem applications as well as in conceptual situations.

- Batteries
- Current
  - Definition
  - Simple circuits
  - Conventional current
  - Direct current (dc)
  - Alternating current (ac)
- Ohm's law and resistors
  - Resistance and resistivity
  - Ohmic and non-ohmic devices
- Power

Teaching Suggestions and Demonstrations

Most people have thrown balls and watched objects fall, so they enter a study of mechanics with everyday experience and observation to build on. Many of your students, however, will not have comparable experiences with circuits. What happens in the wires when a light switch is turned on is invisible to us, so applying common sense and intuition to circuits is difficult. (This may actually be advantageous. Because of friction and air resistance, our "common sense" applied to mechanics is often misleading!) Don't assume previous knowledge, and be sure to use plenty of examples and thought questions to help students develop understanding of the material. Analogies for circuits are particularly helpful.

Sections 18-1 and 18-2

Begin your presentation by drawing a **simple circuit** consisting of a battery and a lightbulb. Draw both a sketch, showing wires from the two battery terminals connected to the bulb, and a circuit diagram, using the symbols for a battery and a resistor. The **battery** maintains a potential difference between its terminals. If a path is available, electrons will flow from one terminal to the other. The flow of electrons is the **current**. Be sure to differentiate between the actual direction the electrons move and the direction of the **conventional current**, as shown in Figure 18-8. Emphasize that **electromotive force** is not a force at all, but rather a potential difference.

➲ **DEMO 18-1** Bring in three or more common household **batteries** (such as D-cells), a bulb, connecting wires, and a voltmeter. (Although voltmeters aren't discussed until the end of this chapter, you can briefly introduce them here.) Place the batteries in series and then in parallel, testing the voltage across the group each time. (Have students predict the results before you actually perform the demonstration.) Connect the batteries (first in parallel and then in series) to the bulb and break the circuit at different points each time. Again, have students predict the results. The bulb goes out no matter where you break the circuit when all the elements are in series, but not when they are all in parallel.

Using a **mechanical analogy for electric circuits**, such as the flow of water, is very helpful to students. The battery is a pump or person raising the water, the current is the flow of the water, and the resistors are waterwheels or restricted pipes. Establish this analogy early in the study of circuits, and you will be able to use it in the discussion of resistors in series and parallel later on.

Sections 18-3 and 18-4

Make sure your students understand the difference between **resistance** and **resistivity**. Example 18-5 provides practice calculating resistance and also justifies the fact that we often ignore the resistance in conducting wires in a circuit. Show samples of color-banded **resistors** used in circuits, but also point out that lightbulbs and toaster ovens act as resistors, since they dissipate electrical energy. Figure 18-12 gives the color code for resistors and explains how to use it.

Use the water-flow analogy to help students derive the relationships expressed by **Ohm's law**. If you keep the pump (voltage) constant but use a narrower pipe (increase the resistance), what happens to the water flow (current)? It decreases. If you keep the pipe size (resistance) the same but turn the pump (voltage) up, what happens to the water flow (current)? It increases. Remind students that Ohm's law is not really a law, but rather a relationship that applies to "ohmic" materials.

➲ **DEMO 18-2** Very simple circuits can be used to demonstrate **Ohm's law**. Bring in a power supply, two resistors with very different resistances, an ammeter, and the necessary connecting wires or circuit board. Put one resistor in series with the power supply and ammeter. Read the current for varying values of voltage. Then, replace the resistor and compare the values of current for the same voltages. In the first case, you increase V and examine I with R constant; then you change R and notice the change in I with V constant.

Discuss the temperature dependence of the resistivity for various materials. Work Example 18-7 and explain the concepts behind and applications for the resistance thermometer.

Sections 18-5 and 18-6

Equations 18-5, 18-6a, and 18-6b each express **electrical power** in terms of two of the three variables in Ohm's law (current, voltage, and resistance). Work Example 18-8 in class. Remind students that a **kilowatt-hour** is a unit of energy, not power. Go through Example 18-9 and talk about ways to conserve energy (and money!).

➲ **DEMO 18-3** Set up a demonstration circuit with a voltage source and a 5-W lightbulb. Include an ammeter and a voltmeter in the circuit and have students **calculate the power dissipated** by the lightbulb. Substitute different wattage bulbs (10 W, 15 W) and calculate the power dissipated by each. (See Resource Information.) This demonstration also provides a nice opportunity to emphasize that a given battery or power supply with constant voltage will not always provide the same current. How much current is drawn from the battery depends on the resistance in the circuit.

There are many excellent examples for calculating power and current flow in household circuits. Go over Figure 18-19 in class and work Example 18-11 and Exercise E.

Section 18-7

Carefully point out the differences between **direct current (dc)** and **alternating current (ac)**. (See Figure 18-21.) **Root mean square values of alternating current and voltage** are often difficult for students to understand at first. The typical household plug provides an rms voltage of 120 V, which corresponds to a maximum or peak voltage of 170 V (see Figure 18-22). The 170-V figure is in a sense not very meaningful; the voltage only reaches that value for brief moments in the cycle. The wall socket *effectively* supplies 120 V, which means that the power dissipated is equivalent to the power that would be dissipated in the same appliance if it were connected to a 120-V direct current source. In fact, rms values for current and voltage are sometimes called effective values. Work Example 18-13 in class.

Sections 18-8 through 18-10

These three sections are optional but contain a great deal of interesting material. Try to cover the microscopic view of current and the basic definition of superconductivity. If you have pre-health professions students, also plan to cover the structure of a neuron and discuss how electrical conduction aids in our perception of the outside world and in the control of our bodies.

Resource Information

Transparencies

T221.	Figure 18-3	Simple electric cell
	Figure 18-4	Diagram of an ordinary dry cell (like a D-cell or AA)
T222.	Figure 18-5	A lightbulb
T223.	Figure 18-8	Conventional current from + to – is equivalent to a negative electron flow from – to +
T224.	Figure 18-9	Graphs of current vs. voltage
T225.	Figure 18-12	The resistance value of a given resistor
T226.	Table 18-1	Resistivity and Temperature Coefficients
T227.	Figure 18-19	Fuses
T228.	Figure 18-20	Connection of household appliances

T229.	Figure 18-21	(a) Direct current (b) Alternating current
	Figure 18-22	Power transformed in a resistor in an ac circuit
T230.	Figure 18-24	Electric field $\vec{E}$ in a wire
	Figure 18-25	Electrons in the volume Al will all pass through the cross section
T231.	Figure 18-27	A simplified sketch of a typical neuron
T232.	Figure 18-28	How a dipole layer of charge forms on a cell membrane
T233.	Figure 18-30	Action potential
T234.	Figure 18-31	Propagation of an action potential along an axon membrane

Suggested Readings

Alper, J., "The Battery: Not Yet a Terminal Case," *Science* (17 May 2002), v. 296, pp. 1224–1226. An interesting article on new developments in battery technology.

Chang, A., "Resistance of a Perfect Wire," *Nature* (3 May 2001), v. 411, pp. 39–40. An interesting discussion of the resistance in electrical contacts.

Cho, A., "New Observations Give Stripes Theory a Lift," *Science* (15 March 2002), v. 295. pp. 1992–1993. A short article on the stripes theory for high-temperature superconductors.

Fisher, K., "Conjugate Relationships in Basic Electricity," *The Physics Teacher* (November 1999), p. 458. Help with new electricity vocabulary and relationships.

Herman, R., "An Introduction to Electrical Resistivity in Geophysics," *American Journal of Physics* (September 2001), pp. 943–952. A nice application of an introductory physics topic to geophysics, including description of a resistivity apparatus.

Service, R., "MgB2 Trades Performance for a Shot at the Real World," *Science* (1 February 2002), v. 295, pp. 786–788. An interesting article on practical superconductors.

Service, R., "Shrinking Fuel Cells Promise Power in Your Pocket," *Science* (17 May 2002), v. 296, pp. 1222–1224. An interesting article on new micro fuel cells.

Voss, D., "Perplexing Compounds Rejoin the Club," *Science* (25 January 2002), v. 295, p. 604. A short article on high temperature superconductors.

Materials

Ammeters and voltmeters are available from a variety of sources. One voltmeter/ammeter/ohmmeter combination is available from Sargent-Welch, model number WLS30710-10.

Notes and Ideas

*Class time spent on material: Estimated:*______ *Actual:*____________

Related laboratory activities:

Demonstration materials:

Notes for next time:

Chapter 19: DC Circuits

Outline

Summary

Basic concepts of electric currents were covered in Chapter 18. Chapter 19 applies these concepts to circuits with combinations of resistors in series and parallel, circuits containing combinations of capacitors, and *RC* circuits. Batteries in combinations in circuits are also treated. At the end of the chapter, devices for measuring current and potential difference in a circuit are discussed.

Major Concepts

By the end of the chapter, students should understand each of the following and be able to demonstrate their understanding in problem applications as well as in conceptual situations.

- EMF
  - Terminal voltage
  - EMFs in series and parallel
- Resistors in series and parallel
- Kirchhoff's rules
- Capacitors in series and parallel
- *RC* circuits
- Ammeters and voltmeters

Teaching Suggestions and Demonstrations

The rules for the behavior of resistors, capacitors, and batteries in circuits are not difficult, but students sometimes have trouble remembering which set of rules applies to which situation. Be sure to go through the derivations of the rules carefully. The students will find the section on electrical hazards informative and interesting.

Sections 19-1 through 19-4

Remind students that **electromotive force (emf)** is not a force at all, but rather a potential difference. Plan to spend a little time on the concept of a **real battery**, complete with **internal resistance**. Figure 19-1 and Example 19-1 will help.

Begin Section 19-2 by introducing the concept of **equivalent resistance for a group of resistors**, then derive the relationships specifically for resistors in **series** and in **parallel**. Equivalent resistance of a group of resistors in series will be fairly intuitive for students; add more resistors and you get more resistance. (See Figure 19-3.) They will likely have more trouble with equivalent resistance of resistors in parallel, when the more resistors you add, the lower the overall resistance becomes. (See Figure 19-4.) Return to the water analogy to help explain this concept. Be sure to do Conceptual Example 19-2 and Example 19-3 or others like them in class.

➲ **DEMO 19-1** Try Conceptual Example 19-6, which compares identical lightbulbs in series and in parallel, as a demonstration. Students should predict the results before you actually connect the circuit. For an extension, you can add more lightbulbs or use an ammeter and a voltmeter to quantify the results.

Reinforce these ideas with plenty of conceptual questions. Consider what happens, for instance, when one more resistor is added to either a series circuit or a parallel circuit. Does the equivalent resistance increase or decrease? What about the current coming from the battery or the current through an individual resistor? What happens to the voltage drop across an individual resistor? You can also briefly discuss **household wiring** at this point. Consider the problems that would be encountered if appliances were connected in series! On the other hand, parallel wiring necessitates fuses or circuit breakers, since adding more branches could increase the total current to dangerous levels.

For **combination circuits**, encourage students to actually sketch the intermediate steps, as is done in Figure 19-10, and to use subscripts on Ohm's law. $V = IR$ is not valid if V refers to the voltage of the battery but I is the current through just one branch.

Begin Section 19-3 by demonstrating **Kirchhoff's rules** applied to a simple circuit. Students can use Ohm's law and equivalent resistance to verify their results. Then move on to a problem such as Example 19-8, for which Kirchhoff's rules are necessary. Students tend to have trouble knowing which sign to use for voltages and recognizing the difference between current direction and loop direction. Emphasize that the current direction is your best guess about what is physically happening in the circuit, and the loop is just a way to traverse a part of the circuit, ending up where you started. Use the water analogy again: in a canoe, you can drift downstream and then paddle back upstream to return to your camping spot, but no matter which way *you* are going, the *river* is flowing downstream. As you drift downstream, you lose potential energy, since you are losing elevation. Likewise, traversing a resistor in the direction of the current also corresponds to a potential drop. Figure 19-12 is a great illustration of this idea. The Problem Solving box in this section is excellent.

➲ **DEMO 19-2** Build a **multiloop circuit** with two or more batteries or power supplies and various resistors. Use an ammeter and a voltmeter to demonstrate the junction theorem and the loop theorem.

Section 19-4 is optional, but should be required of all students who drive a car. Plan to go through Example 19-9 in class. Be sure to include a short discussion on how to jump start a car without the battery exploding.

Sections 19-5 through 19-7

The equations for **equivalent capacitance for capacitors in series and in parallel** can be derived in the same way that the equations for equivalent resistance were derived in Section 19-2. In the case of capacitors, however, a series combination results in an equivalent capacitance that is lower than the capacitance of any one of the individual capacitors, and a parallel arrangement results in a higher capacitance. Remind the students of the physical meaning of capacitance to clarify why this behavior is different from that of resistors. Work several examples in class to reinforce these ideas.

In an ***RC* circuit**, the rate at which the charge on the capacitor changes depends on the amount of charge itself. This results in an exponential relationship. You may need to do a quick review of logarithms in this section. Discuss what happens to both the current and the charge on the plates of the capacitor during charging and discharging. (See Figures 19-20 and 19-21.) The applications of *RC* circuits at the end of Section 19-6 (turn signals, windshield wipers, and heart pacemakers) are quite interesting.

➲ **DEMO 19-3** Set up an ***RC* circuit** with an ammeter in series to demonstrate the time dependence of the charging of a capacitor. After the capacitor is fully charged, open the switch again and watch what happens to the current as the capacitor discharges through the resistors. This demonstration is even more effective if lightbulbs are used as the resistors.

Plan to spend some time on Section 19-7, which covers electrical hazards. This material contains important, practical applications of the concepts you have been discussing in class. Students in general will find the section interesting and useful; it will be of particular value to students with interests in the health professions or in construction, home improvement, or repair.

Section 19-8

This section is marked optional, but you should plan to cover it if possible. If your course has a laboratory component, students will need to be familiar with the operation of **ammeters and voltmeters**. The key differences concern the amount of internal resistance in each and the way each is placed in a circuit. In particular, emphasize that an ammeter, which has a very low internal resistance, must be placed in series with other circuit elements. If placed in parallel, it creates a short circuit that can result in dangerously high currents. The section on how to connect meters is particularly clear.

Resource Information

Transparencies

T235.	Table 19-1	Symbols for Current Elements
T236.	Figure 19-1	Diagram for an electric cell or battery
T237.	Figure 19-3	Resistances connected in series
T238.	Figure 19-4	Resistances connected in parallel
T239.	Figure 19-9	Circuit for Example 19-6
T240.	Figure 19-10	Circuit for Example 19-7, where *r* is the internal resistance of the battery
T241.	Figure 19-11	Currents can be calculated using Kirchhoff's rules
T242.	Figure 19-12	Changes in potential around the circuit in (a) are plotted in (b)
T243.	Figure 19-14	Batteries in series (a) and (b), and parallel (c)
T244.	Figure 19-15	Example 19-9, a jump start
T245.	Figure 19-17	Capacitors in parallel
T246.	Figure 19-18	Capacitors in series
T247.	Figure 19-20	For the *RC* circuit shown, the voltage across the capacitor increases with time
	Figure 19-21	For the *RC* circuit shown, the voltage V_c on the capacitor decreases with time
T248.	Figure 19-23	An *RC* circuit, coupled with a gas-filled tube as a switch
T249.	Figure 19-26	An electric oven
T250.	Figure 19-30	An ammeter is a galvanometer in parallel with a (shunt) resistor
	Figure 19-31	A voltmeter is a galvanometer in series with a resistor with high resistance

Figure 19-32 An ohmmeter
T251. Figure 19-33 Measuring current and voltage

Suggested Readings

Alper, J., "The Battery: Not Yet a Terminal Case," *Science* (17 May 2002), v. 296, pp. 1224–1226. An interesting article on new developments in battery technology.

Amengual, A., "The Intriguing Properties of the Equivalent Resistances of *n* Equal Resistors Combined in Series and in Parallel," *American Journal of Physics* (February 2000), pp. 175–179. An examination of many possible combinations of *n* equal resistors in series and/or in parallel.

Brown, R., "Series and Parallel Resistors and Capacitors," *The Physics Teacher* (November 2003), pp. 483–485. A short article on the addition rules for resistors and capacitors in series and parallel.

Derman, S. and Goykadosh, A., "A Pencil-and-Tape Electricity Experiment," *The Physics Teacher* (October 1999), pp. 400–402. A simple way of verifying the addition rules for resistors in series and parallel.

Denardo, B., Earwood, J. and Sazonova, V., "Experiments with Electrical Resistive Networks," *American Journal of Physics* (November 1999), pp. 981–986. Two large networks of resistors for use in lecture demonstrations or lab.

Hart, F., "Computer-Based Experiments to Measure RC," *The Physics Teacher* (March 2000), pp. 176–177. Computer data acquisition using Vernier equipment.

Henry, D., "Resource Letter TE-1: Teaching Electronics," *American Journal of Physics* (January 2002), pp. 14–23. An examination of the role and content of electronics courses with an excellent list of resources.

Vreeland, P., "Analyzing Simple Circuits," *The Physics Teacher* (February 2002), pp. 99–100. An example of a graphic organizer that helps students correctly apply Ohm's law to series and parallel circuits.

Materials

Ammeters and voltmeters are available from a variety of sources. One voltmeter/ammeter/ohmmeter combination is available from Sargent-Welch, model number WLS30710-10.

Notes and Ideas

Class time spent on material: Estimated:______ *Actual:_____________*

Related laboratory activities:

Demonstration materials:

Notes for next time:

Chapter 20: Magnetism

Outline

Summary

Chapter 20 introduces magnetic fields and magnetic forces. A moving charge will experience a force in a magnetic field if a component of its velocity is perpendicular to the field. Thus, a current-carrying wire may experience a force, and a current-carrying loop may experience a torque in a magnetic field as well. Magnetic fields arise from moving charges. The specific fields due to current-carrying straight wires, wire loops, and solenoids are examined. Magnetism in matter and the Earth's magnetic field are also discussed, as are several applications of magnetic fields in common equipment and scientific instruments.

Major Concepts

By the end of the chapter, students should understand each of the following and be able to demonstrate their understanding in problem applications as well as in conceptual situations.

- Magnets and the magnetic field
  - North and south poles
  - Field lines
  - Earth's magnetic field
- Magnetic force
  - Moving charged particle
  - Magnetic force right-hand rule
  - Current-carrying wire
  - Torque on a current-carrying loop
- Magnetic field
  - Magnetic field right-hand rule
  - Long, straight wire
  - Current loop
  - Solenoid
- Ferromagnetism

Teaching Suggestions and Demonstrations

One main point of Chapter 20 is that electricity and magnetism are intimately connected. In the earlier sections, students find that magnetic fields arise from moving electric charges. The later sections deal with magnetic forces on moving electric charges. None of this is intuitively obvious. Use plenty of conceptual questions and examples to check and reinforce student understanding.

Section 20-1

In studying the electric field, students found that the field direction could be determined by the direction of the force on a positive test particle. In a similar way, the direction of a **magnetic field** can be found using a compass as a "test particle." Point out some differences between the two types of field; electric field lines begin and end on charges, but **magnetic field lines** form closed loops. Although positive and negative electric charges may exist individually, north and south magnetic poles cannot, as illustrated in Figure 20-3.

➲ **DEMO 20-1** Bring in a bar magnet and use a small compass to "test" the **magnetic field** around it. After students have a feel for the shape of the field, place a sheet of paper over the magnet and sprinkle iron filings on top, as shown in Figure 20-4. Alternatively, you can purchase a "magnetic field demonstrator," which consists of a transparent box filled with iron filings suspended in a liquid, with a hole in the center in which a magnet can be inserted. Students will be able to see the filings line up in three dimensions. (See Resource Information.)

Atmospheric auroral displays are caused by the interaction of particles from the Sun with particles in the Earth's atmosphere. The reason these beautiful light shows occur at high latitudes (and are therefore called northern or southern lights) is that the solar particles are directed toward the poles by the **magnetic field of the Earth**, as discussed later in Section 20-4. (See Resource Information.)

Sections 20-2 and 20-3

Section 20-2 discusses the fact that magnetic fields are actually created by moving charges, or current. The **magnetic field right-hand rule** gives the relationship between the current direction and the resulting magnetic field. (See Figures 20-8 and 20-10.) It is sometimes referred to as the right-hand *wire* rule in order to avoid confusion with the magnetic force right-hand rule. Point out to students that **moving charges create a magnetic field**, but that equations for only three specific configurations (long straight wire, current loop, and solenoid) will be discussed in this chapter.

A current-carrying wire will produce a magnetic field. It will also experience a force when placed in an external field. Go through the deduction of the formula for the **force on a current-carrying wire in a magnetic field** and the definition of the magnetic field strength **B**.

➲ **DEMO 20-2** Demonstrate the **force on a current-carrying wire in a magnetic field**. Place a wire connected to a battery or power supply between the poles of a (strong!) horseshoe magnet as shown in Figure 20-11. When current is flowing, the wire will be deflected. Reverse the current direction and notice that the deflection also reverses. Turn the current off to demonstrate that the deflection really is due to the moving charges. (Be sure to experiment with this demonstration ahead of time so you will know how much current is necessary to produce a noticeable result for your magnet.) The students can use the right-hand rule and their observations to determine which end of the magnet is the north pole.

Work several examples like Examples 20-1 and 20-2 in class to reinforce these concepts. Be sure to go over the conventions for drawing a magnetic field pointing into or out of the page.

Sections 20-4 through 20-7

You can begin the somewhat bewildering topic of **magnetic force on moving charges** by asking students what conditions must be met for a particle to experience a force in the other fields encountered so far. All particles experience a force in a gravitational field, but only charged particles do in an electric field. In the case of the magnetic field, the particle must not only be charged, but must also be moving with a component of its velocity perpendicular to the field. The magnitude of the magnetic force is actually defined by the force, charge, speed, and angle.

After introducing the **magnetic force right-hand rule** for the force on a charged particle, draw lots of examples on the board or overhead, and have the class practice determining the direction of the magnetic force. Vary your examples as much as possible. Draw the fields pointing in or out as well as in the plane of the board. Be sure to include a few examples where the velocity and field are parallel or anti-parallel, in which case the force is zero. Also toss in a few negative particles to remind students that the force in this case is opposite the result given by the right-hand rule. Conceptual Example 20-6 is a good one to do together. The Problem Solving box on magnetic fields is excellent; take time to go over it in class.

Point out that the magnetic force acts perpendicular to the direction of the velocity of the particle and therefore does no work on the particle. The magnetic force cannot change the speed of the particle (remember the work–energy theorem!), but what about the velocity of the particle? The magnetic force changes the direction of the velocity vector, but not the magnitude. (See Figure 20-16.)

➲ **DEMO 20-3** Show students photographs of particle tracks from bubble, cloud or spark chambers and point out the spiraling paths. Work Example 20-5 to determine the **radius of the orbit of a moving charged particle in a magnetic field** and discuss the various parameters that can be determined from the experimental data. (See Resource Information.)

Section 20-5 deduces the formula for the **magnetic field strength produced by a straight current-carrying wire**. Go through this deduction carefully, and work Example 20-7.

The **force between two current-carrying wires** is a nice summary of the main points of the chapter. Magnetic fields are produced by moving charges; moving charges in magnetic fields experience forces. Both right-hand rules are necessary to illustrate that two wires carrying currents in the same direction will attract each other, whereas two wires with currents in opposite directions will repel. (See Figures 20-23 and 20-24.) Do Example 20-10 together with students and encourage them to try Example 20-11 on their own for more practice with the direction rules.

➲ **DEMO 20-4** Set up two **parallel wires** and run currents in the same and in opposite directions to illustrate the attractive and repulsive forces that result.

Go over the definitions of the **ampere** and the **coulomb**.

Although the formula for the **field in the center of a current loop of radius *r*** is not derived, you can compare it to the formula for the field the same distance r away from a long straight wire. It will make sense to students that the field in the center of the loop is stronger, since the point in question is surrounded by current on all sides. When discussing the **solenoid**, point out why the magnetic field *outside* is negligible.

Sections 20-8 through 20-12

These sections contain a great deal of important and interesting information, but, except for Section 20-12, they are optional and can be skipped without loss of continuity. If you have time, briefly go over **Ampère's law** and the **torque on a current loop**.

➲ **DEMO 20-5** The easiest way to introduce **torque on a current-carrying loop** is through a demonstration. A wire coat hanger bent into a rectangle or four rulers taped together work fine. Establish a direction in the classroom for an imaginary magnetic field. Place an easily visible marker of some kind on the rectangular loop to indicate current direction. Then hold the loop in the "magnetic field" with the plane of the loop parallel to the field and ask students to use the right-hand rule to determine the direction of the force on each of the four sides. Try different orientations to illustrate maximum torque and zero torque. Note that if the torque causes the loop to "overshoot" the zero position, the new torque attempts to return the loop to the zero position instead of continuing a circular motion. You can discuss the need for a commutator to change current direction in order to produce continuous motion, as in a motor.

Students will find the discussion of the functioning of **galvanometers**, **motors**, and **loudspeakers** informative. The **mass spectrometer** discussed in Section 20-11 is a good example that will be of particular interest to biology majors and students in the pre-health professions. Take time to discuss ferromagnetism and the magnetization of domains within the material. It is a good idea to mention the atomic level structure of "permanent magnets" to show that even these owe their magnetic fields to moving charges.

Resource Information

Transparencies

T252.	Figure 20-2	Like poles of a magnet repel; unlike poles attract
	Figure 20-3	If you split a magnet, two new magnets are produced
T253.	Figure 20-4	Magnetic field lines for a bar magnet
T254.	Figure 20-5	The Earth acts like a huge magnet
T255.	Figure 20-8bc	(b) Magnetic field lines around an electric current in a straight wire (c) Right-hand rule for remembering the direction of the magnetic field
T256.	Figure 20-9	Magnetic field lines due to a circular loop of wire
	Figure 20-10	Right-hand rule for determining the direction of the magnetic field
T257.	Figure 20-11	Force on a current-carrying wire placed in a magnetic field $\vec{B}$
T258.	Figure 20-12	Current-carrying wire in a magnetic field
T259.	Figure 20-14	Force on charged particles due to a magnetic field is perpendicular to the magnetic field direction
T260.	Figure 20-16	Force exerted by a uniform magnetic field on a moving charged particle produces a circular path
T261.	Figure 20-21	Example 20-8 (Magnetic field midway between two currents)
	Figure 20-24	Force between wires carrying parallel and antiparallel currents
T262.	Figure 20-26	Magnetic field of a solenoid
	Figure 20-27	Solenoid used as a doorbell
T263.	Figure 20-31	Cross-sectional view into a solenoid
T264.	Figure 20-32	Calculating the torque on a current loop in a magnetic field $\vec{B}$

T265.	Figure 20-33	Galvanometer
T266.	Figure 20-35	Diagram of a simple dc motor
	Figure 20-36	The commutator-brush arrangement in a dc motor
T267.	Figure 20-38	Loudspeaker
T268.	Figure 20-39	Bainbridge-type mass spectrometer
T269.	Figure 20-40	Domains

Suggested Readings

Baker, B., "Demonstrating Forces between Parallel Wires," *The Physics Teacher* (May 2000), p. 299. A dramatic illustration of the repulsion or attraction between parallel current-carrying wires.

Banerjee, S., "When the Compass Stopped Reversing Its Poles," *Science* (2 March 2001), v. 291, pp. 1714–1715. An accessible, short article on the long-term behavior of the Earth's magnetic field.

Bichsel, E., Wilson, B. and Geerts, W., "Magnetic Domains of Floppy Disks and Phone Cards Using Toner Fluid," *The Physics Teacher* (March 2002), pp. 150–153. A conceptual demonstration using dry and liquid toner to find the magnetic poles at the surfaces of refrigerator magnets and magnetic recording media.

Brown, R., "Demonstrating the Meissner Effect *and* Persistent Current," *The Physics Teacher* (March 2000), pp. 168–169. A demonstration involving levitation of a magnet over a superconductor.

Cavicchi, E., "Experiences with the Magnetism of Conducting Loops: Historical Instruments, Experimental Replications, and Productive Confusions," *American Journal of Physics* (February 2003), pp. 156–167. An interesting study of nineteenth century lab work on electromagnetism.

Chia, C. and Wang, Y., "The Magnetic Field Along the Axis of a Long Finite Solenoid," *The Physics Teacher* (May 2002), pp. 288–289. Investigates the field of a finite solenoid with a multimeter and a pick-up coil.

Erlichson, H., "The Experiments of Biot and Savart Concerning the Force Exerted by a Current on a Magnetic Needle," *American Journal of Physics* (May 1998), pp. 385–391. Descriptions of many of the historically important experiments relating electricity and magnetism.

Gibson, F. and MacInnes, I., "Symmetry in Electromagnetism: A New Magnetic Needle," *The Physics Teacher* (May 2000), pp. 316–317. Simple experiments to demonstrate the relationship between electricity and magnetism.

Grünberg, P., "Layered Magnetic Structures: History, Highlights, Applications," *Physics Today* (May 2001), pp. 31–37. Essay on thin magnetic films with unique properties and their many applications, included electronic data storage.

Schmidt, M., "Investigating Refrigerator Magnets," *The Physics Teacher* (April 2000), pp. 248–249. A nice complement to Walker's discussion of refrigerator magnets.

Sinacore, J. and Graf, E., "Precision Laboratory Apparatus for the Study of Magnetic Forces," *The Physics Teacher* (May 2000), pp. 296–298. A rugged and precise apparatus for introductory magnetism labs.

Stewart, G., "Measuring Earth's Magnetic Field Simply," *The Physics Teacher* (February 2000), pp. 113–114. An easy and popular activity to measure the Earth's magnetic field.

Van Heuvelen, A., Allen, L. and Mihas, P., "Experiment Problems for Electricity and Magnetism," *The Physics Teacher* (November 1999), pp. 482–485. Real-life laboratory problems for introductory electricity and magnetism.

Welsh, G., "Magnetic Therapy in Physics?," *The Physics Teacher* (March 2000), pp. 181–182. A critical thinking and writing assignment on magnetic therapy.

Wright, K., "Seeing the Light," *Discover* (July 2000), pp. 51–57. A very complete discussion of auroras, with great pictures.

Materials

Aurora: Rivers of Light in the Sky, videotape, SkyRiver Films, Anchorage, Alaska, 1-800-248-WILD.

The Fisher Science Deluxe Magnetic Field Apparatus (item # S43059-1) allows students to see iron filings align with the magnetic field in three dimensions.

Bubble chamber photographs can be found on the Web at the Fermi National Accelerator Laboratory Website, http://www.fnal.gov/. Go to the search option and type in "bubble chamber photographs". There is a section with an example photograph and sample questions that can be answered by simple analysis of the picture.

Notes and Ideas

*Class time spent on material: Estimated:*______ *Actual:*______________

Related laboratory activities:

Demonstration materials:

Notes for next time:

Chapter 21: Electromagnetic Induction and Faraday's Law

Outline

Summary

Magnetic fields arise from moving electric charges, as shown in Chapter 20. Chapter 21 looks at the reverse relationship: electric fields are induced by changing magnetic fields. Faraday's law, which quantifies the electromotive force induced by a changing magnetic flux, and Lenz's law, which gives the direction of the induced current, are both described in detail. Electrical and mechanical energy are also discussed. Generators and motors transform between the two types of energy. Inductors store energy in a magnetic field, and transformers convert power at one voltage and current to power at another voltage and current. A generalized notion of resistance, the impedance, is introduced at the end of the chapter and used in analyzing *RLC* circuits. Finally, resonance in an *RLC* circuit and its applications are introduced.

Major Concepts

By the end of the chapter, students should understand each of the following and be able to demonstrate their understanding in problem applications as well as in conceptual situations.

- Magnetic flux
- Induced emf
  - Faraday's law
  - Lenz's law
  - Motional emf
- Mechanical work and electrical energy
  - Generators
  - Motors
- Inductance
  - Inductors and *RL* circuits
  - Energy stored in a magnetic field
  - Transformers
- *LRC* circuits
  - Capacitive reactance
  - Inductive reactance

- Phase relation between current and voltage
- Impedance
- Resonance

Teaching Suggestions and Demonstrations

Electromagnetic induction almost looks like magic. At first encounter it is pretty impressive that merely waving a magnet around in the vicinity of a coil or twisting a loop in an external magnetic field can produce a current. This chapter is fairly long, but don't skimp on the time you spend on Faraday's law and Lenz's law. The remaining sections are essentially extensions or applications of these principles, so ensuring a firm understanding of the early part of the chapter will allow you to move more quickly through the later sections. Mention practical applications of electromagnetic induction that students encounter in their everyday lives, such as credit card readers and magnetic tape recordings.

Sections 21-1 through 21-4

➲ **DEMO 21-1** A nice way to begin the discussion of **induced emf** is through the demonstration shown in Figure 21-2. Connect a solenoid to a sensitive ammeter. Insert a bar magnet into the solenoid while watching the ammeter needle. Also, hold the magnet still inside the coil and then remove it. Repeat, varying the speed of the magnet. As an extension, use a second coil attached to a power supply as the source of the magnetic field. In this case, students will watch the ammeter as the current in the coil is turned on, kept on, and then turned off. Ask students what caused the current registered by the ammeter. Since there was no induced current when the magnet was still or the current in the primary coil was constant, they should recognize that a magnetic field alone doesn't induce a current. Clearly some sort of *change* is necessary. This leads into a discussion of magnetic flux.

The water passing through a fishing net in a stream is a useful analogy for **magnetic flux**. How much water passes through the net depends on the size of the net, the orientation of the net with respect to the flowing water, and how fast the water is moving. Likewise, magnetic flux depends on the surface area of interest, the angle the normal to the surface area makes with the magnetic field direction, and the magnitude of the magnetic field. It is worth spending some time on the concept of the **normal to the surface** and the angle between the normal to the surface and the magnetic field. Remind students to use their common sense. Go over Figures 21-3 and 21-4. If the plane of a loop is perpendicular to the magnetic field, then clearly the flux is a maximum.

To set the stage for **Faraday's law**, try Example 21-2 and Exercise A. Discuss the ways in which the magnetic flux through a surface can change. Changing flux can come from a changing field, a changing area, or a changing orientation in a constant field. (See Figures 21-6 and 21-7.) Show examples of changing area, as in motional emf, as well as changing field. A rotating loop in an external magnetic field is the basis of a generator, which will be discussed in a later section.

Do plenty of conceptual examples for **Lenz's law**. Students catch on to the idea of "opposition to change" fairly easily, but can get confused about just what is being opposed. Include examples in which the original field decreases to emphasize that the induced current creates an induced magnetic field that opposes the *change*, but doesn't necessarily oppose the original field. Go over Conceptual Example 21-4 and Example 21-5 and Exercise B in class.

A moving rod has mechanical energy. In the case of motional emf, as shown in Figure 21-12, this **mechanical energy is converted into electrical energy**. The motion of the rod changes the area of the rectangular loop in the magnetic field, which, consistent with Faraday's law, causes an induced emf. Another way to convert mechanical energy into electrical energy is to rotate a conducting loop in a magnetic field. An emf is induced since the magnetic flux through the loop is changing.

Sections 21-5 and 21-6

➲ **DEMO 21-2** Pull out the bent coat hanger or four rulers taped into a rectangle that you used in the discussion of torque on a current loop in Chapter 20. Now this fancy piece of apparatus will become a **generator**. Establish the direction of the imaginary external magnetic field and rotate the rectangle. Point out to students that you are supplying mechanical energy by turning the loop. Faraday's law states that a current will be induced. To establish the alternating nature of the current, have students apply the magnetic force right-hand rule to the separate sections of the loop for the different positions in the rotation.

The demonstration above convinces students that the current direction alternates. Although calculus is handy for deriving that the alternation is sinusoidal, it isn't actually necessary. Apply the formula for motional emf, $\mathcal{E} = vBl$, but point out that only the component of the velocity perpendicular to the magnetic field contributes, so replace v with $v \cdot \sin\theta$.

Return to the Chapter 20 discussion of torque on a current loop in a magnetic field and contrast **electric motors** with generators. The former convert electrical energy to mechanical work and the latter convert mechanical work to electrical energy. Demonstration motors/generators are available from several scientific equipment companies. (See Resource Information.)

The discussion of back emf is optional but enlightening. Go over this concept, and work through Conceptual Example 21-10 with the class. Eddy currents are covered in this section as well, with very nice applications.

➲ **DEMO 21-3** Demonstrate **back emf** in an inductor in an *RL* circuit using a function generator on square wave mode as the power supply. Use an oscilloscope to examine the voltage across the inductor. (See Resource Information.) The square wave simulates a dc circuit with a switch repeatedly opened and closed. Each time the switch opens or closes, the back emf jumps to a maximum value because the current is changing and thus the magnetic flux through the coil is changing. As the current becomes established at a constant level, the back emf drops to zero because the magnetic flux in the coil becomes constant.

Sections 21-7 through 21-11

A real advantage of alternating current over direct current is the ability of **transformers** to step up or step down the voltage of an ac source. Transformers don't work with direct current because, as Faraday's law states, there must be a changing magnetic flux in order to induce an emf in the secondary coil. Remind students that conservation of energy requires that the power remain constant, so if voltage is stepped up by a factor of 10, current must go down by the same factor. Since losses in power lines depend on the square of the current, stepping current down by a factor of 10 reduces power lost to heating by a factor of 100. Discuss Figure 21-25 with your class and work Example 21-13.

Section 21-8 contains many interesting **applications of the principle of induction**.

Section 21-9 is marked as optional, but you need to cover inductance and inductors if students are going to understand later sections. **Mutual inductance** and **self inductance** are further applications of Faraday's and Lenz's laws. Both inductors and capacitors will be used in the circuits discussed in later sections, so it is a good idea to draw comparisons between them now. **Inductance** is a property of an **inductor** based upon its physical characteristics, just as capacitance depends only on the characteristics of the capacitor and not on the circuit in which the capacitor is placed. When used in a circuit, both store energy in fields—the inductor in a magnetic field and the capacitor in the electric field between its plates. The characteristic decay times for each type of circuit can also be compared. When a dc source is connected (or disconnected) in series with an ***LR* circuit**, the current in the circuit builds up (or decays) in a manner similar to the current in an *RC* circuit. (See Figure 21-34.)

Sections 21-12 through 21-14

All of these sections are optional. We think you should cover them if at all possible. The many new terms introduced in this chapter can be confusing. Spend time clarifying terms such as reactance and impedance and show students how they relate to familiar properties such as resistance.

Root mean square (rms) values of alternating current and voltage are often difficult for students to understand at first. The typical household plug provides an rms voltage of 120 V, which corresponds to a maximum voltage of 170 V (see Section 18-7). The 170-V figure is in a sense not very meaningful; the voltage only reaches that value for brief moments in the cycle. The wall socket *effectively* supplies 120 V, which means that the power dissipated is equivalent to the power that would be dissipated in the same appliance if it were connected to a 120-V direct current source. In fact, rms values for current and voltage are sometimes called effective values.

Phasors and **phasor diagrams** are also very tricky. Introduce them here and convince students that the angular speed ω of the phasor's rotation about the origin is the same as the angular frequency of the sinusoidally alternating voltage. Since the current and voltage of the resistor in an ac circuit are in phase, phasor diagrams aren't really that useful in analyzing circuits that have only resistors. However, they're extremely helpful when capacitors and inductors are in the circuit, so it's best to introduce them on the easier case and let students become familiar with them.

To give students an idea of the goal of these next sections, draw a circuit with a battery and three resistors in series and then a separate circuit with **an ac generator and a resistor, an inductor and a capacitor in series**. Ask students how the voltage drops across the three resistors in the battery circuit are related to the voltage of the battery. They will respond that the resistor voltage drops add up to the total voltage gain from the battery. A reasonable guess for the ac circuit then would be that the voltage drops across the circuit elements add up to the voltage gain across the generator. However, for the ac circuit, the voltages refer to rms values, and they do not add up simply to equal the rms voltage of the generator. The reason is that the different sinusoidal voltages are not in phase. The first task, then, is to discover the **phase relationships**.

Students already know that the voltage and current of the **resistor** are in phase and are related by Ohm's law, which applies to both alternating and direct current circuits as long as V and I correspond to the same type of value, for instance, rms or maximum values.

A **capacitor** in an ac circuit is more complicated. Sketch a sine curve to represent the current in the circuit. Plot the corresponding voltage across the plates of the capacitor by discussing the behavior of the charges. For instance, when the voltage across the plates is zero, the current will be a maximum since there is nothing to oppose it. When the voltage across the plates is a maximum, the current will be zero.

Provide a similar exercise for an **inductor** in an ac circuit. Because of Faraday's law, the voltage across the inductor depends on the *rate of change* of the current, so when the current reaches a maximum, the voltage will be zero. These graphs show that the voltage across the inductor leads the current and the voltage across the capacitor lags the current, each by 90°.

For both capacitor and inductor, note that the **reactance** is a measure of the opposition to current flow. The expressions for the reactances are not derived, but it is a good idea to verify with students that they are at least reasonable. First, have students check units to make sure both are in ohms. Then look at the extreme cases of frequency. When frequency equals zero, which corresponds to the direct current case, the capacitive reactance becomes infinite and the inductive reactance drops to zero, both as expected. Repeat, for the case of a very large frequency.

Now that the students have an understanding of the **leading and lagging** of the various voltages with respect to the current, they will appreciate the usefulness of a **phasor diagram**, such as the one shown in Figure 21-40. Remind them that all the phasors except the generator voltage remain in fixed relationships with each other. (For example, V_L always leads V_R and I by 90°.) The phasor diagram helps students visualize the **phase angle,** which in turn gives them a feel for the **power** dissipated in the circuit.

➲ **DEMO 21-4** Although the rms voltages of the elements in the *RLC* circuit do not add up simply, the **sum of the sine curves** for the voltages of the three elements does represent the sinusoidal voltage of the generator. It is hard to add sine curves freehand, so prepare some computer plots ahead of time. Graph the curves representing V_L, V_R, and V_C and then show their sum. This will be most effective if the numbers correspond to an example you will be doing in class, so the students can see that the sum not only is another sine curve, but also has the expected leading or lagging relationship. If desired, you can show graphs for *RL* and *RC* circuits separately before proceeding to the *RLC* example.

Introduce **resonance** by asking students how to increase current in a given *RLC* circuit. Maximum current will correspond to the minimum impedance. An examination of the formula for impedance indicates that this will occur when $X_L = X_C$. Alternatively, examine the phasor diagram and note that the greatest power results when V_R is a maximum. The largest value V_R can have is the same as the generator voltage, and again, this results when $X_L = X_C$. Figure 21-42 shows current as a function of frequency.

➲ **DEMO 21-5** Set up an ***RLC* series ac circuit** with a variable frequency generator, an inductor, a resistor, and a capacitor. Use an oscilloscope to examine the voltage across the resistor and the generator voltage, displaying both sine curves at once. (Since the resistor voltage is always in phase with the current, it can be used to determine whether the generator voltage is leading or lagging the current.) Begin with frequencies lower than resonance frequency and watch the two curves as you increase to and then beyond the resonance frequency. Students should be able to see both the changing phase relationship and the changing magnitude of V_R. At resonance the two curves will be identical.

Resource Information

Transparencies

T270.	Figure 21-2	Induced current
T271.	Figure 21-3	Determining the flux through a flat loop of wire
	Figure 21-4	Magnetic flux Φ_B
T272.	Figure 21-6	A current can be induced by changing the area of the coil
	Figure 21-7	A current can be induced by rotating a coil in a magnetic field
T273.	Figure 21-9	Example 21-4 (Practice with Lenz's law)

T274.	Figure 21-12	(a) A conducting rod is moved to the right on a U-shaped conductor (b) Upward force on an electron in the metal rod
T275.	Figure 21-15	An ac generator
T276.	Figure 21-17	Induced emf
	Figure 21-18	An ac generator produces an alternating current
T277.	Figure 21-19	Circuit of a motor showing induced back emf
T278.	Figure 21-20	Production of eddy currents in a rotating wheel
T279.	Figure 21-23	Step-up transformer
T280.	Figure 21-25	The transmission of electric power from power plants to homes
T281.	Figure 21-33	*LR* circuit
	Figure 21-34	Growth and decay of current in an *LR* circuit
T282.	Figure 21-35	Resistor connected to an ac source
T283.	Figure 21-36	Inductor connected to an ac source
T284.	Figure 21-37	Capacitor connected to an ac source
T285.	Figure 21-40	Phasor diagram for a series *LRC* circuit
	Figure 21-41	Phasor diagram for a series *LRC* circuit showing the sum vector, V_0
T286.	Figure 21-42	Current in an *LRC* circuit as a function of frequency

Suggested Readings

Backman, P., Murley, C. and Williams, P., "The Driven *RLC* Circuit Experiment," *The Physics Teacher* (October 1999), pp. 424–425. Uses a virtual four-channel oscilloscope installed on a laptop computer to display voltages across all elements in an *RLC* circuit simultaneously.

Baldwin, N., *Edison: Inventing the Century*, Hyperion, New York, 1995. An excellent biography of Edison. Chapter 17 discusses the debate on whether to use alternating or direct current as the electrical standard.

Cervellati, R. and Soldà, R., "An Alternating Voltage Battery with Two Salt-Water Oscillators," *American Journal of Physics* (May 2001), pp. 543–545. Describes the construction and operation of a battery that reverses polarity.

Fay, S. and Portenga, A., "Hey You! Shut the Refrigerator Door!," *The Physics Teacher* (September 1998), pp. 336–338. A great experiment on power usage by electrical appliances.

Higbie, J., "A Novel Approach to Faraday's Law," *The American Journal of Physics* (December 1997), p. 1211. Uses a hypothesis of magnetic monopoles to arrive at Faraday's law.

Ivanov, D., "Another Way to Demonstrate Lenz's Law," *The Physics Teacher* (January 2000), pp. 48–49. Uses falling magnets to illustrate Lenz's law.

Johns, R., "Simplifying AC-Current Measurements," *The Physics Teacher* (May 2001), pp. 314–315. Describes a simple procedure using ac power strips for measuring the current drawn by household appliances.

Kingman, K. and Popescu, S., "Motional EMF Demonstration Experiment," *The Physics Teacher* (March 2001), pp. 142–144. A very visual demonstration of electromagnetic induction.

Kingman, R., Rowland, S. and Popescu, S., "An Experimental Observation of Faraday's Law of Induction," *American Journal of Physics* (June 2002), pp. 595–598. Description of an experimental apparatus demonstrating Faraday's law.

Lan, B. and Lim, J., "Michael Faraday: Prince of Lecturers in Victorian England," *The Physics Teacher* (January 2001), pp. 32–35. An informative historical article on Faraday.

Oliver, J., "Observing Voltage Phases in *RC, RL*, and *RLC* Circuits," *The Physics Teacher* (January 1997), p. 30. A short note on experimental observations of voltage phases.

Ruby, L., "Why DC for Long-Range Power Transmission," *The Physics Teacher* (May 2002), pp. 272–274. A brief history of the debate regarding ac versus dc usage for electrification of the United States, followed by arguments in support of dc.

Sawiki, C., "A Lenz's Law Experiment Revisited," *The Physics Teacher* (October 2000), pp. 439–441. Magnetic braking in a real-world situation.

Silverman, M., "Power, Reaction, and Excitement…in an AC Circuit," *The Physics Teacher* (May 2002), pp. 302–307. Excellent discussion of ac circuits involving fluorescent lamps.

Tanner, P., Loebach, J., Cook, J. and Hallen., H., "A Pulsed Jumping Ring Apparatus for Demonstration of Lenz's Law," *American Journal of Physics* (August 2001), pp. 911–916. Provides instructions for construction and use of a good demonstration of Lenz's law.

Materials

One demonstration motor is available from Fisher Science Education, model number S43580.

Function generators and oscilloscopes are available from a wide range of manufacturers.

An eddy current demonstrator is available from Fisher Science Education, model number S83019. Klinger Educational Products has a demonstration of an eddy-current brake, P3.4.4.

Notes and Ideas

*Class time spent on material: Estimated:*______ *Actual:*____________

Related laboratory activities:

Demonstration materials:

Notes for next time:

Chapter 22: Electromagnetic Waves

Outline

Summary

Chapter 22 brings together concepts from Chapters 16–21. The production and propagation of electromagnetic waves are discussed, and the entire electromagnetic spectrum is presented. The transfer of energy and momentum in electromagnetic waves is explained. The chapter ends with a section on wireless communication.

Major Concepts

By the end of the chapter, students should understand each of the following and be able to demonstrate their understanding in problem applications as well as in conceptual situations.

- Existence of electromagnetic waves
    - Symmetry arguments
    - Generation of electromagnetic waves
- Propagation of electromagnetic waves
    - Speed
    - Direction of propagation
    - The Doppler effect
- The electromagnetic spectrum
- Energy and momentum in electromagnetic waves
    - Energy density
    - Radiation pressure

Teaching Suggestions and Demonstrations

This is an exciting chapter to teach. The discoveries that electromagnetic waves travel at the speed of light and that light is an electromagnetic wave were astounding to scientists of the time. Try to convey the sense of amazement that must have accompanied this realization.

Sections 22-1 through 22-4

Plan to spend a little class time talking about **Maxwell's symmetry arguments** for the existence of magnetic fields produced by changing electric fields (see Figures 22-2 and 22-3) and his theoretical prediction of **electromagnetic waves** that travel at the speed of light. (While this is well-known today, it came as a shock to scientists of the 1800's!) Go through the historical developments sketched at the beginning of Section 22-1 and through the production of an electromagnetic wave by an **antenna** shown in Figures 22-4, 22-5, and 22-6. Point out that electromagnetic waves are **transverse waves** and that the electric and magnetic fields are perpendicular to each other and in phase. (See Figure 22-7.) Show the students how to use the right-hand rule to find the **direction of propagation** of an electromagnetic wave. Be sure students are aware that ***all*** **accelerating charges radiate electromagnetic waves.**

Scientists initially postulated the existence of the **ether** as the medium through which electromagnetic waves travel. The story of the search for the ether is quite interesting; it was a major scientific controversy of the nineteenth century. It does not take long to go over in class, and it gives students the sense of the "real" way science develops—lots of people arguing over the interpretation of data! (See Resource Information.)

All electromagnetic waves in a vacuum travel at the **speed of light**. The speed of light is a large number. Give students several real world estimates to help them understand how fast light travels. The historical material presented in Section 22-4 on the measurement of the speed of light can be left for the students to read if you are short on time.

The inverse relationship between the **wavelength and frequency** of electromagnetic waves may take the students some time to grasp. Give many reminders (energy increases with frequency and decreases with wavelength) and be consistent about how you graph these quantities. Go through the **electromagnetic spectrum**, using Figure 22-8, and giving examples for each region.

➲ **DEMO 22-1** It is a good idea to have a chart of the **electromagnetic spectrum** (see Figure 22-8) to show students as you talk about it. Be sure to point out what a small section of the spectrum is made up of visible light.

Sections 22-5 and 22-6

These two sections are optional, but we recommend that you cover them at least briefly. Electromagnetic waves are made up of electric and magnetic fields, each of which has an **energy density.** Therefore, electromagnetic waves have an energy density and can transport energy from one place to another. Work through the calculation of the total energy density of an electromagnetic wave at the beginning of this section and the relation $E = cB$ for the electric and magnetic fields in an electromagnetic wave. Define the **rms values** of the electric and magnetic fields and the **intensity** of the electromagnetic wave.

Most students will find it easy to understand that electromagnetic waves carry energy. The ideas that these waves also carry **momentum** and can produce **radiation pressure** are much less intuitive. Work through the theory carefully and do several examples, including Example 22-5. Discuss how radiation pressure produces the tail of a comet.

Section 22-7

This section is also optional and can be skipped without loss of continuity, but it contains lots of very interesting information. The discussion of AM and FM radio signals is particularly good.

Resource Information

Transparencies

T287.	Figure 22-2	Ampère's law applied to two different surfaces
	Figure 22-3	A capacitor discharging

T288.	Figure 22-4	Fields produced by charge flowing into conductors
T289.	Figure 22-5	Sequence showing electric and magnetic fields that spread outward
T290.	Figure 22-6	(a) The radiation fields produced by a sinusoidal signal (b) plane waves
T291.	Figure 22-7	Electric and magnetic field strengths in an electromagnetic wave
T292.	Figure 22-8	Electromagnetic spectrum
T293.	Figure 22-11	Electromagnetic wave carrying energy through area A
T294.	Figure 22-12	Block diagram of a radio transmitter
	Figure 22-15	Block diagram of a simple radio receiver
T295.	Figure 22-13	Amplitude modulation (AM)
	Figure 22-14	Frequency modulation (FM)

Suggested Readings

Beaver, J., "The Speed of Light with a Shortwave Radio," *The Physics Teacher* (March 2000), pp. 172–174. Uses shortwave signals from the National Institute of Standards and Technology stations in Colorado and Hawaii to measure the speed of light.

Berman, B., "Sky Lights: Bad Day, Sunshine," *Discover* (February 2001), p. 33. A short article on the depletion of the ozone layer and UV levels on Earth.

Brush, S., "Cautious Revolutionaries: Maxwell, Planck, Hubble," *American Journal of Physics* (February 2002), pp. 119–127. Great article on three influential scientists and their attitudes toward their own remarkable discoveries.

Hosack, H., Marler, N. and MacIsaac, D., "Microwave Mischief and Madness," *The Physics Teacher* (May 2002), pp. 264–266. Lists several short experiments performed by heating objects in a microwave oven.

Kunzig, R., "Trapping Light," *Discover* (April 2001), pp. 72–79. An introductory article on photonics.

Morris, R., *Dismantling the Universe: The Nature of Scientific Discovery*, Simon and Schuster, New York, 1983. A great book to read before teaching a physics class. Chapter 1, pp. 13–43, covers the story of the ether controversy.

Müller, R., "The Ether Wind and the Global Positioning System," *The Physics Teacher* (April 2000), pp. 243–246. Using GPS to verify the Michelson-Morley experiment.

Prather, E. and Slater, T., "Bringing Extra-Solar Planets into the Introductory Physics Classroom," *The Physics Teacher* (February 2001), pp. 120–122. Describes the use of the Doppler shift for light to discover the existence of planets in other star systems.

Rossing, T. and Chiaverina, C., "Resource Letter TLC-1: Teaching Light and Color," *American Journal of Physics* (October 2000), pp. 881–887. Excellent list of resources for teaching optics.

Materials

HeNe lasers are available from a variety of sources, including Edmund Optics, model number NT61-315. Many demonstrations can be done with a laser pointer.

Notes and Ideas

Class time spent on material: Estimated:______ Actual:____________

Related laboratory activities:

Demonstration materials:

Notes for next time:

Chapter 23: Light: Geometric Optics

Outline

23-1 The Ray Model of Light
23-2 Reflection; Image Formation by a Plane Mirror
23-3 Formation of Images by Spherical Mirrors
23-4 Index of Refraction
23-5 Refraction: Snell's Law
23-6 Total Internal Reflection; Fiber Optics
23-7 Thin Lenses; Ray Tracing
23-8 The Thin Lens Equation; Magnification
*23-9 Combinations of Lenses
*23-10 Lensmaker's Equation

Summary

Geometrical optics concerns the reflection and the refraction of light, both of which cause changes in the direction light is moving. The images formed by various types of mirrors and lenses are examined. Combinations of lenses and the lensmaker's equation are two of the applications discussed in this chapter.

Major Concepts

By the end of the chapter, students should understand each of the following and be able to demonstrate their understanding in problem applications as well as in conceptual situations.

- Wave fronts and rays
- Reflection and mirrors
  - The law of reflection
  - Plane mirrors
  - Spherical mirrors—concave and convex
  - Ray tracing and the mirror equation
- Refraction and lenses
  - The law of refraction
  - Total internal reflection
  - Reflection
  - Thin lenses—converging and diverging
  - Ray tracing and the thin-lens equation
  - Combinations of lenses
- The lensmaker's equation

Teaching Suggestions and Demonstrations

Before beginning with the topics of reflection and refraction, give students a brief overview of optics in general. Distinguish between geometrical optics, the topic of this chapter, and physical optics, which will be covered in Chapter 24. To give students a preview of things to come, mention the wave-particle duality of light and emphasize that the current discussion deals with the wave characteristics of light. Throughout this chapter, keep a collection of lenses and mirrors on hand for quick demonstrations. Glass prisms and a laser are also handy for illustrating refraction and total internal reflection.

Sections 23-1 through 23-3

Start by defining **wave fronts** and **rays**. Since parallel rays are used in investigating lenses and mirrors, be sure that students understand that **planar wave fronts** arise when the source is very far away. In geometrical optics, the rays are sufficient for illustrating the phenomena, but it's a good idea to occasionally throw in some wave fronts to remind students what the rays represent.

➲ **DEMO 23-1** The **pinhole camera**, or *camera obscura*, remarkably forms an image without a lens. Make a small viewer by forming a rectangular box without a front or back out of black poster board. For the front, cut another piece of black poster board and put a pinhole in the center. (Alternatively, cut a small square out of the front piece, cover it with aluminum foil and put the pinhole in the foil.) Cover the back with waxed paper. (This is the view screen.) Seal all the edges with black electrical tape. Stand in a darkened room and aim the viewer at the window. An image of the window will form on the waxed paper. Keep the box small so that the image is bright enough to see. Sketch a ray diagram for students to clarify why an image forms and also why the hole is necessarily small.

For safe solar eclipse viewing, poke a small hole in a piece of cardboard and project an image of the sun onto a sheet of white paper. (Remind students to never look directly at the sun!)

Include fronts as well as rays the first time you illustrate the **law of reflection**. One reason is that it helps convince students what a convenient simplification the rays provide, since the diagram with fronts is much more cluttered and harder to follow. Fronts also provide an illustration of the physical basis for the law of reflection. With an angle of incidence other than zero, students can see that one "end" of a front hits the surface before the other, causing a change in the direction of the light.

➲ **DEMO 23-2** Shine a laser onto a rectangular or trapezoidal glass prism at varying angles of incidence. The laser beam essentially acts like a ray, and students can compare the **reflected ray** and the **incident ray** at each surface.

We are so used to looking in **plane mirrors**, it doesn't really occur to students to think about the location of the image. A baby or a puppy, though, when confronted with its own image in a mirror, will try to look behind the mirror to find the playmate! Example 23-1, concerning the minimum height of a "full-length" mirror, is a fun one to do in class. Extend it by asking what other factors determine whether or not you can actually see your whole self. Often, students will think that the distance between the person and the mirror matters. It doesn't, but the vertical placement of the mirror does.

A plane mirror always forms the same type of image. **Convex and concave spherical mirrors** can form different types of images. Point out to students that the image of a convex mirror is always smaller, upright and virtual, but the properties of the image of a concave mirror depend on the object distance as compared to the focal length. Do an example of each case for the concave mirror using the **mirror equation** and the formula for **magnification**. Go over the sign conventions very carefully. Then draw **ray diagrams** to scale to check the results for both image distance and image height. Use the diagrams in Figures 23-12, 23-13, 23-16, and 23-17. Encourage students to practice plenty of ray diagrams on their own, particularly those for virtual images.

➲ **DEMO 23-3** Pass around both **a convex and a concave spherical mirror** and have students examine their own images. For the concave mirror, suggest that they start with the mirror at arm's length and watch the image as they slowly bring it closer and closer. When the object distance equals the focal length, the image will momentarily disappear. Once the object (the student's own face) is inside the focal point, the image will be magnified, upright and virtual.

Students may have trouble with the concepts of **real and virtual images**. If the image can be focused on a screen, then it must be real since the light rays actually pass through it. An image that is made by diverging rays is a virtual image. Light does not actually reach a virtual image, but your brain traces the diverging rays backward to the point from which they appear to have come.

Sections 23-4 through 23-8

Illustrate the **refraction of light** by drawing wave fronts incident on a boundary between two different media. Then add the rays, which are perpendicular to the fronts. Since the light travels at different speeds in different media, one end of each wave front will slow down or speed up before the other, causing the path of the light, or ray, to bend. Once students understand the reason for the bending, drop the wave fronts from the diagrams and work only with the rays, as in Figure 23-19. Every once in a while sketch in a few wave fronts to remind students why the rays behave as they do when entering different media.

Sketch the wave fronts for incident light traveling along the normal. The fact that the light is unbent follows from Snell's law as well as from the sketch. If light is traveling along a normal to the surface boundary, then all parts of the wave front change speed at the same time, resulting in a change of wavelength but not of direction.

➲ **DEMO 23-4** Shine a laser beam onto a rectangular or trapezoidal glass prism with the light incident on one of the parallel sides. Now students can examine the **refracted rays** as well as the reflected rays at each surface. As the light travels from air to glass the ray is bent toward the normal, but as it travels from glass to air on the far side of the prism it is bent away from the normal. Point out also that both rays in air are parallel to each other, but the outgoing ray is slightly deviated from a straight-line extension of the incident ray (see Figure 23-22).

Use both a diagram and Snell's law to help explain why **total internal reflection** cannot occur when light in a medium with a lower index of refraction is incident on a medium with a higher index of refraction. Consider, for example, light traveling from air to glass. Since the ray bends toward the normal, the angle of refraction can never reach 90°. Attempting to solve for the critical angle in this case leaves you trying to find the inverse sine of a number greater than one, which is impossible.

➲ **DEMO 23-5** The trick to demonstrating **total internal reflection** is to use a semicircular prism. Shine the laser in the curved side along a radius. The light will pass straight through that boundary, since the angle of incidence is zero. The angle at which the ray hits the flat side, however, depends upon which radius you choose. Show the path the light follows for angles less than, equal to, and greater than the critical angle. An alternative method consists of placing the prism on a corkboard with a straight pin placed vertically at the center of the flat side of the semicircle. Have students sight through the prism to the pin. As they increase the viewing angle, the pin will suddenly disappear!

➲ **DEMO 23-6** Inexpensive **fiber-optic toys** are available through several distributors. (See Resource Information.) Students enjoy seeing these demonstrated. Also mention biological applications, as shown in Figure 23-28.

The discussion of **converging and diverging thin lenses** is very similar to the treatment of concave and convex mirrors earlier in the chapter. For convenience, ray diagrams are drawn as though the bending of the light occurs in the center of the lens. Of course it doesn't, because there is no boundary between different media at the center. Be sure to draw some rays refracting at the surfaces as they actually do so that students will understand why lenses that are thicker in the middle cause parallel light rays to converge while lenses that are thinner in the middle cause parallel rays to diverge. (See Figures 23-31 and 23-33.)

The equations relating **magnification** and **focal length** to object and image distances are conveniently the same for lenses and mirrors. The **sign conventions** are also the same. The only difference is that the virtual image of a lens is on the same side as the object and a real image is on the opposite side. Light really does pass through a lens, so real objects are formed on the opposite side. Light does not pass through a mirror, so virtual images are on the opposite side. Make sure students understand that the sign of the image distance corresponds to whether or not the image is real or virtual, not the side of the lens or mirror on which it appears. Emphasize the importance of the sign conventions and of familiarity with the types of images formed. Be sure to work several examples of objects far away from the lens and close to the lens for both primary kinds of lenses.

➲ **DEMO 23-7** Use various **converging lenses** to project the image of a light source onto a screen. Begin with a very large object distance, which will result in a tiny image that is formed very close to the focal point. Have students watch what happens to the image as the distance between the light and the lens decreases. Once the light is at a distance less than the focal length, it will be impossible to project the image onto a screen.

➲ **DEMO 23-8** Students can also look through both **converging and diverging lenses** and describe and compare the images they see for different object distances. A converging lens used with an object inside the focal point is simply a magnifying glass. The resulting image is upright, magnified and virtual.

➲ **DEMO 23-9** An optics set that includes a ray projector and various lenses and mirrors is an excellent tool for demonstrating **geometrical optics**. Many sets come with magnets or suction cups that allow you to attach the elements to the board. You can use the set to demonstrate the laws of reflection and refraction and the behavior of light rays as they interact with mirrors and lenses. (See Resource Information.)

➲ **DEMO 23-10** Students often mistakenly believe that research-grade astronomical telescopes are so large because magnification depends on the **lens diameter**. Find two converging lenses with the same focal length but different diameters. Dim the room lights and use each to project an image of a window and the scene outside it on the opposite wall. First, the students will notice that all the trees and cars framed by the window and, indeed, the window frame itself, are upside down. Now have students compare the two different images. The magnification will be the same, but the image formed by the larger diameter lens will be brighter and thus show more detail.

Sections 23-9 and 23-10

These two sections are optional and can be skipped without loss of continuity. However, if you have time, it is useful to go through the material in Section 23-9 on combinations of lenses. Figures 23-41 and 23-42 are particularly clear and helpful. An important idea to get across to the students is that the image of the first lens acts as the object for the second lens. Demonstrate finding the image by both the thin-lens equation and by using a ray diagram.

Resource Information

Transparencies

T296.	Figure 23-2	Law of reflection
T297.	Figure 23-3	Diffuse reflection from a rough surface
T298.	Figure 23-4	Reflection from (a) white paper, and (b) a mirror
T299.	Figure 23-6	Formation of a virtual image by a plane mirror
T300.	Figure 23-7	Seeing oneself in a mirror, Example 23-1
T301.	Figure 23-12	Rays parallel to the principle axis of a spherical mirror
T302.	Figure 23-13	Rays leave point O′ on the object (an arrow)
T303.	Figure 23-14	Diagram for deriving the mirror equation
T304.	Figure 23-16	Object placed within the focal point F
T305.	Figure 23-17	Convex mirror
T306.	Table 23-1	Indices of Refraction
T307.	Figure 23-19	Refraction
T308.	Figure 23-24	Total internal reflection
	Figure 23-25	Example 23-8 (View up from under water)
	Figure 23-26	Total internal reflection of light by prisms in binoculars
T309.	Figure 23-31	Parallel rays are brought to a focus by a converging thin lens
	Figure 23-33	Diverging lens
T310.	Figure 23-34	Finding the image by ray tracing for a converging lens
T311.	Figure 23-36	Finding the image by ray tracing for a diverging lens
T312.	Figure 23-37	Deriving the lens equation for a converging lens
	Figure 23-38	Deriving the lens equation for a diverging lens
T313.	Figure 23-40	An object placed within the focal point of a converging lens
T314.	Figure 23-41	Two lenses, A and B, used in combination, Example 23-12
T315.	Figure 23-42	Determining the focal length of a diverging lens, Example 23-13

Suggested Readings

Chakravarti, S. and Siegel, P., "Visualizing the Thin-Lens Formula," *The Physics Teacher* (September 2001), pp. 342–343. Uses a graph of $1/i$ versus $1/o$ to help students investigate and verify the thin-lens formula.

Gatland, I., "Thin Lens Ray Tracing," *American Journal of Physics* (December 2002), pp. 1184–1186. Presents a ray-tracing approach to thin lens analysis.

Greenslade, T., "A Quick Experiment on Reflection from Concave Mirrors," *The Physics Teacher* (April 2000), p. 206. Describes how to make inexpensive concave mirrors and use them for simple experiments.

Helfgott, H. and Helfgott, M., "A Noncalculus Proof That Fermat's Principle of Least Time Implies the Law of Refraction," *American Journal of Physics* (December 2002), pp. 1124–1125. An algebraic proof that shows how Fermat's principle of least time implies Snell's law.

Houser, B., "Demonstrating the Decreased Wavelength of Light in Water," *The Physics Teacher* (April 2001), pp. 228–229. A good demonstration of what is and is not constant about light as it moves from one medium to another.

Johnson, A. and Hitz, C., "Career Opportunities in Optics," *Physics Today* (May 2000), pp. 25–28. An interesting article about advances and applications in the area of optical communications and the resulting increase in career possibilities.

Kakaes, K., "Theorists Doubt Claim for Perfect Lens," *Science* (24 May 2002), v. 296, pp. 1380–1381. A short, interesting article on materials with negative indices of refraction.

Lawrence, R., "Magnification Ratio and the Lens Equations," *The Physics Teacher* (March 2000), pp. 170–171. Presents a method of quickly visualizing relationships of object and image distances to magnification.

Newburgh, R., Rueckner, W., Peidle, J. and Goodale, D., "Using the Small-Angle Approximation to Measure the Index of Refraction of Water," *The Physics Teacher* (November 2000), pp. 478–479. A nice variation of a standard introductory physics lab.

O'Connell, J., "Optics Experiments Using a Laser Pointer," *The Physics Teacher* (October 1999), pp. 445–446. Nine simple demonstrations of geometrical and physical optics.

Shelby, R., Smith, S. and Schultz, S., "Experimental Verification of a Negative Index of Refraction," *Science* (6 April 2001), v. 292, pp. 77–79. An advanced article on a material with a negative index of refraction.

Tape, C., "Aquarium, Computer, and Alaska Range Mirages," *The Physics Teacher* (May 2000), pp. 308–311. Photographs and descriptions of mirages.

Thompson, G. and Mathieson, D., "The Mirror Box," *The Physics Teacher* (November 2001), pp. 508–509. Detailed instructions for constructing a mirror box suitable for class demonstrations of the law of reflection and for magic tricks.

Materials

The Blackboard Optics Basic Set from Klinger (item KO4100 or KO4100M) is one example of an assembly of lenses, mirrors and ray projectors that can attach right to your blackboard. An additional accessory set (KO4100A or KO4100AM) provides components for more demonstrations and experiments in optics.

Edmund Optics can be found online at www.edmundoptics.com. It is a great source for lenses, prisms, and other optical components.

Fun fiber-optic demonstrations include “lumirods” from Fisher Science Education (item SA1678-1, for example) and fiber-optic light wands (item CR30317-71) from Edmund Scientific.

Notes and Ideas

*Class time spent on material: Estimated:*______ *Actual:*_____________

Related laboratory activities:

Demonstration materials:

Notes for next time:

Chapter 24: The Wave Nature of Light

Outline

Summary

Chapter 24 discusses two major properties of light that illustrate its wave characteristics, interference and diffraction. Superposition of waves is applied to electromagnetic waves in this chapter. Double-slit interference, single-slit diffraction, and thin-film interference are the main topics covered. The chapter ends with a discussion of polarization.

Major Concepts

By the end of the chapter, students should understand each of the following and be able to demonstrate their understanding in problem applications as well as in conceptual situations.

- Superposition
- Interference
  - Young's two-slit experiment
  - Air wedge
  - Newton's rings
  - Thin-film interference
- Diffraction
  - Single-slit diffraction
  - Resolution
  - Diffraction gratings
- Polarization
  - Transmission axis
  - Law of Malus

Teaching Suggestions and Demonstrations

Students won't necessarily be able to visualize interference patterns until they see them, so bring out the laser and do lots of short demonstrations of double-, single-, and multiple-slit interference throughout the chapter. Continually remind students that they are seeing wave phenomena. Experiments discussed in Chapters 27 and 28 will illustrate the particle nature of light. Students will probably ask how light can be both a particle and a wave, and how it knows when to act like a particle and when to act like a wave. Begin these discussions of wave-particle duality now and continue them in the chapters on modern physics.

Sections 24-1 through 24-3

Have students compare Figure 24-2, showing the **diffraction** of waves around an obstacle and through a slit, to Figure 24-1, showing **Huygen's principle**. Light diffracts as it passes through the narrow slits. Emphasize that diffraction is a wave phenomenon and thus illustrates the wave nature of light. The idea of wave-particle duality bears repeating, since it is completely nonintuitive.

Section 24-2 is optional. The illustration on highway mirages at the end of the section is well worth taking time to discuss.

Students should already be familiar with **wave superposition and interference** from their study of waves on a string and sound waves in Chapters 11 and 12. Take a little time at the beginning of this chapter to review these concepts (see Figure 24-8) and then apply them to electromagnetic waves. The ideas of **phase** and **phase difference** are critical to the understanding of the interference of light waves, so be sure students have a firm grasp of these terms.

The conditions necessary for interference effects of light to be noticeable are also important to discuss. Most students understand the idea of **monochromatic** light fairly easily. However, the concept of **coherent light sources** is a little trickier. Point out that the two sources don't necessarily need to be in phase, but they do need to maintain a constant phase difference. Overlapping light waves will interfere no matter what, but if the sources are incoherent the interference will vary rapidly and randomly so that no overall interference pattern will be noticeable.

Use Figures 24-5, 24-6, and 24-7 to talk about **Young's double-slit experiment**. Point out that Young performed his experiment before lasers were available to produce coherent light, and so he had to use an initial single slit to assure a coherent light source. Next, ask students what the particle model of light would predict from the experiment. If light were a stream of tiny particles, you would expect them to pile up on the screen directly across from each slit instead of resulting in the alternating bands of dark and light. Go over the derivations for the **conditions for bright and dark fringes** carefully. In particular, make sure students understand the concept of **path difference** and that the relationship between the path difference and the wavelength is what determines whether constructive or destructive interference occurs.

➲ **DEMO 24-1** A set of slides for **demonstrating interference and diffraction** is well worth purchasing. (See Resource Information.) Sets usually include single slits, pairs of slits, and multiple slits as well as circular apertures. Shine a laser through a double slit to project the interference pattern on a wall. Point out to students that you don't need the initial slit as in Young's original experiment because your source is already coherent. In actuality, the pattern will be a combination of the double-slit pattern with a single-slit pattern superimposed, because the two slits are each of finite width. You can mention this and return to the demonstration when you discuss single-slit diffraction later. After students have examined the pattern, ask them to predict what will happen if you use a smaller slit separation. Using other slides from the set, demonstrate both smaller and larger slit separations.

Section 24-4 through 24-7

"Blue bends best; red refracts rotten" is a handy saying to help students remember how the index of refraction for a given material depends on the color of the light. Students will be familiar with the **dispersion** caused by a triangular prism. Draw a diagram of white light incident on a slab of glass with parallel faces and show them that in this case, too, the colors are dispersed. However, the rays representing the different colors all leave the glass parallel to one another and are so slightly displaced that we can't distinguish the colors. The reason a prism separates the colors of light is that the rays are diverging when they leave the glass.

Use Figure 24-16 to talk about how rainbows are formed. Ask the students to think about where to look for a rainbow in the sky—toward the Sun or away from the Sun?

It makes sense to students that light from two different slits will interfere, but the idea that light passing through a **single slit** will interfere with itself to produce dark and light bands is more difficult. Examine Figure 24-20 with students and use it to derive Equation 24-3b, the condition for a dark fringe in single-slit interference.

➲ **DEMO 24-2** Shine a laser through a **single slit** and examine the pattern produced. Have students try to find the relationship between the width of the slit and the width of the central maximum. The result that a narrower slit produces a wider central maximum is important and not necessarily obvious. Compare patterns formed by single slits of different widths, return to the double-slit interference pattern, and show students the effect of the individual slit widths on the overall pattern.

Extend the discussion of single and double slits to **multiple slits** and you have a diffraction grating. Point out the advantages of diffraction gratings over prisms for spreading light into its colors.

➲ **DEMO 24-3** Use the laser and a slide with multiple slits to demonstrate **diffraction gratings**. Have students predict and then examine what happens as the number of slits and the slit spacing change.

➲ **DEMO 24-4** **Diffraction grating glasses** are inexpensive to buy and lots of fun. (See Resource Information.) Students put on the glasses and view different light sources. Have them look at extended sources, such as the lights in the classroom, as well as narrow sources. In Chapter 27, students can use the glasses again to view spectral lines of different gases.

Section 24-8 and 24-9

By now students will understand the importance of path difference to the interference of two light waves. For the **interference of reflected waves**, emphasize that a second factor must also be taken into account—namely, the possible **phase change due to reflection**. The combination of these two factors will determine if the interference is constructive or destructive.

Thin-film interference can be confusing for students because the condition for destructive interference in some cases is the condition for constructive interference in others. The key is whether a phase shift on reflection occurs at only one or at both interfaces, which depends on the relative indices of refraction. To clarify the difference, first derive the interference conditions for films of negligible thickness. In this case, only phase shifts matter. Then add thickness to the film and take into account the path difference as well, with which students are already familiar from Young's two-slit experiment. Point out that when comparing path difference to wavelength, the wavelength *in the medium*, not in air, is what matters. The colors in soap bubbles and oil slicks are two very common examples of thin-film interference. In addition, the interference of reflected waves can be used for practical applications, as discussed for Newton's rings, air wedges, and CD players.

➲ **DEMO 24-5** Students certainly will have seen the **colors of soap bubbles** before, but it's fun to pass out a few jars of bubble liquid and blowers and have them take a closer look. As the thickness of a bubble changes from point to point, so do the colors. If the humidity is right and the bubbles last a while before popping, you can even see what happens as the soap film becomes negligibly thin at the top of the bubble.

➲ **DEMO 24-6** Get a brand-new package of microscope slides and separate them in pairs. Tape each pair together near the short edges. A very thin film of air will be trapped between the slides. If you press down in the center with a dull pencil point, you will see **Newton's rings** surrounding the point of pressure.

Section 24-9 is a discussion of the Michelson interferometer. It is worth taking a few minutes to go over this very important instrument.

Sections 24-10 through 24-12

Begin the section on the **polarization** of light by discussing the orientation of the electric field in an electromagnetic wave. Point out that in most light sources, the light is generated randomly by many atoms. Each bit of light is produced separately and each has its own polarization direction. Such a source is called **unpolarized**, because the combination of all the bits of light has no overall polarization. (See Figure 24-40.) You can then discuss how a polarizer allows electric fields with certain orientations through and stops others. Each polarizer has a **transmission axis**.

➲ **DEMO 24-7** Get a large piece of cardboard (at least 0.5 m square) and cut a slit in it a little wider than a demonstration Slinky. Use it and the Slinky to demonstrate the **action of a polarizer**, as shown in Figure 24-39. (Be sure to point out that the direction of the slit in the cardboard corresponds to the transmission axis of the polarizer and is perpendicular to the direction of the long molecular chains.)

➲ **DEMO 24-8** It is well worth the cost to purchase several sheets of **polarizing material** large enough to cover the top of an overhead projector. (See Resource Information.) Use the projector as the incandescent light source. Place one of the polarizer sheets over the top. Students will notice a decrease in the intensity of the light. (See Figure 24-42.) Now place a second polarizer on top of the first, with the transmission axes aligned. Rotate the top polarizer slowly through 90 degrees and show the resultant decrease in the intensity of transmitted light. Use this to introduce the **Law of Malus**, $I = I_0\cos^2\theta$.

Work several examples (Example 24-11 and others).

➲ **DEMO 24-9** Go through the exercise described in Conceptual Example 24-12 with two crossed polarizers and a third polarizer inserted in the middle. Discuss the result. Repeat using a piece of waxed paper (an "unpolarizer") in the middle instead of the third polarizer sheet.

➲ DEMO **24-10** You can easily do several dramatic examples of **photoelastic stress analysis** with the overhead projector and two crossed polarizers. Again, use the projector as the incandescent light source. Place almost any clear plastic object between the polarizers. (Clear plastic protractors and rulers work quite well, as do clear plastic cups and forks.) As you rotate the polarizers, a brilliant color display shows the places on the object with greatest stress.

If you have time, go through the discussion of **liquid crystals** in optional Section 24-11. Point out that liquid-crystal displays are much more energy efficient than LED displays, a critical consideration in battery-operated devices such as calculators. Finish up this chapter with a discussion of why the sky is blue (optional Section 24-12) and why polarized sunglasses are better for driving than regular sunglasses (see the end of Section 24-10).

Resource Information

Transparencies

T316.	Figure 24-1	Huygen's principle, used to determine wave front CD
	Figure 24-2	Huygen's principle is consistent with diffraction
T317.	Figure 24-5	Young's double-slit experiment
	Figure 24-6	If light is a wave
T318.	Figure 24-7	The wave theory
T319.	Figure 24-8	(a) Constructive interference (b) Destructive interference
T320.	Figure 24-9	Interference fringes produced by a double-slit experiment
	Figure 24-10	Examples 24-1 and 24-2
T321.	Figure 24-12	The spectrum of visible light
	Figure 24-14	Index of refraction as a function of wavelength for various transparent solids
	Figure 24-15	White light dispersed by a prism into the visible spectrum
T322.	Figure 24-20	Analysis of diffraction pattern formed by light passing through a narrow slit
	Figure 24-21	Intensity in the diffraction pattern of a single slit
T323.	Figure 24-24	Diffraction grating
	Figure 24-25	Intensity as a function of viewing angle θ
T324.	Figure 24-26	Spectra produced by a grating
T325.	Figure 24-28	Line spectra for the gases indicated and spectrum from the Sun showing absorption lines
T326.	Figure 24-30	Light reflected from a thin film
	Figure 24-32	Reflected ray changes phase
T327.	Figure 24-42	Unpolarized light has equal intensity vertical and horizontal components
	Figure 24-43	Crossed Polaroids completely eliminate light.
	Figure 24-45	Example 24-12. (Three Polaroids.)
T328.	Figure 24-51	“Twisted” form of liquid crystal and molecules disoriented by electric field

Suggested Readings

Ambrose, B., Heron P., Vokos, S. and McDermott, L., "Student Understanding of Light as an Electromagnetic Wave: Relating the Formalism to Physical Phenomena," *American Journal of Physics* (October 1999), pp. 891–898. Development and modification of tutorials to address student difficulty with the wave nature of light.

Ambrose, B., Shaffer P., Steinberg, R. and McDermott, L., "An Investigation of Student Understanding of Single-Slit Diffraction and Double-Slit Interference," *American Journal of Physics* (February 1999), pp. 146–155. Discusses evidence of student misunderstanding of both the wave nature of light and the photon, with implications for curriculum development.

Benenson, R., "Light Polarization Experiments with a Diode Laser Pointer," *The Physics Teacher* (January 2000), pp. 44–46. Introductory polarized light measurements with a diode laser pointer.

Easton, D., "Transmission through Crossed Polaroid Filters," *The Physics Teacher* (April 2001), pp. 231–233. Describes modifications to digital lux meters for use in Law of Malus and other polarization experiments.

Ellenstein, M., "The Broken-Protractor Prediction Trick," *The Physics Teacher* (January 2002), p. 52. An example of using polarizing sheets to investigate stress lines in molded plastic objects. Includes photographs.

Gallis, M., "Automating Microwave Optics Experiments," *The Physics Teacher* (April 2002), pp. 217–219. Describes two types of microwave experiments that demonstrate interference phenomena.

Hernández-Andrés, J., Valero, E., Nieves, J. and Romero, J., "Fizeau Fringes At Home," *American Journal of Physics* (July 2002), pp. 684–688. A description of an easy, inexpensive method to obtain Fizeau fringes using a flatbed scanner and a personal computer.

Kovács, A., Varjú, K., Osvay, K. and Bor, Z., "On the Formation of White-Light Interference Fringes," *American Journal of Physics* (November 1998), pp. 985–989. An experiment to investigate white-light interference fringes using a Michelson interferometer and a spectrograph.

Middleton, A. and Sampere, S., "Color Mixing via Polarization," *The Physics Teacher* (February 2001), pp. 123–124. Description of an apparatus for using polarizers to mix colors.

Moloney, M., "Homemade Interference," *The Physics Teacher* (November 1999), pp. 504–505. Uses the interference of two tiny light sources viewed through a cloth grating to determine the wavelength of light.

Ouseph, P., Driver, K. and Conklin, J., "Polarization of Light by Reflection and the Brewster Angle," *American Journal of Physics* (November 2001), pp. 1166–1168. Describes an experiment to determine the Brewster angle and to study polarization of reflected light using equipment available from PASCO Scientific.

Perkalskis, B. and Freeman, J., "Demonstrating Induced Birefringence in Stressed Glass," *American Journal of Physics* (September 2000), pp. 871–872. Description of apparatus for demonstrating birefringence.

Perkalskis, B. and Freeman, R., "Herschel's Interference Demonstration," *The Physics Teacher* (March 2000), p. 142. Apparatus for using two prisms to demonstrate multiple interference.

Rueckner, W. and Papaliolios, C., "How to Beat the Rayleigh Resolution Limit: A Lecture Demonstration," *American Journal of Physics* (June 2002), pp. 587–594. Describes an experiment to demonstrate the effect of instrument aperture size on resolution, as well as the use of interferometry to detect the presence of more than one object even when they cannot be resolved by the instrument. Includes photographs.

Sawicki, C., "Easy and Inexpensive Demonstration of Light Interference," *The Physics Teacher* (January 2001), pp. 16–19. Uses interference fringes to measure the thickness of plastic wrap.

Sawicki, M. and Sawicki, P., "Supernumerary Rainbows," *The Physics Teacher* (January 2000), p. 19. Beautiful photograph of multiple rainbows.

Wein, G., "A Video Technique for the Quantitative Analysis of the Poisson Spot and other Diffraction Patterns," *American Journal of Physics* (March 1999), pp. 236–240. Presentation of an approach for the observation and analysis of diffraction phenomena using a video camera.

Materials

The Introductory Optics System from Pasco (item OS-8500) is a complete set including equipment for demonstrating geometrical optics as well as interference and diffraction.

The Diffraction Optics Kit (Pasco item OS-8531) includes only those components necessary for demonstrating diffraction.

The Slide Set and Gratings available from Fisher Science Education (item S42539) contains six different slides with various numbers of slits and slit separations.

The Deluxe Laser Accessories Kit (Fisher item S42701ND) is a much more comprehensive set, including circular apertures, a prism, various slits and diffraction gratings, a hologram, other accessories, and a book detailing 30 experiments.

Edmund Scientific has diffraction-grating film available for purchase by the roll (item CR30521-16).

Diffraction-grating glasses can be ordered from Fisher Science Education. They are listed as diffraction viewing glasses, catalog number NCS48814 for individual glasses, or catalog number NCS48814A for a package of 10.

Edmund Scientific sells various sizes and types of polarizing sheets. (See, for instance, item CR30384-93.) A polarizer/analyzer kit (item CR30815-70) that can be used with a microscope is also available.

Notes and Ideas

*Class time spent on material: Estimated:*______ *Actual:*_____________

Related laboratory activities:

Demonstration materials:

Notes for next time:

Chapter 25: Optical Instruments

Outline

Summary

The human eye is a remarkable optical instrument. Chapter 25 applies the geometric optics from Chapter 23 to explain how the eye creates an image. Optical instruments that correct or extend our vision, such as eyeglasses, contact lenses, magnifying glasses, microscopes, and telescopes, are explored. Finally, lens aberrations and ways to correct for them are discussed.

Major Concepts

By the end of the chapter, students should understand each of the following and be able to demonstrate their understanding in problem applications as well as in conceptual situations.

- The human eye and the camera
  - Focusing and image formation
  - The *f*-number
  - Nearsightedness
  - Farsightedness
- The magnifying glass
- Instruments with lens combinations
  - The compound microscope
  - The telescope
- Lens aberrations and corrections

Teaching Suggestions and Demonstrations

You very likely will not have much time to spend on this chapter, but try to at least hit the highlights. Students, especially those with an interest in biology and/or those who wear glasses, typically find the study of the human eye fascinating. Likewise, cameras, telescopes and microscopes provide a wonderful opportunity to apply geometrical optics to practical instruments with which the students are familiar. The end sections on X-ray diffraction and computed tomography are quite interesting.

Sections 25-1 and 25-2

A simple **camera** is diagramed in Figure 25-1; Figures 25-5, 25-6, and 25-7 give additional information about focusing and the **single-lens reflex camera**. Section 25-1 has a nice discussion of how the **CCD sensor** in a **digital camera** works. The **parts of the eye**, **image formation**, and **accommodation** are illustrated in Figures 25-8 and 25-9. Point out the difference in focusing between an eye and a camera. The muscles in the eye change the shape of the lens and thus its focal length. The focal length of a camera lens is fixed, so focusing is accomplished by moving the entire lens toward or away from the film.

➲ **DEMO 25-1** Show students a simple **camera** and have them compare it to an eye. In addition to the focusing differences mentioned above, compare the camera's adjustable aperture to a person's pupil, and the film to the retina.

A "normal" human eye can focus on any object between about 25 cm away (the near point) and infinity (the far point). That's quite an impressive range! Of course, many people don't have this range and therefore wear glasses or contact lenses to correct their **nearsightedness** or **farsightedness**. In either case, the external lens and the eye work together as two lenses in combination. The external lens forms an image between the near and far points, and then the eye uses that image as its object. The **refractive power** of a lens can be confusing. Point out that it is just the inverse of the focal length and therefore its units, diopters, are simply inverse meters. If you are short on time, skip the quantitative discussion of corrective lenses and just illustrate with diagrams such as those in Figures 25-10, 25-12, and 25-13.

➲ **DEMO 25-2** There are no rods or cones at the location of the optic nerve, so if an image happens to fall there, no message gets sent to the brain and you can't see the object in question. To demonstrate this **blind spot**, have students do the following. Mark two dark X's 10–15 cm apart on a piece of paper. With the paper stretched out in front of you at arm's length, close your left eye and look with your right eye at the left-hand X. (The right-hand X will be visible farther to the right with your peripheral vision.) Now slowly bring the paper toward your face, focusing on the left-hand X but noticing the right-hand X out of the corner of your eye. The right-hand X disappears at the point when its image falls on your blind spot.

Sections 25-3 through 25-5

A **magnifying glass** is a simple convex lens like those discussed in Chapter 23. It works essentially by allowing the viewer to move the object closer to the eye but still focus it. Discuss the definition of **angular magnification** and compare it to the definition of magnification given in Chapter 23.

Both **compound microscopes** and **refracting telescopes** consist of an objective lens and an eyepiece. The eyepiece is a magnifier, which uses the image of the objective as its object. Go over the arrangements of the lenses as shown in Figures 25-19 and 25-23 and compare the two instruments, pointing out the difference in object distance. Again, if you are short on time, concentrate on the conceptual diagrams and skip the quantitative analysis.

➲ **DEMO 25-3** **Telescopes** can be demonstrated quite easily with a couple of lenses, a meter stick and some clay. Place the lenses on the horizontal meter stick and hold them in place vertically with the clay. Figure 25-19 will help students determine how far apart to put the lenses. Then have students hold the telescope up and sight through the pair of lenses at an object across the room or outside. It takes some practice but they should be able to view magnified images. The magnification is hard to estimate; counting bricks on a brick wall or examining some other regular pattern helps. Repeat using lenses with different focal lengths and compare. Also try a diverging lens as the eyepiece (see Figure 25-22).

A good example of a lens-mirror combination is the **reflecting telescope**, which has a mirror as its objective and a lens as its eyepiece. Figure 25-21 shows one possible arrangement of the two optical elements. For telescopes, a larger objective is better since most objects of astronomical interest are dim and larger apertures collect more light. Large mirrors are easier to make than large lenses, so the best astronomical research telescopes in the world are reflectors. An ingenious method of making telescope mirrors, called "spin casting," involves melting glass in a huge, spinning oven. The rotation helps form the curvature of the mirror.

Section 25-6

This section is optional and can be omitted, but the two main types of **lens aberrations**, chromatic and spherical, can be covered very briefly. Students may have observed chromatic aberration while looking through a telescope eyepiece. Note that chromatic aberration is not a problem with mirrors, since the angle of reflection is not wavelength dependent.

Sections 25-7 through 25-9

➲ **DEMO 25-4** To introduce **resolution** and **Rayleigh's criterion**, shine a laser through a circular aperture and note the resulting circular diffraction pattern, as shown in Figure 25-28. In addition, try making a poster of pairs of dots with different separations. Have students start far away from the poster and see which pairs they can resolve. As they move closer, they will be able to resolve the more closely spaced pairs.

Go over Figure 25-30 carefully, then work Example 25-10 and 25-11. Have the students pay attention to the resolution of approaching car headlights at night, and estimate how far away they can tell the difference between a motorcycle (with one headlight) and a car or truck (with two headlights).

➲ **DEMO 25-5** Look on the Web for images from the Hubble Space Telescope and for images of your area from satellites (see Resource Information). Show these to the students and discuss the importance of resolution in the making and interpreting of the images.

Sections 25-10 through 25-12

These sections are optional and can be omitted without loss of continuity. The material is very interesting (phase contrast microscopes, X-ray diffraction, and CT scans), but unless you are teaching a specialized course you will not have time to cover these in class.

Resource Information

Transparencies

T329.	Figure 25-1	A simple camera
	Figure 25-7	Single-lens reflex (SLR) camera
T330.	Figure 25-2	Portion of a typical CCD sensor
T331.	Figure 25-8	Diagram of a human eye
	Figure 25-9	Accommodation by a normal eye

T332.	Figure 25-10	Correcting eye defects with lenses
	Figure 25-11	A cylindrical lens forms a line image of a point object
T333.	Figure 25-12	Lens of reading glasses (Example 25-5)
	Figure 25-13	Example 25-6 (Nearsighted eye)
T334.	Figure 25-15	When the same object is viewed at a shorter distance
T335.	Figure 25-19	Astronomical telescope (refracting)
T336.	Figure 25-22	Terrestrial telescopes that produce an upright image
T337.	Figure 25-23a	Compound microscope
T338.	Figure 25-24	Spherical aberration
	Figure 25-26	Chromatic aberration
T339.	Figure 25-30	The *Rayleigh criterion*

Suggested Readings

Andereck, B. and Secrest, S., "The Magic Magnifier," *The Physics Teacher* (May 2001), pp. 301–302. Describes the construction of a simple telescope out of PVC pipe and lenses.

Graf, E., "How Do *You* Use a Magnifying Glass?," *The Physics Teacher* (May 2001), pp. 298–300. Compares standard explanations of the use of magnifying glasses to the way people really use them.

Henderson, H., "A Dialog in Paradise: John Milton's Visit with Galileo," *The Physics Teacher* (March 2001), pp. 179–183. An essay about Milton, his literary works, and his visit with Galileo.

Huebner, J., Gibbs, D. and Ryan, P., "Projecting Chromatic Aberrations," *American Journal of Physics* (September 2000), pp. 869–870. Describes apparatus for demonstrating chromatic aberration to a large class using an overhead projector.

Knight, J. and Russell, P., "New Ways to Guide Light," *Science* (12 April 2002), v. 296, pp. 276–277. A good short article on new optical fibers.

McLellan, J., Marcos, S., Prieto, P. and Burns, S., "Imperfect Optics May Be the Eye's Defense Against Chromatic Blur," *Nature* (9 May 2002), v. 417, pp. 174–176. An advanced but accessible treatment of the optics of the eye.

Miller, D., "Retinal Imaging and Vision at the Frontiers of Adaptive Optics," *Physics Today* (January 2000), pp. 31–36. Good discussion of the potential for adaptive optics to correct for aberrations, diffraction and the eye's defects.

Milonni, P., "Resource Letter: AOA-1: Adaptive Optics for Astronomy," *American Journal of Physics* (June 1999), pp. 476–485. A guide to the basic concepts and the literature on adaptive optics for astronomy.

Porter, N., *Physicists in Conflict*, Institute of Physics Publishing, Bristol and Philadelphia, 1998. Chapters 3 and 4 discuss Galileo and Kepler.

Salinas, J. and Sandoval, J., "Geometrical Optics and Visual Perception," *The Physics Teacher* (October 2001), pp. 420–423. An examination of the relationship between images perceived by an observer and the curvature of the light wave fronts reaching the eye.

Thomas, G., Ackerman, D., Prucnal, P. and Cooper, S., "Physics in the Whirlwind of Optical Communications," *Physics Today* (September 2000), pp. 30–36. Basics of digital optical communications as well as descriptions of new innovations.

Wagner, D. and Walkiewicz, T., "When the Eye Meets the Lens," *The Physics Teacher* (November 2000), pp. 474–475. Describes experiments to clarify the confusion between the observer-independent magnification of a lens and the magnifying power as seen by a particular observer.

Wardle, D., "The Time Delay in Human Vision," *The Physics Teacher* (October 1998), pp. 442–444. A nice demonstration of reaction time and human vision.

Materials

Images from the Hubble Space Telescope are available at http://hubble.nasa.gov/.

Images of the Earth from satellites are available from a variety of sites, some free and some for a fee. One site is http://edcsns17.cr.usgs.gov/EarthExplorer/.

Notes and Ideas

*Class time spent on material: Estimated:*______ *Actual:*____________

Related laboratory activities:

Demonstration materials:

Notes for next time:

Chapter 26: The Special Theory of Relativity

Outline

Summary

Chapter 26 begins the study of "modern" physics. Relativity governs the behavior of systems moving at large velocities, and quantum mechanics, the subject of Chapter 27, deals with microscopic systems. All of these new laws must reduce to the more familiar Newtonian mechanics as the systems enlarge to macroscopic size and slow down to everyday speeds. The postulates of special relativity are presented, and their consequences, including time dilation, length contraction and mass increase, are derived. Einstein's famous equation $E = mc^2$ is also discussed. The chapter ends with a short discussion of the correspondence principle.

Major Concepts

By the end of the chapter, students should understand each of the following and be able to demonstrate their understanding in problem applications as well as in conceptual situations.

- Special relativity
  - The two postulates
  - Simultaneity
  - Time dilation
  - Length contraction
  - Mass increase
  - Relativistic addition of velocities
- Mass-energy equivalence and $E = mc^2$
  - Matter and anti-matter
  - Relativistic kinetic energy
- Correspondence Principle

Teaching Suggestions and Demonstrations

Relativity is an incredibly fascinating (and difficult) subject to teach. Students will most likely not believe you at first and exchange "let's just humor the professor" looks as you attempt to convince them of time dilation and length contraction. Be sure to include details of experimental evidence verifying relativity throughout your presentation of this chapter. You can begin by making some general points regarding relativity to set the stage and address misconceptions students may have. Relativity is about the nature of space and time. Relativity is not difficult mathematically, but it is very difficult conceptually because we do not have any day-to-day experiences with macroscopic objects traveling near the speed of light. Relativity is not an optical illusion created by high-speed travel.

Sections 26-1 and 26-2

As an overview of the next few sections, tell students that you will present the two **postulates of special relativity**, see what consequences result from them, and then look for experimental verification of those consequences. The idea is not to prove the postulates, since postulates are by definition statements accepted without proof, but to accept them and then see where they lead. Any "proof" of relativity will come as experiments are compared to predictions. You will need to spend some time going over the postulates. For the first, emphasize that if you are in an inertial reference frame, there is no experiment you can perform to show if you are moving or at rest. For the second, point out that nothing can travel faster than c.

Give the students a hint of what is to come by examining the postulates a little further. Notice that the speed of, say, a baseball depends on the reference frame of the observer. The batter will see a different speed than someone racing by the ball field in a fast car. However, this does not apply to light; the speed of light is the same for all observers. Distance equals speed multiplied by time, so if the speed stays constant, weird things must happen to distance and time.

Sections 26-3 through 26-5

Begin by defining the concept of **simultaneity**. Then you can introduce the consequences that follow from the postulates of special relativity by deriving Equation 26-1a for **time dilation** carefully, using Figure 26-6. Point out explicitly when you are using the postulates. Note that the math isn't difficult, but the results are hard to believe! Do a number of examples to help students recognize which time is the **proper time** in each situation. Point out that "proper" does not mean "correct." Students will be tempted to think that the observer in the rest frame of the event is right and the other observer is wrong, which is not the case; both are right. Another common misconception arising from the phrase "moving clocks run slow" is that somehow the mechanisms in clocks don't work right when they are speeding along. There is nothing wrong with the clocks; time itself is dilated.

Do a problem involving hypothetical high-speed travel such as Example 26-2 and discuss the results. The discussion of the twin paradox after Example 26-2 is quite clear. Be sure to define the **light-year** and take a moment to calculate just how far a light-year is in kilometers. Of course, we are not capable of accelerating spaceships to anywhere near the speed of light, so it is important to do a realistic example regarding muons (see Example 26-1) or other elementary particles as well. Discuss actual experiments, such as those involving muons performed at CERN.

There are two optional parts of these sections, one on the variable γ and one on applications of relativity to **global-positioning systems**. Both of these discussions will be interesting to the students and useful for later study. We recommend covering them.

After deriving the expression for **length contraction** (Equations 26-3a and 26-3b) go over the concept of **proper length**. Then work Example 26-5 to help students understand the direction dependence of the length contraction.

Sections 26-6 through 26-11

Section 26-6 is optional, but useful to go over conceptually; it leads nicely into the rest of the sections of the chapter. Discuss the difference between **relativistic mass** and **rest mass** and how the increase of mass with speed is consistent with the result that nothing can be accelerated to a speed greater than the speed of light. The result is particularly important in particle accelerators.

Even students who have never had any physics will come to your class having heard of Einstein's famous equation, $\boldsymbol{E = mc^2}$. Do a few quantitative examples to emphasize just how big the number c^2 is. Mention the difference between fusion and fission, which will be covered in more detail in Chapter 31. Many students may think that **antimatter** is an invention of science fiction, so spend a little time discussing it. The positron appears again in the nuclear decay equations of Chapter 30. Work through Examples 26-9 and 26-10 in class to give students an idea of the differences in energy of nuclear and chemical reactions.

Take time to work through Example 26-12 on the **relativistic addition of velocities**. Students generally have trouble with this concept. Remind them that in this chapter "low speed" means low with respect to c. Even one-tenth the speed of light, though, is incredibly fast in everyday terms, so we don't notice relativistic effects in our everyday lives. Students sometimes think that Newton's laws apply to slow speeds and Einstein's relativity to fast speeds, and then wonder where to draw the line. Emphasize that the relativistic equations apply to all speeds, but when $v \ll c$, they reduce to the equations of classical physics. This is the **correspondence principle**.

Resource Information

Transparencies

T340.	Figure 26-2	A coin is dropped by a person in a moving car
T341.	Figure 26-3	A moment after lightning strikes at points A and B
	Figure 26-5	Thought experiment on simultaneity
T342.	Figure 26-6	Time dilation
T343.	Figure 26-10	Rocket 2 is fired from rocket 1 with speed $u' = 0.60c$

Suggested Readings

Ashby, N., "Relativity and the Global Positioning System," *Physics Today* (May 2002), pp. 41–47. Explains the importance of taking fundamental relativistic principles into account for the precise operation of the GPS.

Berlinski, D., "Einstein and Gödel," *Discover* (March 2002), pp. 38–42. An excellent historical article on the friendship between these two men.

Berman, B., "Sky Lights: Curved Space," *Discover* (August 2001), p. 28. A good short article on the bending of light by matter.

Berman, B., "Sky Lights: The Hole Story," *Discover* (October 2001), p. 36. A short article on black holes.

Blackman, E., "Astrophysical Perspectives in Teaching Special Relativity," *The Physics Teacher* (March 1998), p. 177. A nice astrophysical example for special relativity.

Brown, H., "The Origins of Length Contraction," *American Journal of Physics* (October 2001), pp. 1044–1054. An analysis of the null result of the Michelson–Morley experiment and its interpretation by both Lorentz and FitzGerald.

Carlip, S., "Kinetic Energy and the Equivalence Principle," *American Journal of Physics* (May 1998), pp. 409–413. A reanalysis of experimental data to test the equivalence principle for the kinetic energy of atomic electrons.

Field, J., "Space–Time Exchange Invariance: Special Relativity as a Symmetry Principle," *American Journal of Physics* (May 2001), pp. 569–575. An advanced article on relativity, with a useful introduction.

Folger, T., "Does the Universe Exist If We're Not Looking?" *Discover* (June 2002), pp. 44–49. Reflections from John Wheeler on observation and experiment.

Greenslade, T., "Relativistic Meter Sticks," *The Physics Teacher* (May 2000), p. 315. A visual display of length contraction.

Gülmez, E., "Measuring the Speed of Light with a Fiber Optic Kit: An Undergraduate Experiment," *The American Journal of Physics* (July 1997), pp. 614–618. A description of an experiment using a commercially available kit in which time-delay is measured to determine the speed of light.

Hecht, E., "From the Postulates of Relativity to the Law of Inertia," *The Physics Teacher* (November 2000), pp. 497–498. Musings on the connections between the postulates of relativity and the law of inertia.

Irion, R., "B-Meson Factories Make a 'Number From Hell'," *Science* (23 February 2001), v. 291, p. 1471. A short article on the potential imbalance of matter and antimatter in the universe.

Kunzig, R., "Black Holes Spin?" *Discover* (July 2002), pp. 32–39. A good general article on new information on black holes.

Lamoreaux, S., "Testing Times in Space," *Nature* (25 April 2002), v. 416, pp. 803–804. An accessible summary of a *Physical Review* article on space-based testing of the constancy of universal constants.

Naddy, C., Dudley, S. and Haaland, R., "Projectile Motion in Special Relativity," *The Physics Teacher* (January 2000), pp. 27–29. Revisiting projectile motion using special relativity considerations.

Osborne, I., Rowan, L., and Coontz, R., " Spacetime, Warped Branes, and Hidden Dimensions," *Science* (24 May 2002), v. 296, p. 1417. A short introduction to a collection of articles on spacetime.

Scheider, W., *Maxwell's Conundrum: A Serious but Not Ponderous Book About Relativity*, Cavendish Press, Michigan, 2000. An algebraic introduction to relativity.

Seife, C., "Crystal Stops Light in Its Tracks," *Science* (11 January 2002), v. 295, p. 255. A short report on efforts to slow down light.

Seife, C., "The Intelligent Noncosmologist's Guide to Spacetime," *Science* (24 May 2002), v. 296, pp. 1418–1421. An excellent short history on the development of the theory of spacetime.

Seife, C., "Relativity Goes Where Einstein Sneered To Tread," *Science* (10 January 2003), p. 185. A short, accessible note on quantum entanglement and relativity.

Tegmark, M., "Measuring Spacetime: From the Big Bang to Black Holes," *Science* (24 May 2002), v. 296, pp. 1427–1433. A more advanced article on measurements of spacetime.

Wang, L., Kuzmich, A. and Dogariu, A., "Gain-Assisted Superluminal Light Propagation," *Nature* (20 July 2000), v. 406, pp. 277–279. Results of an experiment demonstrating superluminal light propagation in atomic cesium gas.

Will, C., "Gravitational Radiation and the Validity of General Relativity," *Physics Today* (October 1999), pp. 38–43. A discussion of how observations of gravity waves would test Einstein's theories.

Notes and Ideas

*Class time spent on material: Estimated:*______ *Actual:*_____________

Related laboratory activities:

Demonstration materials:

Notes for next time:

Chapter 27: Early Quantum Theory and Models of the Atom

Outline

Summary

Chapter 27 discusses the beginning of quantum physics, which treats the behavior of microscopic systems. Quantization, the photon and its interactions with matter, wave–particle duality and its implications for subatomic particles, early models of the atom, de Broglie's hypothesis, and atomic spectra are presented. The chapter ends with information on the early quantum mechanical model of hydrogen. Data are presented that are inconsistent with the earlier atomic models but support the quantum model.

Major Concepts

By the end of the chapter, students should understand each of the following and be able to demonstrate their understanding in problem applications as well as in conceptual situations.

- Blackbody radiation
  - Energy distribution
  - Wein's displacement law
- Quantization of energy
  - Planck's constant
- Photons
  - Energy
  - Momentum
- Photoelectric effect
  - Work function
  - Cutoff frequency
- Compton effect
- Wave–particle duality
- The de Broglie hypothesis
- Early models of the atom
  - Thomson model
  - Rutherford model

- Bohr model
  - Hydrogen spectrum
  - Assumptions
  - Orbits and energies

Teaching Suggestions and Demonstrations

This chapter contains many new ideas, some of which were very difficult for those who first developed them to accept. This material is conceptually complex. Stress the experimental proof of these ideas and do lots of example calculations. If you are running short on time, go over the models of the atom qualitatively.

Sections 27-1 through 27-4

The discovery of the **electron** and its properties gave the first hint of quantization. Thomson's measurement of the **charge-to-mass ratio of the electron**, and Millikan's **oil drop experiment** will both be interesting to students and are worth covering in class.

Go over the definition of a **blackbody**. Students may have trouble with this concept. Have them think of a hot summer day. Would they rather walk barefoot on a white concrete sidewalk or the black asphalt of the street? Why? Use their comments as a starting point for your description of a blackbody as a **perfect absorber** and a **perfect emitter**. Use Figure 27-4 to introduce **Wien's displacement law**. Talk about Examples 27-1 and 27-2 and Exercise A on the relationship between temperature and color.

If you have time, step the students through the historical development of the **ultraviolet catastrophe**. (It's another of the great stories of physics!) It also helps the students understand how Planck could have proposed the **quantization of energy** when he didn't really think it was true. Go over the concepts of the **quantum number** and the energy $E = nhf$ of transitions between states.

Einstein was the first to apply the idea of quantization of energy to light. Discuss the differences between the classical model and Einstein's model of what happens in the **photoelectric effect**. Ask the students to tell you which model fits the data best. Go through Example 27-4 in class.

➲ **DEMO 27-1** Use a photoelectric cell to demonstrate the **photoelectric effect**. (See Resource Information.) Vary the intensity of the light and the frequency of the light to show students how each affects the current produced. Determine the **threshold frequency** for the cell and the **work function** of the metal.

Be sure to point out to the students that Einstein won the Nobel prize for his work on the photoelectric effect, not relativity.

Discuss the properties of **photons**. Mention that the photon rest mass is zero, that photons travel at the velocity of light, and that photons carry momentum. The definition of **photon momentum** follows from the description of the relativistic energy. Go through this derivation and work Exercise 27-6.

Sections 27-5 and 27-6

If you are short on time, you can skip Section 27-5 entirely. If you have time to cover it, the **Compton effect** is a great example of the conservation of energy and momentum. Work out the derivation of

Equation 27-7 in class and then go through Example 27-8. Go over the photon interactions listed in Section 27-6 briefly.

Sections 27-7 through 27-9

Students are usually interested in the **de Broglie hypothesis**, particularly when they learn that de Broglie was a graduate student when he suggested it. Go over the de Broglie relation (Equation 27-8) and calculate the wavelength of several macroscopic objects (like baseballs) traveling at typical speeds. Discuss the reasons why we don't have to be concerned with the wave nature of large objects moving at normal speeds. Then work Example 27-11 for electrons.

Experimental proof of the de Broglie hypothesis first came from **electron diffraction** by crystals. (See Figure 27-12.) Remind the students that all particles have a wave nature, just as all electromagnetic waves have a particle nature. The discussion of various types of electron microscopes in optional Section 27-9 is quite informative.

Sections 27-10 and 27-11

Discuss the two early models of the atom, the **Thomson model** and the **Rutherford model**, in terms of hypothesis, collection and interpretation of data, and reformulation of hypothesis. (See Figures 27-17, 27-18, and 27-19.) The development of these models makes an exciting story; try to convey some of that sense of discovery to the students. Point out the flaws in each model and how the study of the flaws led to the next model.

One of the major problems with the Rutherford model of the atom was its prediction of a continuous atomic spectrum, since all charged particles traveling in a circular path are accelerating and all accelerated charges emit radiation. **Line spectra** characteristic of individual elements were already known (see Figures 27-20 and 27-21) but not well understood. Go over the difference between **emission spectra** and **absorption spectra**. Be sure the students understand that the original **Balmer formula** was empirical, not derived. Go over Figures 27-22 and 27-23 and show how they are connected. Discuss the **Lyman** and **Paschen series** briefly in class.

➲ **DEMO 27-2** Use a hydrogen spectral tube and the diffraction-grating glasses from Chapter 24 to let students view the Balmer series in hydrogen. (See Resource Information.) Have them identify the *n* value corresponding to each visible line. Let the students look at spectral lines from other elements as well. If you don't have a selection of spectral tubes, you can improvise by sending students across campus or out into the community to look at "neon" signs (which are mostly *not* neon) with the diffraction-grating glasses. Sodium and mercury parking lot lights also work well!

Sections 27-12 and 27-13

The **Bohr model of the atom** is the model most people in modern society use. It is easy to imagine, explains many of the properties of atoms, and uses a minimum number of assumptions. Go over the four basic assumptions and derive the equations for the radii of the **Bohr orbits** and the speed of an electron in these orbits. (See Figures 27-25 and 27-26.) Calculate the energies of electrons in the first several orbits and construct an **energy-level diagram**, as in Figure 27-27. Use the difference in energy of two levels to derive the Balmer equation. The derivation of the Balmer equation and the definition of the **Rydberg constant** in terms of fundamental constants were the triumphs of the Bohr model. Do Example 27-13. Remind students that the wavelengths calculated in this example can be **emitted** as an electron moves from a higher energy level to a lower one, or **absorbed** as an electron moves from a lower energy level to a higher one.

If you have time, go through Section 27-13, which connects the Bohr orbits to **standing de Broglie waves**. Use Figures 27-28, 27-29, and 27-30 as illustrations. Discuss with the class how the success of the Bohr model led physicists to take matter waves seriously. Talk qualitatively about the **Schrödinger equation** and its importance in quantum mechanics.

Resource Information

Transparencies

T344.	Figure 27-1	Discharge tube
	Figure 27-2	Cathode rays deflected by electric and magnetic fields
T345.	Figure 27-4	Measured spectra of wavelengths and frequencies emitted by a blackbody
T346.	Figure 27-6	The photoelectric effect
	Figure 27-7	Photoelectric effect
T347.	Figure 27-9	The Compton effect
	Figure 27-10	Pair production
T348.	Figure 27-13	Transmission electron microscope
T349.	Figure 27-14	Scanning electron microscope
	Figure 27-16	Probe tip of scanning tunneling electron microscope
T350.	Figure 27-17	Plum-pudding model of the atom
	Figure 27-19	Rutherford's model of the atom
T351.	Figure 27-18	Experimental setup for Rutherford's experiment
T352.	Figure 27-21	Emission spectra of gases
T353.	Figure 27-23	Line spectrum of atomic hydrogen
T354.	Figure 27-26	Possible orbits in the Bohr model of hydrogen
	Figure 27-27	Energy-level diagram for the hydrogen atom
T355.	Figure 27-28	An ordinary standing wave compared to a circular standing wave
	Figure 27-30	Standing circular waves for two, three, and five wavelengths

Suggested Readings

Birnbaum, J. and Williams, R., "Physics and the Information Revolution," *Physics Today* (January 2000), pp. 38–42. Discusses physical and economic limits to semiconductor devices and options for advancing computing involving quantum physics.

Chen, C. and Zhang, C., "New Demonstration of the Photoelectric Effect," *The Physics Teacher* (October 1999), p. 442. A modification of an introductory physics photoelectric-effect demonstration.

Collins, D., "Video Spectroscopy—Emission, Absorption, and Flash," *The Physics Teacher* (December 2000), pp. 561–562. Uses a diffraction grating attached to a color video camera to display spectra.

Fitzgerald, R., "What Really Gives a Quantum Computer Its Power," *Physics Today* (January 2000), pp. 20–22. An investigation of quantum computing with NMR.

Isenberg, C., "Laser Diffraction Experiments with Pseudoliquids and Pseudosolids," *The Physics Teacher* (October 2000), pp. 411–413. Analog experiments using laser light to illustrate X-ray and neutron scattering.

Laloë, F., "Do We Really Understand Quantum Mechanics? Strange Correlations, Paradoxes, and Theorems," *American Journal of Physics* (June 2001), pp. 655–701. A comprehensive article on the present understanding of quantum mechanics.

Levi, B., "Researchers Stop, Store, and Retrieve Photons—or at Least the Information They Carry," *Physics Today* (March 2001), pp. 17–18. New developments in the study of photons.

Mermin, N., "The Contemplation of Quantum Computation," *Physics Today* (July 2000), pp. 11–12. A self-described old physicist and young philosopher presents thought-provoking issues of quantum computation.

Mohrhoff, U., "What Quantum Mechanics Is Trying to Tell Us," *American Journal of Physics* (August 2000), pp. 728–745. An interesting and novel interpretation of quantum mechanics; includes a thorough glossary of terms.

Palmquist, B., "Interactive Spectra Demonstration," *The Physics Teacher* (March 2002), pp. 140–142. A creative method for modeling emission, absorption, and continuous spectra that involves throwing balls and jumping on chairs!

Peercy, P., "An Eye for Impurity," *Nature*, (25 April 2002), v. 416, pp. 799–800. A short article on the use of scanning transmission electron microscopy in locating the dopant atoms in semiconductors.

Smalley, E., "Future Tech: Hack-Proof Chatting," *Discover* (May 2002), pp. 26–27. A practical application of quantum principles.

Voss, D., "Quantum Engine Blasts Past High Gear," *Science* (18 January 2002), v. 295, p. 425. A short article on a theoretical "quantum afterburner" for heat engines.

See also the theme issue of the *American Journal of Physics*, March 2002, on quantum mechanics.

Materials

A photoelectric cell is available on its own or as part of the Complete h/e System (item AP-9370) from Pasco. The complete package includes the apparatus and instructions for performing photoelectric-effect experiments.

Eight different spectral tubes (items SE-9461 through SE-9468) and a spectral tube power supply and mount (item SE-9460) are available from Pasco.

Ordering information for the diffraction-grating glasses is given at the end of Chapter 24.

Notes and Ideas

Class time spent on material: Estimated:______ *Actual:_____________*

Related laboratory activities:

Demonstration materials:

Notes for next time:

Chapter 28: Quantum Mechanics of Atoms

Outline

Summary

Chapter 28 extends the quantum model to a more comprehensive theory that is able to explain the structure and predict the spectra of more complex atoms. The Heisenberg uncertainty principle is presented. Quantum numbers are defined. The periodic table is built using the Pauli exclusion principle. The chapter ends with a discussion of various quantum phenomena, including lasers and holography.

Major Concepts

By the end of the chapter, students should understand each of the following and be able to demonstrate their understanding in problem applications as well as in conceptual situations.

- The quantum model
  - Wave functions
  - Schrödinger equation
  - Electron probability clouds
- Heisenberg uncertainty principle
- The quantum mechanical model of hydrogen
  - Quantum numbers
  - Pauli exclusion principle
- Periodic table
- Lasers

Teaching Suggestions and Demonstrations

The first exposure to the quantum model is exciting but difficult. Students will find the concepts of wave functions and probability clouds odd and not very believable. Talk about the philosophical foundation of the quantum model, but emphasize the experimental evidence that supports it. After all, we use the quantum model because it correctly predicts the results of experiments! Students may have particular trouble with the uncertainty principle. If you have time, draw connections between the world of physics in the early part of the twentieth century and other cultural events and movements of the time.

Sections 28-1 through 28-5

Start with a little review of the correspondence principle, wave properties, and the de Broglie formula, then introduce the concept of the wave function. The discussion in Section 28-2 is quite good and should be very helpful. Go over the double-slit experiment with electrons to help convince students that the wave function description of the electron predicts the outcome of experiments correctly.

The **Heisenberg uncertainty principle** is difficult to explain at the introductory level. One way to make it seem more reasonable is to tie it to measurement. If we want to see a billiard ball rolling across a table at night, we can shine a flashlight beam on it. While we now know that the beam will create a radiation pressure on the ball, the force generated is negligible. The ball's path is undisturbed by the beam. We can calculate both the position of the ball and its momentum at some time t.

If we want to measure the position of an electron, what do we use as a probe? If we shine light on the electron, the photons in the light beam will interact with the electron and cause it to change position and/or momentum. It is possible to devise an experiment in which the position can be measured, or one in which the momentum can be measured. It is *not* possible to devise an experiment in which both can be measured simultaneously. In fact, it is not possible for both of these quantities to have precise values at the same time. Go through the qualitative argument given in Section 28-3 for the form of the Heisenberg uncertainty principle for position and momentum. Work Examples 28-1 and 28-2. Discuss briefly the other forms of the uncertainty principle.

Sections 28-4 and 28-5 treat the **philosophical shift in the world view** of physicists that resulted from the acceptance of the quantum theory. We think it is very important to take some time in class to discuss this. The last paragraph in Section 28-4 is particularly good and is a nice lead-in to the idea of **electron clouds** in Section 28-5. Talk qualitatively about the **wave function** of an electron in a hydrogen. Use Figures 28-6 and 28-9. Point out that the electrons do not really travel in orbits, as in the Bohr model, but that there are certain regions of space where the electrons are more likely to be found. Don't expect the students to grasp all the details of quantum theory at this level.

Sections 28-6 through 28-8

Skim briefly through the **four quantum numbers** (n, ℓ, m_ℓ, and m_s) required by the **quantum mechanical model** of the hydrogen atom. Pay particular attention to the **magnetic quantum number** and the **spin quantum number**, these quantities will be very different from anything the students have seen before. The magnetic quantum number can be discussed as a quantized projection of the angular momentum vector. (See Figure 28-7.) Be sure students understand that **spin** is an intrinsic property of the electron (and of other particles) and has no direct mechanical analog. Use the definitions of the quantum numbers to define a **quantum state** of hydrogen. (See Table 28-1.) Go through Conceptual Example 28-3 in class. If you have time, briefly discuss forbidden transitions and how they can occur.

Use Figure 28-10, which shows energy level diagrams of helium, lithium, and sodium, to begin talking about the **shell model** of atomic structure. Remind students that as you add more protons and neutrons to the nucleus and more electrons to the atom, the interactions between all of the parts become very complex. It is quite amazing that the simple shell model works as well as it does for so many atoms. Define the **Pauli exclusion principle** and discuss its implications for the filling of atomic **shells** and **subshells**. Go over the shorthand notation for the **electronic configuration** of an atom and do several examples. Chemistry students will be particularly interested in this topic. Go through Conceptual Example 28-5 and Exercise D with the class.

➲ **DEMO 28-1** Get a large copy of the **periodic table** of the elements (you can borrow one from your colleagues in chemistry if you don't have one) and point out how the properties of the elements change with their electronic structure. Discuss why the table is set up in its present configuration.

Sections 28-9 through 28-12

These sections are all optional. It is important to talk briefly about the source of X-rays and the different types of **X-ray** imaging available today. If you have time, the history of the medical use of X-rays is very interesting. (See Resource Information.) Go through the mechanisms involved in **fluorescence** and **phosphorescence**. Discuss the relative energies of the photons given off as X-rays with those given off in visible-light fluorescence.

Lasers have many more technological applications than X-rays. Plan to spend some time discussing how a laser works and how the light emitted from a laser has special characteristics. Medical applications of lasers are many and varied. If you have a class full of biology majors or students in the pre-health professions, they will find this material fascinating.

➲ **DEMO 28-2** Bring a HeNe **laser** to class. Talk about the color of the light and how it is determined by the electronic structure of neon. Also discuss the fact that laser light is **coherent**. Ask the class to name uses for the laser that they see every day. Obtain a **hologram** that is visible with the laser, and give students a chance to view it.

Resource Information

Transparencies

T356.	Figure 28-3	Parallel beam, of light or electrons, falls on two slits
	Figure 28-4	Young's double-slit experiment done with electrons
T357.	Figure 28-6	Electron cloud for the ground state of the hydrogen atom
T358.	Figure 28-7	Quantization of angular momentum direction for $l = 2$
T359.	Figure 28-8	When a magnetic field is applied, an $n = 3$, $l = 2$ energy level is split
T360.	Table 28-1	Quantum Numbers for an Electron
T361.	Figure 28-9	Electron cloud, or probability distribution, for $n = 2$ states in hydrogen
T362.	Figure 28-10	Energy level diagram
T363.	Table 28-2	Value of l
T364.	Table 28-3	Electron Configuration of Some Elements
T365.	Figure 28-17	Two energy levels for a collection of atoms
	Figure 28-18	Laser diagram, showing excited atoms stimulated to emit light
T366.	Figure 28-20	Energy levels for He and Ne
T367.	Figure 28-22	Reading a CD or DVD

Suggested Readings

Borbat, P., Costa-Filho, A., Earle, K., Moscicki, J. and Freed, J., "Electron Spin Resonance in Studies of Membranes and Proteins," *Science* (12 January 2001), v. 291, pp. 266–269. A review of electron spin resonance techniques used in biology.

Chu, S., "Cold Atoms and Quantum Control," *Nature* (14 March 2002), v. 416, pp. 206–210. An excellent overview article on the history of laser cooling of atoms, followed by five advanced articles.

Colson, W., Johnson, E., Kelley, J. and Schwettman, H., "Putting Free-Electron Lasers to Work," *Physics Today* (January 2002), pp. 35–41. Describes a new type of laser that is tunable over a wide range of wavelengths.

Day, C., "Ultrafast X-ray Diffraction Tracks Molecular Shape-Shifting," *Physics Today* (March 2001), pp. 19–20. A discussion of X-ray diffraction that can look at structural changes in molecules.

Folger, T., "Quantum Shmantum," *Discover* (September 2001), pp. 36–43. An interesting article on the "'many worlds" interpretation of quantum mechanics.

Gottesman, D. and Lo, H., "From Quantum Cheating to Quantum Security," *Physics Today* (November 2000), pp. 22–27. Discusses the use of quantum phenomena to distribute code keys.

Horsewill, A., Jones, N. and Caciuffo, R., "Evidence for Coherent Proton Tunneling in a Hydrogen Bond Network," *Science* (5 January 2001), v. 291, pp. 100–103. An advanced article on evidence for quantum tunneling in a cyclic network of hydrogen bonds.

Laloë, F., "Do We Really Understand Quantum Mechanics? Strange Correlations, Paradoxes, and Theorems," *American Journal of Physics* (June 2001), pp. 655–701. A comprehensive article on the present understanding of quantum mechanics.

Lindley, D., *Boltzmann's Atom: The Great Debate That Launched a Revolution in Physics*, Free Press, NY, 2001. An interesting historical account of the work of Boltzmann.

Mould, R., *A Century of X-rays and Radioactivity in Medicine*, Institute of Physics Publishing, Bristol and Philadelphia, 1993. Wonderful photographs of early X-ray photographs and X-ray equipment.

Peyser, L., Vinson, A., Bartko, A. and Dickson, R., "Photoactivated Fluorescence from Individual Silver Nanoclusters," *Science* (5 January 2001), v. 291, pp. 103–106. An advanced article on the storage and retrieval of data using nanoparticles and fluorescence.

Stevens, R., "A Simulation of Optical Polarization," *The Physics Teacher* (April 2000), pp. 222–223. A demonstration that can be used to model polarization states of light as well as particle spin.

See also the theme issue of the *American Journal of Physics*, March 2002, on quantum mechanics.

Materials

Eight different spectral tubes (items SE-9461 through SE-9468) and a spectral tube power supply and mount (item SE-9460) are available from Pasco.

The laser accessories kit from Fisher listed in the resource section of Chapter 28 includes a hologram. Kits for making holograms are also available from most major suppliers. See, for instance, the Complete Holography Kit (item S42800) from Fisher.

Ordering information for the diffraction-grating glasses is given at the end of Chapter 24.

Notes and Ideas

*Class time spent on material: Estimated:*______ *Actual:*____________

Related laboratory activities:

Demonstration materials:

Notes for next time:

Chapter 29: Molecules and Solids

Outline

Summary

Chapter 29 focuses on applications of quantum mechanics to molecules and solids. Molecular bonding and spectra, bonding between molecules in solids, the band theory of solids, and semiconductor devices are all discussed.

Major Concepts

By the end of the chapter, students should understand each of the following and be able to demonstrate their understanding in problem applications as well as in conceptual situations.

- Molecular bonding
  - Covalent bonds
  - Ionic bonds
- Bonding between molecules
  - Potential energy diagrams
  - Van der Waals bonds
- Molecular spectra
  - Rotational levels
  - Vibrational levels
- Solid lattices
- Band theory of solids
  - Conductors
  - Insulators
  - Semiconductors
  - Doping
- Semiconductor devices
  - Diodes
  - Transistors
  - Integrated Circuits

Teaching Suggestions and Demonstrations

This entire chapter is optional, although it contains a great deal of interesting and useful information. Unless you are teaching a three-semester or three-quarter course, you will probably need to skip it or cover (very quickly!) a limited number of concepts. Many students will get some of this material in chemistry courses.

Sections 29-1 through 29-4

Discuss the difference between **covalent bonds** and **ionic bonds**. Use Figures 29-1 and 29-2 to show the students how the electron clouds distribute themselves between the two nuclei in a hydrogen bond, and Figures 29-3, 29-4, and 29-5 to talk about ionic bonds. Many bonds have both ionic and covalent character; Figure 29-6 illustrates the charge distribution in the water molecule, which is a **polar molecule**. Go over the two **potential energy diagrams** for molecules in Figures 29-8 and 29-9 and discuss how the **activation energy** might affect molecular bonds.

➲ **DEMO 29-1** Use a track for toy cars or a track for marbles to represent the potential energy diagrams for molecules. First build a track like Figure 29-8, and show how a small car or marble will be "attracted" to the bottom of the potential energy curve. Then build a track like Figure 29-9, and use it to demonstrate the need for activation energy. Discuss the stability of the bond in each case.

Section 29-3, on **van der Waals bonds**, will be interesting for biology or biochemistry students. Use Figures 29-11, 29-12, and 29-13 to discuss the role of van der Waals bonding in the structure and replication of **DNA** and **RNA**.

Section 29-4 contains an excellent treatment of **rotational and vibrational energy levels** in molecules. Go through Example 29-2 with the class to illustrate rotational energy transitions; CO is an important molecule in biology and in environmental studies. Example 29-3 considers vibrational energy transitions in hydrogen. Compare the difference in the energies of the rotational levels and the vibrational levels and talk about its significance.

Sections 29-5 through 29-9

Begin these sections by defining the difference between **crystalline and amorphous solids**. You can then move on to discussing the various types of crystalline structures. (See Figure 29-19.) Models are very helpful here (see Resource Information). Talk about the role of ionic and covalent bonding in solids, and define **metallic bonding**. Use Figures 29-21 and 29-22 to introduce the band theory. Define the terms **valence band**, **conduction band**, and **band gap**, and use them to distinguish between **conductors**, **insulators**, and **semiconductors**. (See Figure 29-23.) Work through Example 29-4 and Conceptual Example 29-5 with the class. Section 29-7 on the **doping** of semiconductors is very clear and leads into the treatment of **semiconductor devices** in the remaining sections of the chapter.

➲ **DEMO 29-2** Set up the **diode** circuits shown in Figures 29-29 and 29-30. Use a multimeter or an oscilloscope to show the class the input and output voltages.

Resource Information

Transparencies

T368.	Figure 29-1	Electron probability distribution for two H atoms when their spins are the same
	Figure 29-2	Electron probability distribution for two H atoms when their spins are opposite
T369.	Figure 29-3	Probability distribution for last electron of Na in NaCl

	Figure 29-4	In a neutral sodium atom, the 10 inner electrons shield the nucleus
	Figure 29-5	Neutral chlorine atom
T370.	Figure 29-8	Potential-energy diagram for H_2 molecule
	Figure 29-9	Potential-energy diagram for a bond requiring an activation energy
T371.	Figure 29-12	Section of a DNA double helix
T372.	Figure 29-13	Protein synthesis
T373.	Figure 29-14	Energy levels of atoms and molecules
T374.	Figure 29-16	Rotational energy levels and allowed transitions
	Figure 29-18	Allowed vibrational energies for a diatomic molecule
T375.	Figure 29-19	Arrangement of atoms in a crystal
	Figure 29-20	Diagram of an NaCl crystal, showing the "packing" of atoms
T376.	Figure 29-21	The splitting of 1*s* and 2*s* atomic energy levels
	Figure 29-22	Energy bands for sodium
T377.	Figure 29-23	Energy bands for (a) a conductor, (b) an insulator, and (c) a semiconductor
T378.	Figure 29-24	Two-dimensional representation of a silicon crystal
	Figure 29-25	A *p*-type semiconductor, gallium-doped silicon
	Figure 29-26	Impurity energy levels in doped semiconductors
T379.	Figure 29-27	Schematic diagram showing how a semiconductor diode operates
	Figure 29-28	Current through a diode as a function of applied voltage
	Figure 29-29	A simple (half-wave) rectifier circuit using a semiconductor diode and ac source voltage
T380.	Figure 29-31	Schematic diagram and symbols of *npn* and *pnp* transistors
	Figure 29-32	An *npn* transistor used as an amplifier

Suggested Reading

Tokura, Y., "Correlated-Electron Physics in Transition-Metal Oxides," *Physics Today* (July 2003), pp. 50–55. A look at some new crystal growth technologies that may provide the basis for a new type of electronics.

Materials

Toy car and marble tracks are available at most toy and discount stores.

A Matter Model is available from Pasco (ME-9825).

Notes and Ideas

*Class time spent on material: Estimated:*_______ *Actual:*______________

Related laboratory activities:

Demonstration materials:

Notes for next time:

Chapter 30: Nuclear Physics and Radioactivity

Outline

Summary

Chapter 30 focuses on the nucleus, the center of the atom. The structure and properties of the nucleus and its decay patterns are discussed. Conservation laws and half-life are presented. Decay series are defined. The chapter ends with a brief description of the detection of radiation.

Major Concepts

By the end of the chapter, students should understand each of the following and be able to demonstrate their understanding in problem applications as well as in conceptual situations.

- Nuclei
  - Nucleons
  - Atomic number
  - Mass number
  - Isotopes
  - Binding energy
  - Stability
- Nuclear forces
  - Strong force
  - Weak force
- Radioactivity
  - Alpha decay
  - Beta decay
  - Gamma decay
  - Conservation laws
  - Activity and decay rates
  - Half-life
  - Radioactive decay series
  - Radioactive dating
- Methods for radiation detection

Teaching Suggestions and Demonstrations

There are issues in this chapter about which students need to be informed so they can be educated decision makers for their own health and for the health of their communities. If you have only a single class to spend on these topics, try to describe nuclei, radioactivity, and the detection of radiation qualitatively and go on to Chapter 31.

Sections 30-1 and 30-2

These first two sections contain lots of new vocabulary. Students will generally know about protons and neutrons, and if they have already taken chemistry, they may be familiar with more of the terms. It is a good idea to make a poster, or write on the board in a semipermanent place, a list of the definitions of terms such as **nucleon**, **atomic number**, **neutron number**, and **mass number**. Go over **standard nuclear notation** ${}_{Z}^{A}X$, which will likely be unfamiliar to most students. Write several examples on the board. Go over the definition of **isotopes**, and remind the students that atoms with the same number of protons in their nuclei are the same element. Define the **unified atomic mass unit *u*** in terms of kilograms and MeV/c^2. Table 30-1 lists the masses of the proton, neutron, and electron in various units.

Work Example 30-1, which estimates the **radius of the nucleus**. Point out that the radius of the nucleus varies from element to element (Equation 30-1) and that the size of an average nucleus is on the order of one **fermi**. Go through Example 30-2 and calculate the average **density of a nucleus**. Try to give students an idea of how large this number is! Emphasize that although the radius of the nucleus depends on the mass number, the density does not.

If the protons in the nucleus all repel each other, what holds the nucleus together? Ask the students this question and use their answers to introduce the **strong nuclear force**. Go over the properties of this force. Discuss the competing influences of strong nuclear attraction and electrostatic repulsion and their effects on the stability of the nucleus. Discuss the **mass deficit** and the **total binding energy** of the nucleus. Example 30-3 and Exercise A are nice exercises to help students connect this material to what they already know. Go over Figure 30-1, the **binding energy per nucleon curve**, and Figure 30-2, the plot of N versus Z for stable nuclides, and discuss the implications of each for nuclear stability. Work Example 30-4 in class.

Sections 30-3 through 30-7

Talk about the **radioactive decay** of unstable nuclei. Discuss the main types of particles (**alpha particles**, **electrons**, **positrons**, and **photons**) that are emitted during the decay of a nucleus. Go over the properties of the positron and point out that electrons and positrons are often grouped together and called **beta particles**. Mention that other particles can be emitted by a nucleus, but that these are the most common.

➲ **DEMO 30-1** If you have access to a **Geiger counter** or other radiation detector and some low-level sources of radiation (alpha, beta, and gamma), bring them to class. (See Resource Information.) Use a source of each type of radiation in turn. Start with the detector close to the source and slowly move it farther and farther away. This will demonstrate the **range** of the radiation in air. Then hold the detector at a constant distance and place barriers between the source and the detector. Use paper, aluminum sheets, and lead. Begin with one piece, then add more. This will demonstrate the **penetrating ability** of the various types of radiation. Make a chart on the board and begin to talk about the three main types of decay.

Point out that an **alpha particle** is just a helium nucleus, which has a particularly stable structure. Discuss what quantities change in a nucleus when it emits an alpha particle. Be sure that the students understand that the **parent nucleus** and the **daughter nucleus** are different elements. Go through Examples 30-6 and 30-7 in class and talk about how the excess energy is carried off in alpha decay. The Physics Applied element on smoke detectors at the end of Section 30-4 will be of interest to the students.

Go over the basic reaction (neutron conversion to a proton) involved in **beta decay** and work Example 30-6. Discuss the distribution of kinetic energy in beta decay and how measurement of this energy led to the discovery of the **neutrino** and the **antineutrino**. Define the processes that take place in **electron capture** and **gamma decay**. Point out that in gamma decay the parent and daughter nuclei are the same isotope. Be sure to mention the difference between X-rays and gamma rays.

Sections 30-8 through 30-13

As you begin to talk about **half-life**, point out that radioactive decay is a statistical process. We can make predictions about when a percentage of the nuclei in a sample will decay, but we cannot predict when a particular nucleus will decay. You can discuss this in relation to predictions on the average age at which people get married, or in relation to life insurance studies of death rates in the general population. If most of the students in your class are comfortable with logarithms, go through the derivation of Equation 30-4.

➲ **DEMO 30-2** Give each student in the class the same number of pennies. (This works best if the total number of pennies is close to 1000. You can use any kind of disk with two different sides. Pennies are common and easy to obtain.) The total number of pennies for the class will be N_0. Have all of the students flip the pennies at the same time. Each time they flip, one half-life has passed. Have them report the number of heads they get and discard the pennies that came up tails. Repeat this process about 10 times, or until most of the students have no pennies left. Record the results of each flip on the board or overhead. Plot the result on rectangular graph paper and on semilog paper. (Make a transparency of a piece of each type of graph paper and do this on an overhead.) The exponential decay curve is quite dramatic and usually quite smooth! It will be easy to see that the number of pennies left decreases by half in equal amounts of time.

Talk about the relationship between **activity** and half-life. Do Examples 30-9 and 30-11 in class. Choose one **radioactive decay series** (Figure 30-11 describes the decay of $^{238}_{92}U$) and go through it with the students. Comment on how decay series can be used to determine an estimate of the age of the Earth. Define the **activity** of a sample and its units, the **curie** and the **becquerel**.

The students will find **carbon-14** dating interesting. Work Example 30-13 and talk about the limits of the procedure. Section 30-12 is optional and can be skipped without loss of continuity.

➲ **DEMO 30-3** Ask the students to think about how we detect radioactivity. If you have access to low-level radioactive mineral samples, bring one to class, along with other nonradioactive samples, and ask the students to tell which one is radioactive just by looking at it. Of course, they can't. Then bring out the Geiger counter and talk about how radiation is detected.

You can't see, smell, hear, or feel nuclear radiation. How do we measure it? Go over the diagram of the **Geiger counter** in Figure 30-13 and the **photomultiplier tube** in Figure 30-14 and discuss the limitations of each. If you have time, talk about **semiconductor detectors** and **cloud and bubble chambers**.

Resource Information

Transparencies

T381.	Table 30-1	Rest Masses in Kilograms, Unified Atomic Mass Units, and MeV/c^2
T382.	Figure 30-1	Binding energy per nucleon as a function of mass number *A* for stable nuclei
T383.	Figure 30-2	Number of neutrons versus number of protons for stable nuclides
T384.	Figure 30-8	Energy-level diagram showing how $^{12}_{5}\text{B}$ can decay
T385.	Table 30-2	The Three Types of Radioactive Decay
T386.	Figure 30-10	The number *N* of parent nuclei in a given sample
T387.	Figure 30-11	Decay series beginning with $^{238}_{92}U$
T388.	Figure 30-13	Diagram of a Geiger counter
	Figure 30-14	Scintillation counter with a photomultiplier tube

Suggested Readings

Aberg, S., "Weighing Up Nuclear Masses," *Nature* (30 May 2002), v. 417, pp. 499–501. A short article on the discrepancy between calculated and measured nuclear masses.

Franklin, A., "The Road to the Neutrino," *Physics Today* (February 2000), pp. 22–28. Historical description of the discovery and subsequent study of the neutrino.

Garrison, E., "Physics and Archeology," *Physics Today* (October 2001), pp. 32–36. Outlines important contributions of physics to archeology, including radioisotope dating as well as many newer physics-based techniques.

Haxton, W. and Holstein, B., "Neutrino Physics," *American Journal of Physics* (January 2000), pp. 15–32. Technical and thorough treatment of the basic concepts of neutrino physics.

Macklis, R., "Scientist, Technologist, Proto-Feminist, Superstar," *Science* (1 March 2002), v. 295, pp. 1647–1648. An excellent short article on Marie Curie.

Mak, S., "Radioactivity Experiments for Project Investigation," *The Physics Teacher* (December 1999), pp. 536–539. Discusses the use of potassium chloride in half-life estimation and other radioactivity experiments.

Seife, C., "Atom Smasher Probes Realm of Nuclear 'Gas'," *Science* (25 January 2002), v. 295, pp. 603–604. Discusses new developments in the liquid drop model of the nucleus.

Materials

A basic radiation meter is available from Fisher Science Education, model number S44126B.

A radioactive source set is also available from Fisher Science Education, model number S440975.

A basic nuclear lab station, including radioactive sources, model number S46860IBM, and an intermediate nuclear lab station, model number S46863IBM, are also available from Fisher Science Education. These stations come in MAC versions as well.

Notes and Ideas

Class time spent on material: Estimated:______ *Actual:_____________*

Related laboratory activities:

Demonstration materials:

Notes for next time:

Chapter 31: Nuclear Energy; Effects and Uses of Radiation

Outline

Summary

Chapter 31 presents information on nuclear reactions, including fission and fusion. The interactions of radiation with matter and methods for measuring those interactions are discussed. Biomedical applications of nuclear processes, including nuclear imaging techniques, are covered in optional sections at the end of the chapter.

Major Concepts

By the end of the chapter, students should understand each of the following and be able to demonstrate their understanding in problem applications as well as in conceptual situations.

- Nuclear reactions
  - Q-value
  - Fission
  - Liquid drop model
  - Compound nuclear model
  - Fusion
- Nuclear reactors
- Interaction of radiation with matter
- Dosimetry and units
- Medical applications of nuclear physics

Teaching Suggestions and Demonstrations

By the time you reach this chapter, the end of the semester or quarter will be near. It is important to find time to discuss the interaction of radiation with matter, especially if your course is aimed at promoting science literacy. Knowledge of the nature of radiation is key to making informed decisions about a variety of modern issues in medicine and society in general. If you have only a single class to spend on these topics, try to describe nuclear reactions qualitatively and spend time discussing dosimetry and the applications of nuclear physics to medicine and other fields.

Sections 31-1 through 31-3

Begin these sections by defining **transmutation**. Point out the differences between a **nuclear reaction** and a chemical reaction. Work Conceptual Example 31-1 and Example 31-3 in class. Be sure to talk about **threshold energy** at the end of Example 31-3.

Remind students of the **binding energy per nucleon curve** from Chapter 30 (repeated as Figure 31-11). Point out the fission region (to the right of ^{56}Fe) and the fusion region (to the left of ^{56}Fe). Talk about the **fission** and **fusion** processes (see Figure 31-2). Be sure to discuss the energy per nucleus released in each process and the reaction products. If you have time, go over **chain reactions** (Figure 31-4), **critical mass** (Figure 31-6), and the basic structure of a **nuclear reactor** (Figure 31-7). Do try to find time to talk about the **proton-proton cycle** (Equations 31-6 a, b, and c) and the Sun. The discussion of the **tokamak** and **magnetic confinement** at the end of Section 31-3 is very clear.

Sections 31-4 and 31-5

All students find a discussion of the **biological effects of radiation** interesting. The units tend to be confusing, so go over the definitions of the **curie**, the **becquerel**, the **roentgen**, the **rad**, the **gray**, the **RBE**, and the **rem** carefully. Go over Table 31-1, which lists **quality factors** for different types of radiation. Work Examples 31-10 and 31-12 in class. There are Web sources of average radiation doses for common activities. (See Resource Information.)

➲ **DEMO 31-1** If you didn't do this demonstration in Chapter 30, do it now. Use a **Geiger counter** or other radiation detector and some low-level sources of radiation (alpha, beta, and gamma) to demonstrate how distance and shielding can protect humans from radiation exposure. (See Resource Information.) Use a source of each type of radiation in turn. Start with the detector close to the source and slowly move it farther and farther away. This will demonstrate the **range** of the radiation in air. Then hold the detector at a constant distance and place barriers between the source and the detector. Use paper, aluminum sheets, and lead. Begin with one piece, then add more. This will demonstrate the **penetrating ability** of the various types of radiation. Talk about the difference between the damage caused by different types and energies of radiation and about the difference between whole-body dose and exposure to particular parts of the body.

Sections 31-6 through 31-9

Several medical applications of nuclear physics are given at the end of Chapter 31 in these optional sections. These discussions are very clear and interesting. Pre-health professions students in your class will be quite interested in them. If you have time, talk to the class about risks versus benefits for these procedures.

Resource Information

Transparencies

T389.	Figure 31-2	Fission of a $^{235}_{92}U$ nucleus after capture of a neutron
T390.	Figure 31-4	Chain reaction
	Figure 31-6	Critical mass
T391.	Figure 31-7	A nuclear reactor

T392.	Figure 31-12	Tokamak configuration
T393.	Table 31-1	Quality Factor (QF) of Different Kinds of Radiation

Suggested Readings

Aberg, S., "Weighing Up Nuclear Masses," *Nature* (30 May 2002), v. 417, pp. 499–501. A short article on the discrepancy between calculated and measured nuclear masses.

Badash, L., "Marie Curie: In the Laboratory and on the Battlefield," *Physics Today* (July 2003), pp. 37–43. A treatment of Marie Curie's contribution to the medical use of X-rays.

Drell, S., "Andrei Sakharov and the Nuclear Danger," *Physics Today* (May 2000), pp. 37–41. A look at some of the ideas of 1975 Nobel peace prize winner Andrei Sakharov as they apply in the post–Cold War world.

Feder, T., "Food Meets Physics at Irradiation Facility," *Physics Today* (April 2002), p. 24. A short article on food irradiation studies.

Mak, S., "Radioactivity Experiments for Project Investigation," *The Physics Teacher* (December 1999), pp. 536–539. Discusses the use of potassium chloride in half-life estimation and other radioactivity experiments.

Sawicki, M., "A Note on Quantum Teleportation," *The Physics Teacher* (November 1999), p. 499. An analogy illustrating the essence of quantum teleportation experiments.

Seife, C., "Atom Smasher Probes Realm of Nuclear 'Gas'," *Science* (25 January 2002), v. 295, pp. 603–604. Discusses new developments in the liquid drop model of the nucleus.

Steiger, W., "A Radioactive Tracer in Medicine," *The Physics Teacher* (October 1999), pp. 408–409. Of particular interest to students in the pre-health professions.

Stone, R., "Living in the Shadow of Chernobyl," *Science* (20 April 2001), v. 292, pp. 420–426. A description of the Chernobyl accident and its long-term aftermath, with interviews.

Tigner, M., "Does Accelerator-Based Particle Physics Have a Future?," *Physics Today* (January 2001), pp. 36–40. A discussion of fundamental questions in particle physics and the experimental facilities that can address them.

The Environmental Protection Agency Division of Radiation Protection is an excellent source of information and data on radiation exposure. The Web site address is: http://www.epa.gov/radiation/.

Materials

A basic radiation meter is available from Fisher Science Education, model number S44126B.

A radioactive source set is also available from Fisher Science Education, model number S440975.

A basic nuclear lab station, including radioactive sources, model number S46860IBM, and an intermediate nuclear lab station, model number S46863IBM, are also available from Fisher Science Education. These stations come in MAC versions as well.

Notes and Ideas

Class time spent on material: Estimated:______ *Actual:_____________*

Related laboratory activities:

Demonstration materials:

Notes for next time:

Chapter 32: Elementary Particles

Outline

Summary

This chapter begins with descriptions of various types of particle accelerators and moves on to a discussion of exchange particles and the four fundamental forces in nature. Basic categories of elementary particles and conservation laws are defined. The chapter ends with a brief description of the Standard Model and the Grand Unified Theory.

Major Concepts

By the end of the chapter, students should understand each of the following and be able to demonstrate their understanding in problem applications as well as in conceptual situations.

- Particle accelerators
    - Cyclotron
    - Synchrotron
    - Linear accelerators
    - Colliding beams
- Particles and antiparticles
    - Feynman diagrams
    - Field particles
    - Conservation laws
    - Classification schemes
- Quarks
- Standard Model
    - Quantum chromodynamics
    - Electroweak theory
- Grand Unified Theory

Teaching Suggestions and Demonstrations

At this point in the semester you will probably be out of time. Try to take a class period (or half a period) to discuss the concepts in this chapter qualitatively. Many of the students will find this material quite interesting and will want to spend some time discussing it.

Sections 32-1 through 32-6

Point out to the students that of all the **elementary particles** they have studied so far, only electrons and neutrinos are fundamental particles. Go over the various types of **particle accelerators** briefly. Move on to a general overview of the **four fundamental forces** in nature and the concept of the **field particle** (see Figure 32-6 and discussion). Show the class how Feynman diagrams can be used to illustrate this type of interaction. Introduce antiparticles and discuss their properties and the proofs of their existence. Go over the definition of a **lepton** and talk about which particles fall into this category. Then briefly discuss **hadrons**. (See Table 32-2.) The basic conservation laws are important—take time to go through them in class.

Sections 32-7 through 32-12

It is fun to talk about **quarks** and their properties. The names of the properties are interesting, and the math is actually very easy. Go over Table 32-3 and work through Conceptual Example 32-7 with your class. If you have time, describe the general ideas of the **Standard Model** and the **Grand Unified Theory**. Students will run into these ideas in the popular press; it is nice if they have some sense of what they mean.

Resource Information

Transparencies

T394.	Table 32-1	The Four Forces in Nature
T395.	Table 32-2	Particles (stable under strong decay)
T396.	Table 32-3	Properties of Quarks

Suggested Readings

Fortson, N., Sandars, P. and Barr, S., "The Search for a Permanent Electric Dipole Moment," *Physics Today* (June 2003), pp. 33–39. Describes experiments looking for the electric dipole moment of the neutron.

Franklin, A., "The Road to the Neutrino," *Physics Today* (February 2000), pp. 22–28. Historical description of the discovery and subsequent study of the neutrino.

Haxton, W. and Holstein, B., "Neutrino Physics," *American Journal of Physics* (January 2000), pp. 15–32. Technical and thorough treatment of the basic concepts of neutrino physics.

Joshi, C. and Katsouleas, T., "Plasma Accelerators at the Energy Frontier and on Tabletops," *Physics Today* (June 2003), pp. 47–53. Describes a new form of particle accelerator that is powerful but compact.

Lubkin, G., "Nobel Prize to 't Hooft and Veltman for Putting Electroweak Theory on Firmer Foundation," *Physics Today* (December 1999), pp. 17–19. Discussion of the recipients of the 1999 Nobel prize in physics and their work.

Riordan, M., Rowson, P. and Wu, S., "The Search for the Higgs Boson," *Science* (12 January 2001), v. 291, pp. 259–260. A short article on the search for the Higgs boson, an important piece of the Standard Model.

Sawicki, M., "A Note on Quantum Teleportation," *The Physics Teacher* (November 1999), p. 499. An analogy illustrating the essence of quantum teleportation experiments.

Seife, C., "Muon Experiment Challenges Reigning Model of Particles," *Science* (9 February 2001), v. 291, p. 958. A short article on experimental evidence of the potential failure of the Standard Model.

Tigner, M., "Does Accelerator-Based Particle Physics Have a Future?," *Physics Today* (January 2001), pp. 36–40. A discussion of fundamental questions in particle physics and the experimental facilities that can address them.

Notes and Ideas

*Class time spent on material: Estimated:*______ *Actual:*____________

Related laboratory activities:

Demonstration materials:

Notes for next time:

Chapter 33: Astrophysics and Cosmology

Outline

Summary

Chapter 33 introduces the ideas of astrophysics and cosmology. Stellar evolution, the measurement of astronomical distances, and general relativity are discussed. The Big Bang theory and the early history of the universe are presented. The chapter ends with a brief description of dark matter and energy and the large-scale structure of the universe.

Major Concepts

By the end of the chapter, students should understand each of the following and be able to demonstrate their understanding in problem applications as well as in conceptual situations.

- Astrophysics
  - Light-year
  - Galaxies and nebula
  - Absolute luminosity and apparent brightness
  - Hertzsprung–Russell diagrams
  - Stellar evolution and nucleosynthesis
  - Novae and supernovae
  - Distance measurements
- General relativity
- Hubble's law and the red shift
- Big Bang
- Cosmic microwave background
- Early history of the universe
- Dark matter and energy
- Large scale structure of the universe

Teaching Suggestions and Demonstrations

The material in this chapter is not often presented as part of an introductory physics course, but it is fascinating, with great opportunities for discussions of how we know what we know. If you have a long enough semester or quarter and you have time for this material, you will find the students interested and enthusiastic.

Sections 33-1 through 33-3

These first three sections contain new vocabulary: the **light-year**, **galaxies**, **nebula**, and **star and galaxy clusters**. Many students will have heard these terms before but may not know their precise scientific definitions. Be sure to go over them carefully. The difference between **absolute luminosity** and **apparent brightness** is not obvious to students. Use Example 33-3 to clarify these concepts.

The relationship between color, absolute luminosity, and mass of stars is presented in the **Hertzsprung–Russell diagram**. Go over Figure 33-6, pointing out the main sequence of stars and the red giants and white dwarfs. Students will find the story of **stellar evolution** and **nucleosynthesis** very interesting. The formation of a **nova** or a **supernova** is also of interest to most students.

Distance measurements are extremely important in astronomy and astrophysics. Distances clearly cannot be measured directly, but there are many ways to measure them indirectly. Go over these and discuss the advantages and sources of inaccuracy for each.

➲ **DEMO 33-1** Use the parallax method (see Figure 33-11 and Example 33-6) to determine the distance from one landmark on your campus to another. Then measure the distance with a long tape measure. Comment on the degree of accuracy you are able to achieve.

Section 33-4

General relativity, which applies to accelerated frames of reference and to gravitation, can be covered briefly. Discuss the **principle of equivalence**, using the elevator examples as shown in Figures 33-12 and 33-13. Be sure to give experimental verification of relativity, such as Eddington's experiment measuring the displacement of stars during a solar eclipse (see Figure 33-14).

➲ **DEMO 33-2** Stretch a balloon or a piece of rubber over as large a cookie tin as you can find. Fasten it securely. The flat surface represents empty space. If you place a very light glass marble on the surface it will not move. Now place a steel marble in the center of the surface. The distortion of the surface represents space-time as curved by the presence of matter. If you place the light glass marble back on the surface, it will be "attracted" by the heavy steel marble and roll down the gravitational "well". (See Figure 33-18.)

If you have time, discuss **black holes** and the **Schwarzschild radius** and define the **event horizon**.

Sections 33-5 through 33-10

Do plan to spend time on the **Doppler effect** for electromagnetic waves, even if you just discuss it qualitatively. The applications to police and weather radar are fascinating. Talk about the **red shift** and what information it gives about the motions of stars in the universe. Work Example 33-7. Introduce **Hubble's Law** (Equation 33-6) and how it can be used to measure distance and estimate the age of the universe.

➲ **DEMO 33-3** Use a balloon to demonstrate the **cosmological principle**. Start with an uninflated balloon (light colors work best). This will represent the universe. Mark some "stars" on it with a permanent marker. As you blow up the balloon, the universe expands, and all of the stars will move away from one another.

Plan to discuss the **Big Bang** theory and the **cosmic microwave background**, then go through the **early history of the universe**, as shown in Figure 33-25. The idea of a flat universe expanding at an accelerating rate is relatively new; define **dark matter** and **dark energy** and talk about how they contribute to the expansion of the universe and the recently discovered **large-scale structure of the universe**.

Resource Information

Transparencies

T397.	Figure 33-6	Hertzsprung–Russell (H–R) diagram
T398.	Figure 33-14	(a) Three stars in the sky (b) Gravity bends light from one star
T399.	Figure 33-18	Rubber-sheet analogy for space-time curved by matter
T400.	Figure 33-25	Compressed graphical representation of the development of the universe

Suggested Readings

Brumfiel, G., "Cosmology Gets Real," *Nature* (13 March 2003), pp. 108–110. An accessible update on modern cosmology.

Harwit, M., "The Growth of Astrophysical Understanding," *Physics Today* (November 2003), pp. 38–43. A nice history of the development of the ideas of astronomy and astrophysics.

Notes and Ideas

*Class time spent on material: Estimated:*______ *Actual:*____________

Related laboratory activities:

Demonstration materials:

Notes for next time:

Instructor Notes on ConcepTest Questions

Cornelius Bennhold • Gerald Feldman

Physics

Principles with Applications

Giancoli

Instructor Notes on ConcepTests
Table of Contents

Instructor Notes on ConcepTests
Table of Contents (continued)

Introduction

A. Facilitating the paradigm shift: From Knowledge Delivery to Skill Training

Teaching by telling is an ineffective mode of instruction for most students.
L.C. McDermott

In most science and engineering courses, a primary goal of the course is to help the students develop their competencies in critical thinking, problem solving and analytical reasoning. However, many instructors have come to recognize that, in spite of their best intentions, many students emerge from their study of these technical fields with serious shortcomings regarding their level of ability in these cognitive skills and thus with significant gaps in their understanding of important topics. As a consequence, there is a discernible shift from "content-driven" learning toward a more "skill-centered" type of instruction. In particular, there is consensus that "doing science" does not just mean acquiring a body of knowledge of scientific facts and formulas — it also involves a specific set of skills to mediate the application of this knowledge. In order to achieve a reasonable level of competence, as with most physical and mental activities, this enhancement of thinking skills cannot be achieved without extensive exercise and many hours of practice.

In the last two decades, physicists have begun to approach this problem from a scientific perspective by conducting detailed systematic studies on the learning and teaching of physics [McD99]. Research in Physics Education, which focuses on empirical evidence of student understanding of science content, complements traditional education research that has a stronger emphasis on educational theory and methodology. From the perspective of science departments, discipline-based education research conducted by science faculty has proven to be an effective approach for improving student learning in science [McD01].

Guided by such recent results in Physics Education research, many physics departments have set out to reinvent their introductory physics courses to create a forum in which students can significantly improve their conceptual understanding of physical principles and better develop their skills in critical thinking, analytical reasoning and problem solving. However, reallocating classroom time from lecturing to interactive student engagement requires a fundamental shift in teaching philosophy. It changes the focus from the subject content to the intellectual activity of the students; thus, to some degree, knowledge transfer is being replaced by the development and practice of such cognitive skills.

B. Practicing Physics: The Socratic Method and Peer Instruction

Research in the field of Physics Education has revealed a basic ineffectiveness in the conventional lecture course [Cr01]. This refers to a "typical" classroom course, in which the students attempt to absorb information that is presented by the lecturer. In fact, it is the passivity of the student audience that is at the heart of the problem. The notion that the "traditional" lecture communicates little to students is a conclusion not readily accepted in the physics community. Indeed, the idea that a very clear explanation suffices for complete student comprehension may appear sensible. However, numerous empirical studies have shown this not to be the case [Me96]. On the other hand, Physics Education research has uncovered convincing evidence that an "interactive engagement" (IE) approach, in which the students are active participants in the classroom environment, is considerably more effective in the learning process [Ha98]. The main point is that the students spend more of the classroom time practicing the material, thereby exercising their minds and developing their skills, which allows them to assimilate the basic concepts in a more natural manner. Further studies in general higher education have shown that students develop complex reasoning skills most effectively when actively engaged with the material and have found that cooperative activities are an excellent way to engage students [Jo91].

As a first step toward engaging the students in the classroom, it is recommended that the entire philosophy of the classroom period be altered. Rather than repeat the same material that is already covered in the textbook, we begin practicing the material as soon as possible and as often as possible. Thus, we require the students to read the class material in advance, and they are expected to have some familiarity with the concepts to be covered that day. The material presentation can be limited to the highlights in order to get to the major classroom activity, which is the implementation of a technique known as Peer Instruction.

Peer Instruction is a Socratic teaching style pioneered by Eric Mazur at Harvard University in which students actively participate in the learning process through an interactive sequence of questions and answers, called *ConcepTests* [Ma97]. Largely abandoning the principles of behavioral learning theory, Peer Instruction is based on Piagetian learning principles that promote students constructing their understanding of concepts rather than rote memorization.

Peer Instruction replaces the traditional lecture format by partitioning the class period into 10 – 15 minute "mini-lectures," each of which is terminated by a series of short multiple-choice conceptual questions (*ConcepTests*) designed to challenge the students to immediately apply the material. The students think for 1 – 2 minutes about the questions and then commit to individual answers. They then spend 2 – 3 minutes attempting to convince their neighbors about the accuracy of their answer. After this period, students are asked to reconsider their original answer and record it again. Another quick poll is taken, after which the instructor indicates the correct answer. The immediate feedback reveals to the students their level of understanding, leading them to ask more questions, while the instructor, based on the student answers, can decide whether to move on to the next subject or to supplement the current one with additional practice.

The objective of Peer Instruction is to foster active student learning, even (or especially) in a large lecture environment. By encouraging student participation and interaction during the lecture, Peer Instruction encourages students to think critically about the arguments being developed and to discuss their ideas and insights with other students. Engaging students in the learning process has been shown to improve student learning of and attitude toward the subject, increase classroom attendance and reduce attrition. It is important to point out that the *ConcepTests* are not numerical questions that require a calculator or other manipulation. Mazur found that students understood concepts better and performed more successfully on conventional problems in this environment, as reflected in scores on the Force Concept Inventory [He92] (a nationally distributed assessment instrument) and final exams in the course.

It is helpful (but not essential) to supplement Peer Instruction with an electronic student response system that records the student responses to the *ConcepTests.* Most such systems display the composite class responses for each question as a histogram and records all of the responses for each individual student during the class period. Depending on the system, students may login to the system at the beginning of each class with a valid ID number, so the keypad responses are unambiguously linked to each student in a unique fashion. Student authentication permits complete tracking of each student's activity throughout the class period; such systems allow giving credit for participation and permit far more analysis possibilities than the simple display of a response histogram during class.

It is during the discussion with neighbors that the primary learning can take place. Students sometimes have their own way of thinking or explaining that can often differ from the approach taken by the instructor. The "student way" of thinking may be different in such a way as to be more attuned to the other students. The instructor is still a primary resource, but generally during the discussion time, the instructor is surveying the discussion and listening to the arguments of the students. In that sense, the role of the instructor is more akin to that of a "coach" who facilitates and advises while the student "players" work at the desired activity. Following the second round of responses, the correct answer and a full explanation are provided, and possible reasons why students might have selected a particular wrong answer are offered. Often a particular misconception can be uncovered in this manner. Those students who have not changed to the correct answer after discussion are forced to confront their own deeply seated beliefs, and the "shock" of getting the wrong answer can be an effective means of dislodging this misconception, or at least, getting them to reconsider what they had previously believed.

In our five years of experience of using such a system, certain aspects of the system and its use as an interactive engagement tool have become apparent, such as:

- Students enjoy using the system and readily engage in the in-class activities.
- Attendance in class is typically at the 90% level or higher for *every* class.
- Answering questions in class and discussion with neighbors breaks up the monotony of a traditional lecture.
- Anonymity during class responses permits students to answer honestly.
- Discussion among the students often reveals hidden misconceptions or elicits interesting questions that pertain to the entire class.
- Students reach a higher level of motivation with active in-class participation.

Even relatively simple concepts that are "covered" in a few minutes of conventional lecture time turn out to be profoundly confusing to students (*e.g.,* "The only force [neglecting air resistance] acting on a projectile in flight is gravity"). Furthermore, ideas which instructors may consider too trivial to warrant more than a passing reference have been found to stump many students when they are asked to make use of them in problems (*e.g.,* "The total momentum of a pair of objects sitting at rest is zero"). The improvement in student comprehension and retention from the interactive engagement methods is a measurable effect, obtained from experimental data subjected to a rigorous statistical analysis. A landmark study by Richard Hake [Ha98] with a data set of over 6000 students in 62 independent introductory physics courses identified an increase in the normalized gain (defined as the ratio of the average gain to the maximum possible gain) on the Force Concept Inventory [He92] by a factor of two for courses employing IE techniques, compared to traditional lecture courses.

C. From *ConcepTests* to *ConcepModules*

As discussed above, the term *ConcepTest* refers to a single conceptual multiple-choice question that challenges the student to immediately apply certain material and thereby develop critical thinking and analytical reasoning skills. Since Mazur's publication [Ma97], many new such questions have been developed and tested within the physics education community. With extensive implementation of Peer Instruction, some disadvantages of an isolated *ConcepTest* have become apparent, among them [Fe99]:

- The question may be too difficult, thus students have little chance to arrive at the correct answer, even after extensive discussion.
- The question may be too easy and thus appear gratuitous to the students.
- The students may not see the connection between the question and the material being discussed, thus it "comes out of nowhere."
- The students may not realize how discussing a particular conceptual question helps them solve numerical homework problems, thus students may question the purpose of engaging in the activity.
- The students may be left with the feeling that, after answering a particular question, they have exhausted all open questions pursuant to this line of inquiry.

We therefore advocate using a refinement of the *ConcepTest* in class, the **Conceptual Module**, or *ConcepModule*. This instrument contains the following features:

1. A sequence of 2 – 4 *ConcepTests*, usually ordered by level of difficulty or ranked according to some other pedagogical reason. This allows students to "climb the mountain slowly" when a difficult concept is being practiced.
2. A follow-up numerical problem to be solved by the instructor in class after or in between several *ConcepTests*. This relates the conceptual question(s) to the numerical problems the students will need to solve for homework and in exams and thus establishes a linkage between different knowledge representations.
3. An accompanying demo (or movie of a demo) that further elucidates the concept and accommodates different student learning styles.
4. Open-ended follow-up questions for the last *ConcepTest* in the sequence that emphasize that each answer to a *ConcepTest* invites another question.

The use of *ConcepModules* in class, rather than isolated *ConcepTests*, integrates the exploration of certain material, especially for the average or below-average student who may otherwise perceive each class activity in isolation. Clearly, not all *ConcepModules* contain all the above features, and many instructors may wish to continue using individual *ConcepTests* without the accompanying features of a module. As an example, we show below a conceptual module from Chapter 3 (the assignment of the difficulty level is discussed in the next section):

Conceptual Module 3.10 (three parts)

ConcepTest 3.10a: Shoot the Monkey I
Difficulty Level: 3

This series deals with variations of the "monkey hunter" problem. The first question deals with the relatively simple case of the shooter and the monkey on the same level.

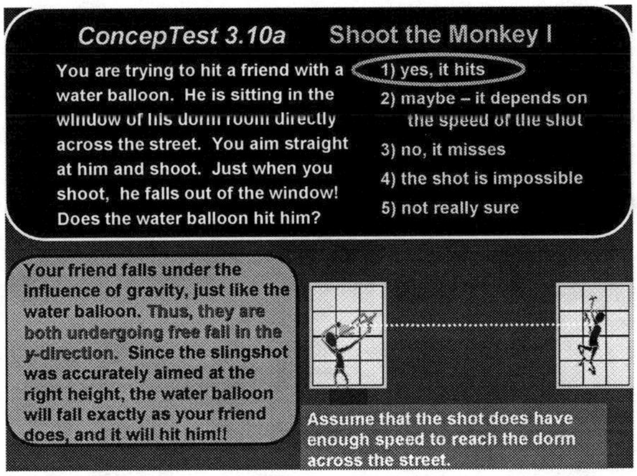

ConcepTest 3.10b: Shoot the Monkey II
Difficulty Level: 4

This is the standard formulation of the "monkey hunter" problem, in which the shooter is on the ground and the monkey is located at some elevation. It is hoped that the students will be able to extrapolate their experience from the first question to this one.

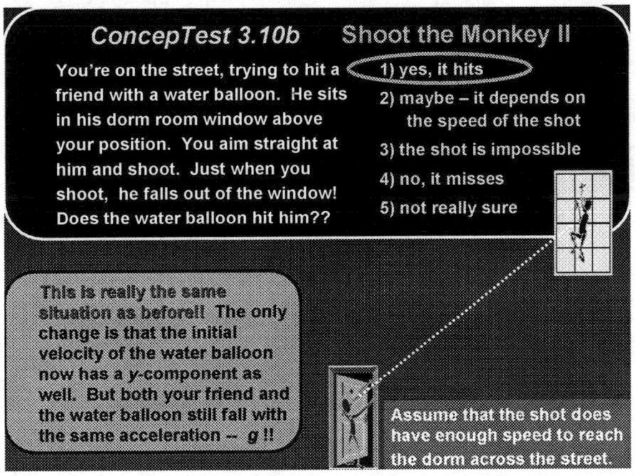

ConcepTest 3.10c: Shoot the Monkey III
Difficulty Level: 4
Suggested follow-up conceptual question:

- **When would they *not* hit each other?**

In this case, two projectiles initially aimed at each other are fired. The fact that both projectiles are only under the influence of gravity means that they will indeed hit. This can be more easily seen for the case of projectiles that are launched at the same level.

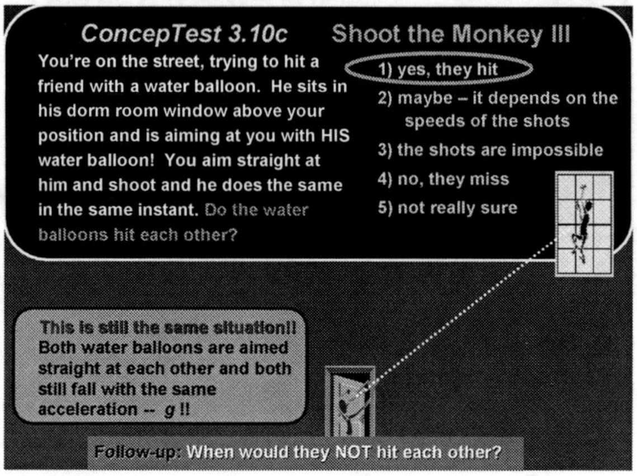

D. Classification of question difficulty according to *Bloom's Taxonomy of Thinking Skills*

As an introductory science course moves from the mere delivery of knowledge to the acquisition and training of thinking skills, the question arises, what performance indicators are instructors to use to determine if the learning objectives of the course have been achieved? As a prerequisite, problems the student uses to train with must be classified according to some scale. Usually, for numerical end-of-chapter problems, most standard textbooks tend to use an "easy – medium – hard" distinction that is based on the experience of the textbook author, rather than any rigorous analysis of the level of thinking skills needed to solve a particular problem. Conceptual questions are usually not rated at all.

To clarify our educational objectives, we have found it helpful to work with a classification scheme referred to as Bloom's Taxonomy [An94], which establishes a hierarchy of cognitive skills, listed below:

Level 1: Knowledge

Knowledge in this context refers to the ability to recall previously learned material. This may cover a wide range, from specific facts to complete theories, but all that is required is remembering the appropriate information. Knowledge represents the lowest level of thinking skills in the cognitive domain.

Examples of general learning objectives at this level are: know common terms, know specific facts, know methods and procedures, know basic concepts and principles.

Physics questions at this level would simply require familiarity with specific facts. Reading quizzes that test students having read the book before class would use this level.

Level 2: Comprehension

Comprehension is defined as the ability to grasp the meaning of material. This may be shown by translating material from one form to another (words to numbers) or by interpreting material (explaining or summarizing). These learning outcomes go one step beyond the simple remembering of material and represent the lowest level of understanding.

Examples of general learning objectives at this level are: understand facts and principles, interpret verbal material, interpret charts and graphs, translate verbal material to mathematical formulae, and estimate the future consequences implied in data.

Physics questions at this level would test the lowest level of understanding of a concept beyond the mere remembering of a fact, but without expecting the student to apply it in a given problem or situation.

Level 3: Application

Application refers to the ability to use learned material in concrete situations. This may include the application of such things as rules, methods, concepts, principles, laws, and theories. Learning outcomes in this area require a higher level of understanding than those under comprehension.

Examples of general learning objectives at this level are: apply concepts and principles to new situations, apply laws and theories to practical situations, solve simple mathematical problems, construct graphs and charts, and demonstrate the correct usage of a method or procedure.

Physics problems at this level could require simple numerical calculations. It is thus an important step, since it constitutes the first link between theory (concepts) and practice (applications). Conceptual questions would require applying a concept to a specific new situation or problem. Simple laboratory procedures would fall into this category.

Level 4 (higher order): Analysis

Analysis refers to the ability to break down material into its component parts so that its organizational structure may be understood. This may include the identification of parts, analysis of the relationship between parts, and recognition of the organizational principles involved. Learning outcomes here represent a higher intellectual level than comprehension and application because they require an understanding of both the content and the structural form of the material.

Examples of general learning objectives at this level are: recognize unstated assumptions, recognize logical fallacies in reasoning, distinguish between facts and inferences, evaluate the relevancy of data, and analyze the organizational structure of a work.

Physics problems at this level involve taking apart the elements of a more complex exercise and independently manipulating the components. Usually, multi-step numerical problems (end-of-chapter exercises) fall into this category. Another aspect of the analysis level would include processing of data in lab exercises with the objective of drawing conclusions from those data.

Level 5 (higher order): Synthesis

Synthesis refers to the ability to put parts together to form a new whole. This may involve items like a plan of operations (research proposal) or a set of abstract relations (scheme for classifying information). Learning outcomes in this area stress creative behaviors, with major emphasis on the formulation of new patterns or structure.

Examples of general learning objectives at this level are: write a well-organized funding proposal, propose a plan for an experiment, integrate learning from different areas into a plan for solving a problem, formulate a new scheme for classifying objects (or events, or ideas).

The physics problems at this level normally involve analyses between categories or topics that are not alike, requiring the students to reassemble component parts of these categories in new ways. Problems at this level are seldom found in standard textbooks.

Level 6 (higher order): Evaluation

Evaluation is concerned with the ability to judge the value of material (statement, novel, poem, research report) for a given purpose. The judgments are to be based on definite criteria. These may be internal criteria (organization) or external criteria (relevance to the purpose), and the student may determine the criteria or have them be given. Learning outcomes in this area are highest in the cognitive hierarchy because they contain elements of all the other categories, in addition to performing value judgments based on clearly defined criteria.

Examples of general learning objectives at this level are: judge the logical consistency of written material, judge the adequacy with which conclusions are supported by data, judge the value of a work by the use of internal criteria or of external standards of excellence.

The physics problems in this level should require students to make judgment calls as to the validity of the information provided and place material in a larger context of other knowledge. Problems at this level are very rarely found in standard textbooks.

E. Student Response Histograms

Many conceptual questions will be presented with a *student response histogram (SRH)* [Gu01] that displays student responses when the question was asked in class. It allows an instructor to anticipate the potential student response and may influence the decision to use this question in class. Clearly, student responses to individual questions depend on many factors, some of which will be discussed in the context of each question. Normalizing the student response to the overall quality of the student body is a difficult issue, which is addressed below.

The complete SRH consists of two graphs, the *response histogram* and the *change graph.* The *response histogram* displays five entry columns, where each entry reports the percentage of students giving one possible answer in the ConcepTest. The number of students giving a particular answer in round 1 is shown as red (light gray) bars. The blue (dark gray) bars display the number of students who have chosen this answer in round 2, after discussion with peers. The correct answer is displayed in the PowerPoint slide and is shown as a circle below the correct column.

To further evaluate and anticipate the effectiveness of a given *ConcepTest*, the change graph displays five separate groups of students:

- C-C (Correct → correct): This group of students ("sustainers") chose the correct answer both before and after discussion. This group constitutes the best students who are confident about having chosen the correct answer and stick to it.
- I-C (Incorrect → correct): This group of students ("improvers") chose an incorrect answer before discussion but the correct answer after discussion. This group has obviously benefited from the discussion process and either was persuaded by a line of reasoning or just followed a group consensus.
- I-S (Incorrect → same incorrect): This group of students ("insisters") chose the same incorrect answer both before and after discussion. This group clearly felt confident about their wrong answer, and the group discussion could not change their conviction.
- I-D (Incorrect → different incorrect): This group of students ("switchers") chose an incorrect answer before discussion and then switched to a different incorrect answer after discussion. This group may have randomly selected an answer the first time and may follow an incorrect line of argument during discussion.
- C-I (Correct → incorrect): This group of students ("doubters") chose the correct answer before discussion and then switched to an incorrect answer after discussion. This group may either have chosen the correct answer accidentally before discussion or had a low level of confidence in their answer.

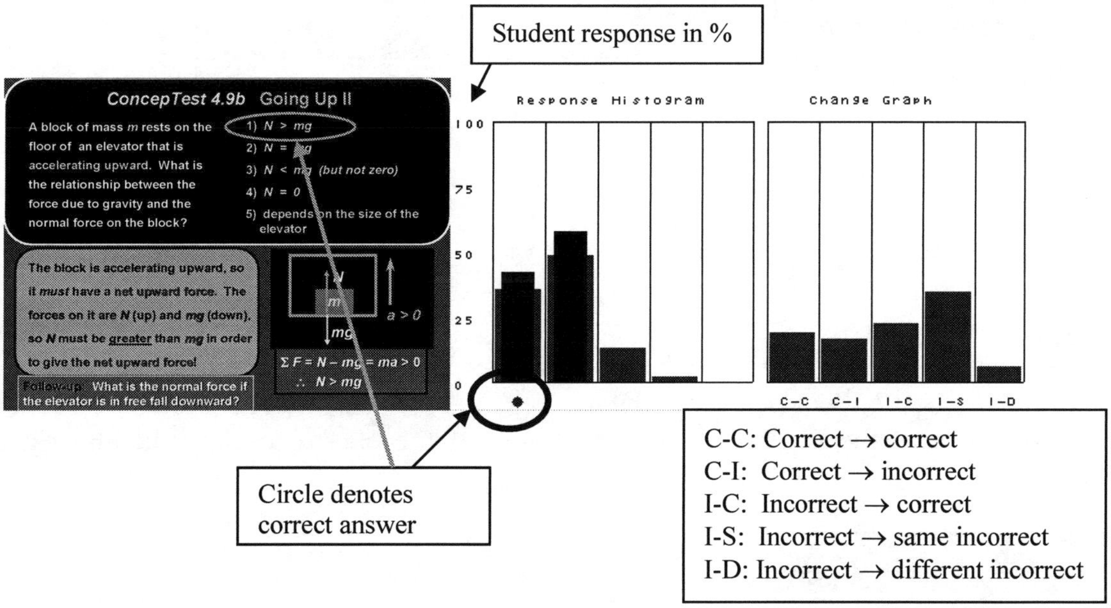

F. Examples of Common Histogram Types

- The "softball" question:
  Such questions are usually easy for students, indicating the large number of correct responses before discussion. The purpose of asking these questions is to either perform a quick reality check for the instructor – that the students have indeed acquired this level of competency – or as a warm-up exercise with harder questions to follow.

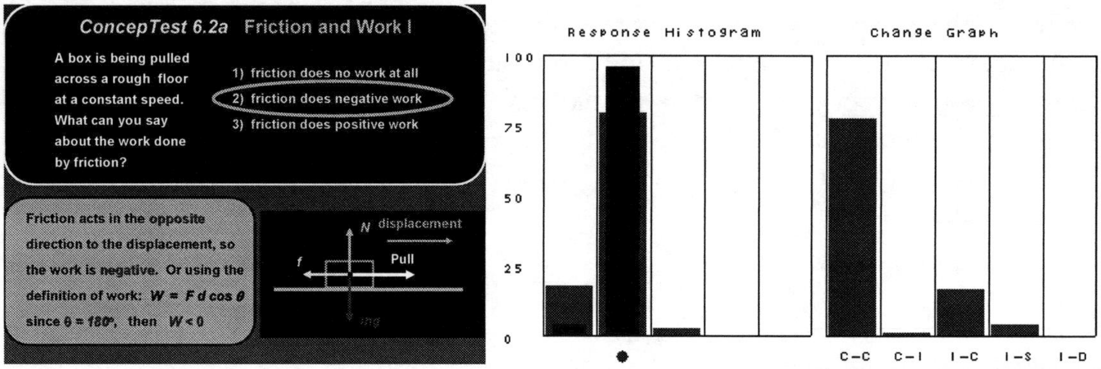

- The "perfect" question:
  Questions with such histograms display an ideal *ConcepTest*. Around half the students answered the question correctly by themselves. After discussion, enough students convinced others of the correct answer that most of the class ends up with the right answer.

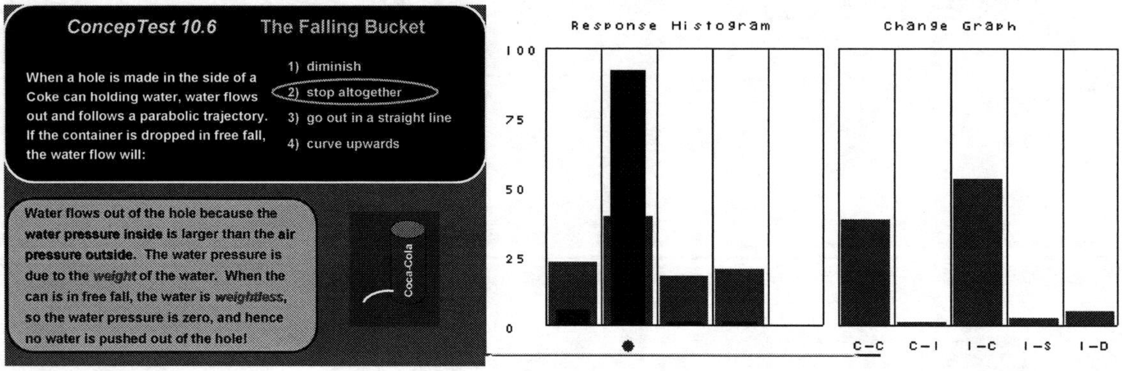

- <u>The "corrected misconception" question:</u>
  Questions of this type indicate that initially most students follow their instincts when answering. Upon further deliberation within groups, most students recognize that the correct answer may be counterintuitive and change their initial choice, with a significant percentage moving toward the correct answer.

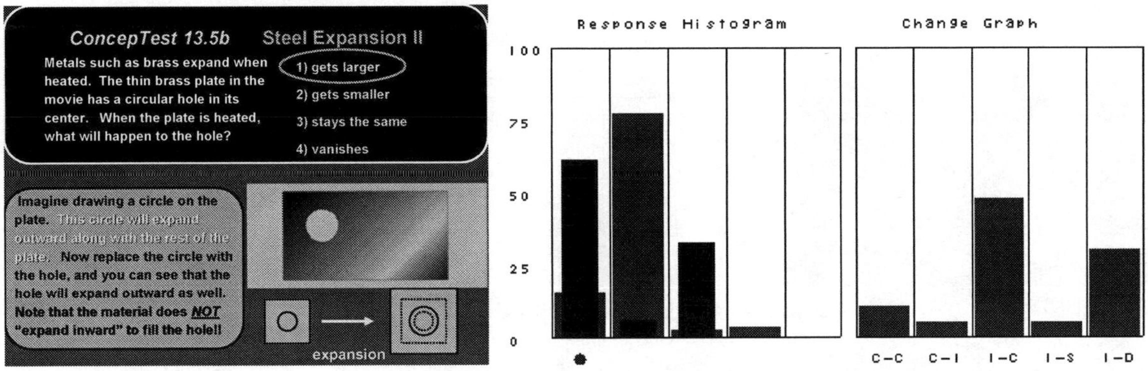

- <u>The "stubborn misconception" question:</u>
  Questions of this type reveal misconceptions held by students that are not easily resolved, even after group discussion with peers. Most students hold on to their initial choices, even though they realize that there is a fairly significant spread. Such responses definitely require further intervention by the instructor.

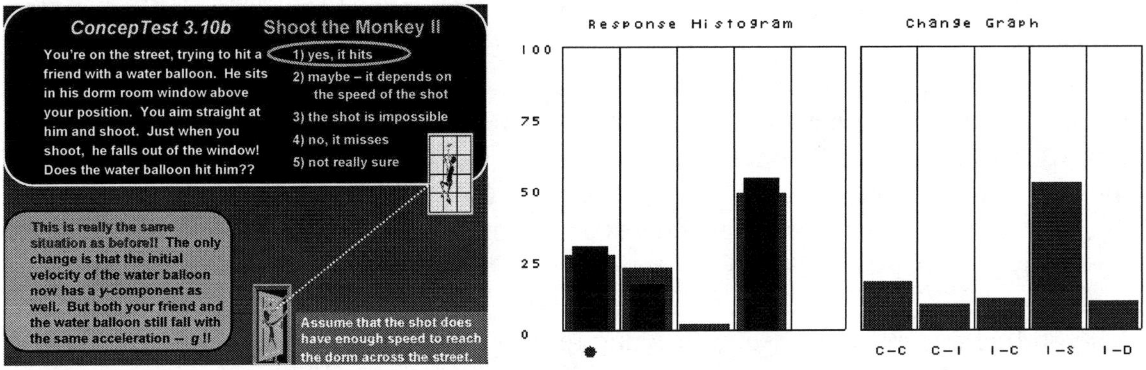

- The "killer" question:
  Every once in a while a very tough question comes along that will throw the entire class for a loop. When histograms of this nature are displayed in class after discussion with peers, they can incite a riot-like atmosphere. This can be useful for certain purposes (such as dampening overflowing student self-confidence), but the instructor needs to be prepared for it! Aim to spend about 20-30 minutes of class time to discuss the solution to such questions.

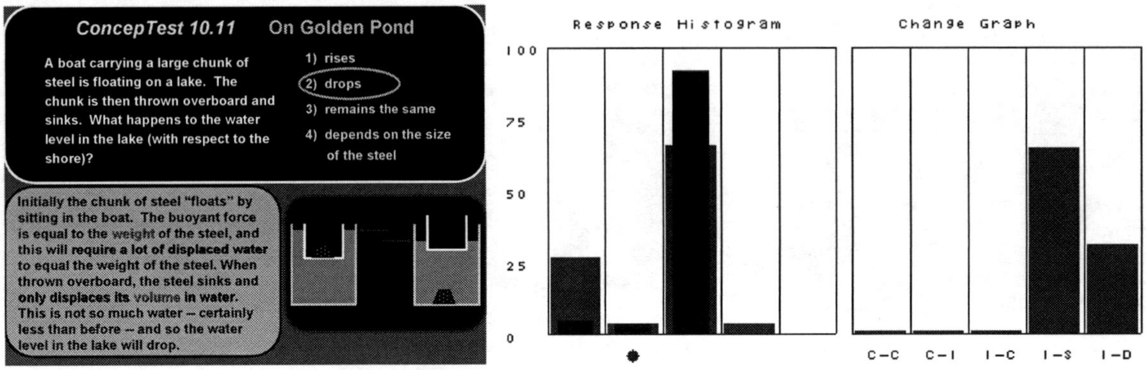

G. Normalizing Student Responses through FCI Scores

The student response histograms have to be used with caution since there can be many factors that lead to a particular student response, besides the nature of the question. One important overall factor is clearly the level of student preparation. Students attending Harvard University can on the whole be expected to have higher correct scores (a higher number of "sustainers" and "improvers") than students in two-year community colleges. To provide a benchmark for instructors we recommend giving the class the FCI as a pretest at the beginning of the semester. The SRHs displayed in this manual were obtained at The George Washington University, where the pretest FCI score has been about **33 ± 3 %** (10 ± 1 correct questions out of the 30 FCI questions) over a number of years. Thus, our SRH should be understood based on that background of student preparation. A lower pretest FCI score would lead to a smaller student group of "sustainers" and "improvers." To provide a different reference point, Mazur reported a pretest FCI score of Harvard students of around 66% [Ma97]. Ideally, one would like to see SRHs analyzed as a function of their pretest FCI scores, a project far beyond the scope of this manual.

We hope that you find this manual a useful teaching resource as you implement *ConcepTests* and *ConcepModules* in the classroom, and we welcome all questions, comments, and corrections.

October, 2004

Cornelius Bennhold
bennhold@gwu.edu

Gerald Feldman
feldman@gwu.edu

References

[An94] Lorin W. Anderson, Lauren A. Sosniak (Editors), *Bloom's Taxonomy: A Forty-Year Retrospective* (University of Chicago Press, 1994).

[Cr01] C.H. Crouch and E. Mazur, Am. J. Phys. **69**, 970 (2001).

[Fe99] G. Feldman and C. Bennhold, AAPT Announcer **29**, 114 (1999).

[Gu01] N. Guttenberg, G. Feldman and C. Bennhold, AAPT Announcer **31**, 137 (2001).

[Ha98] R.R. Hake, Am. J. Phys. **66**, 64 (1998).

[He92] D. Hestenes, M. Wells and G. Swackhammer, Physics Teacher **30**, 141 (1992).

[Jo91] D.W. Johnson, R.T. Johnson and K.A. Smith, *Active Learning: Cooperation in the College Classroom* (Interaction Book Co., Edina, MN, 1991).

[Ma97] E. Mazur, *Peer Instruction: A User's Manual* (Prentice-Hall, Upper Saddle River, NJ, 1998).

[McD99] L.C. McDermott and E.F. Redish, Am. J. Phys. **67**, 755 (1999).

[McD01] L.C. McDermott, Am. J. Phys. **69**, 1127 (2001).

[Me96] D.E. Meltzer and K. Manivannan, Physics Teacher **34**, 72 (1996).

Conceptual Module 2.1

ConcepTest 2.1: Walking the Dog
Difficulty Level: 2
Suggested follow-up conceptual question:

- **Have you and your dog traveled the same distance?**

This question examines the difference between displacement and distance. These concepts are often confused by students, and this ConcepTest is intended to help emphasize the main distinction between these two quantities.

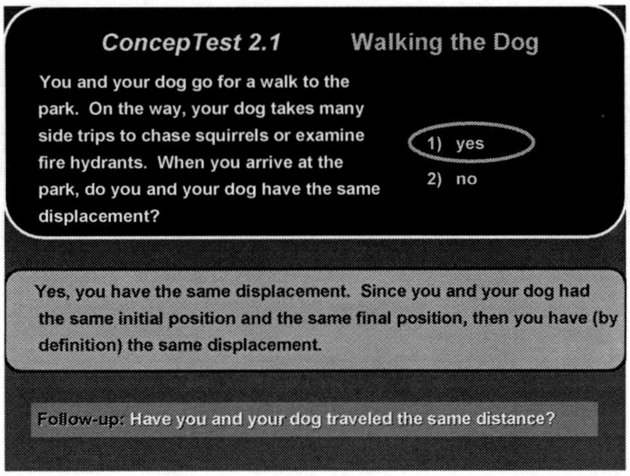

Conceptual Module 2.2

ConcepTest 2.2: Displacement
Difficulty Level: 2

The displacement of an object does *not* depend on the origin, because the displacement is the difference between two positions. The *absolute position* of an object *will* depend on the origin, but when you take the difference between two positions, the origin drops out.

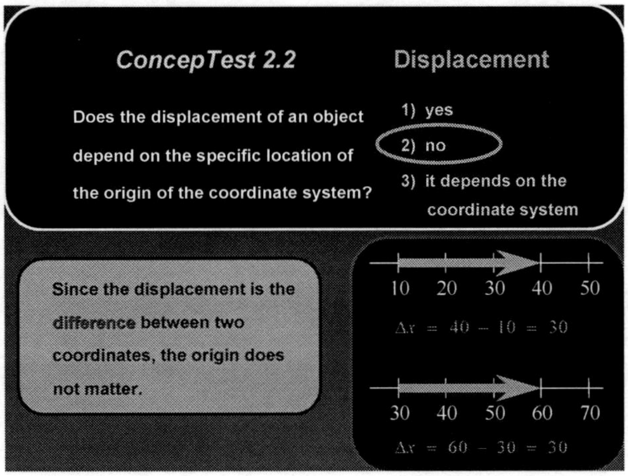

Conceptual Module 2.3

ConcepTest 2.3: Position and Speed
Difficulty Level: 2

The position of an object and its speed have no relation. An object can have any speed, in principle, at any position.

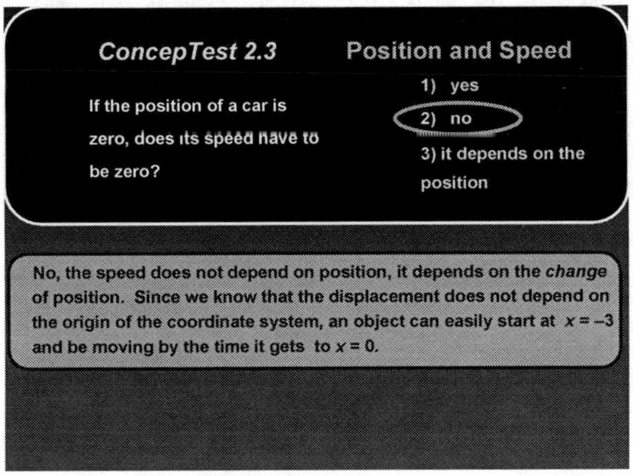

Conceptual Module 2.4

ConcepTest 2.4: Odometer
Difficulty Level: 1
Suggested follow-up conceptual question:

- **How would you measure displacement in your car?**

This question again addresses the distinction between distance and displacement.

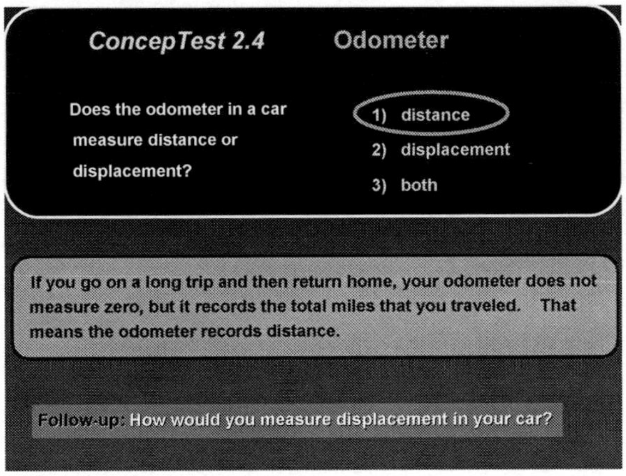

Conceptual Module 2.5

ConcepTest 2.5: Speedometer
Difficulty Level: 1
Suggested follow-up conceptual question:

- **How would you measure velocity in your car?**

This question examines the difference between speed and velocity. One is a vector while the other is a scalar. These concepts are often confused by students, and this ConcepTest is intended to help emphasize the main distinction between these two quantities.

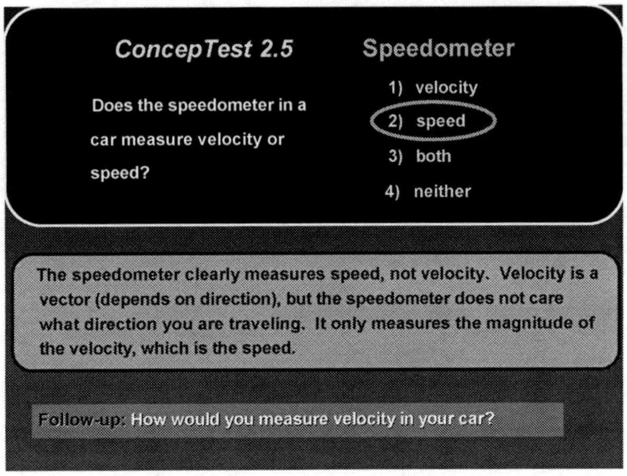

Conceptual Module 2.6 (two parts)

ConcepTest 2.6a: Cruising Along I
Difficulty Level: 3

The definition of average speed is the ratio of total distance over total time. In this ConcepTest, different speeds apply to two identical time intervals, so the average speed is straightforward to determine.

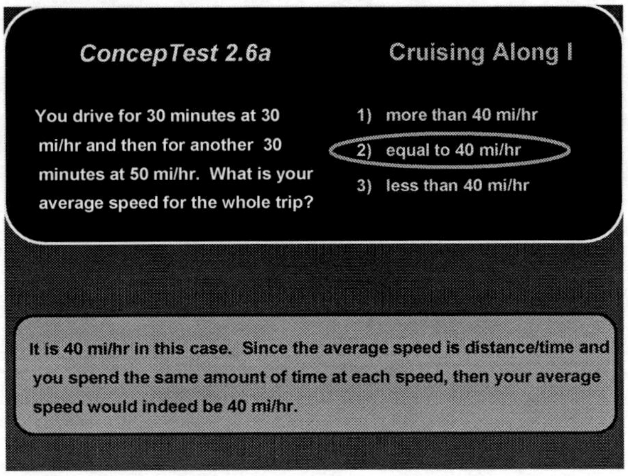

ConcepTest 2.6b: Cruising Along II
Difficulty Level: 4
Suggested follow-up conceptual question:

- **How much further would you have to drive at 50 mi/hr in order to get back your average speed of 40 mi/hr?**

This ConcepTest is different from the previous one in the sense that different speeds now apply to two identical distances. In this case, determining the average speed is not quite as straightforward as in the previous ConcepTest.

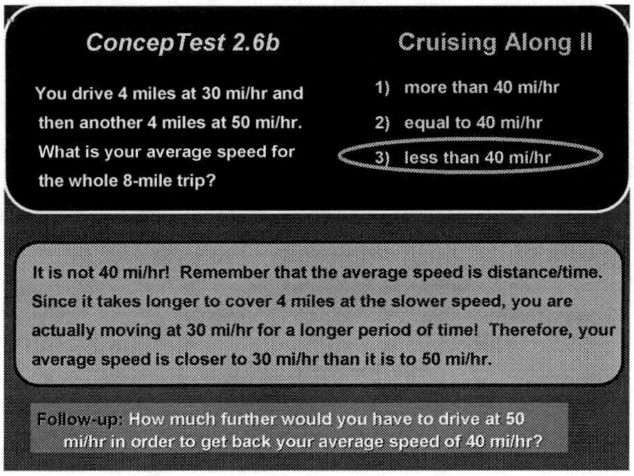

Conceptual Module 2.7

ConcepTest 2.7: Velocity in One Dimension
Difficulty Level: 2

This ConcepTest addresses the concepts of average velocity and instantaneous velocity.

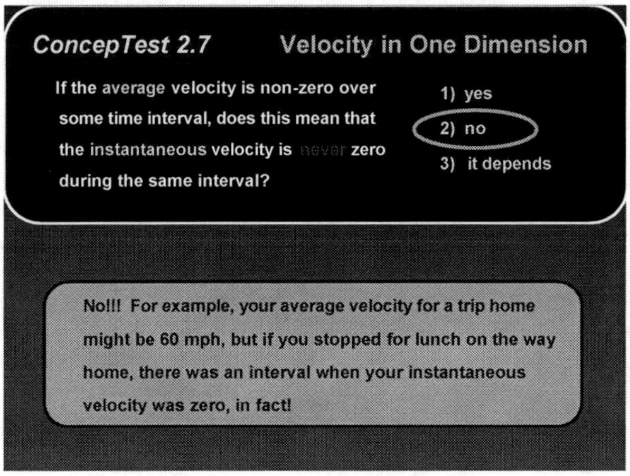

Conceptual Module 2.8 (two parts)

ConcepTest 2.8a: Acceleration I
Difficulty Level: 2

Acceleration is an especially difficult concept for students. This question is intended to give a feeling that "acceleration is not the same as velocity," which is a notion that many students seem to hold.

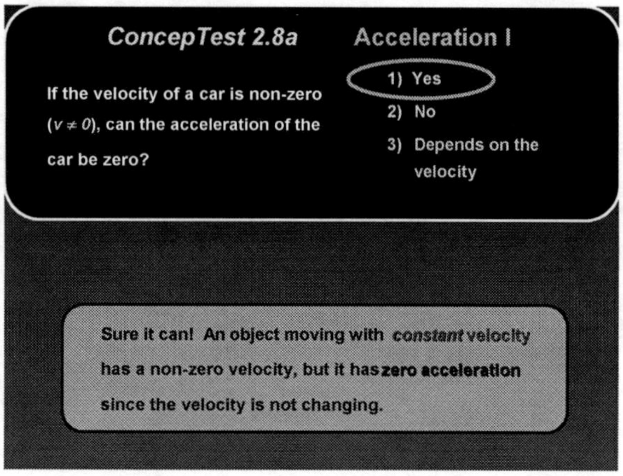

ConcepTest 2.8b: Acceleration II
Difficulty Level: 3
Suggested follow-up conceptual question:

- **...and the value of *a* is?**

This illustrates an example that is the opposite of the previous question. That is, in this case, the velocity is zero but the acceleration is not. The idea that an object that is not moving can actually have a non-zero acceleration is very tricky for students.

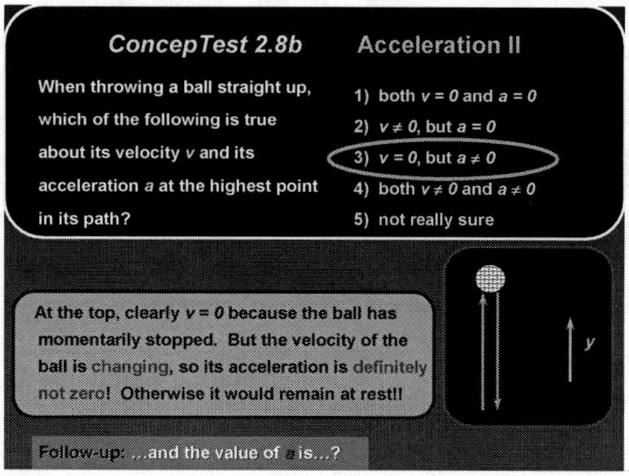

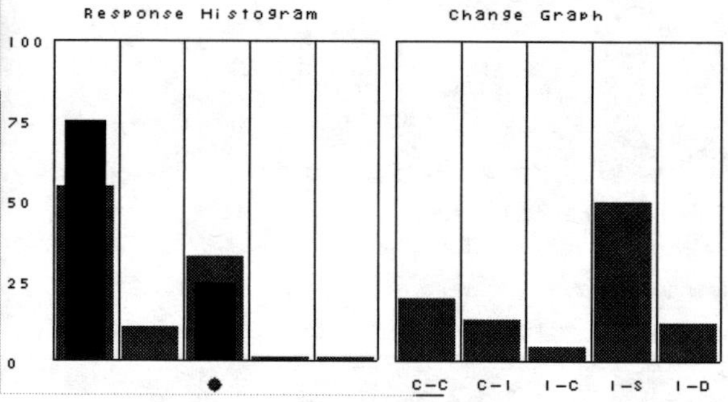

Conceptual Module 2.9 (two parts)

ConcepTest 2.9a: Free Fall I
Difficulty Level: 2

This question is intended to make students realize that the acceleration of a freely falling object is always $-g$, independent of the initial velocity of the object.

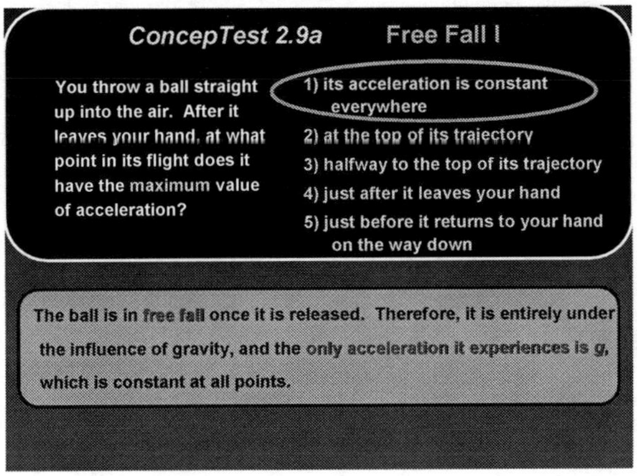

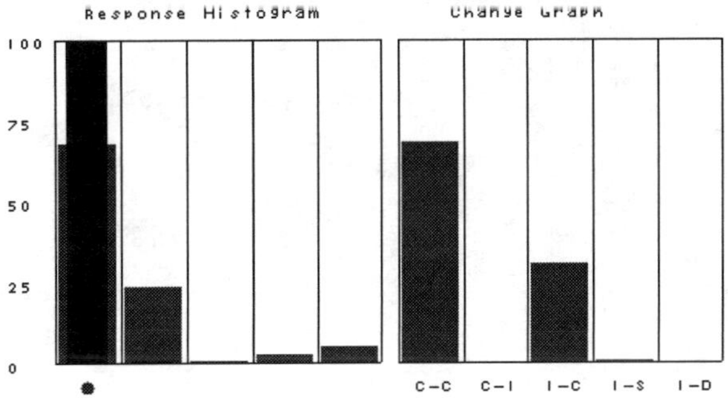

ConcepTest 2.9b: Free Fall II
Difficulty Level: 3
Suggested follow-up conceptual question:

- **Which one has the greater velocity when they hit the ground?**

This question is intended to make students realize that the acceleration of a freely falling object is always $-g$, independent of the initial velocity of the object.

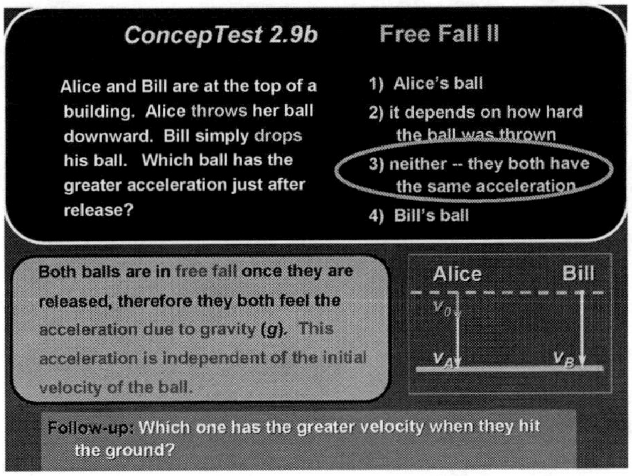

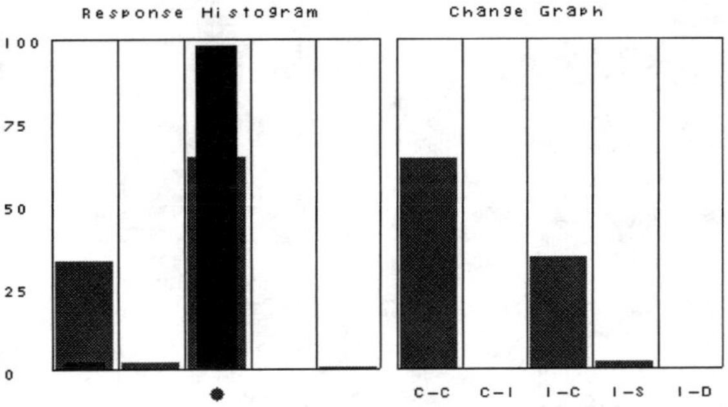

Conceptual Module 2.10 (two parts)

ConcepTest 2.10a: Up in the Air I
Difficulty Level: 3

The symmetry of the rising and falling ball is illustrated in this ConcepTest. Since the acceleration of the freely falling ball is $-g$, on the way up and on the way down, its downward trajectory is just the reverse of its upward trajectory.

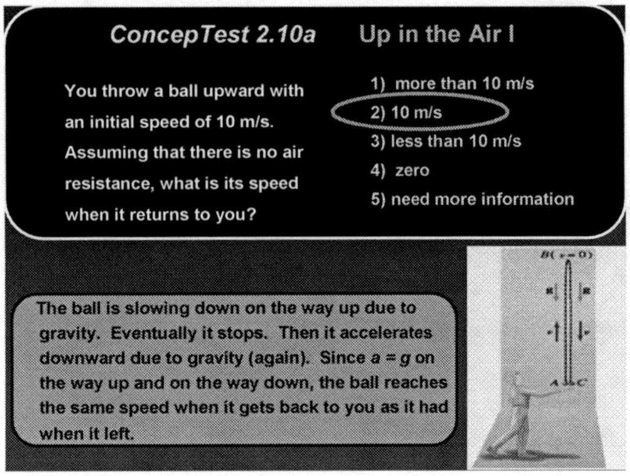

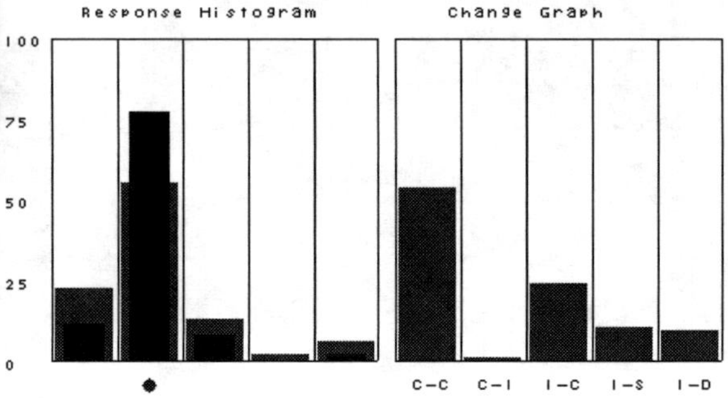

ConcepTest 2.10b: Up in the Air II
Difficulty Level: 4
Suggested follow-up conceptual question:

- **What happens if there is air resistance?**

This ConcepTest continues to build on the idea addressed in the previous question. By recalling the previous answer, students should realize that the two cases depicted in this question are identical, in terms of the behavior of the ball *below* the thrower's position.

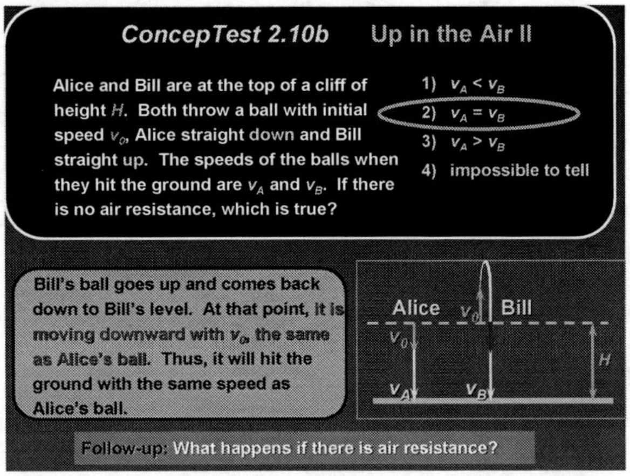

Conceptual Module 2.11

ConcepTest 2.11: Two Balls in the Air
Difficulty Level: 5
Suggested follow-up conceptual question:

- **How could you calculate where they meet?**

Students will have a tendency to say that the balls will meet in the middle. This is not true, however, due to the different initial velocities of the balls. Realizing that the balls will meet after traveling for similar time periods (not similar distances) requires a bit of sophisticated thinking on the part of the students.

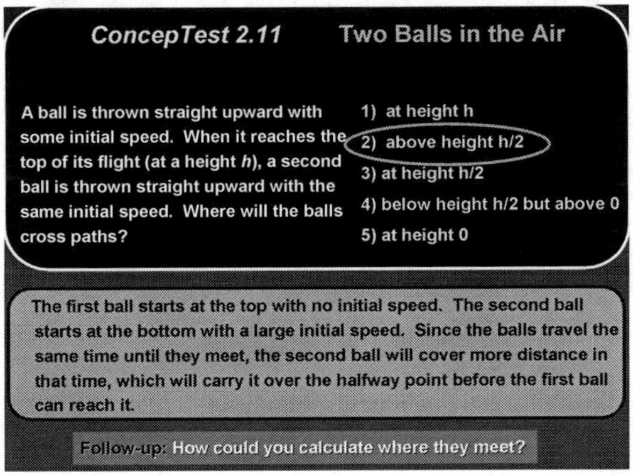

Conceptual Module 2.12

ConcepTest 2.12a: Throwing Rocks I
Difficulty Level: 4

Students will recognize that both rocks have the same acceleration, but the rocks obviously have different velocities. While the velocity difference will remain the same (see next question), the position difference will always increase.

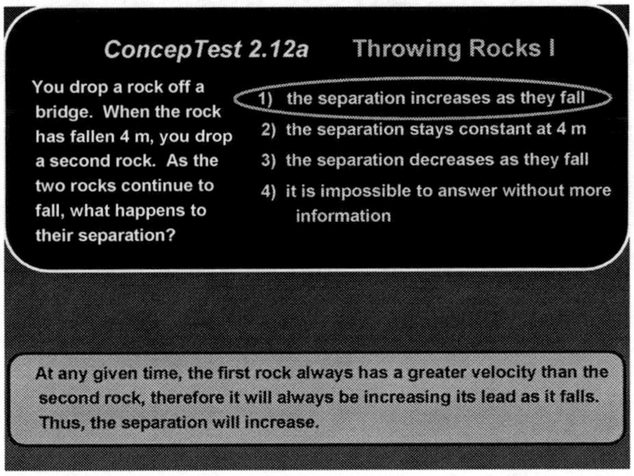

ConcepTest 2.12b: Throwing Rocks II
Difficulty Level: 4
Suggested follow-up conceptual question:

- **What happens when air resistance is present?**

This question moves to the next step, which is asking about the velocities. Here the notion of constant acceleration (and what it means) is emphasized.

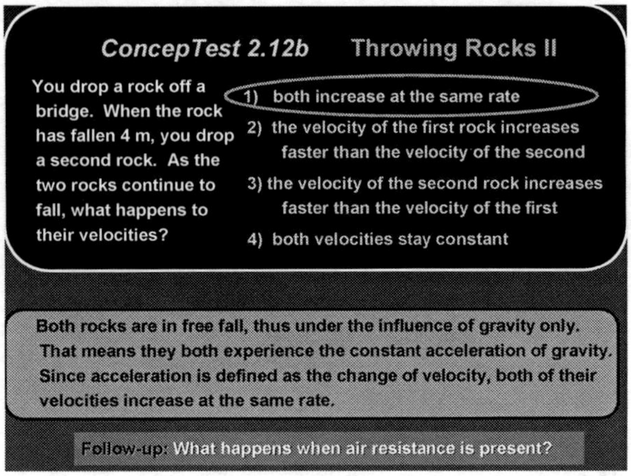

Conceptual Module 2.13 (two parts)

ConcepTest 2.13a: Graphing Velocity I
Difficulty Level: 2

Understanding graphical representations is the theme of this question and the questions that immediately follow. Here the slope of the x vs. t plot is constant.

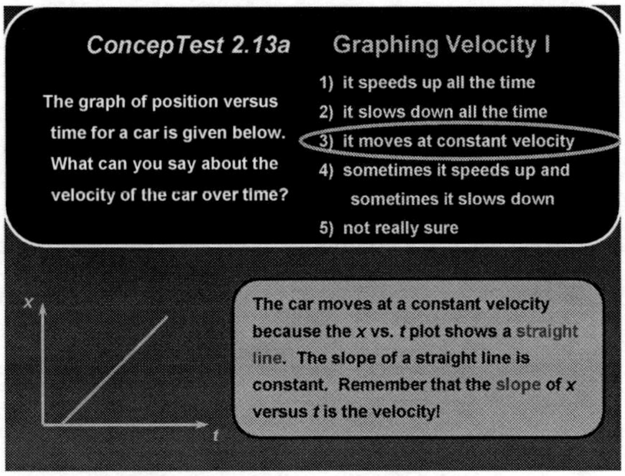

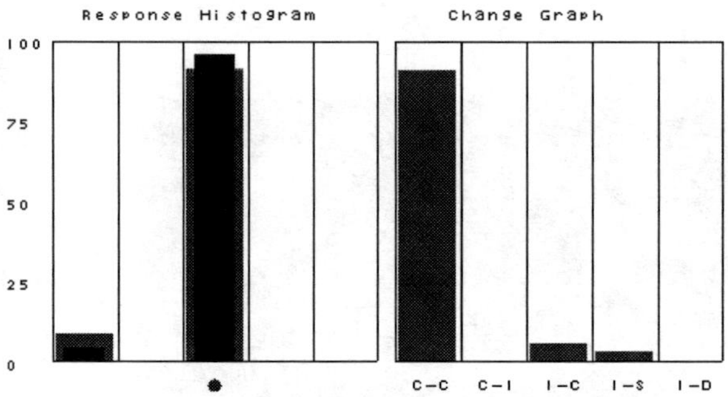

ConcepTest 2.13b: Graphing Velocity II
Difficulty Level: 2

In this question, the *x* vs. *t* plot has a changing slope, which requires students to explain the connection of the slope to the concept of velocity.

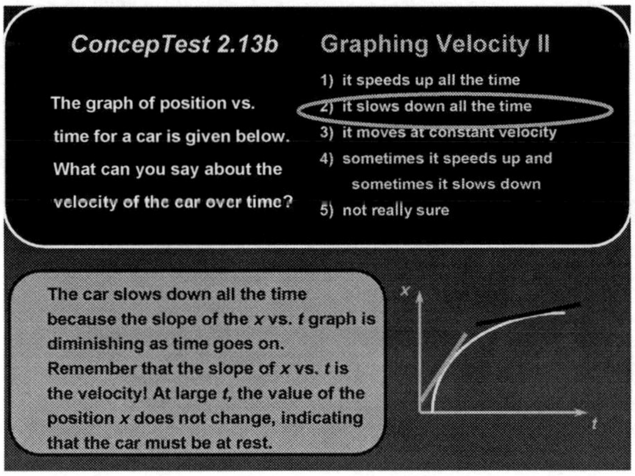

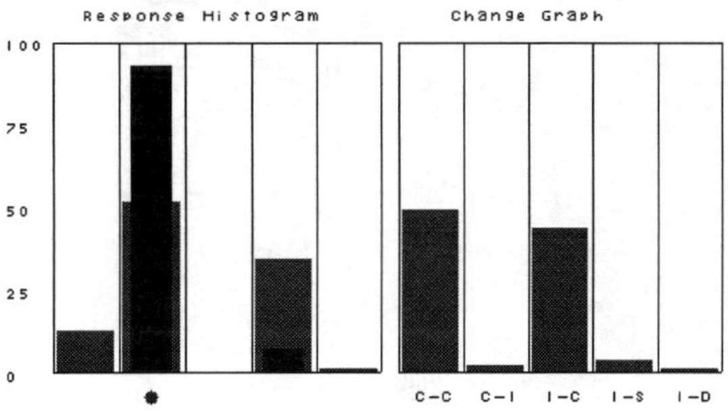

Conceptual Module 2.14 (two parts)

ConcepTest 2.14a: *v* versus *t* graphs I
Difficulty Level: 2

This question moves on to the understanding of the *v* vs. *t* graph. This question has nothing to do with the slope of the graph, but rather the actual velocity values that can be read off the *y*-axis (i.e., the velocity axis).

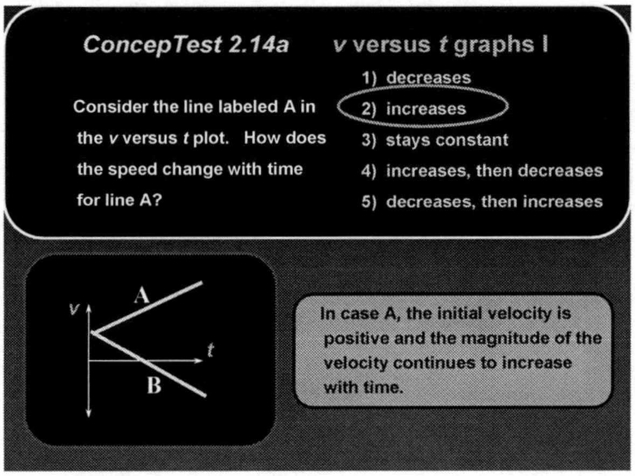

ConcepTest 2.14b: *v* versus *t* graphs II
Difficulty Level: 3

This question continues to examine the v vs. t graph. In this case, the velocity values decrease, pass through zero, and then become negative (but increasing in magnitude). Thus, the plot of line B actually describes a complicated motion, despite the simplicity of the actual plot.

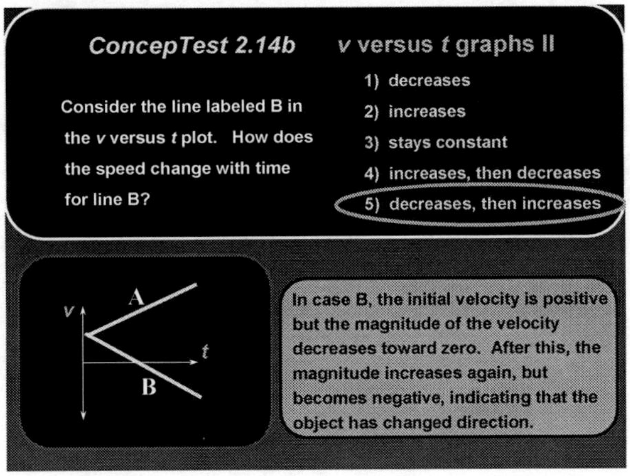

Conceptual Module 2.15 (three parts)

ConcepTest 2.15a: Rubber Balls I
Difficulty Level: 2

In this set of questions, the students must formulate their own view of the motion graphs, based on a description of the physical motion. In this case, the ball is simply dropped and falls down. The correct v vs. t plot has a negative slope that begins at $v = 0$.

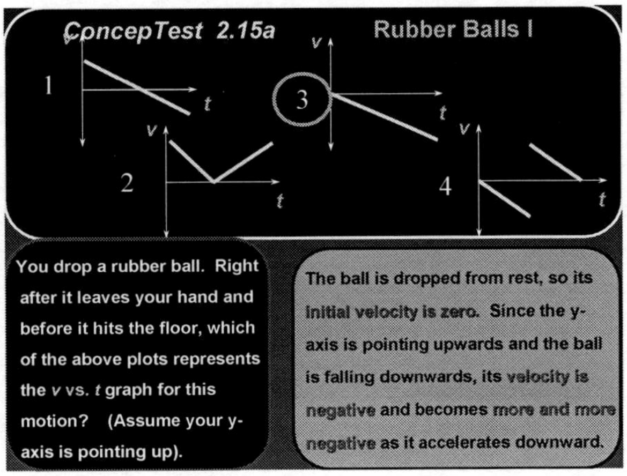

ConcepTest 2.15b: Rubber Balls II
Difficulty Level: 3

In this set of questions, the students must formulate their own view of the motion graphs based on a description of the physical motion. In this case, the ball is tossed up in the air. The correct v vs. t plot has a negative slope that begins at $v > 0$ and then crosses $v = 0$ and goes to negative values of v.

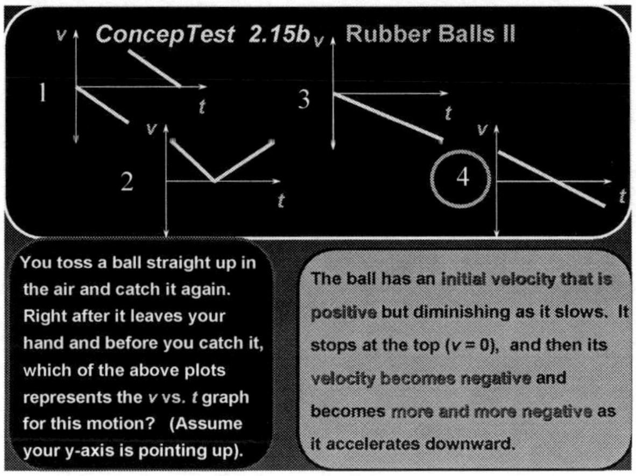

ConcepTest 2.15c: Rubber Balls III
Difficulty Level: 4

In this set of questions, the students must formulate their own view of the motion graphs, based on a description of the physical motion. In this case, the ball is dropped from rest, but bounces off the floor and rebounds up in the air. The bounce leads to a discontinuity in the v vs. t graph, but the acceleration is everywhere the same (except for the period during which the ball is in contact with the floor).

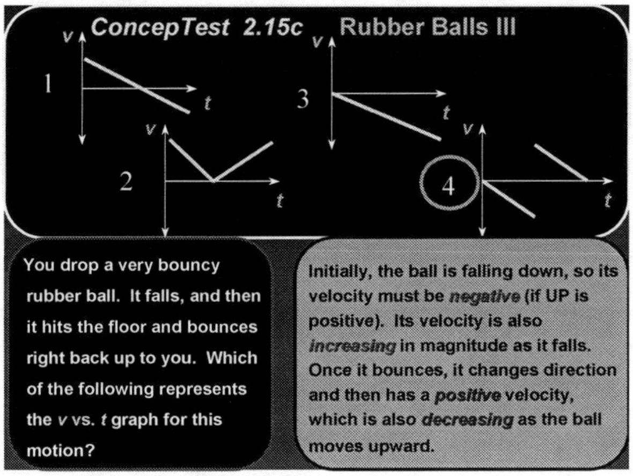

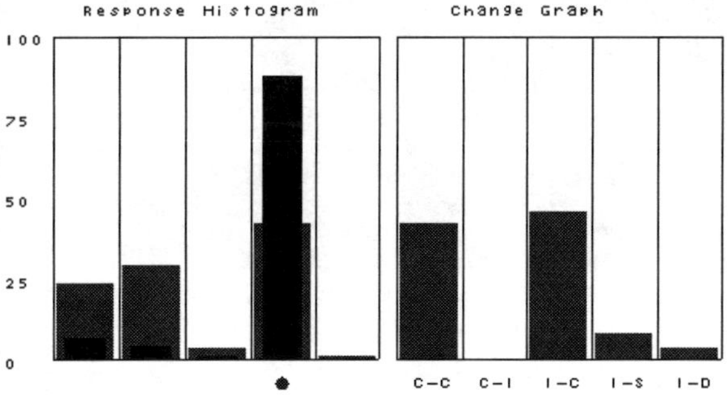

Conceptual Module 3.1 (three parts)

ConcepTest 3.1a: Vectors I
Difficulty Level: 2

This is a series of questions about basic characteristics of vectors. The questions are intended to give the students a feel for combining vectors. This first ConcepTest deals with two vectors that add up to zero.

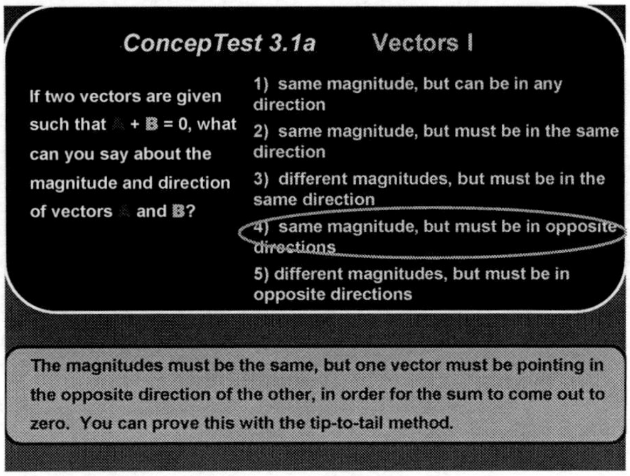

ConcepTest 3.1b: Vectors II
Difficulty Level: 3

In this case, students must recognize that when two vectors satisfy the Pythagorean theorem, then they must be perpendicular.

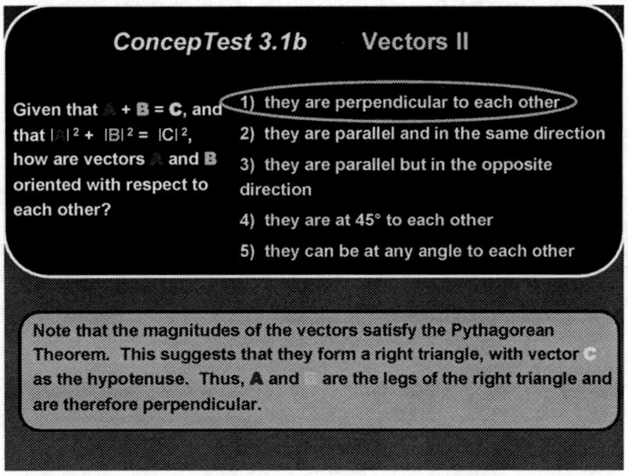

ConcepTest 3.1c: Vectors III
Difficulty Level: 3

In this case, students must recognize that when two vectors add to their maximum resultant, then they must be parallel.

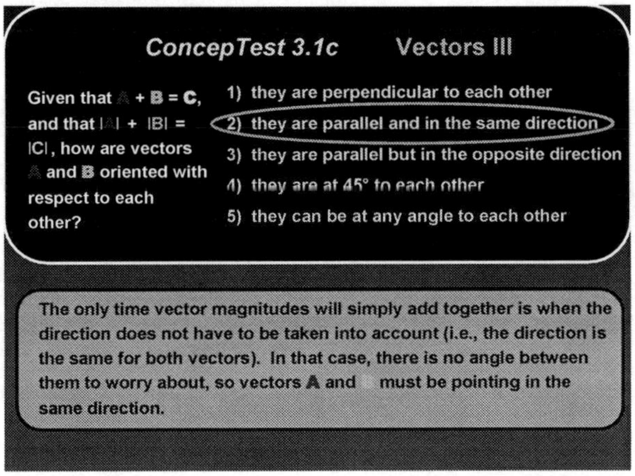

Conceptual Module 3.2 (two parts)

ConcepTest 3.2a: Vector Components I
Difficulty Level: 2
Suggested follow-up conceptual question:

- **If you double one component and not the other, how would the angle change?**

This series of questions deals with the components of a vector and how they relate to the angle of the vector. In this case, the components are multiplied by a factor, and the effect on the angle is sought.

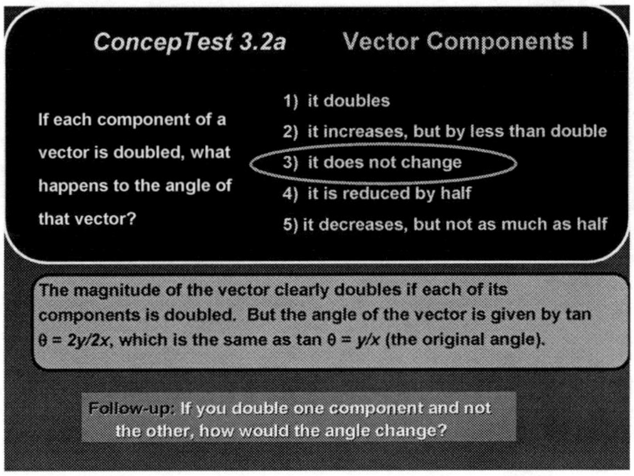

ConcepTest 3.2b: Vector Components II
Difficulty Level: 2

In this case, the angle of a vector with equal components is examined.

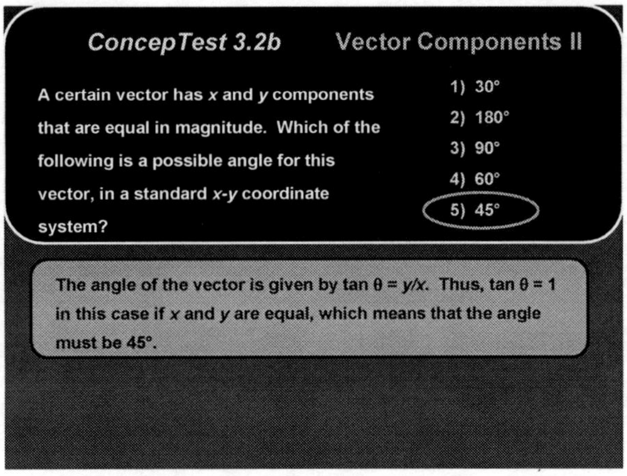

Conceptual Module 3.3

ConcepTest 3.3: Vector Addition
Difficulty Level: 3

Two vectors being added can have a range of resultants, but this range is well defined. This ConcepTest asks the students to identify the range of validity for the resultant of two vectors.

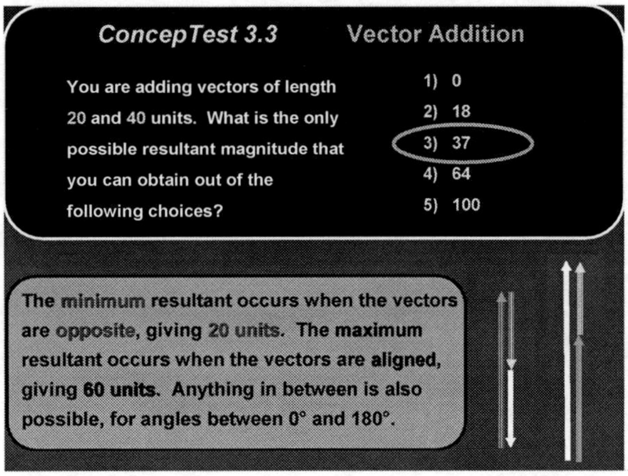

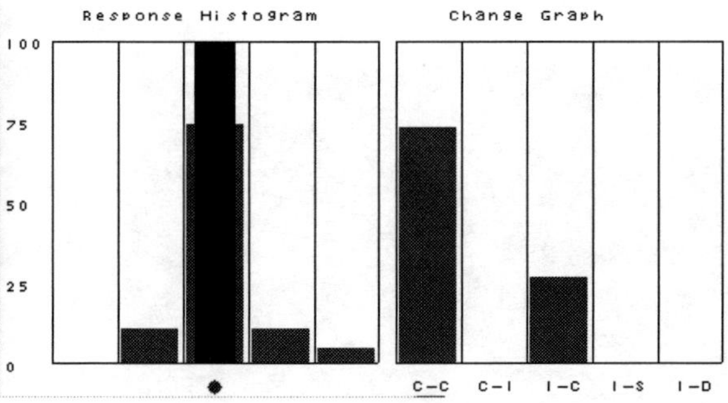

Conceptual Module 3.4 (three parts)

ConcepTest 3.4a: Firing Balls I
Difficulty Level: 3

Treating the x and y motions independently tends to be a problem for students. This series of ConcepTests explores this independence of the two motions. In this case, the cart is moving with a constant speed.

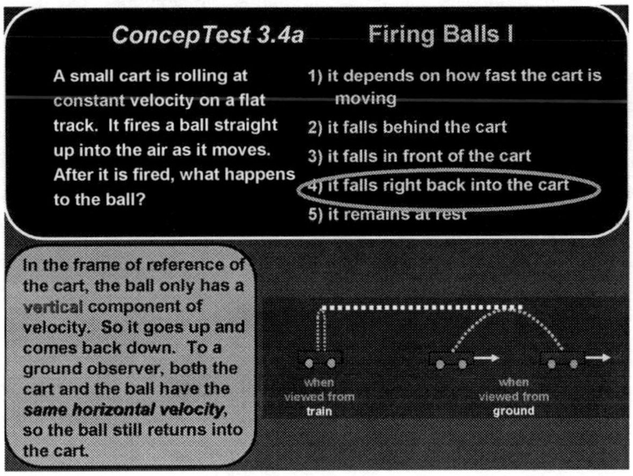

ConcepTest 3.4b: Firing Balls II
Difficulty Level: 4

In this variation of the preceding question, the cart is now being accelerated by an external force as it moves along the track.

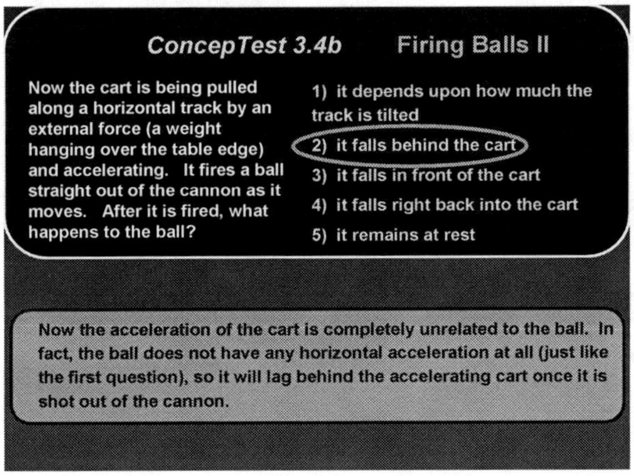

ConcepTest 3.4c: Firing Balls III
Difficulty Level: 5

In this final variation of the series, the cart is now accelerating down an inclined track, under the influence of gravity.

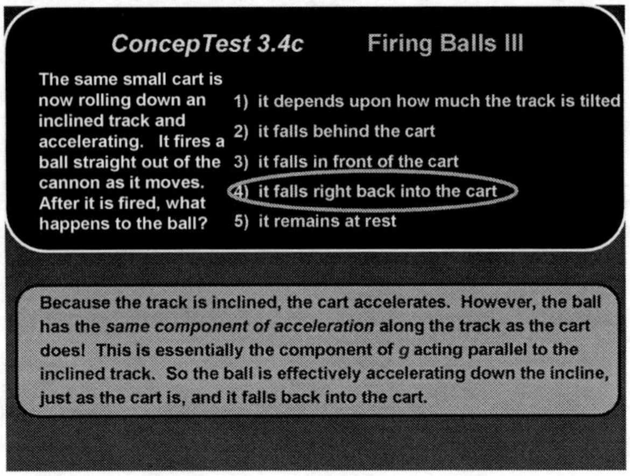

Conceptual Module 3.5

ConcepTest 3.5: Dropping a Package
Difficulty Level: 2
Suggested follow-up conceptual question:

- **What would happen if air resistance is present?**

This question again addresses the independence of the *x* and *y* motions. Since the horizontal motions of the package and the plane remain the same after the drop, the package will "pace" the plane and therefore remain directly under the plane as it falls.

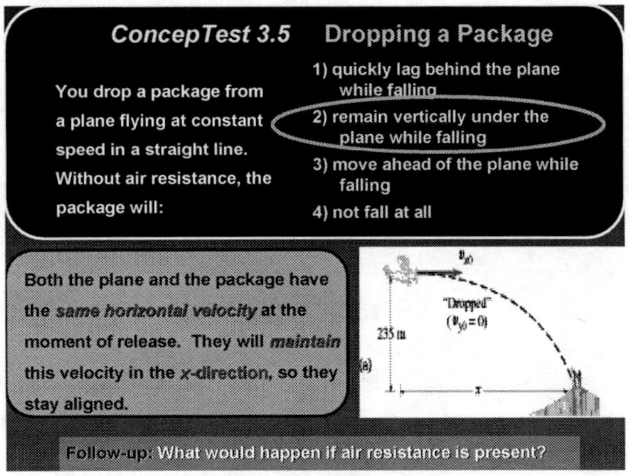

Conceptual Module 3.6 (three parts)

ConcepTest 3.6a: Dropping the Ball I
Difficulty Level: 3
Suggested follow-up conceptual question:

- **Is that also true if there is air resistance?**

This series of ConcepTests continues to examine the *x* and *y* motions. While they are independent, some quantities (such as the range) depend on both motions, while others (such as flight time) only depend on one. This question addresses the flight time.

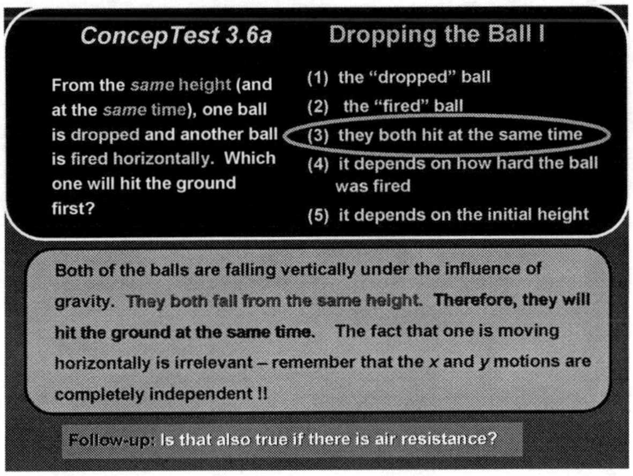

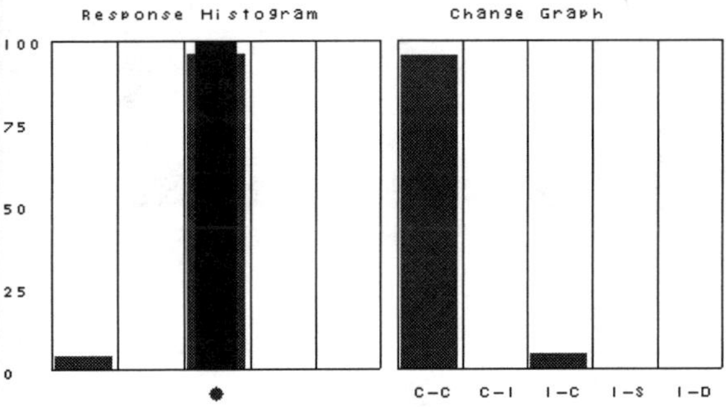

ConcepTest 3.6b: Dropping the Ball II
Difficulty Level: 3
Suggested follow-up conceptual question:

- **What would you have to do to have them both reach the same final velocity at the ground level?**

The second question in this series deals with the final velocity, which involves the combination of both the x and y motions (velocities).

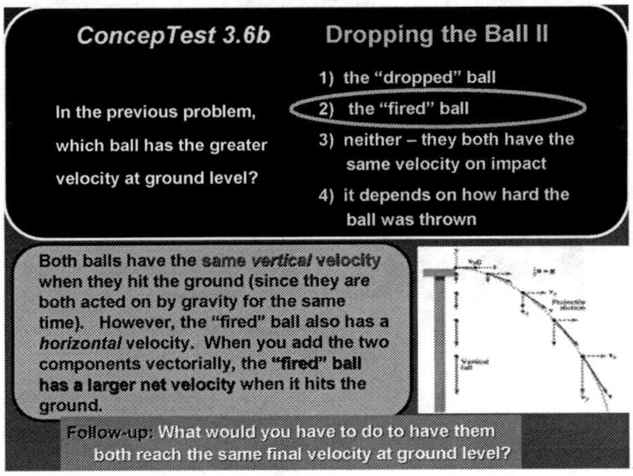

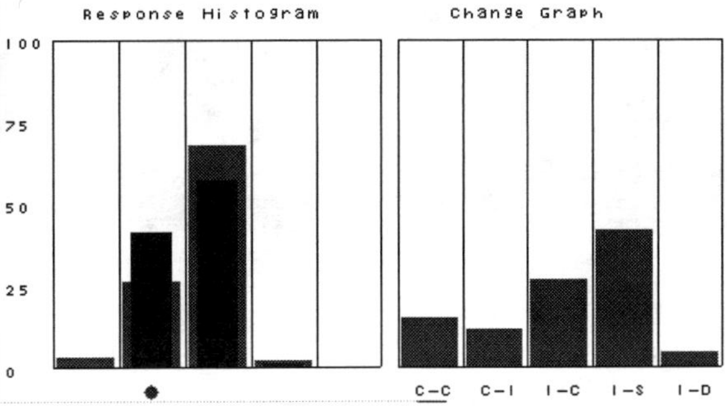

ConcepTest 3.6c: Dropping the Ball III
Difficulty Level: 3

The total speed of a projectile is changing because the *y* velocity is changing (while the *x* velocity remains constant). The total speed is the resultant of *x* and *y* velocities at any given moment.

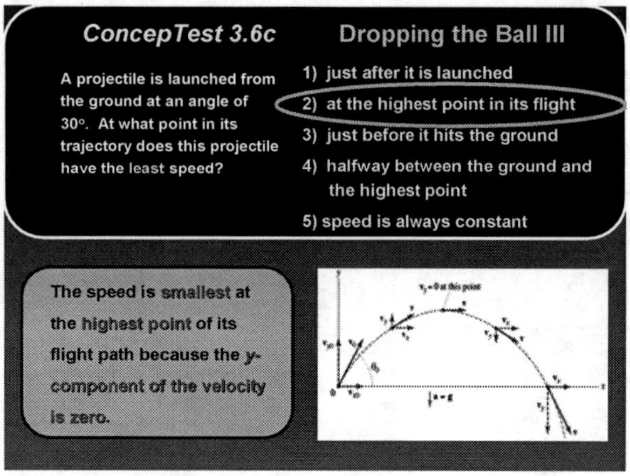

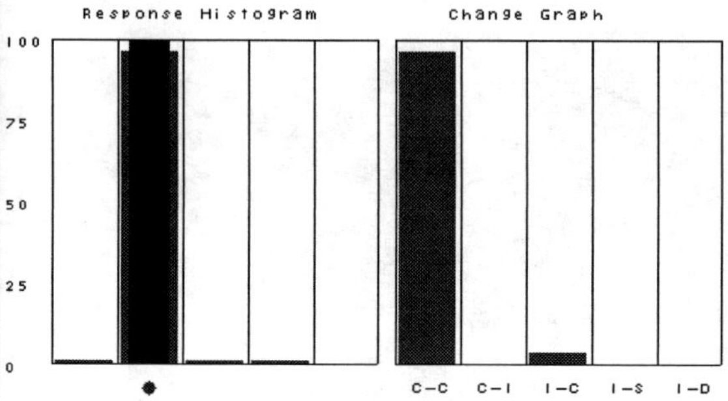

Conceptual Module 3.7 (two parts)

ConcepTest 3.7a: Punts I
Difficulty Level: 3
Suggested follow-up conceptual question:

- **Which one had the greater initial velocity?**

The time in the air is determined by the *y* motion, not the *x* motion. The students must recognize that these punts all have the same initial *y* velocity, and so they all spend the same time in the air.

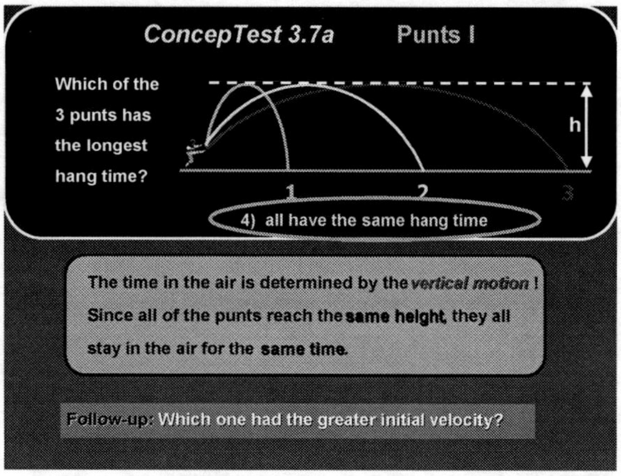

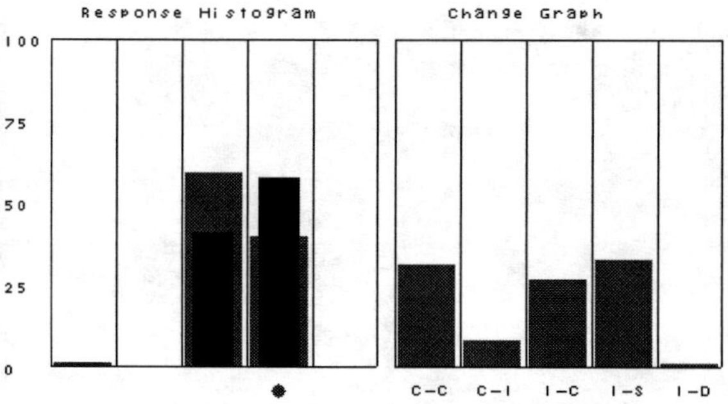

ConcepTest 3.7b: Punts II
Difficulty Level: 4
Suggested follow-up conceptual question:

- **Which one traveled the greater distance?**

Here again, the students must recognize that the flight time is determined by the *y* motion, which is independent of the *x* motion.

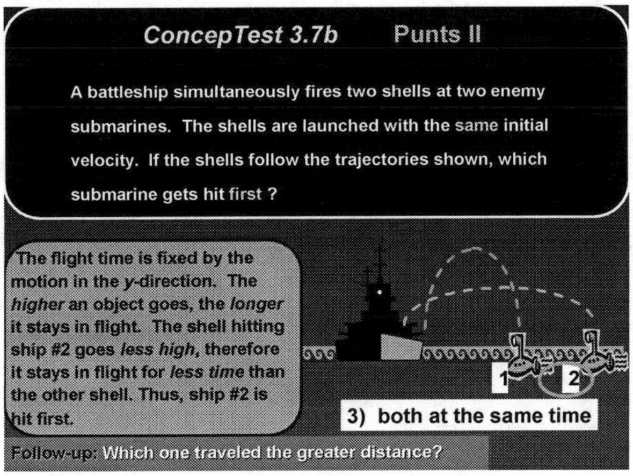

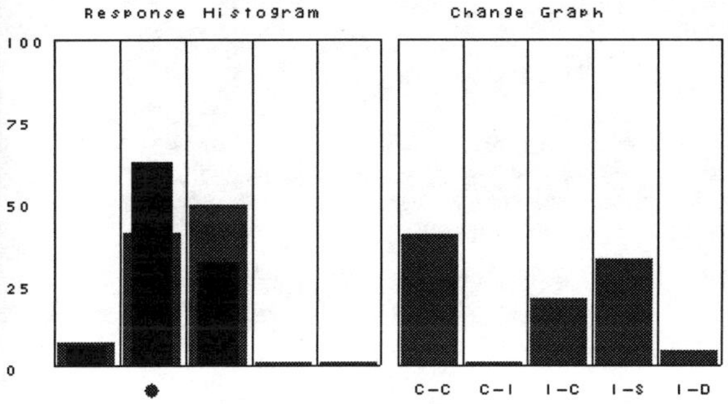

Conceptual Module 3.8

ConcepTest 3.8: Cannon on the Moon
Difficulty Level: 3
Suggested follow-up conceptual question:

- **Which path would it take in outer space?**

In this case, the acceleration due to gravity has changed, and the students must realize what effect that will have on the subsequent motion. This variation cannot be related directly to an earthbound experience, so it requires a certain level of understanding.

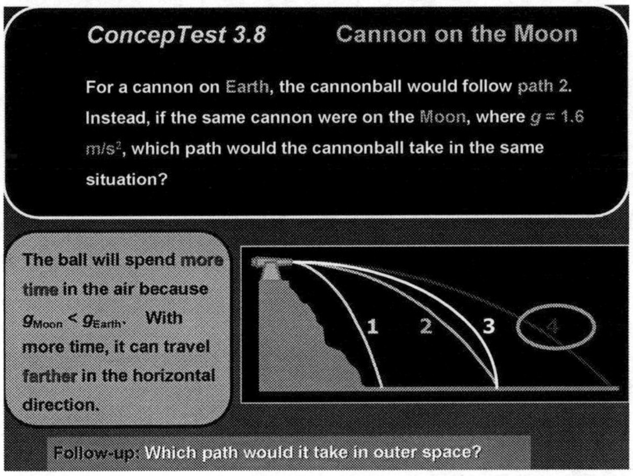

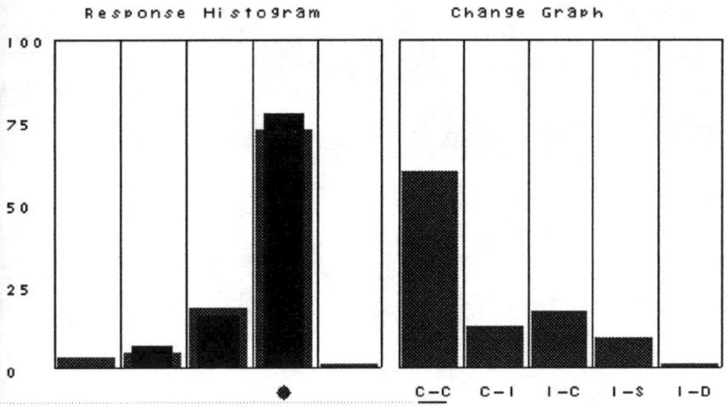

Conceptual Module 3.9
ConcepTest 3.9: Spring-Loaded Gun
Difficulty Level: 1

The question of maximum range is fairly standard. In this case, however, the students may wish to derive the range formula to prove that 45° is the optimal angle.

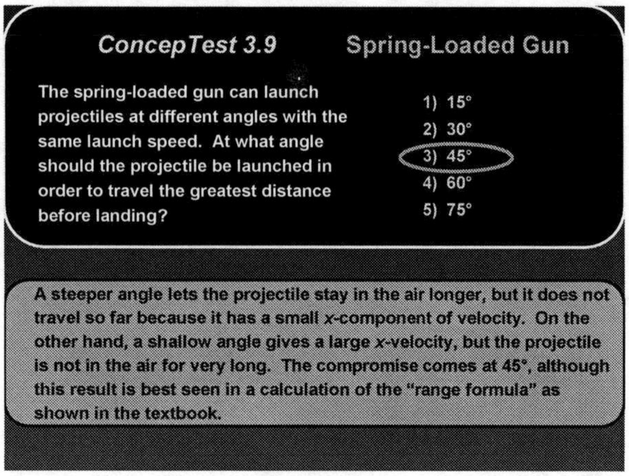

Conceptual Module 3.10 (three parts)
ConcepTest 3.10a: Shoot the Monkey I
Difficulty Level: 3

This series deals with variations of the "monkey hunter" problem. The first question deals with the relatively simple case of the shooter and the monkey on the same level.

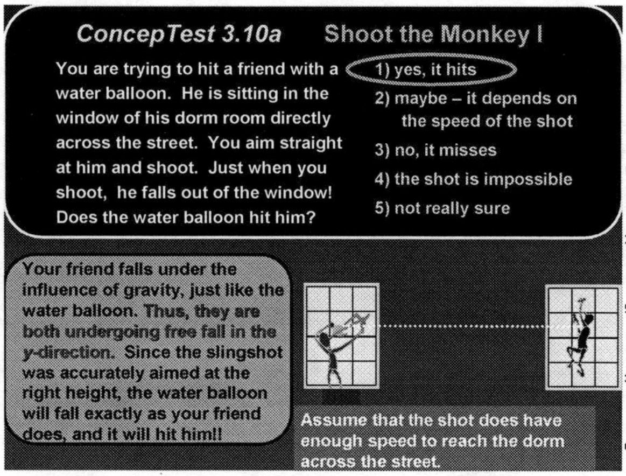

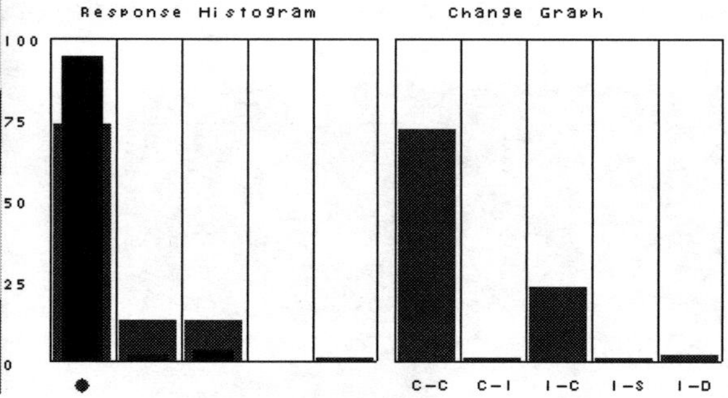

ConcepTest 3.10b: Shoot the Monkey II
Difficulty Level: 4

This is the standard formulation of the "monkey hunter" problem, in which the shooter is on the ground and the monkey is located at some elevation. It is hoped that the students will be able to extrapolate their experience from the first question to this one.

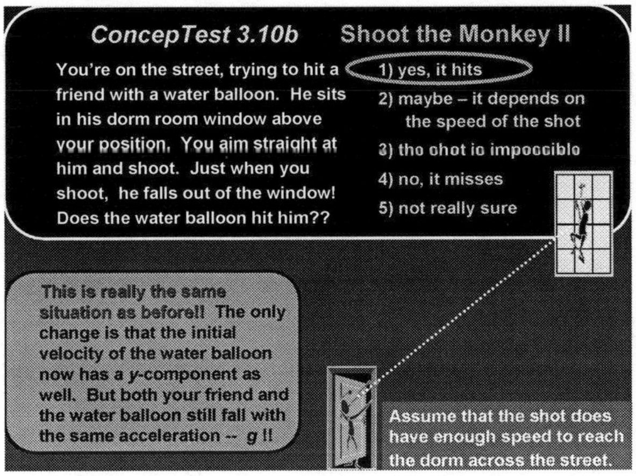

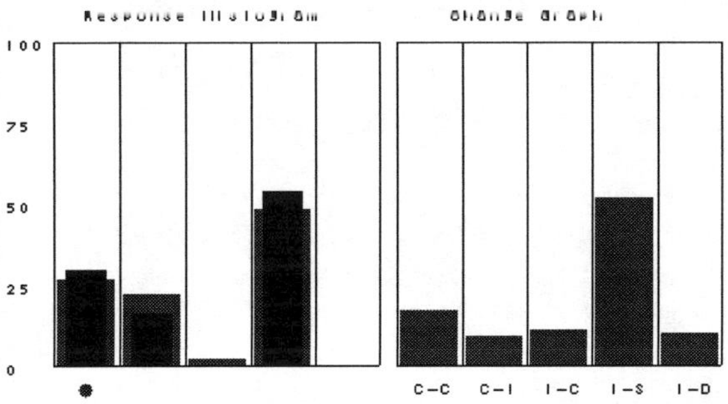

ConcepTest 3.10c: Shoot the Monkey III
Difficulty Level: 4
Suggested follow-up conceptual question:

- **When would they not hit each other?**

In this case, two projectiles initially aimed at each other are fired. The fact that both projectiles are only under the influence of gravity means that they will indeed hit. This can be more easily seen for the case of projectiles that are launched at the same level.

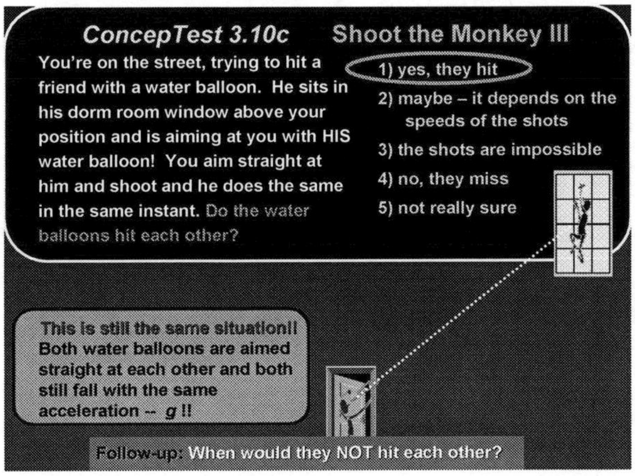

Conceptual Module 4.1 (four parts)

ConcepTest 4.1a: Newton's First Law I
Difficulty Level: 2

This series of questions examines Newton's First Law. In the first question, students must realize that just because an object is at rest, it does not mean that there are no forces acting on it at all (*i.e.*, no net force does not necessarily mean no force at all).

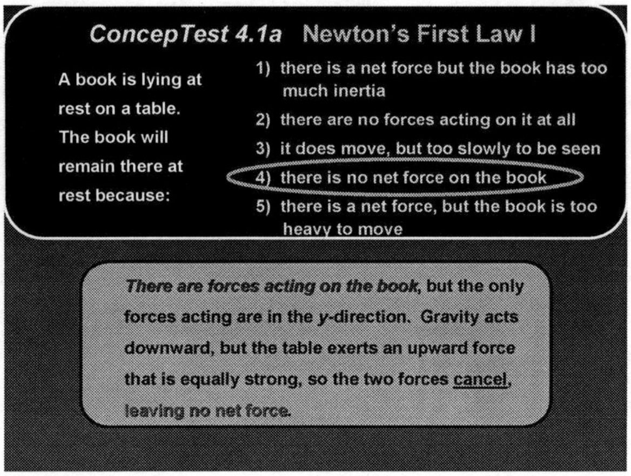

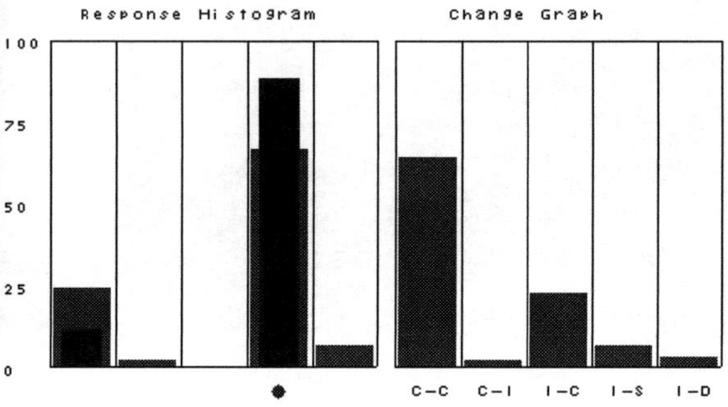

ConcepTest 4.1b: Newton's First Law II
Difficulty Level: 2
Suggested follow-up conceptual questions:

- **Are there any forces acting on the puck? What are they?**

A moving object does not need a force to keep it in motion. This misconception is addressed in this case, where an object moves at constant velocity.

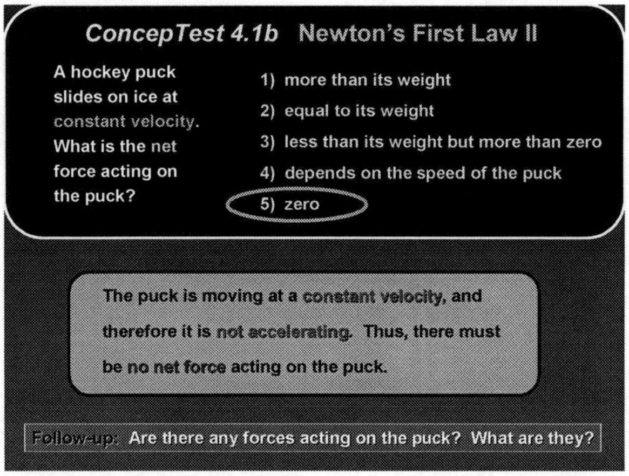

ConcepTest 4.1c: Newton's First Law III
Difficulty Level: 3
Suggested follow-up conceptual question:

- **What is the force that usually keeps the book on the seat?**

This question probes the students' ability to recognize that the book is simply following its natural tendency to remain in motion (unless acted upon by an outside force, which in this case, it is not). There can be some confusion possible due to different reference frames, since the students are actually in a non-inertial frame.

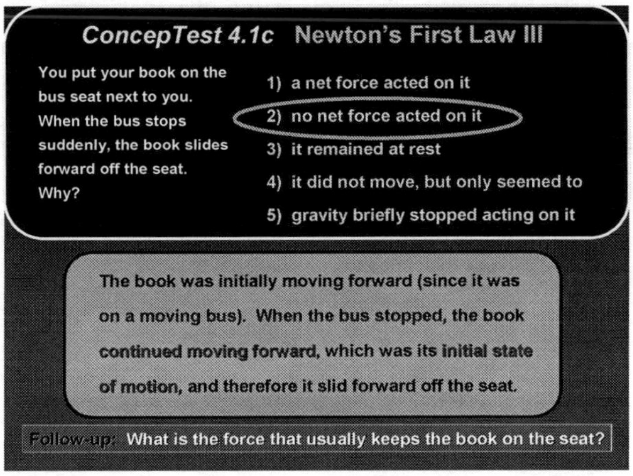

ConcepTest 4.1d: Newton's First Law IV
Difficulty Level: 3
Suggested follow-up conceptual question:

- **What would you have to do to keep the stone moving?**

Concluding with Newton's First Law, a common situation is presented in which an object slows and eventually comes to rest. This apparent contradiction to the First Law arises due to the presence of an external force acting on the object (in this case, friction).

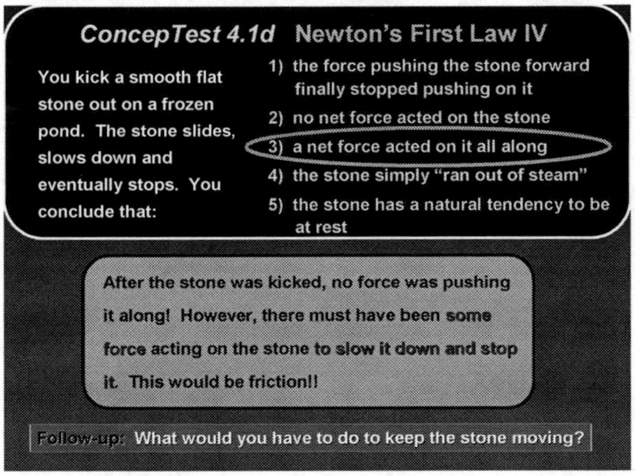

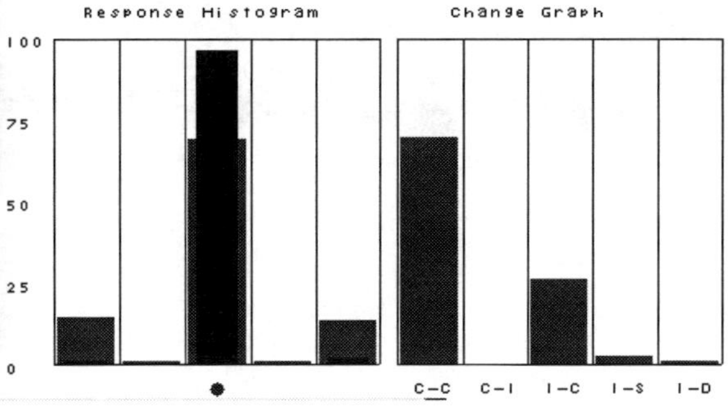

Conceptual Module 4.2 (two parts)
ConcepTest 4.2a: Cart on Track I
Difficulty Level: 2

This question serves as a reminder of Newton's First Law. It is important for students to recognize that the question is dealing with the motion *after* the push, not during the push. This distinction becomes important in the next question.

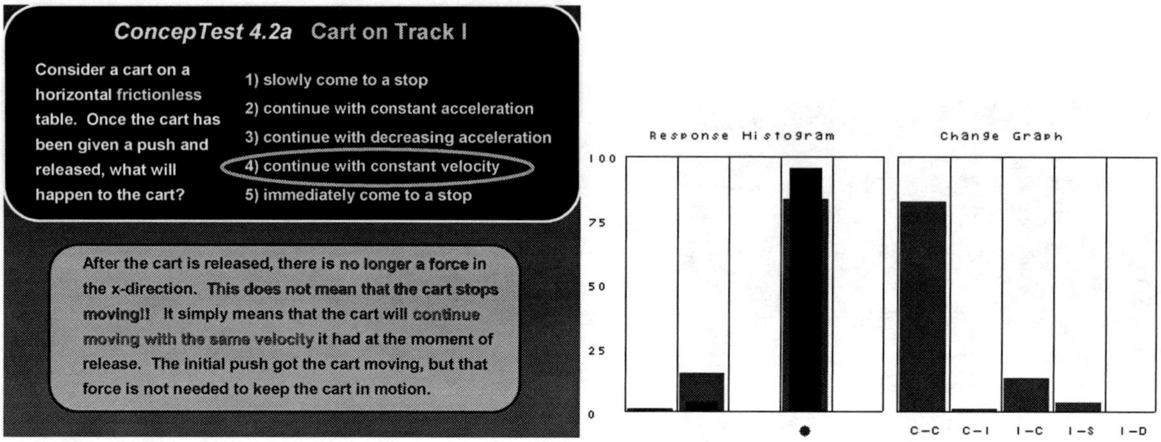

ConcepTest 4.2b: Cart on Track II
Difficulty Level: 2

Getting into Newton's Second Law now, the students should realize that in order to have a constant acceleration, a constant force must be *continuously* applied during the motion.

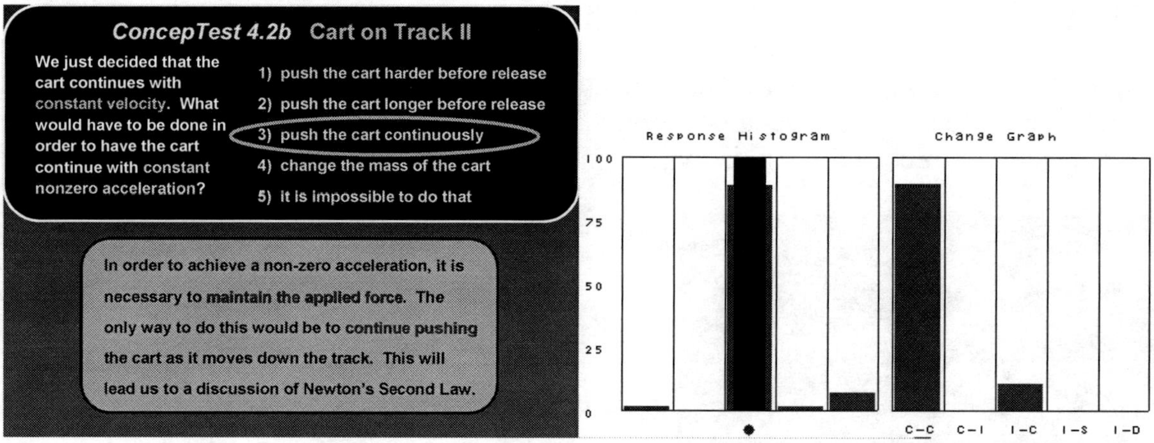

Conceptual Module 4.3

ConcepTest 4.3: **Truck on Frozen Lake**
Difficulty Level: **3**
Suggested follow-up conceptual question:

- **What is the truck doing 5 minutes after the fly hit it?**

In this case, an impulsive force gives the truck a small but finite velocity, which it will maintain until something else happens. After the collision, the truck will continue moving with that same velocity, even though the fly is no longer in the picture.

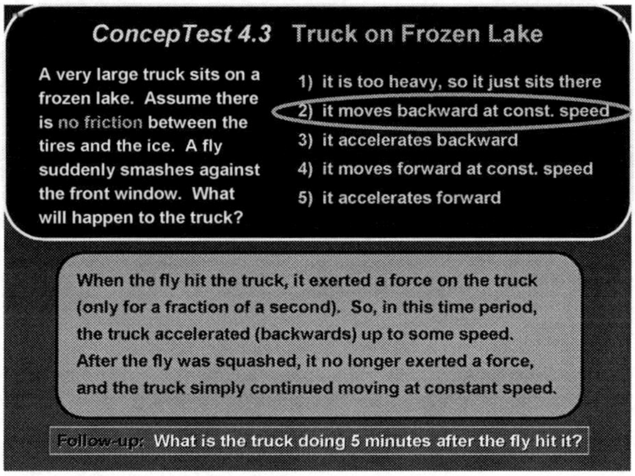

Conceptual Module 4.4 (four parts)

ConcepTest 4.4a: **Off to the Races I**
Difficulty Level: **3**

Kinematics plays a role in the following series of ConcepTests. In this first case, students must determine the acceleration (from the force) and then use it to obtain the time required to achieve the same final speed as before.

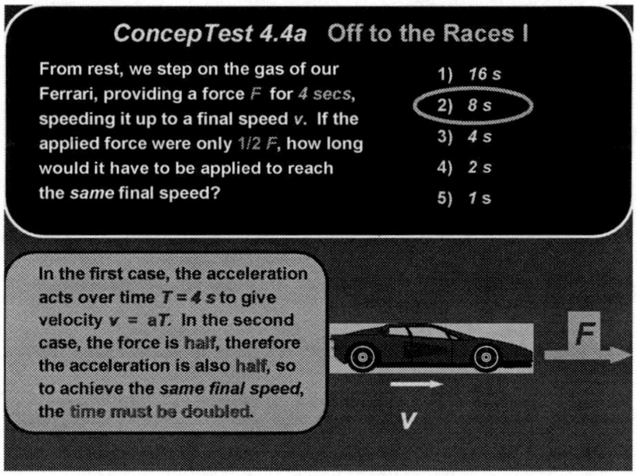

ConcepTest 4.4b: Off to the Races II
Difficulty Level: 3

In this case, students must determine the distance traveled when the same force is applied over a different time interval than before.

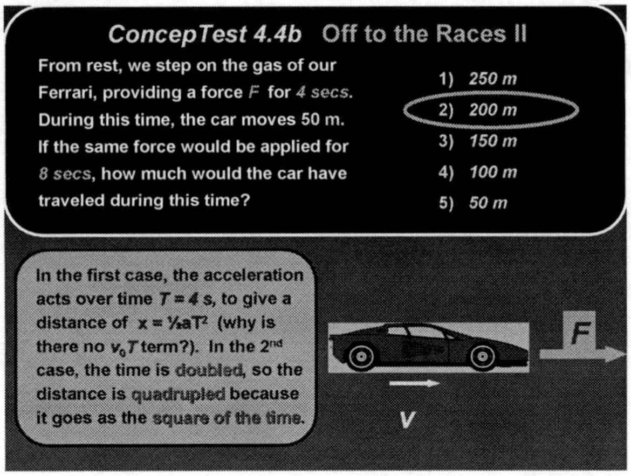

ConcepTest 4.4c: Off to the Races III
Difficulty Level: 4

This question is the "reverse" of the previous one, in the sense that in this case, the car is slowing down due to the applied force (rather than speeding up). The answer to this question is not nearly as intuitive as the previous one.

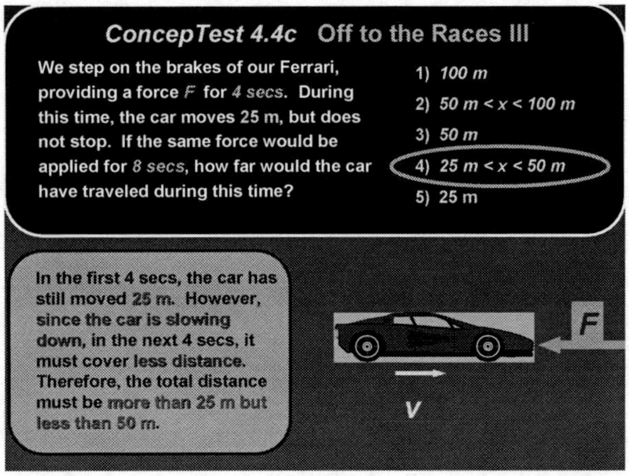

ConcepTest 4.4d: Off to the Races IV
Difficulty Level: 4

Here the students examine a different relationship. In this case, the same force is applied over a different distance and the question asks for the new final speed. In all of these questions, the students must be aware of the kinematics relations that govern the motion.

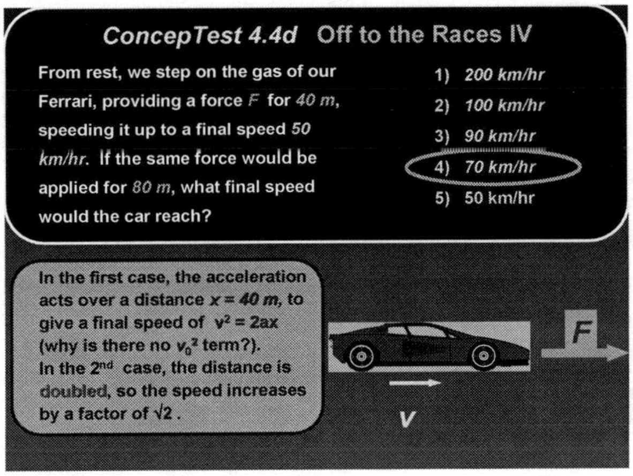

Conceptual Module 4.5

ConcepTest 4.5: Force and Mass
Difficulty Level: 3
Suggested follow-up conceptual question:

- **What would you have to do to get 2M to reach speed *v*?**

This is another exercise in linking the concepts of force and acceleration (from Newton's Second Law) with the kinematics equations that are used to calculate the motion of an object.

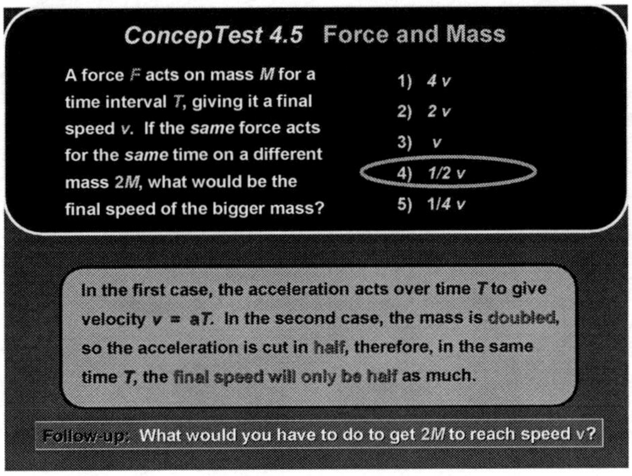

Conceptual Module 4.6

ConcepTest 4.6: Force and Two Masses
Difficulty Level: 4

This is an applied example of Newton's Second Law. Students must learn to break the problem down into steps, using what they are given to determine what they need for a subsequent part of the problem.

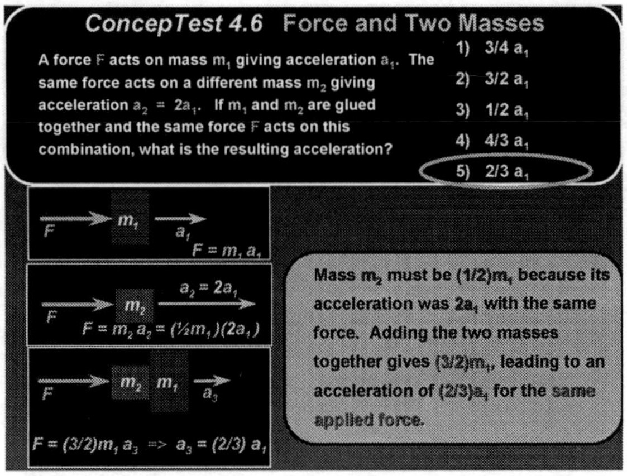

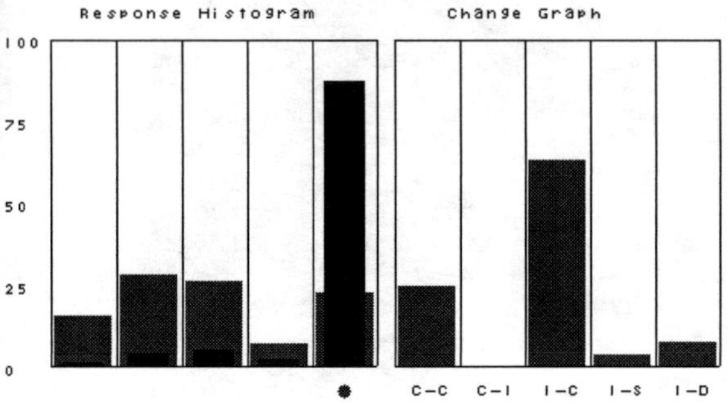

Conceptual Module 4.7 (two parts)

ConcepTest 4.7a: Gravity and Weight I
Difficulty Level: 1

Issues relating to gravity can often confuse students. While heavy and light objects do indeed have different forces of gravity on them (weight), the acceleration of these objects is in fact the same! This first question establishes that the students recognize the force (weight) difference between two objects.

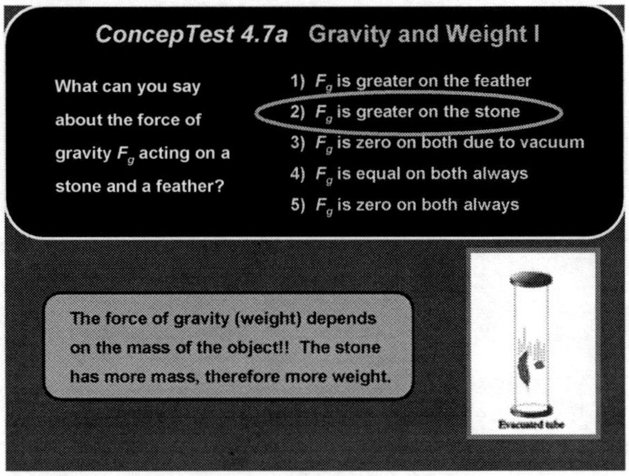

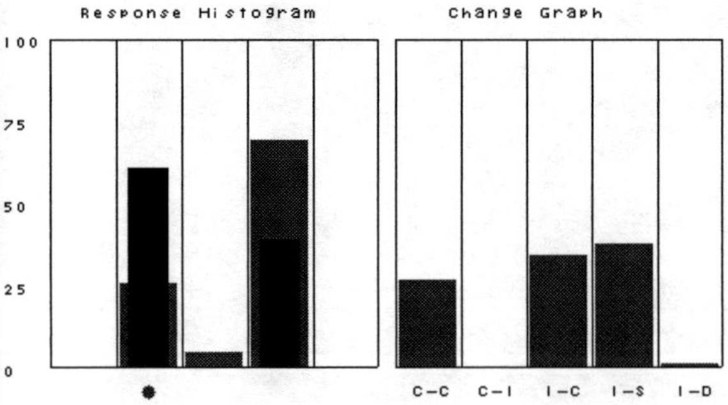

ConcepTest 4.7b: Gravity and Weight II
Difficulty Level: 2
Suggested follow-up conceptual question:

- **Which one hits the bottom first?**

In this second question, however, the acceleration is asked for. The students must realize that they must divide out the mass ($F=ma$) to obtain the acceleration, and that leaves them with the same value (g) for each of the objects.

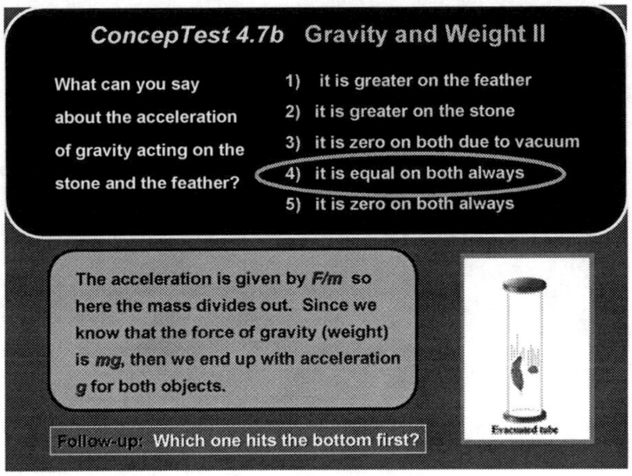

Conceptual Module 4.8

ConcepTest 4.8: On the Moon
Difficulty Level: 2
Suggested follow-up conceptual question:

- **What is different about the bowling ball on the Moon?**

This is a classic question that is often misunderstood. The extent to which a body resists acceleration is characterized by the *mass*, not the weight. It is this issue that is relevant in this question, and so the fact that the bowling ball is lighter (weighs less) on the Moon does not alter the answer.

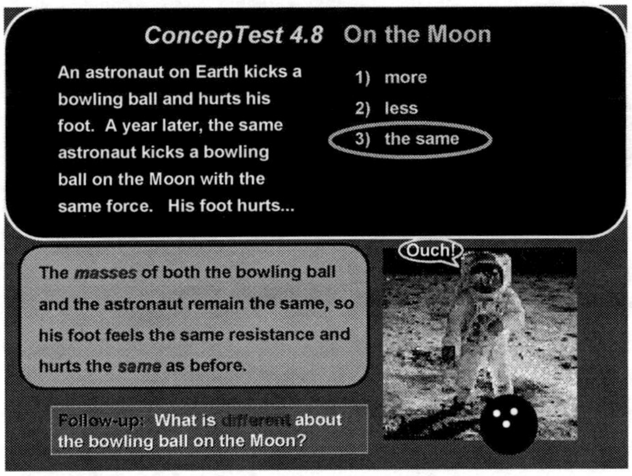

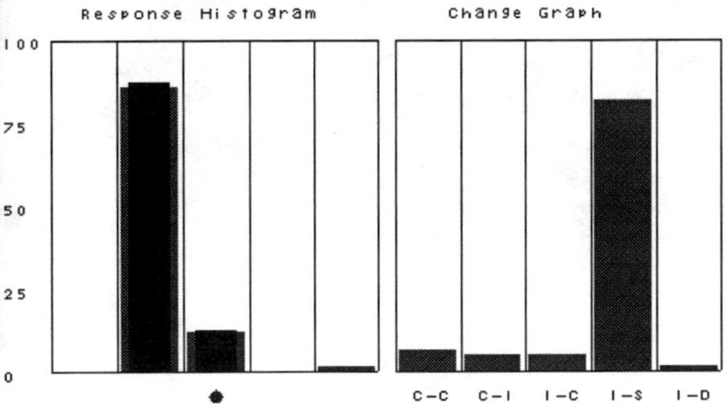

Conceptual Module 4.9 (two parts)

ConcepTest 4.9a: Going Up I
Difficulty Level: 2

If the elevator were at rest, this would be a trivial question ($N = mg$). But in this case, the elevator is moving at constant speed. Students must recognize that this changes nothing from the "at rest" case, so the normal force will still simply be the weight of the block.

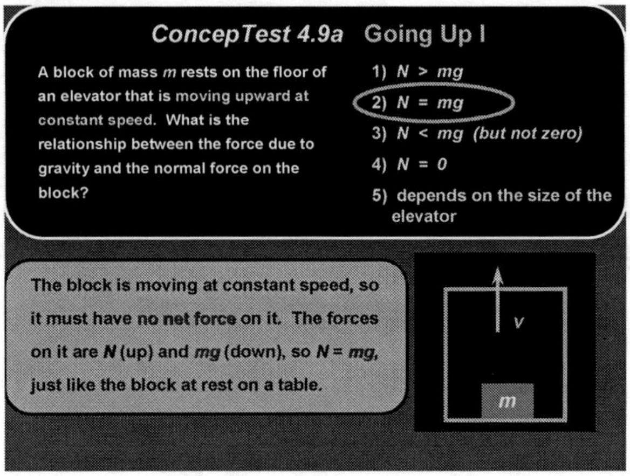

ConcepTest 4.9b: Going Up II
Difficulty Level: 3
Suggested follow-up conceptual question:

- **What is the normal force if the elevator is in free fall downward?**

In this follow-up, the elevator is accelerating upward, so there must be a net upward force. So in this case, the normal force must exceed the weight, for a net upward force to be exerted on the block.

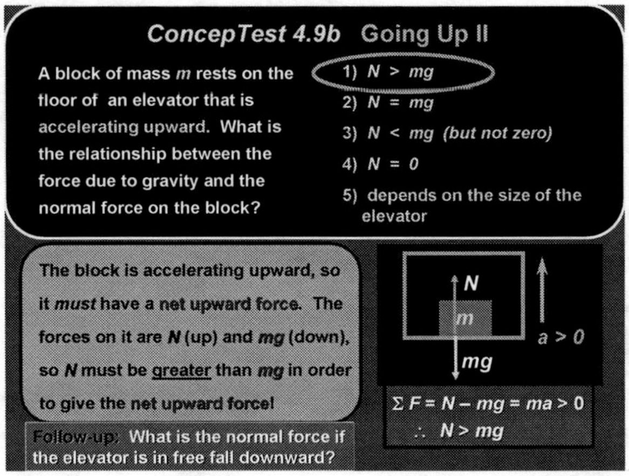

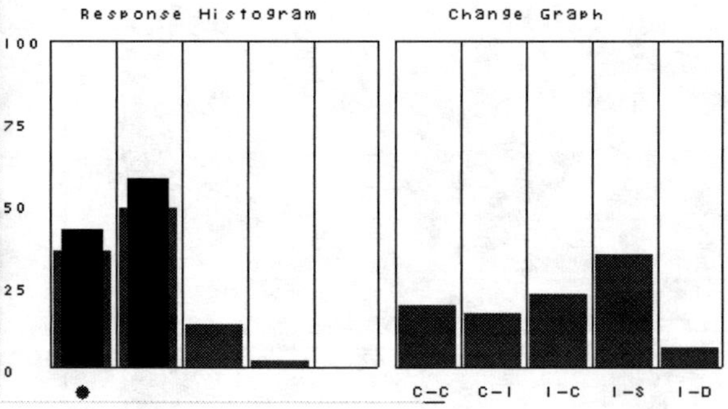

Conceptual Module 4.10
ConcepTest 4.10: Normal Force
Difficulty Level: 2

The normal force is often confused with the weight of the object, but the normal force must be determined by considering all forces applied to the object. In this case, the person pushing on the sled is providing an additional vertical component that will affect the normal force on the sled. This will later be revisited in the case of friction.

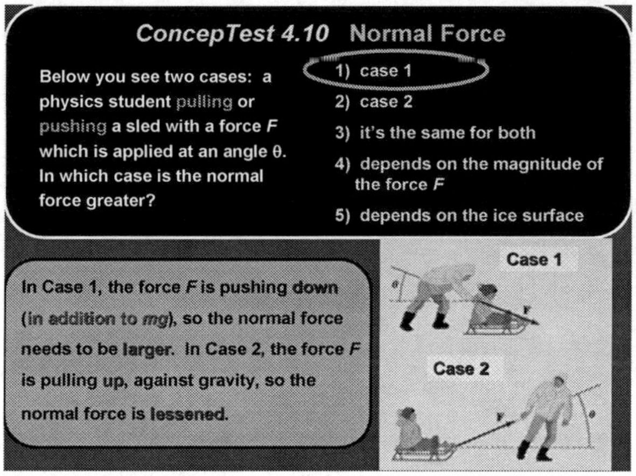

Conceptual Module 4.11
ConcepTest 4.11: On an Incline
Difficulty Level: 4

It is important for the students to get used to the fact that the normal force is perpendicular to a surface, and is not necessarily in a vertical direction. Here the normal force in case B is equal to the component of the weight perpendicular to the plane (which is not the same as the weight itself!).

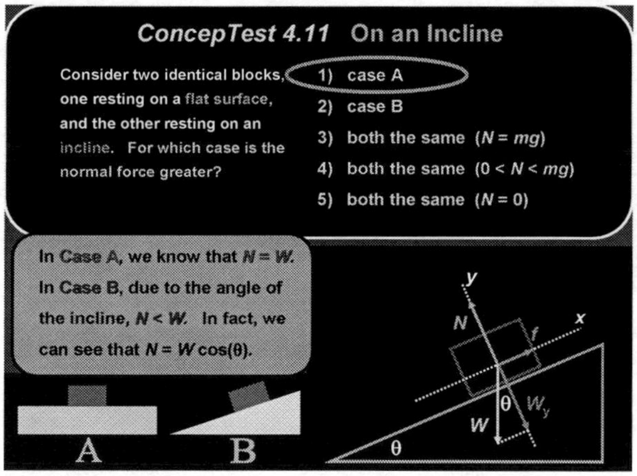

Conceptual Module 4.12

ConcepTest 4.12: Climbing the Rope
Difficulty Level: 2

This is an introduction to Newton's Third Law. The notion that your pulling down on the rope somehow helps you ascend may seem a bit strange when students first think about it in those terms, but this is the essence of Newton's Third Law.

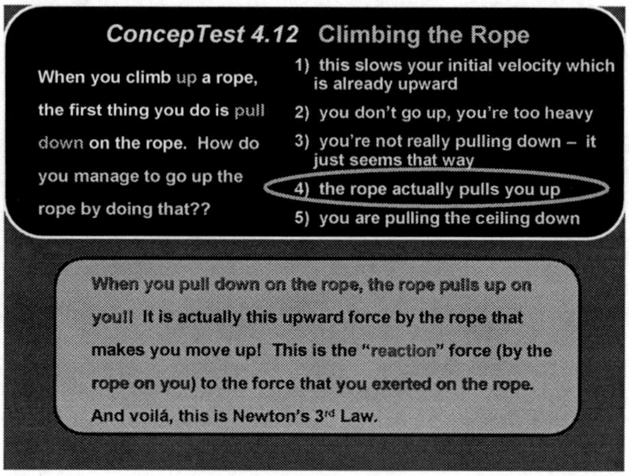

Conceptual Module 4.13 (two parts)

ConcepTest 4.13a: Bowling vs. Ping-Pong I
Difficulty Level: 2

This is a quick check of the level of understanding of Newton's Third Law. When the two objects in question are so different in size and mass, the students often have trouble convincing themselves that the forces are indeed still equal and opposite.

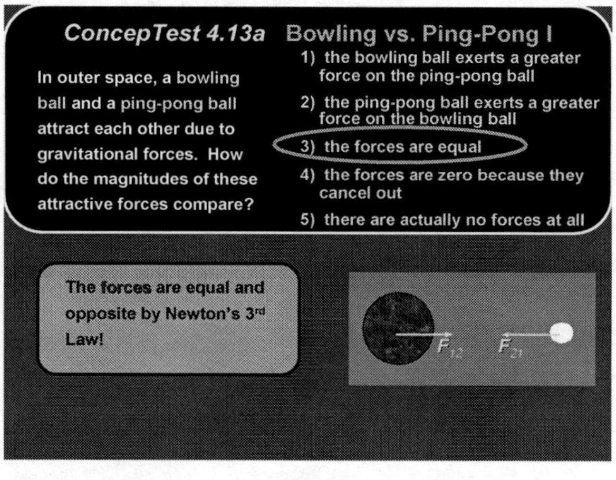

ConcepTest 4.13b: Bowling vs. Ping-Pong II
Difficulty Level: 3
Suggested follow-up conceptual question:

- **Where will the balls meet if they are released from this position?**

Following up on the previous question, now the impact of the different masses is revealed. While the forces are equal and opposite by Newton's Third Law, the accelerations are not equal, and are inversely proportional to the masses. This is a more intuitive result that the students are likely to be more comfortable with, and it emphasizes where the effect of the different masses comes in.

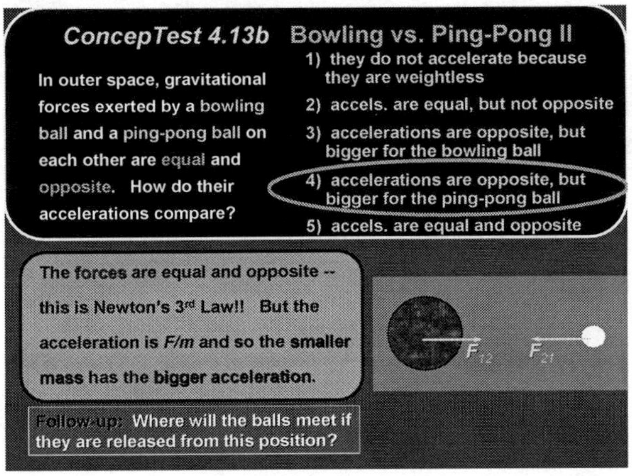

Conceptual Module 4.14 (two parts)

ConcepTest 4.14a: Collision Course I
Difficulty Level: 2

This question is similar to ConcepTest 4.13a, where two rather different objects are seen to exert the same force on each other, due to Newton's Third Law.

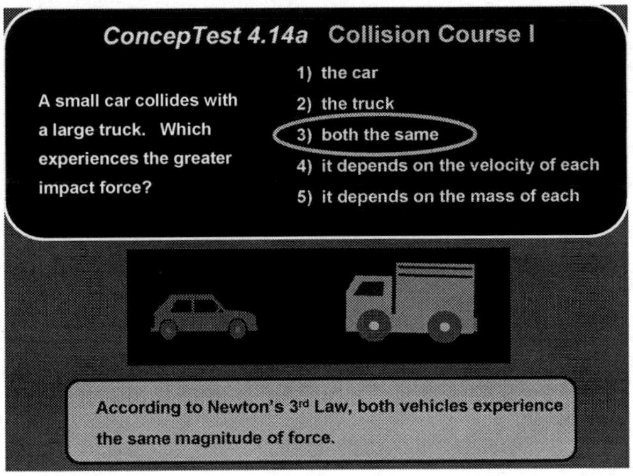

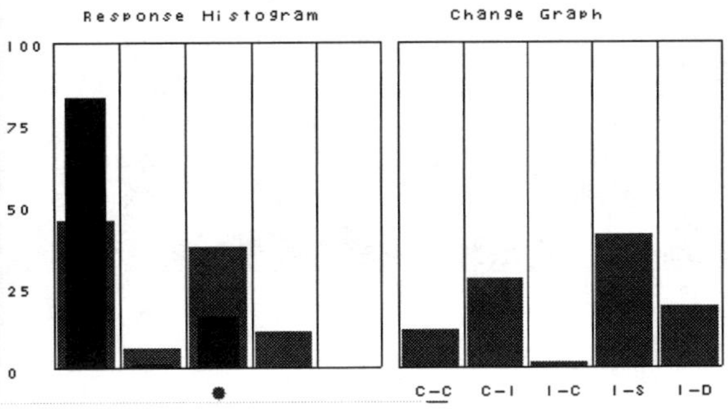

ConcepTest 4.14b: Collision Course II
Difficulty Level: 3

This question is similar to ConcepTest 4.13b, where the accelerations between the two different objects are shown to be different, but the forces were the same. It is here (acceleration) that the mass enters.

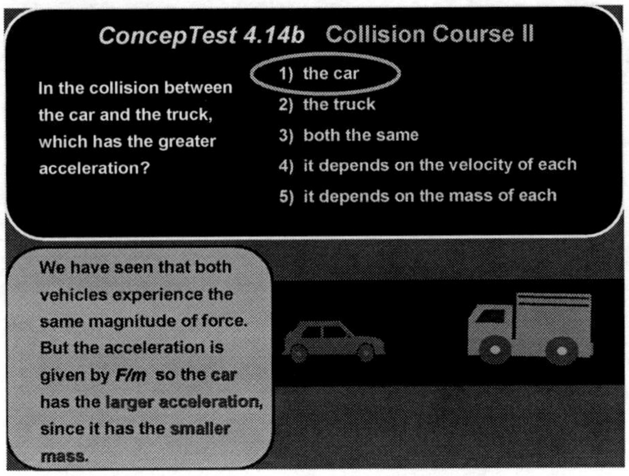

Conceptual Module 4.15 (two parts)

ConcepTest 4.15a: Contact Force I
Difficulty Level: 4

In this case, the problem must be broken down into parts in order to be able to address the question of the contact force acting on one of the blocks. Using the acceleration obtained by treating the combined system, the contact force can be determined from the individual masses and the acceleration.

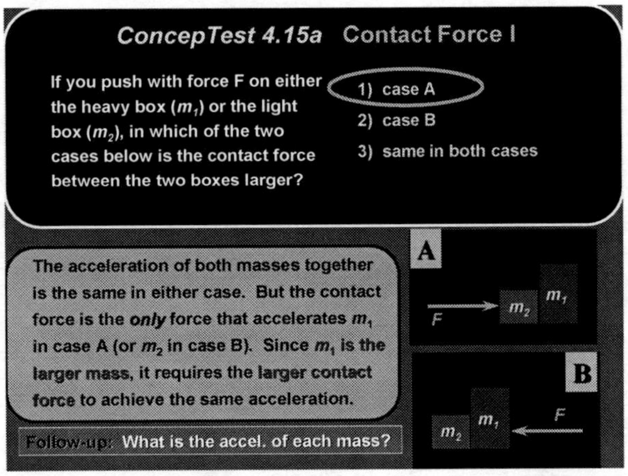

ConcepTest 4.15b: Contact Force II
Difficulty Level: 5
Suggested follow-up conceptual question:

- **What is the acceleration of each mass?**

This is another question about contact forces.

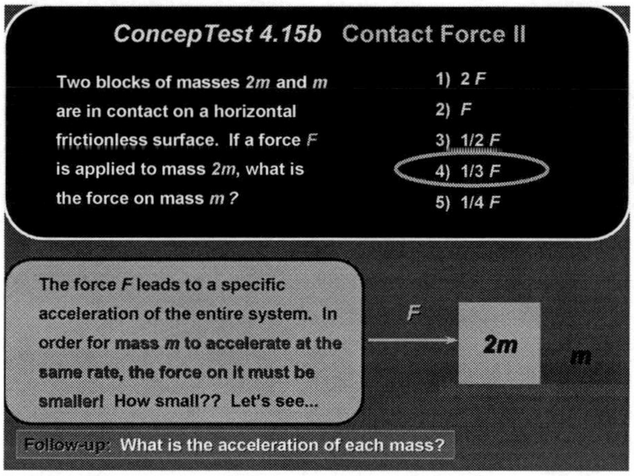

Conceptual Module 4.16 (three parts)

ConcepTest 4.16a: Tension I
Difficulty Level: 2

The tension in the rope is the force that any segment of the rope experiences on it at each end. The force applied at one end of the entire rope is "transmitted" down the length of the rope, segment by segment (by Newton's Third Law), so this is the force that each segment feels.

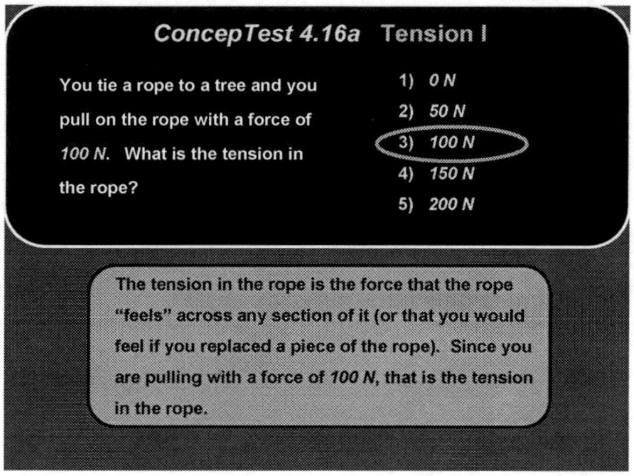

ConcepTest 4.16b: Tension II
Difficulty Level: 2

There is a tendency for students to believe that the tension is twice the force applied by each team. However, this is not the case at all. This situation is exactly the same as the previous question!

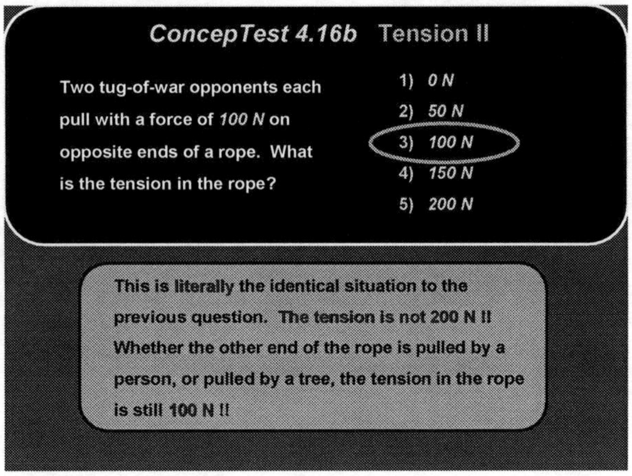

ConcepTest 4.16c: Tension III
Difficulty Level: 3

Using the experience gained by the previous two questions, the students should now be able to recognize that if they use a tree to provide a "passive" force to balance their own applied force, then they can achieve better results by pulling on the same end.

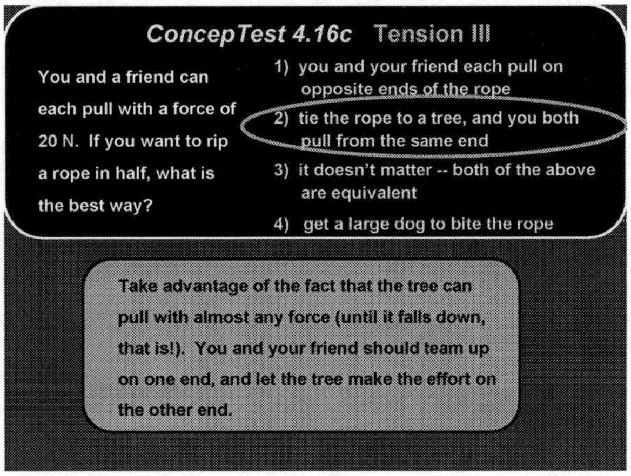

Conceptual Module 4.17

ConcepTest 4.17: Three Blocks
Difficulty Level: 4
Suggested follow-up conceptual question:

- **What is T_1 in terms of m and a?**

This problem involves a basic knowledge of Newton's Second Law. The acceleration of each block is the same, since the combination of blocks moves together. Each string effectively is pulling all of the blocks to its left, so the largest tension would be for the case of string #1, and so on.

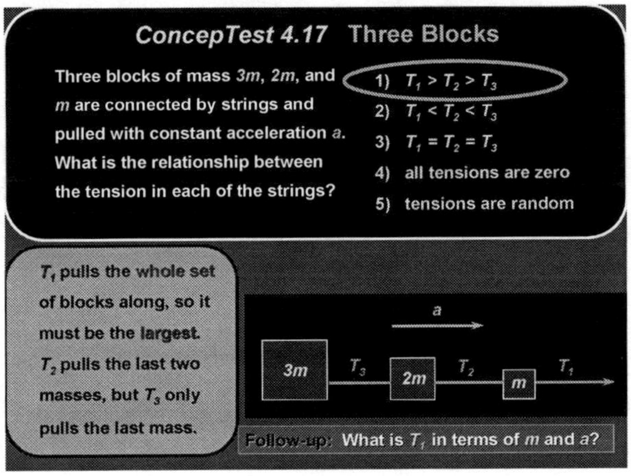

Conceptual Module 4.18

ConcepTest 4.18: Over the Edge
Difficulty Level: 4

This question requires the students to determine the tension in the rope. In case #2, it is simply the applied force, but in case #1, it is not *mg* of the falling block! Since the block is accelerating downward, the tension must be *less* than *mg* in order to have a net downward force. This distinction is not so subtle, but the students must pay careful attention to realize this.

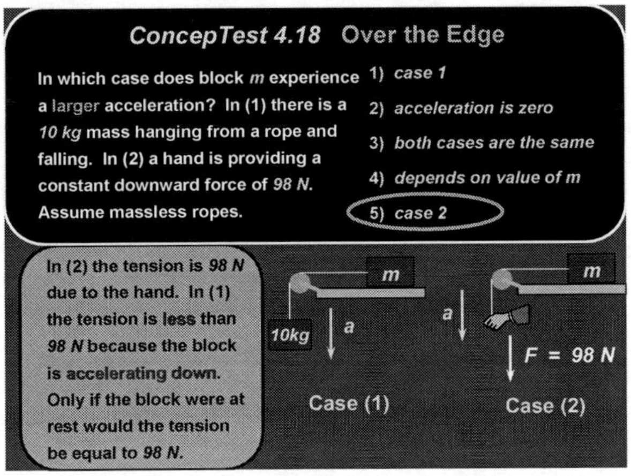

Conceptual Module 4.19

ConcepTest 4.19: Friction
Difficulty Level: 2

This question revisits Newton's First Law. The box is not "pushed" off the back of the truck, it has simply stayed in one place while the truck moved away. If the box were to move with the truck, then a net force would be needed, and that would most likely be supplied by friction.

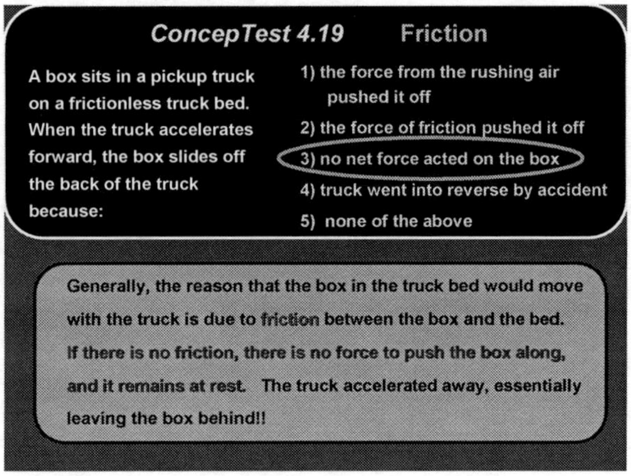

Conceptual Module 4.20

ConcepTest 4.20: Antilock Brakes
Difficulty Level: 2

Static friction can in principle exert larger forces than kinetic friction, due to the larger value of the coefficient of static friction. When the tires are making good contact with the road (not skidding), they are experiencing static friction. Students must appreciate that this is therefore better for braking, which is the logic behind antilock brakes (as well as maintaining control of the car).

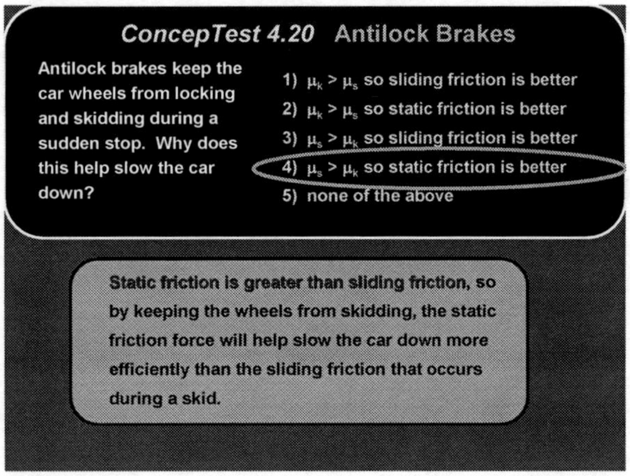

Conceptual Module 4.21

ConcepTest 4.21: Going Sledding
Difficulty Level: 3

This question relates to the same issue as ConcepTest 4.10, which dealt with the normal force. Since the frictional force is proportional to the normal force, the situation below that is easiest is the one with the smallest normal force.

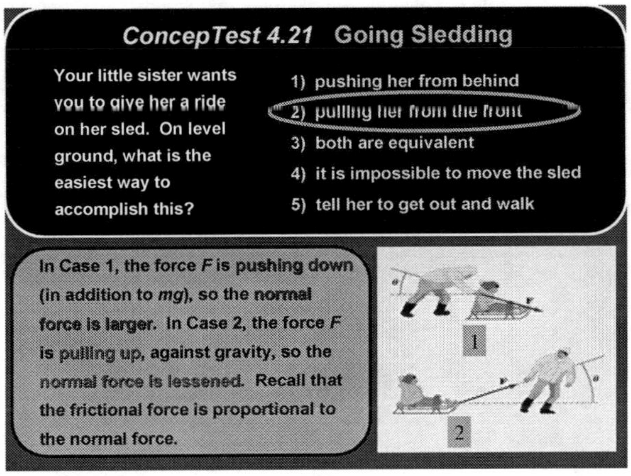

Conceptual Module 4.22

ConcepTest 4.22: Will It Budge?
Difficulty Level: 3
Suggested follow-up conceptual questions:

- **What happens if the tension is 35 N? What about 45 N?**

This is an exercise in determining the force of static friction and realizing that such a force can vary up to a maximum value (beyond which the object will slip). In this case, the maximum static friction force is 40 N, and so any force less than or equal to that value will not move the block.

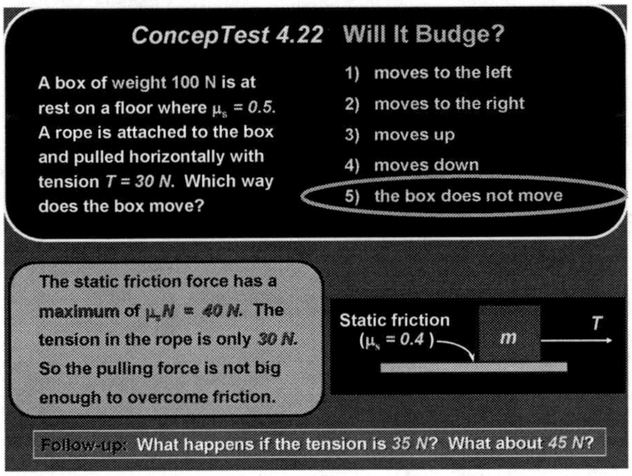

Conceptual Module 4.23 (two parts)

ConcepTest 4.23a: Sliding Down I
Difficulty Level: 4

This question addresses the concept of friction and, more generally, of the forces acting on a body. The motion of the box is determined by the forces acting parallel to the plane, which include a friction force and a component of gravity. Both of these change when the angle varies. For the friction force, it is the changing normal force that causes the variation.

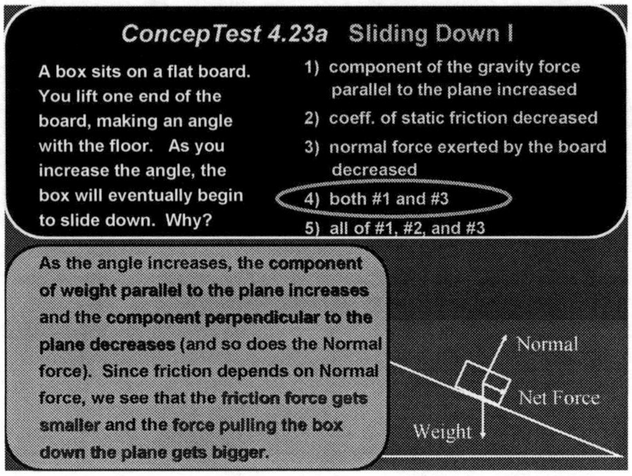

ConcepTest 4.23b: Sliding Down II
Difficulty Level: 5

In this case, the angle is not changed, but the mass on the plane is changed. By considering the same forces parallel to the plane as were examined in the previous question, students must come to the conclusion that both parallel forces change by the same amount in this case, so nothing is different.

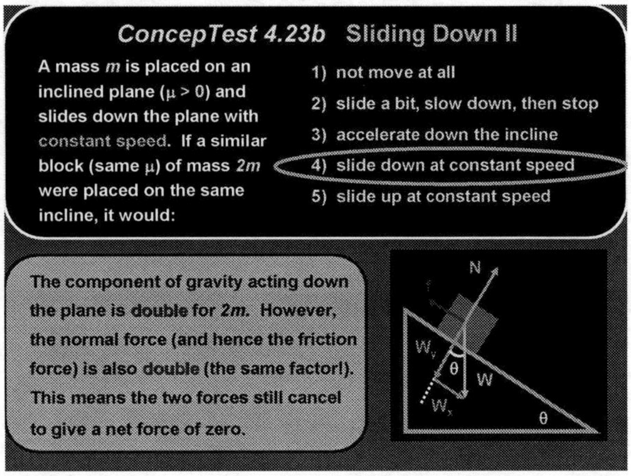

Conceptual Module 5.1

ConcepTest 5.1: Tetherball
Difficulty Level: 3

For uniform circular motion, there must be an inward force (centripetal force) that keeps the ball in motion. In this case, that inward force is supplied by the horizontal component of the tension. The other forces (vertical component of tension and the weight) are balanced.

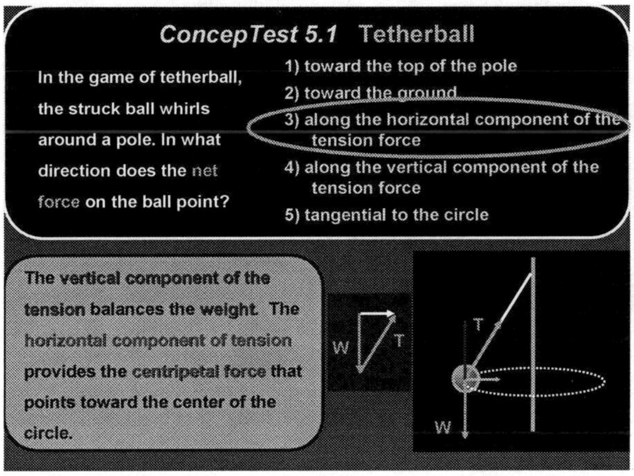

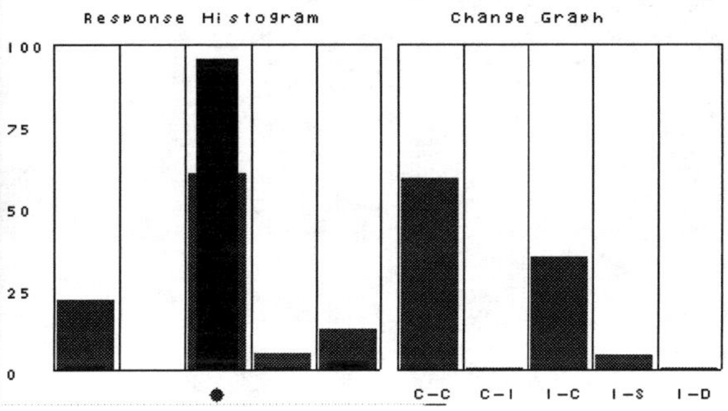

Conceptual Module 5.2 (three parts)

ConcepTest 5.2a: Around the Curve I
Difficulty Level: 2

This is an experience that most students will have had – that of being "thrown" into the car door when a car executes a sharp turn. Students must realize that they are not actually being thrown, but rather they are simply trying to continue moving in a straight-line path.

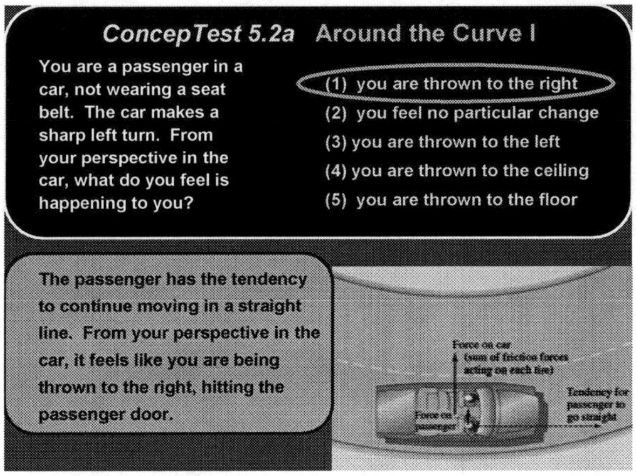

ConcepTest 5.2b: Around the Curve II
Difficulty Level: 2

This follow-up asks the students to interpret what is actually happening to them during the sharp turn in the car. They must realize that without the leftward force exerted by the door, they would continue moving in a straight line and therefore fall out of the car when it turns.

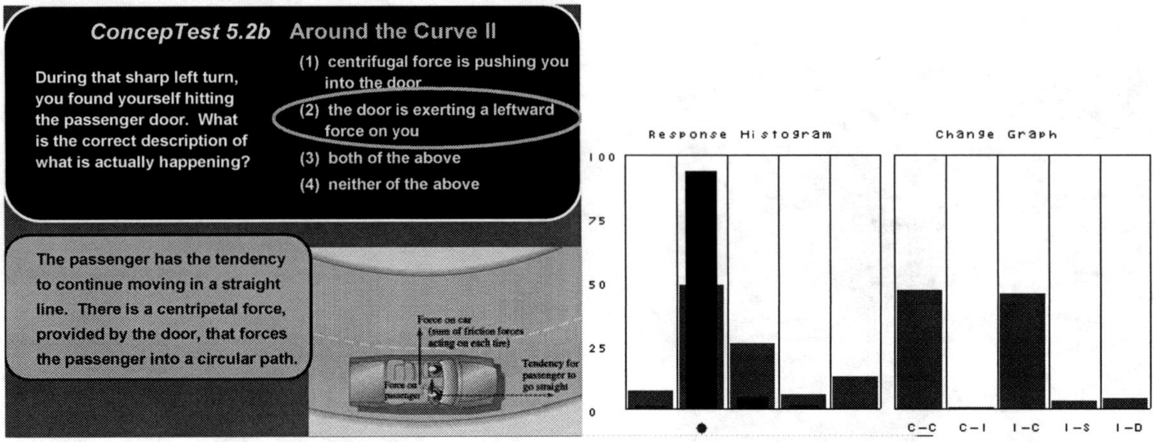

ConcepTest 5.2c: Around the Curve III
Difficulty Level: 3
Suggested follow-up conceptual question:

- **What could be done to the road or car to prevent skidding?**

This question emphasizes the tendency of a moving object to continue in a straight line (by Newton's First Law) by describing a skid. It is in fact the friction between the road and the tires that provides the centripetal force to make the car move in a circle. The lack of sufficient friction leads to the skid.

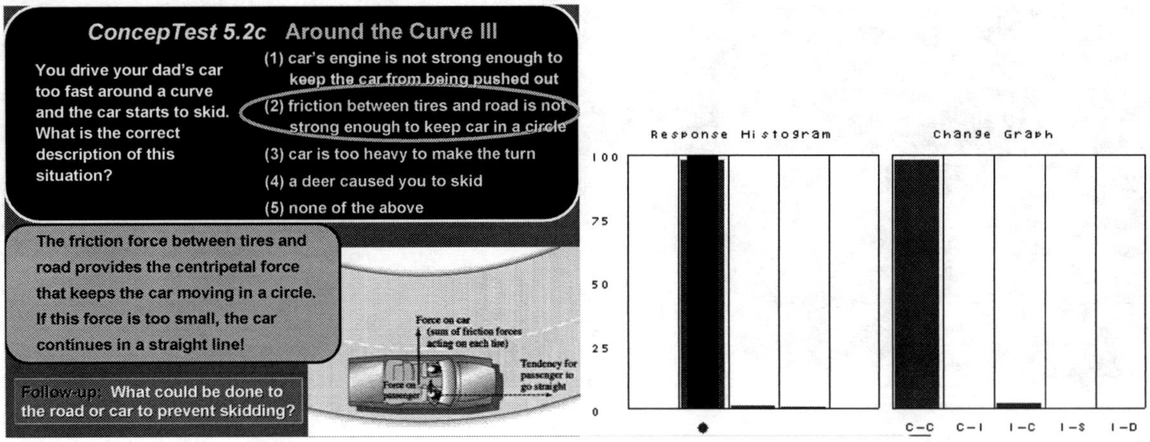

Conceptual Module 5.3

ConcepTest 5.3: Missing Link
Difficulty Level: 3
Suggested follow-up conceptual question:

- **What physical force provides the centripetal force?**

This question again addresses Newton's First Law and circular motion. In this case, the normal force of the tube wall is what provides the centripetal force to keep the ball moving in a circle. Once that force is removed, the ball will then continue in a straight line, as shown by path #2. Students might tend to think that the ball will "remember" its curved path and therefore choose path #1.

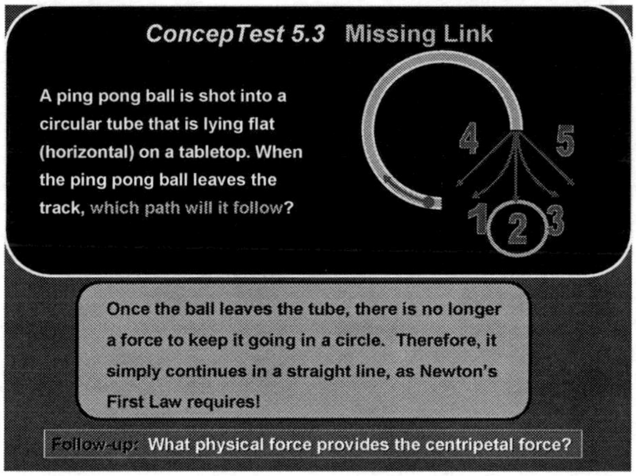

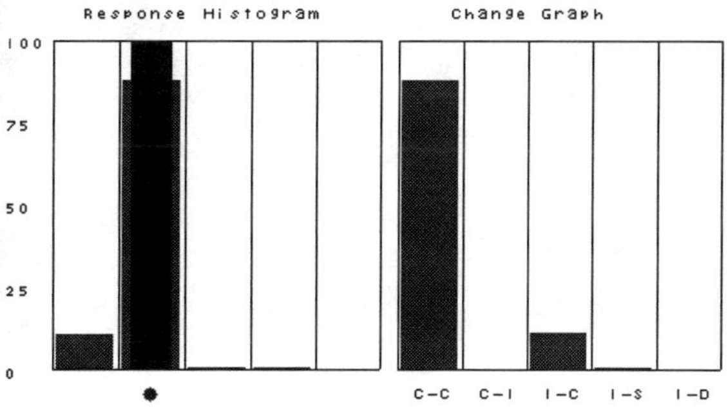

Conceptual Module 5.4

ConcepTest 5.4: Ball and String
Difficulty Level: 5

This is a multiple-step problem, demanding that the students use their knowledge of the relationship between mass, velocity and radius in the centripetal force and in the period. Relating these quantities will lead the students to the appropriate relationship between the tensions in the two cases.

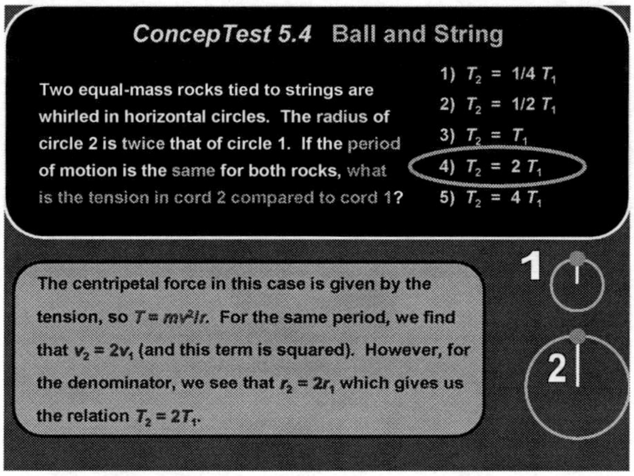

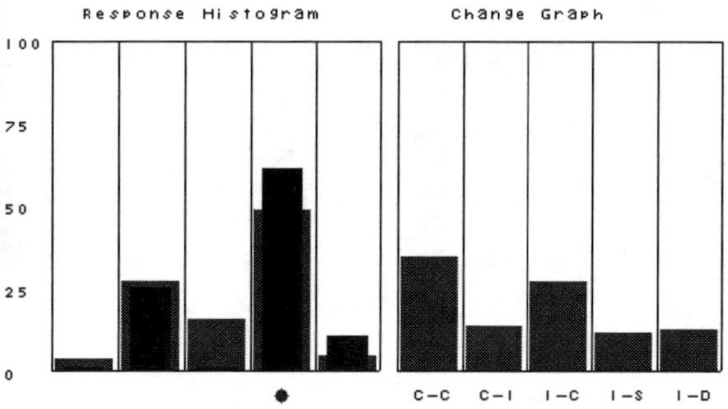

Conceptual Module 5.5

ConcepTest 5.5: Barrel of Fun
Difficulty Level: 3
Suggested follow-up conceptual question:

- **What happens if the rotation of the ride slows down?**

This is an exercise in visualizing the free-body diagram for a case of circular motion. The example relates to a familiar amusement park ride, in which case the centripetal force is provided by the normal force of the wall. In the figures, the students must realize that the net force must point inward only and that condition is satisfied by picture #1 and no other picture.

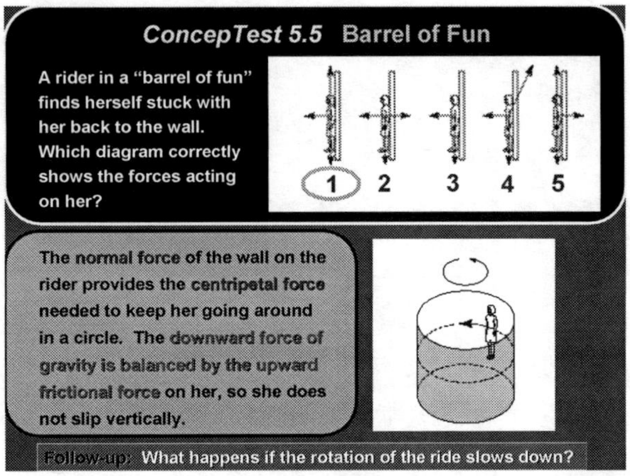

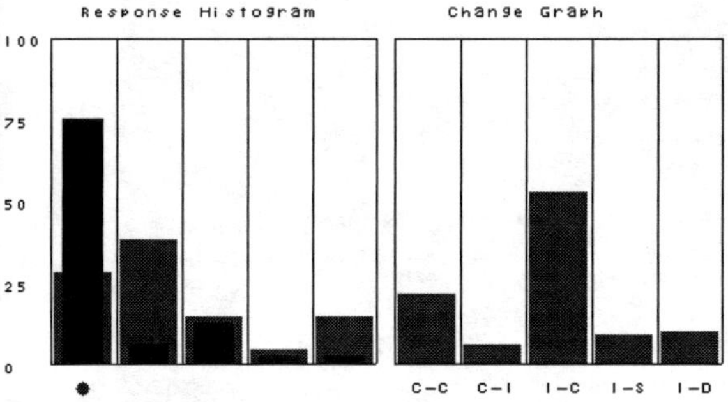

Conceptual Module 5.6 (three parts)

ConcepTest 5.6a: Going in Circles I
Difficulty Level: 4
Suggested follow-up conceptual question:

- **Where is *N* larger than *mg*?**

Again, the net force must be inward (which is downward at the top of the Ferris wheel). Thus, at this position, the downward force (*mg*) must exceed the upward force (*N*).

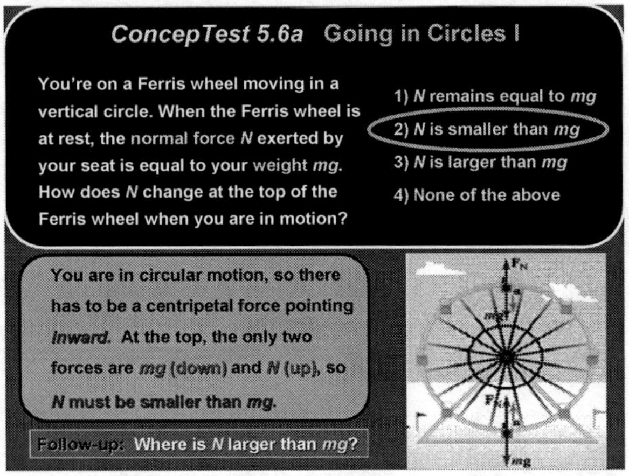

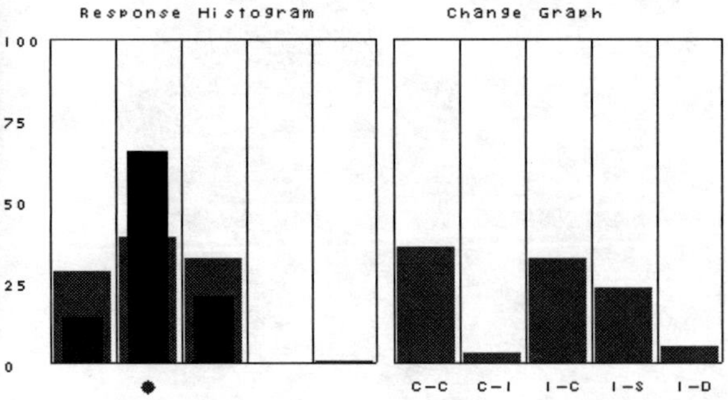

ConcepTest 5.6b: Going in Circles II
Difficulty Level: 4
Suggested follow-up conceptual question:

- **What happens when the skier goes into a small dip?**

In this case, as in the previous question, the centripetal force must be directed in toward the center of the circle (which is downward at the top of the hill). At that position, the weight and normal force are in opposite directions.

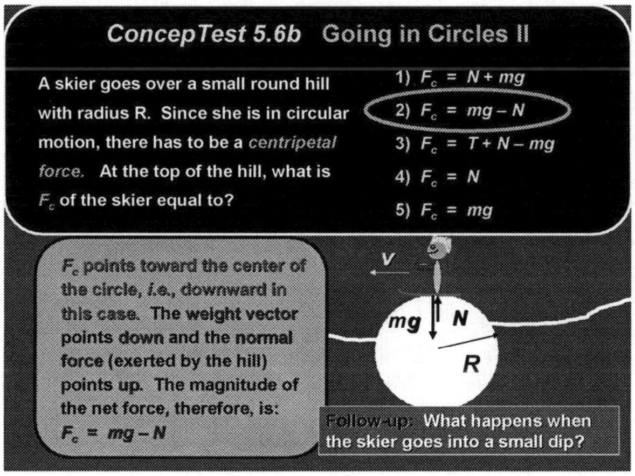

ConcepTest 5.6c: Going in Circles III
Difficulty Level: 4
Suggested follow-up conceptual question:

- **What is F_c at the bottom of the ball's path?**

In this case, as in the previous questions, the centripetal force must be directed in toward the center of the circle (which is downward at the top of the ball's path). At that position, the weight and tension force are in the same direction.

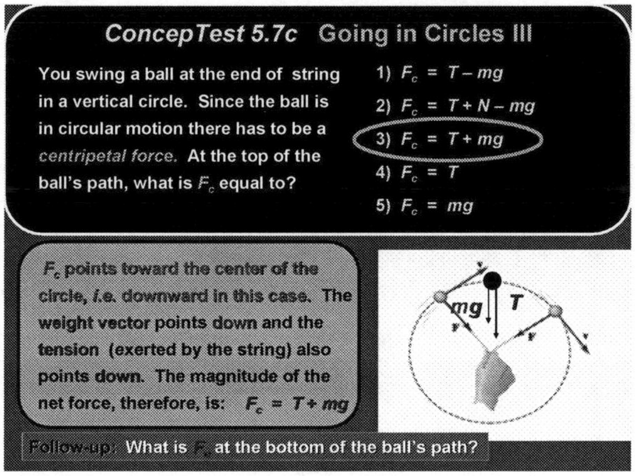

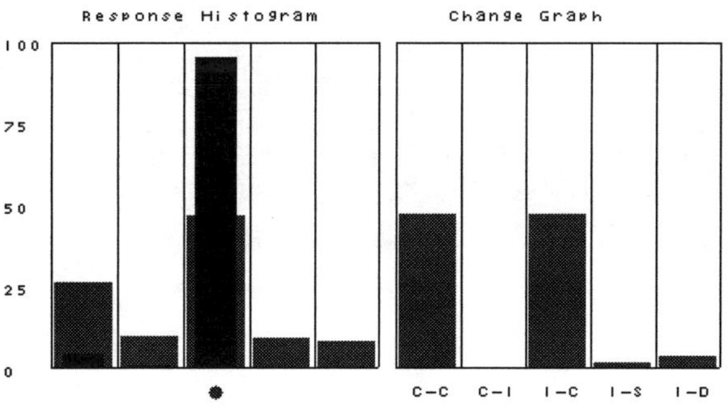

Conceptual Module 5.8 (two parts)

ConcepTest 5.8a: Earth and Moon I
Difficulty Level: 2

This is a review of Newton's Third Law, as it applies to the gravitational attraction between two bodies.

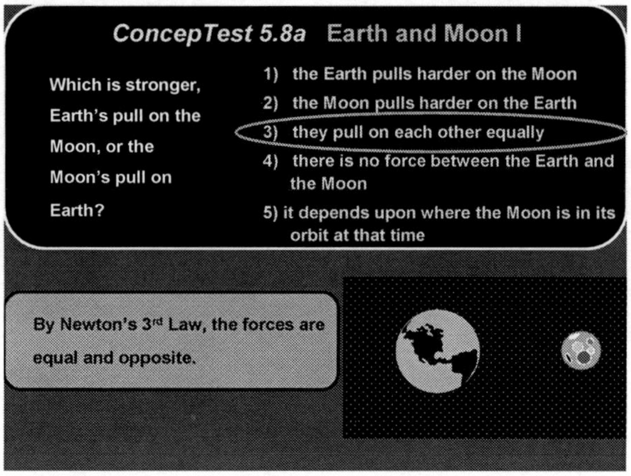

ConcepTest 5.8b: Earth and Moon II
Difficulty Level: 2
Suggested follow-up conceptual question:

- **What distance would increase the force by a factor of 2?**

This question examines the expression for Newton's Law of Gravitation. In particular, it is requiring students to recognize the inverse square nature of the force law.

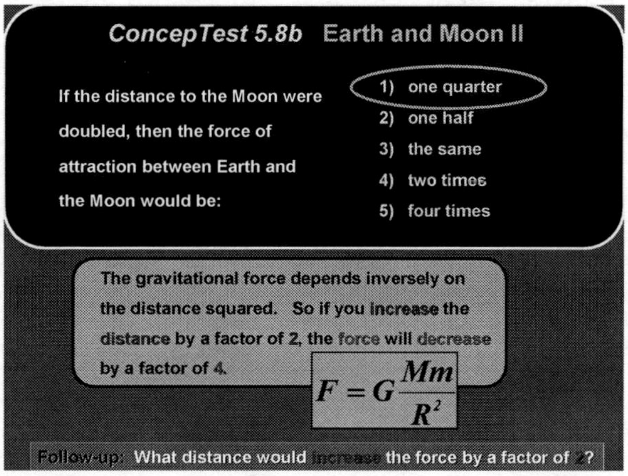

Conceptual Module 5.9
ConcepTest 5.9: Fly Me Away
Difficulty Level: 3

Students must appreciate that the gravity they feel on the surface of Earth is simply a manifestation of the Law of Gravitation. If they were to increase their distance from the center of Earth (as in an airplane or on a mountaintop), then there would be a reduction in the effective force of gravity acting on them (i.e., their weight).

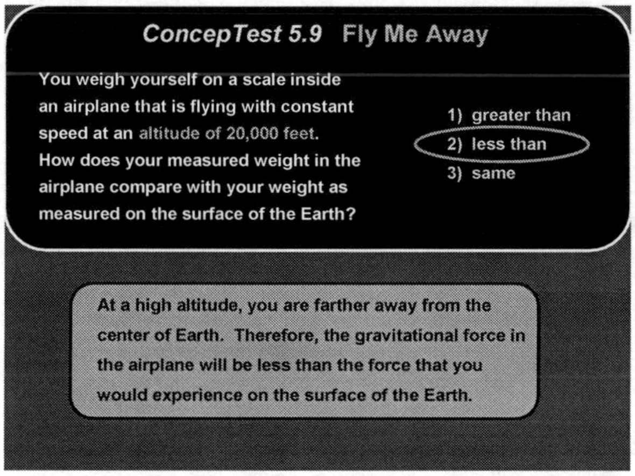

Conceptual Module 5.10
ConcepTest 5.10: Two Satellites
Difficulty Level: 5

At first glance, this question may appear to be asking about something (centripetal force) that has nothing to do with gravity. However, students must recognize that the centripetal force is provided by gravity in this case (just as tension provides the centripetal force for a mass being whirled on a string). Realizing this, the answer is then a simple matter of comparing the square of the distances from the center of Earth.

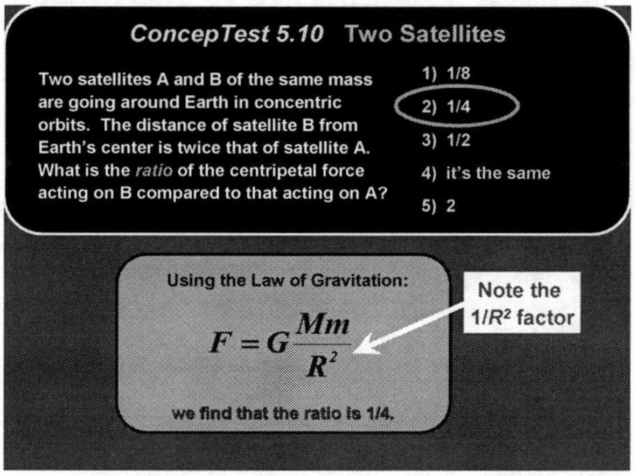

Conceptual Module 5.11

ConcepTest 5.11: Averting Disaster
Difficulty Level: 4
Suggested follow-up conceptual question:

- **What happens to a satellite orbiting Earth as it slows?**

The notion of gravity as the "physical force" that provides the centripetal force for circular motion is critically important for students to appreciate. Many have probably not pondered the question of why the Moon doesn't fall into Earth, and it could be quite illuminating to realize that this is a consequence of the Moon's high orbital speed.

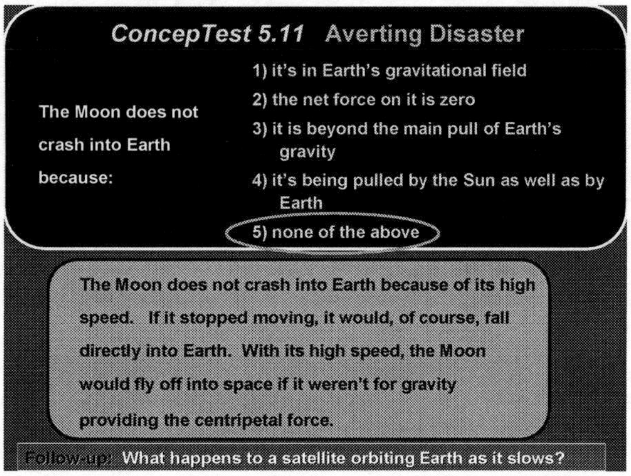

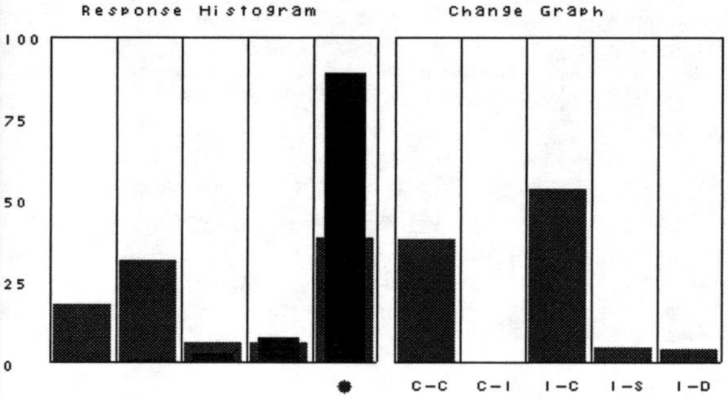

Conceptual Module 5.12

ConcepTest 5.12: In the Space Shuttle
Difficulty Level: 4
Suggested follow-up conceptual question:

- **How weak is the value of *g* at an altitude of 300 km?**

There is a tendency to believe that astronauts float because they are not affected by gravity. In fact, this is far from the case – the value of *g* is only reduced by about 10% at an orbital height of 300 km. The "apparent" weightlessness arises due to the fact that the astronauts are in free fall.

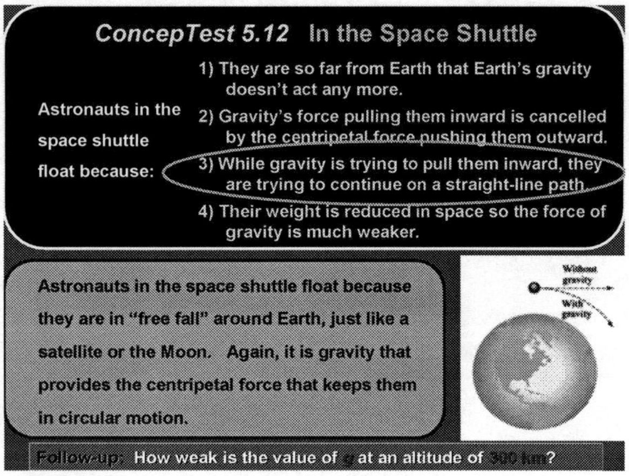

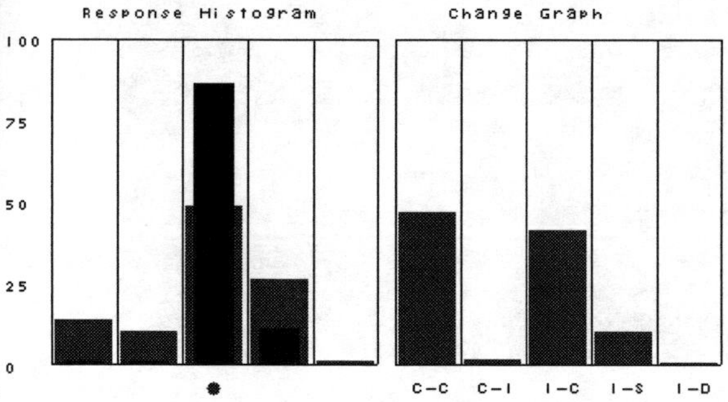

Conceptual Module 5.13
ConcepTest 5.13: Guess My Weight
Difficulty Level: 4

This question addresses the fact that Earth is rotating and therefore all objects on the surface of Earth are in circular motion. This implies a net inwardly directed force, so when comparing the weight *mg* and the normal force *N*, the latter must be slightly less (and it is the normal force that a scale reads).

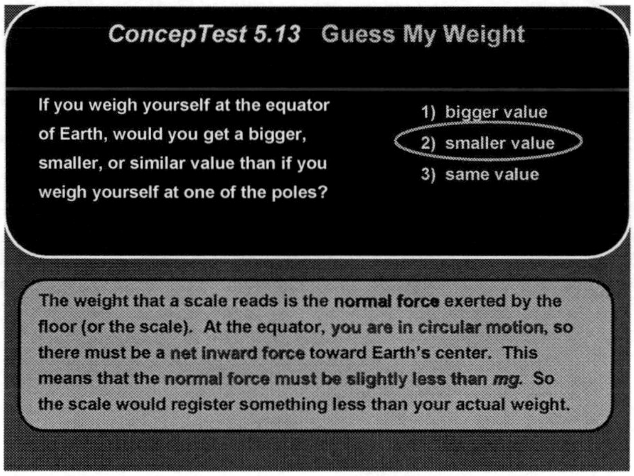

Conceptual Module 5.14
ConcepTest 5.14: Force Vectors
Difficulty Level: 3

This question addresses the vector nature of forces, coupled with the comprehension of the Law of Gravitation. The force vectors for the attraction to *m* and *2m* must be estimated and then added vectorially to get the resultant.

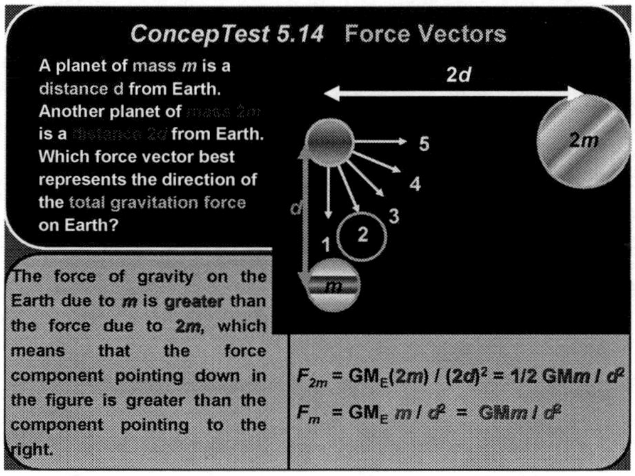

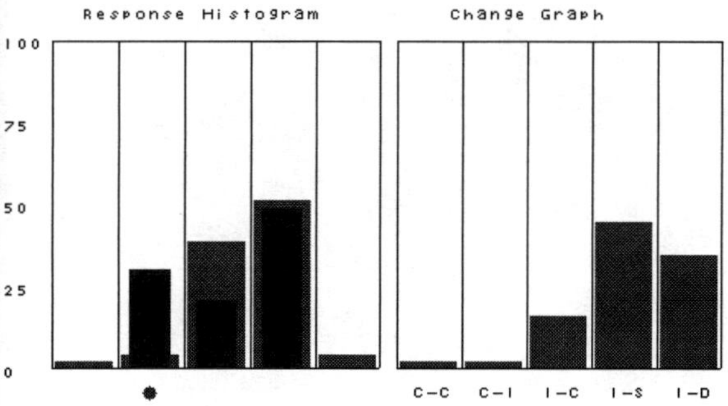

Conceptual Module 6.1

ConcepTest 6.1: To Work or Not to Work
Difficulty Level: 1

Work is a novel concept to the students. There is a belief that simply applying a force to an object constitutes work. This question emphasizes that the object must also have a certain displacement in order for work to be done.

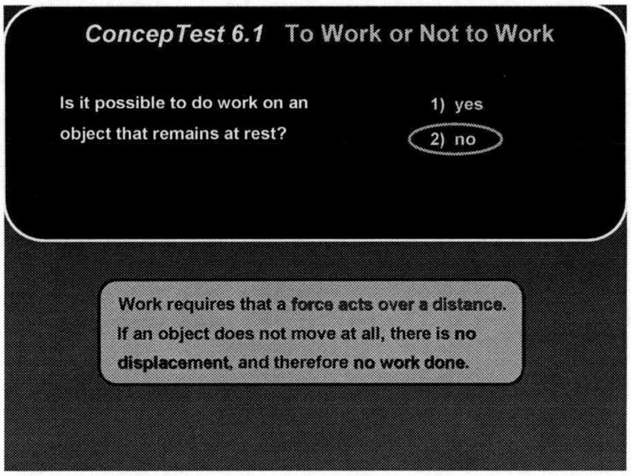

Conceptual Module 6.2 (four parts)

ConcepTest 6.2a: Friction and Work I
Difficulty Level: 2

This question asks students to examine the relationship between the direction of the applied force and the direction of the displacement. Understanding the meaning of positive work or negative work is a distinction that the students should be able to make.

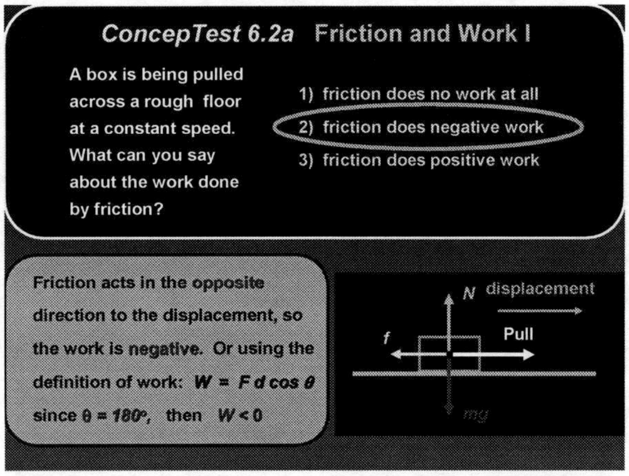

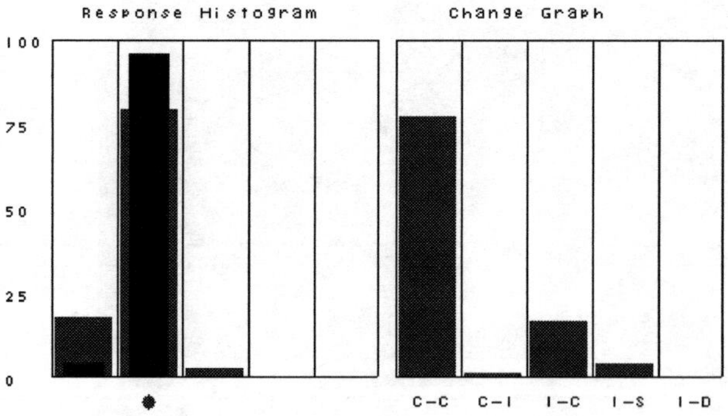

ConcepTest 6.2b: Friction and Work II
Difficulty Level: 3

The notion that friction "resists" motion may lead to the conclusion that friction can only do negative work. Examples of positive work by friction include a box moving on a conveyor belt or a crate in the back of a moving pickup truck.

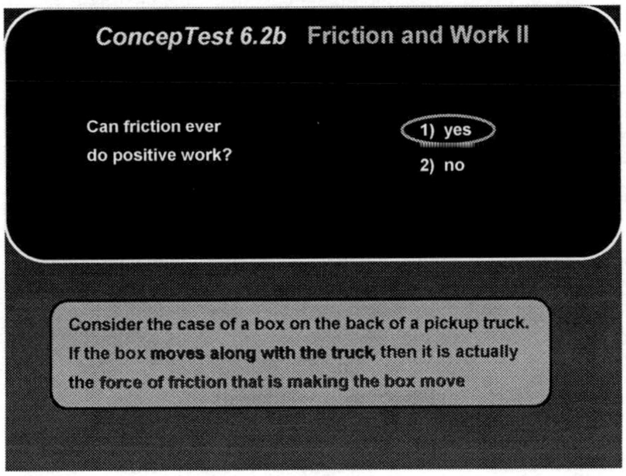

ConcepTest 6.2c: Play Ball!
Difficulty Level: 2
Suggested follow-up conceptual question:

- **What about the work done by the ball on the catcher?**

This again deals with the directions of the force and the displacement in order to determine the sign of the work.

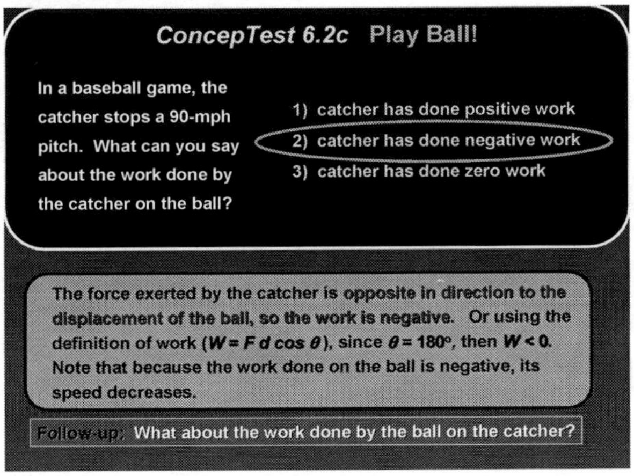

ConcepTest 6.2d: Tension and Work
Difficulty Level: 3
Suggested follow-up conceptual question:

- **Is there a force in the direction of the velocity?**

Finally, we examine the case where the applied force is perpendicular to the displacement, leading to a situation where zero work is done.

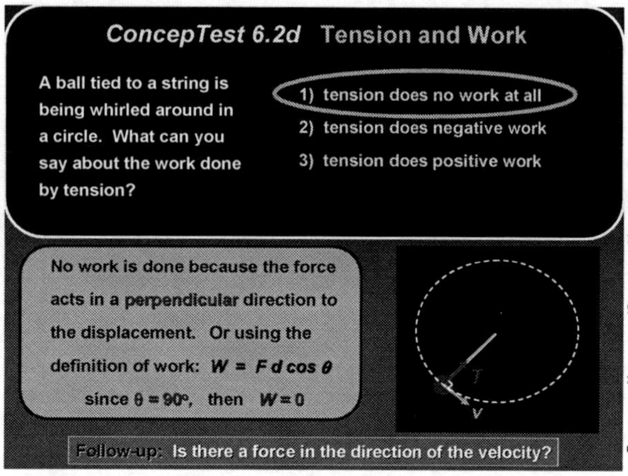

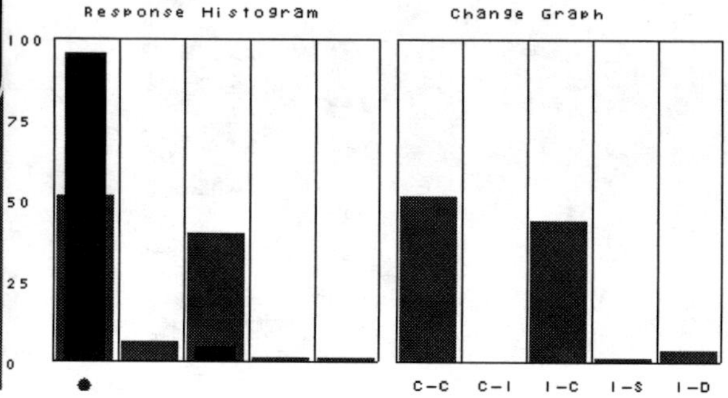

Conceptual Module 6.3

ConcepTest 6.3: Force and Work
Difficulty Level: 4

This question requires a careful free-body diagram in order to address the forces on the box and which forces do non-zero work. Note that not all of the forces do non-zero work.

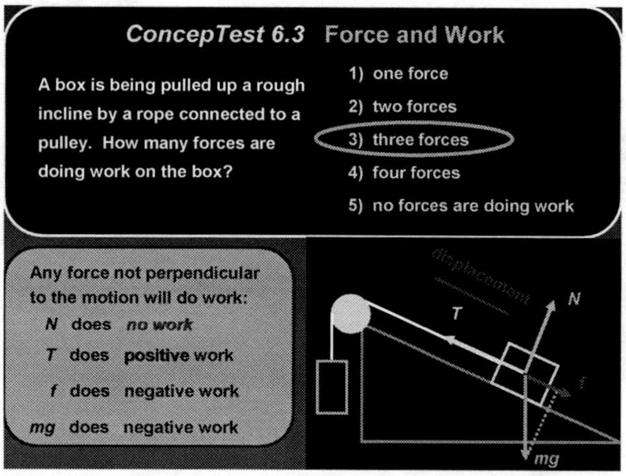

Conceptual Module 6.4

ConcepTest 6.4: Lifting a Book
Difficulty Level: 3
Suggested follow-up conceptual question:

- **What would happen if F_{HAND} was greater than mg?**

Here the net work on an object is examined. Various forces may do work (positive or negative) on the object, but the net work is the sum of those separate components. Or, alternatively, the net work relates directly to the net force. Either way, if no net work is done, then the kinetic energy will not change.

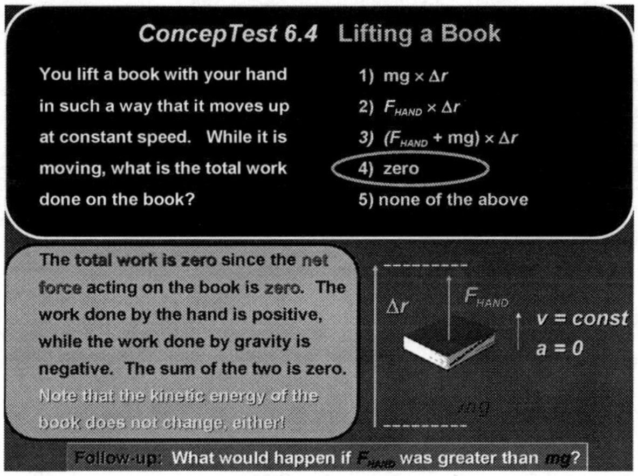

Conceptual Module 6.5 (two parts)

ConcepTest 6.5a: Kinetic Energy I
Difficulty Level: 2
Suggested follow-up conceptual question:

- **How would you achieve a KE increase of a factor of 2?**

This series of questions is intended to be a gentle introduction to kinetic energy. Here students are expected to recognize the fact that velocity is squared in the kinetic energy.

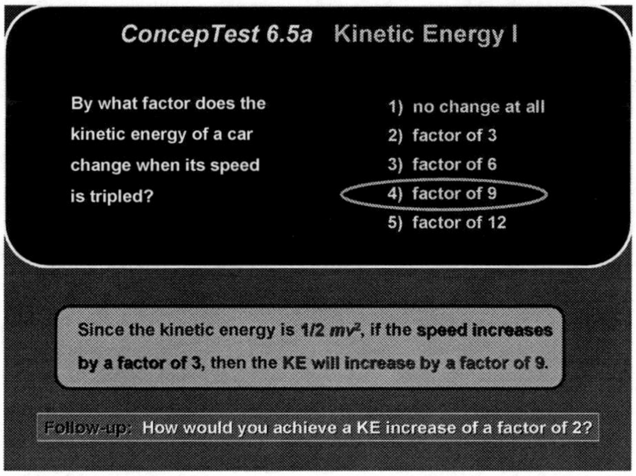

ConcepTest 6.5b: Kinetic Energy II
Difficulty Level: 3

In this case, the relation between mass and velocity in the expression for kinetic energy is being examined.

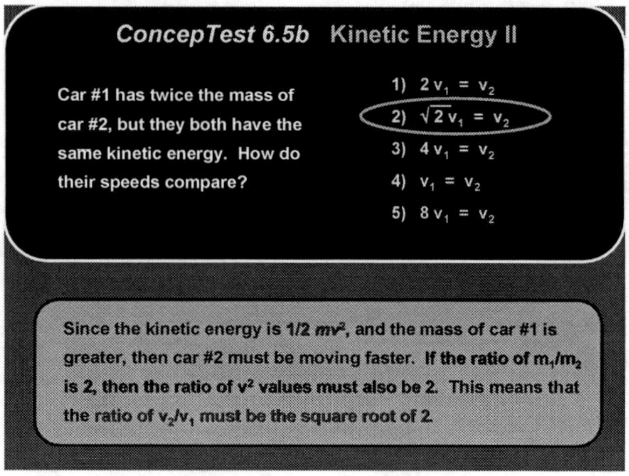

Conceptual Module 6.6 (two parts)

ConcepTest 6.6a: Free Fall I
Difficulty Level: 2
Suggested follow-up conceptual question:

- **How do the initial values of gravitational PE compare?**

In this case, the students must find the work done by gravity and then recognize that the only difference in the two cases is the mass. This also gives them a preview of the expression for gravitational potential energy.

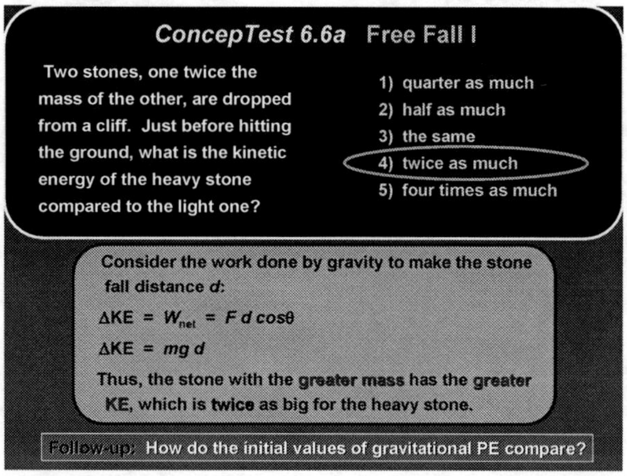

ConcepTest 6.6b: Free Fall II
Difficulty Level: 2

In this case, the students must realize that the final velocities of the two objects will be the same. Therefore, the kinetic energy ratio is directly related to the mass ratio, which independently confirms the result from the previous ConcepTest.

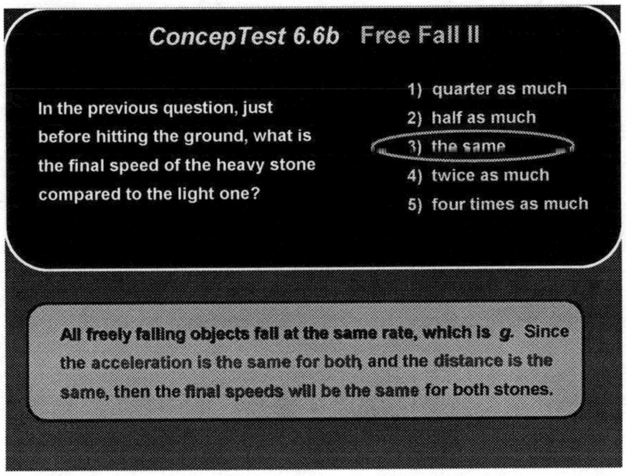

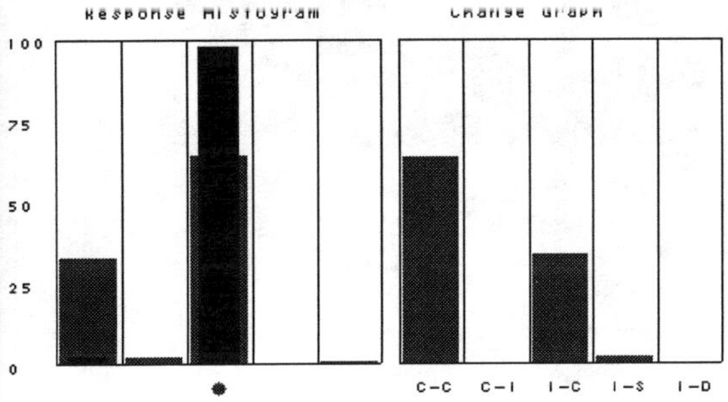

Conceptual Module 6.7

ConcepTest 6.7: Work and KE
Difficulty Level: 2
Suggested follow-up conceptual question:

- **What does it mean for negative work to be done on the child?**

The net work done by external forces leads to a change in kinetic energy. This relation is the so-called “work-kinetic energy” theorem and is crucial for students to understand.

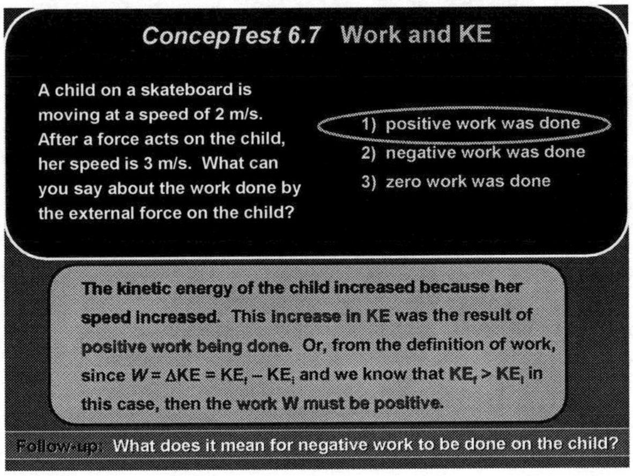

Conceptual Module 6.8 (three parts)

ConcepTest 6.8a: Slowing Down
Difficulty Level: 3

The students are asked to relate work to the change in kinetic energy in order to calculate stopping distance. This question also has definite real-world implications, just to give a sense of how much more difficult it is to stop safely when moving at higher speeds.

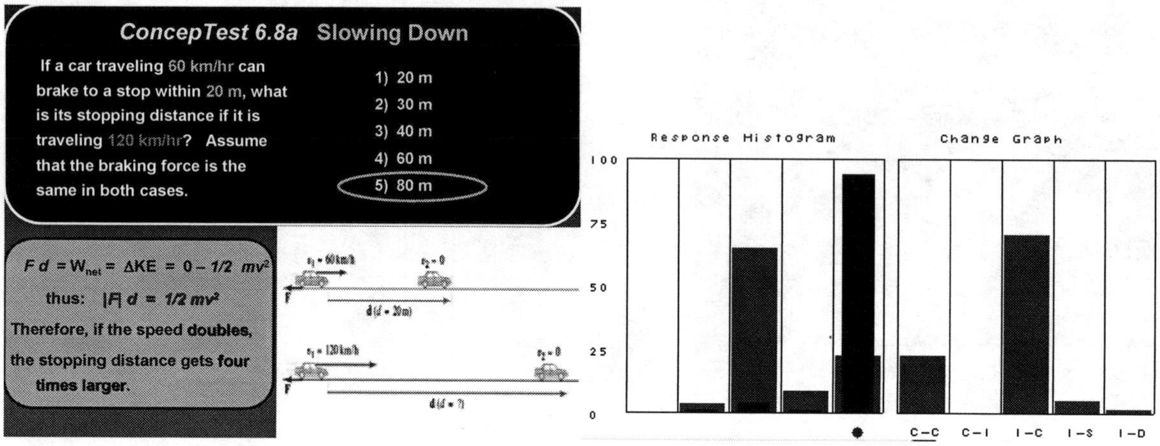

ConcepTest 6.8b: Speeding Up I
Difficulty Level: 4
Suggested follow-up conceptual question:

- **How much energy is required to stop the 60 mph car?**

In this case, a more careful assessment of the "work-kinetic energy" theorem is required. The intuitive notion that both accelerating stages take the same amount of energy is not true, and this is dictated largely by the square of the velocity in the kinetic energy.

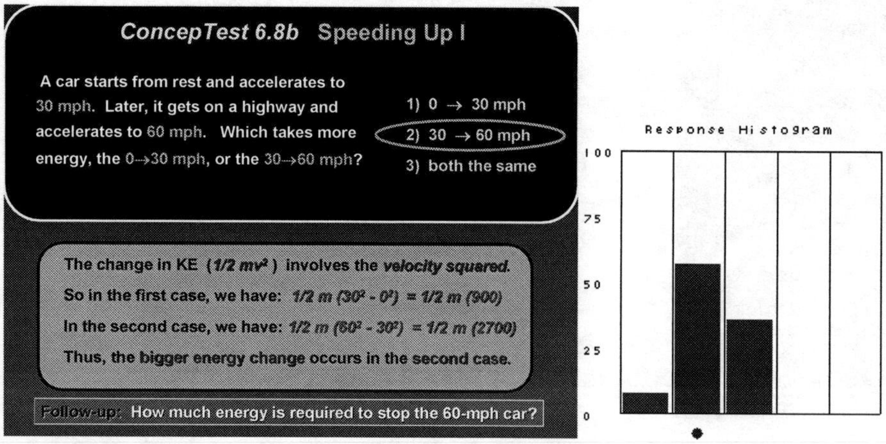

ConcepTest 6.8c: Speeding Up II
Difficulty Level: 4
Suggested follow-up conceptual question:

- **How much work is required to stop the 150 km/hr car?**

This question asks the students to quantify the energy difference between the two stages of acceleration. They must apply what they have learned from the previous ConcepTest to address this issue semi-quantitatively.

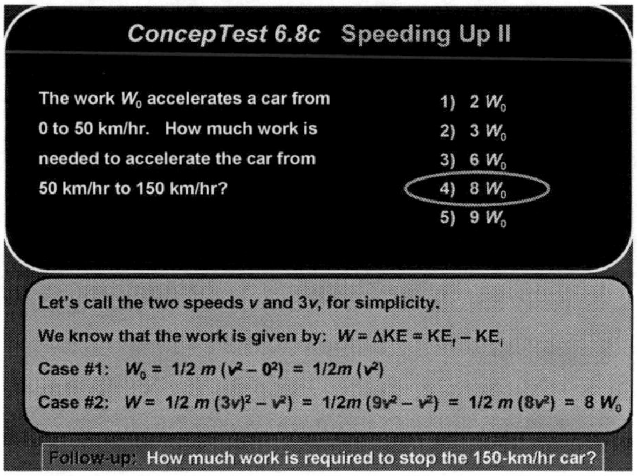

Conceptual Module 6.9 (two parts)

ConcepTest 6.9a: Work and Energy I
Difficulty Level: 4
Suggested follow-up conceptual question:

- **Which block has the greater magnitude of acceleration?**

This question addresses the relationship between work and kinetic energy in the case where the work is done by friction. The simplification that the blocks start with the same kinetic energy allows the students to focus on the determination of frictional work.

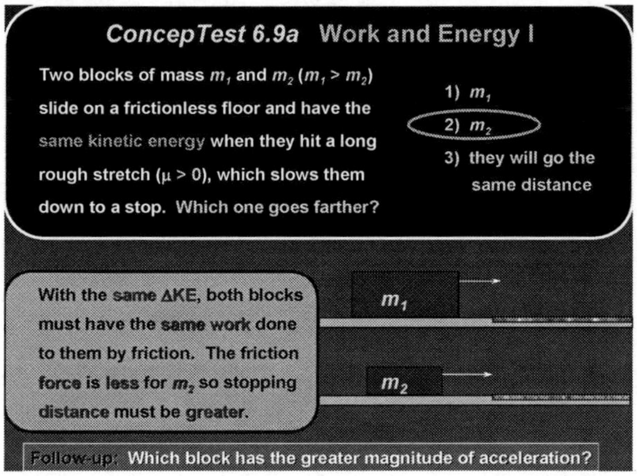

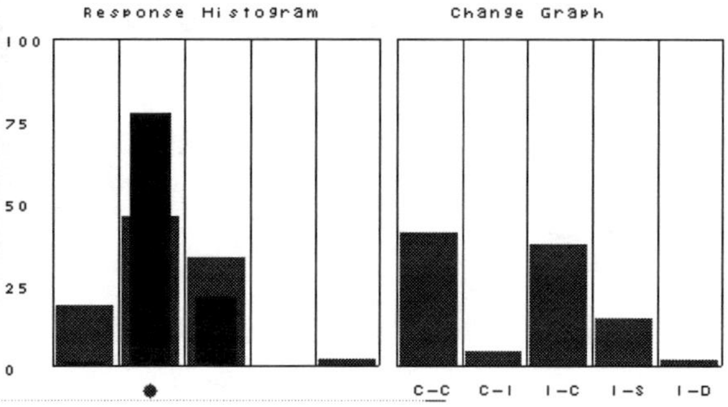

ConcepTest 6.9b: Work and Energy II
Difficulty Level: 3

In this case, the frictional force is constant, and the students must examine the kinetic energy part of the problem. Here the velocity dependence plays the major role.

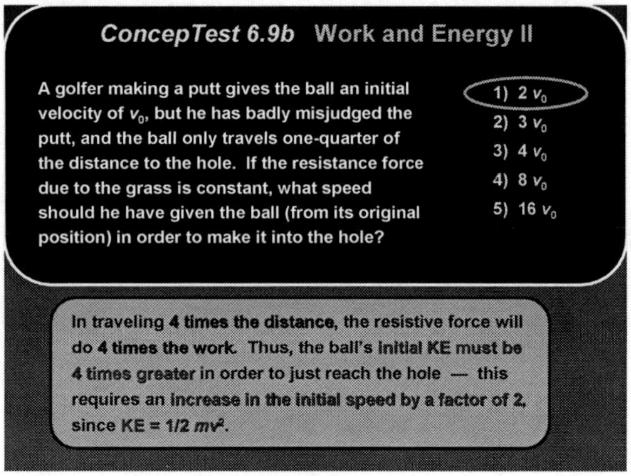

Conceptual Module 6.10

ConcepTest 6.10: Sign of the Energy I
Difficulty Level: 1

This is a simple question relating to kinetic energy.

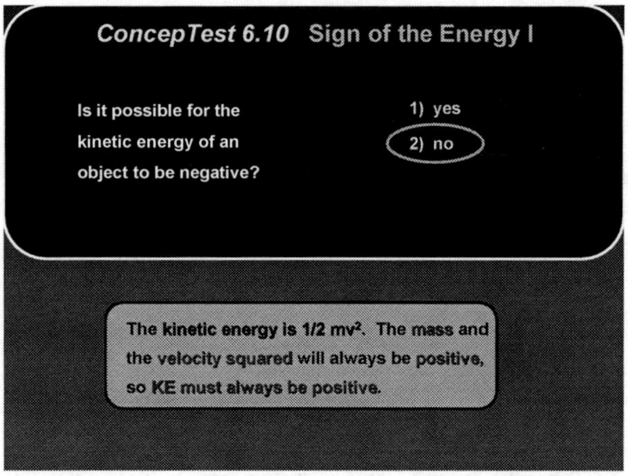

Conceptual Module 6.11
ConcepTest 6.11: Sign of the Energy II
Difficulty Level: 2

It is important for students to understand that the *actual value* of the gravitational potential energy is not that important – what is physically relevant is the *difference* between initial and final values (whatever they may be). The zero reference level for PE is arbitrary, and so it is certainly possible for the PE to be negative.

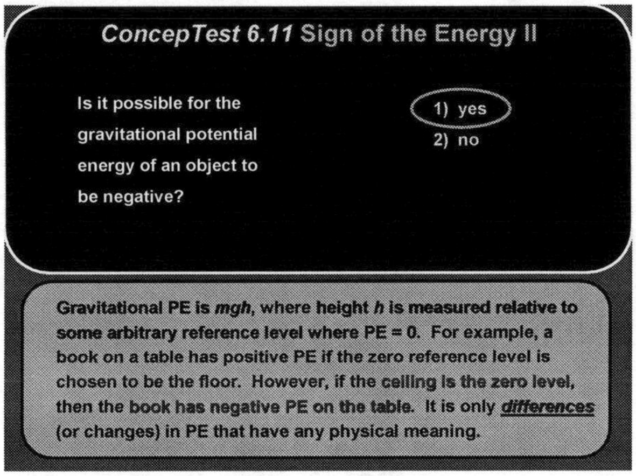

Conceptual Module 6.12
ConcepTest 6.12: KE and PE
Difficulty Level: 2
Suggested follow-up conceptual question:
- **Does anything change physically by the choice of $y = 0$?**

This question continues to address the issue of the zero reference level for gravitational potential energy. Here students are asked to recognize that while the value of PE depends on the reference level, the ΔPE does not depend on that (and therefore KE does not).

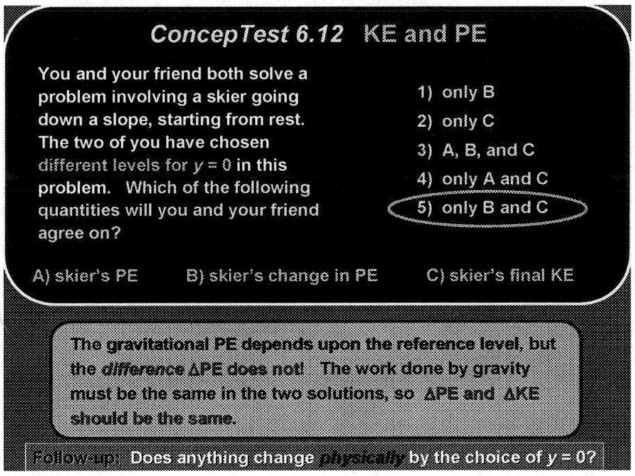

Conceptual Module 6.13

ConcepTest 6.13: Up the Hill

Difficulty Level: 2

Suggested follow-up conceptual questions:

- **How much more work do you do in taking the steeper path?**
- **Which path would you rather take? Why?**

The students must realize that the change in gravitational potential energy depends on the height difference between initial and final levels and is independent of the path taken to go from one point to another (because gravity is a conservative force). What does depend on the path is the force and the distance (related to work), while the work is unchanged.

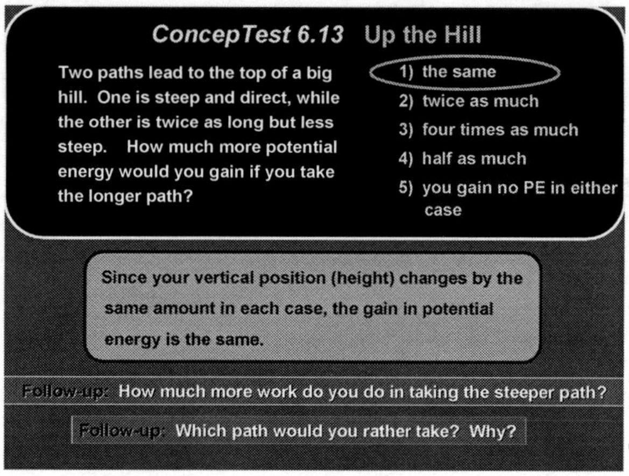

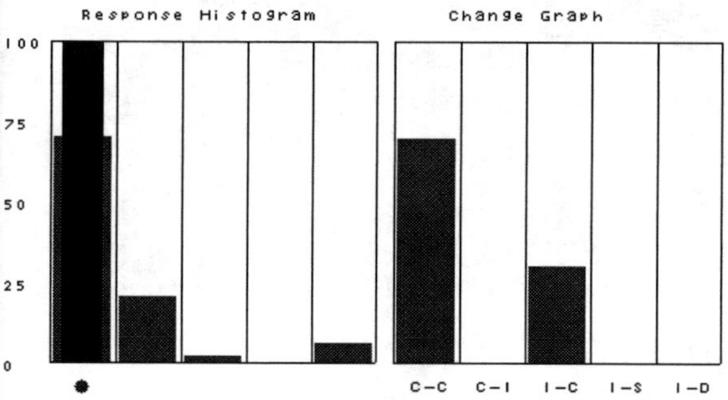

Conceptual Module 6.14

ConcepTest 6.14: Elastic Potential Energy

Difficulty Level: 2

The work has been defined as the change in potential energy. Using the relation for elastic potential energy, students are asked to determine the work for two different displacements, keeping in mind that elastic PE depends on the square of displacement.

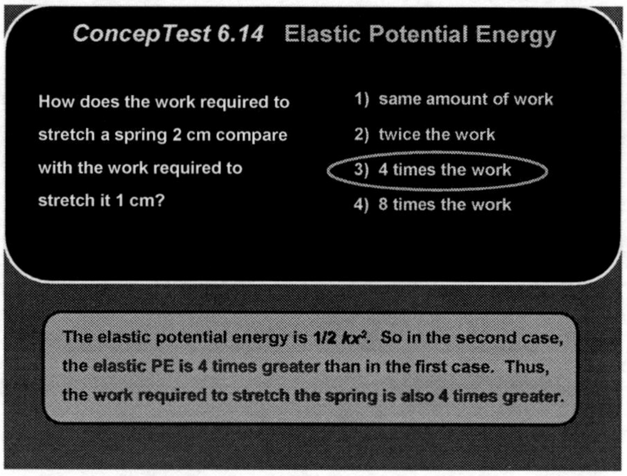

Conceptual Module 6.15

ConcepTest 6.15: Springs and Gravity
Difficulty Level: 2

This question is a basic check on changes in potential energy, both gravitational and elastic. In the case of a vertical spring, both of these come into play.

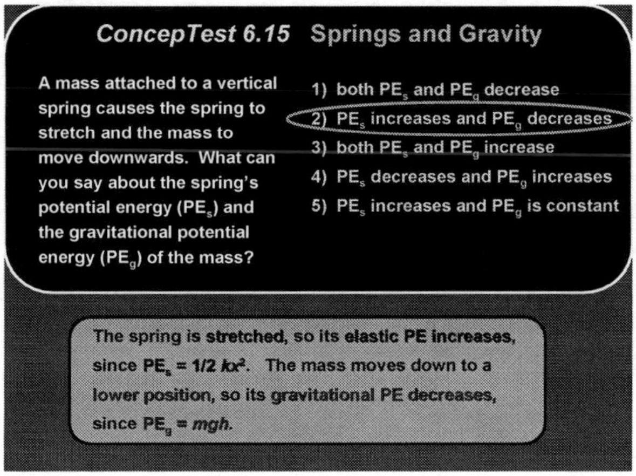

Conceptual Module 6.16

ConcepTest 6.16: Down the Hill
Difficulty Level: 2
Suggested follow-up conceptual question:

- **Which ball takes longer to get down the ramp?**

The kinetic energy is related to the change in potential energy. Here the gravitational PE depends on the height of the ramp, which is the same in all cases, so the KE must end up the same as well (and therefore, the speed).

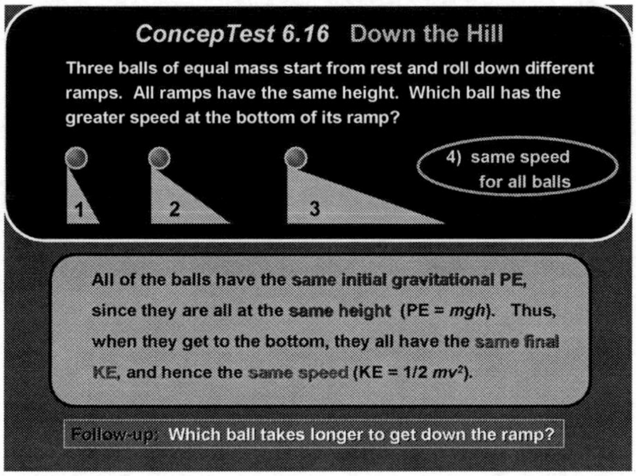

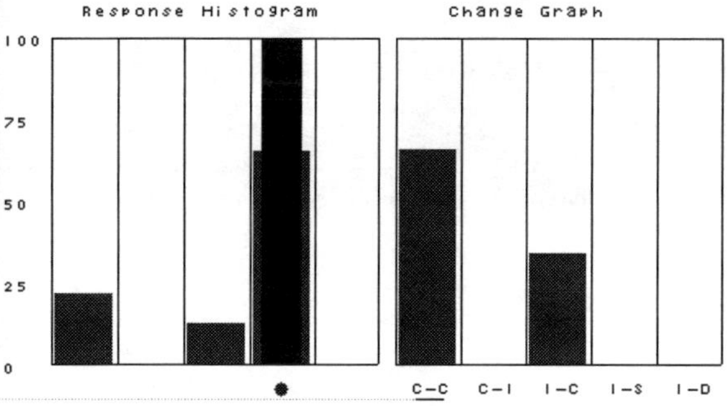

Conceptual Module 6.17 (two parts)

ConcepTest 6.17a: Runaway Truck
Difficulty Level: 3

This question examines conservation of energy, where potential energy is converted into kinetic energy. The dependence of these quantities on the relevant variables (in this case, height and velocity) is the principal concern here.

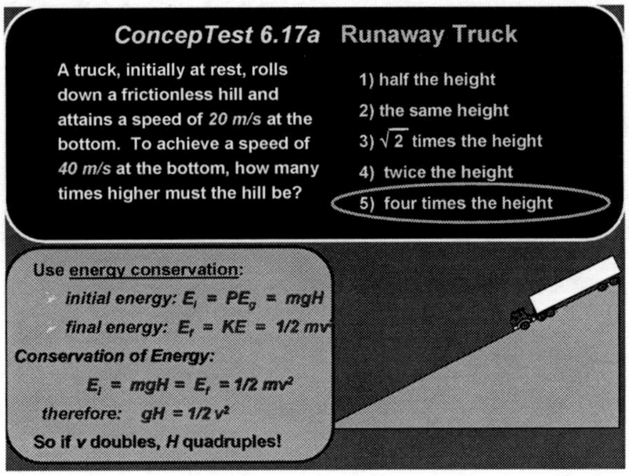

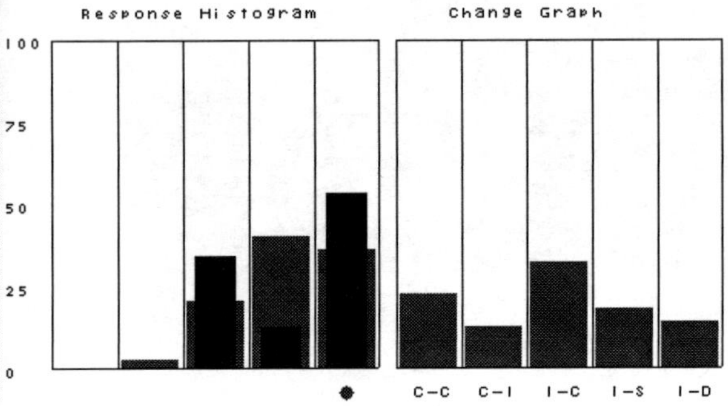

ConcepTest 6.17b: Runaway Box
Difficulty Level: 3

This question follows the same line of reasoning as the previous question, except that in this case, the elastic potential energy (instead of gravitational PE) is being related to kinetic energy.

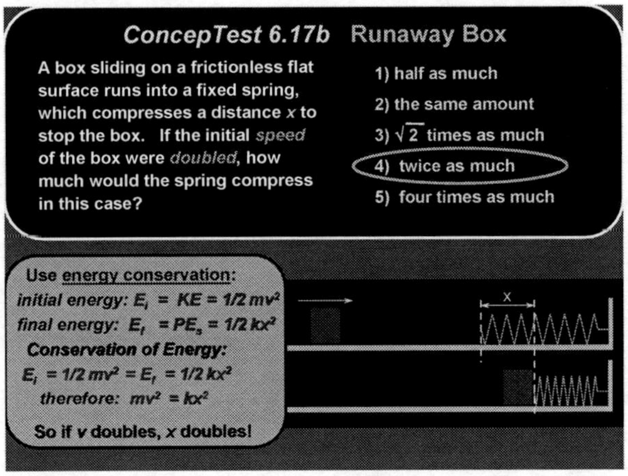

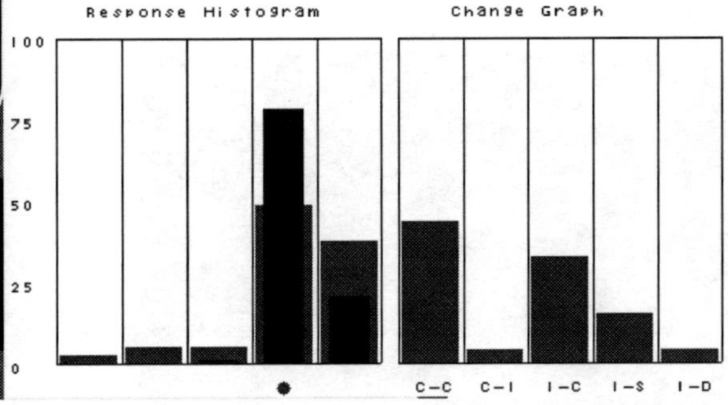

Conceptual Module 6.18 (two parts)

ConcepTest 6.18a: Water Slide I
Difficulty Level: 3

This question emphasizes the path independence of the potential energy. Both of the kids start at the same height, so their final velocities will be the same. The velocity does not depend on the shape of the slide (i.e., the path taken).

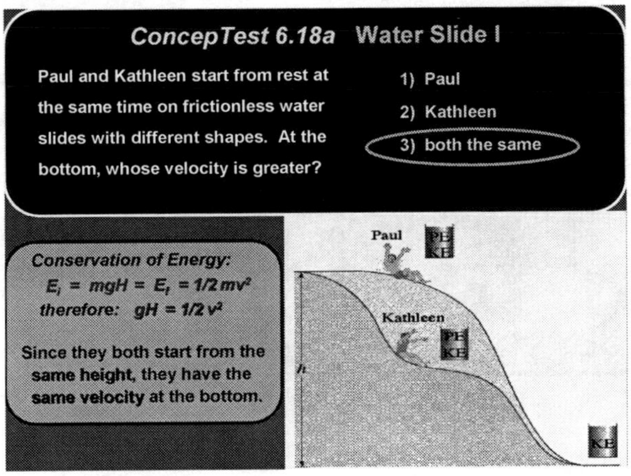

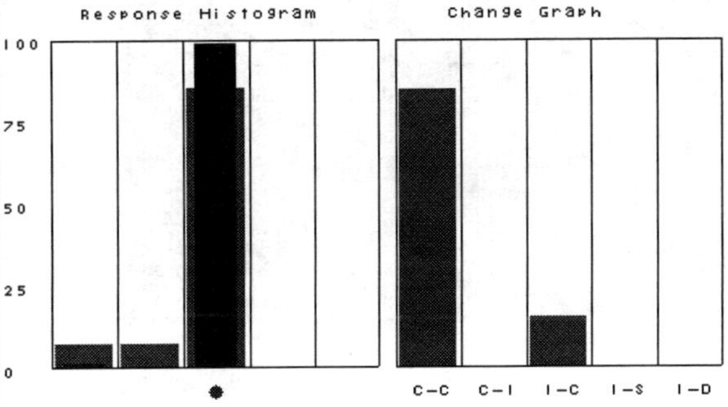

ConcepTest 6.18b: Water Slide II
Difficulty Level: 5

This follow-up question is somewhat tricky. In this case, when determining the travel time, the path actually does matter. Kathleen converts her gravitational potential energy into kinetic energy more quickly, which means that she has a greater velocity for most of the ride. Thus, she arrives at the bottom earlier.

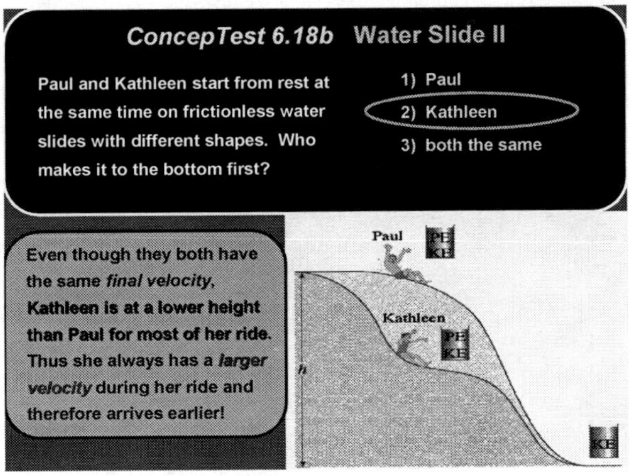

Conceptual Module 6.19
ConcepTest 6.19: Cart on a Hill
Difficulty Level: 4

This question requires a careful consideration of energy conservation. The two velocities in the question cannot simply be added to obtain the answer.

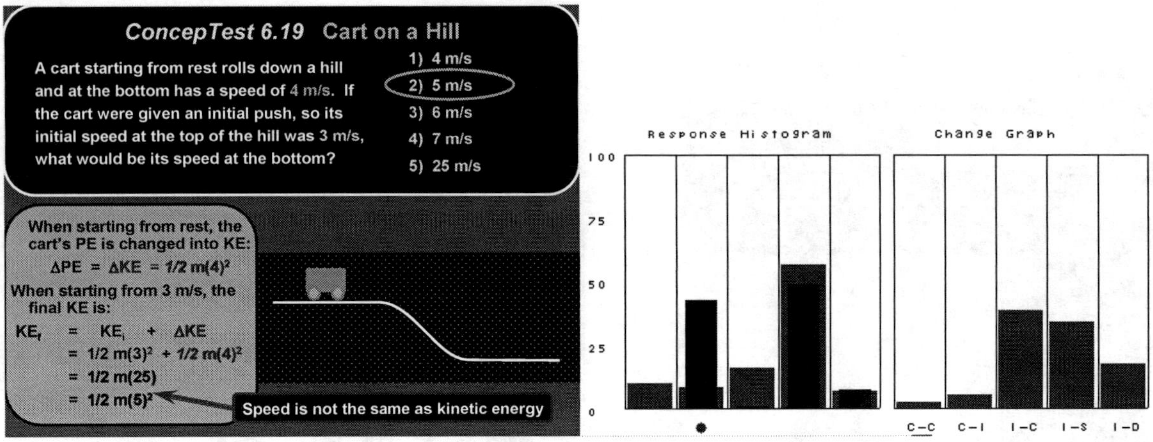

Conceptual Module 6.20 (two parts)
ConcepTest 6.20a: Falling Leaves
Difficulty Level: 4
Suggested follow-up conceptual questions:
- **What happens to the leaf's KE as it falls? What net work is done?**

In this case, the work done by non-conservative forces is involved. The leaf is falling at constant speed because there is a frictional force (air resistance); this constitutes the non-conservative force in the problem. Therefore, by the work-energy theorem, the final energy will be less than the initial energy.

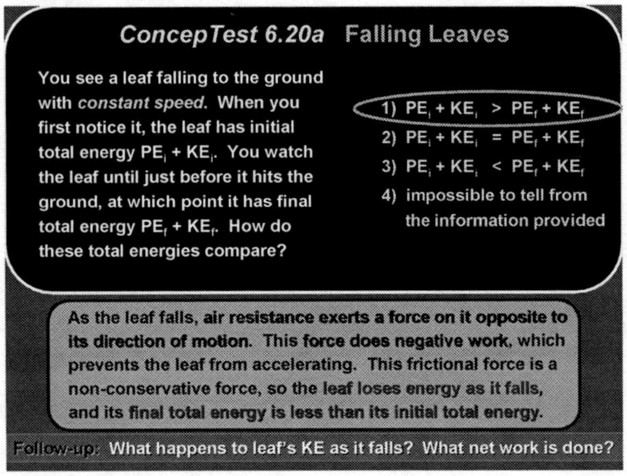

ConcepTest 6.20b: Falling Balls
Difficulty Level: 5
Suggested follow-up conceptual question:

- **How does the force of air resistance compare to gravity when the ball reaches terminal velocity?**

In this case, again, a non-conservative force (air resistance) is playing a role. Since the ball has "lost" energy due to work by non-conservative forces, it will be moving slower at each point on the way down than it was at those same points on the way up. Thus, the downward trip will take a longer time.

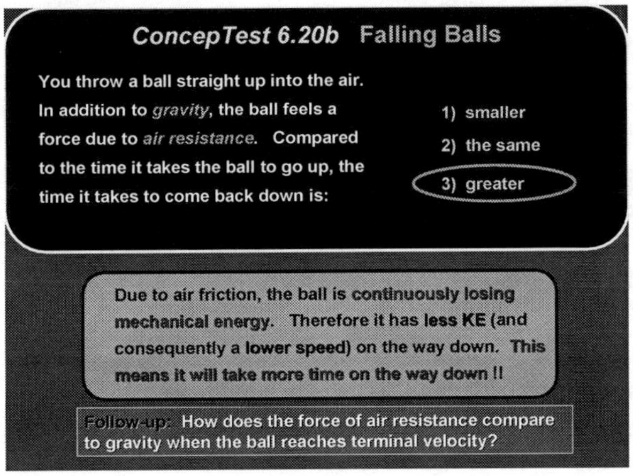

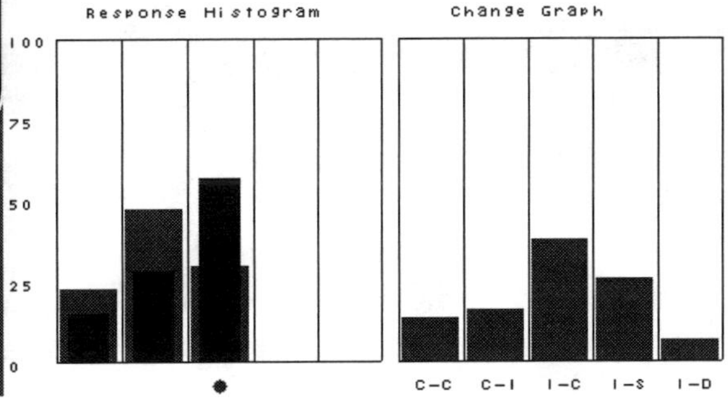

Conceptual Module 6.21 (three parts)

ConcepTest 6.21a: Time for Work I
Difficulty Level: 2

As a precursor to introducing power, a comparison is done between two cases where the time of application of the force is different. This does not affect the work at all.

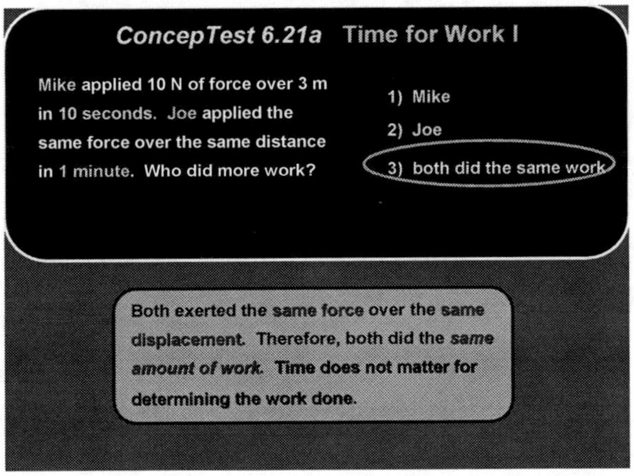

ConcepTest 6.21b: Time for Work II
Difficulty Level: 3

In this question, the relationship between work, time and power is examined.

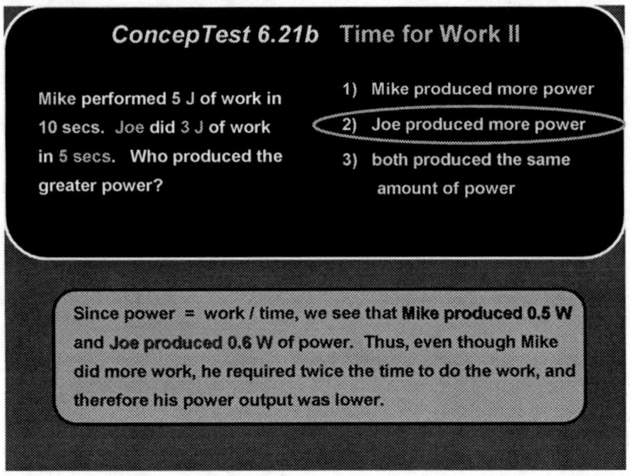

ConcepTest 6.21c: Power
Difficulty Level: 2

Power and work are not the same quantities. They are related, but not the same. Students often confuse the two concepts, and the distinction must be made clear.

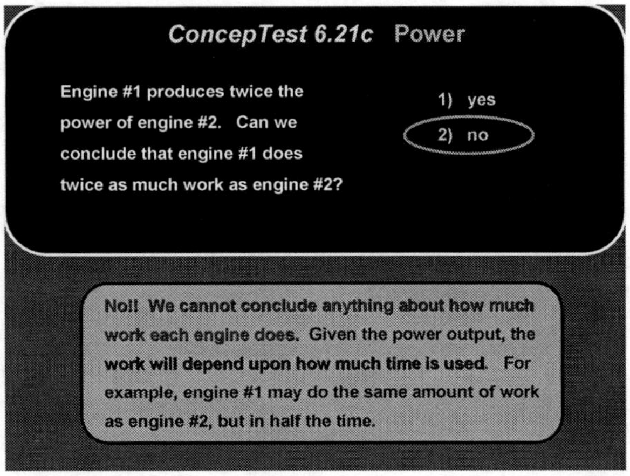

Conceptual Module 6.22 (two parts)

ConcepTest 6.22a: Electric Bill
Difficulty Level: 2

Power and work are not the same quantities. They are related, but not the same. Students often confuse the two concepts, and the distinction must be made clear. In this case, the units of energy (from an electric bill) carry the information to lead students to realize that they are paying for energy, not power.

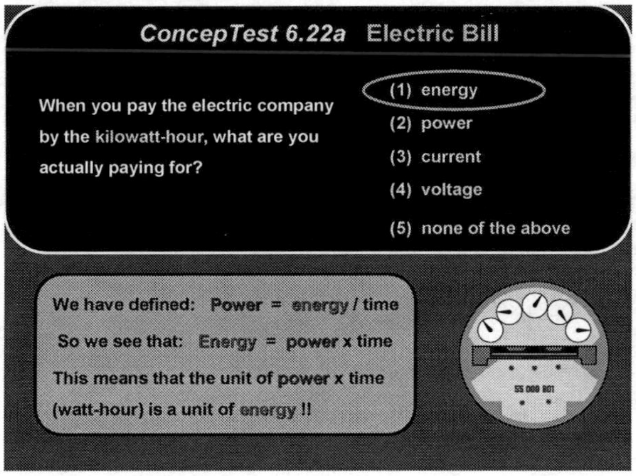

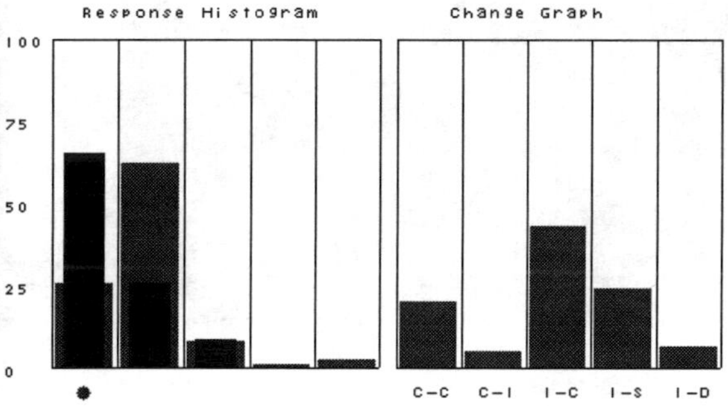

ConcepTest 6.22b: Energy Consumption
Difficulty Level: 2

Here again, the distinction between power and energy is emphasized. The total energy expended depends on the time that an appliance is in operation. Obviously, a higher-power device will use energy *faster*, but the total energy still depends on the time.

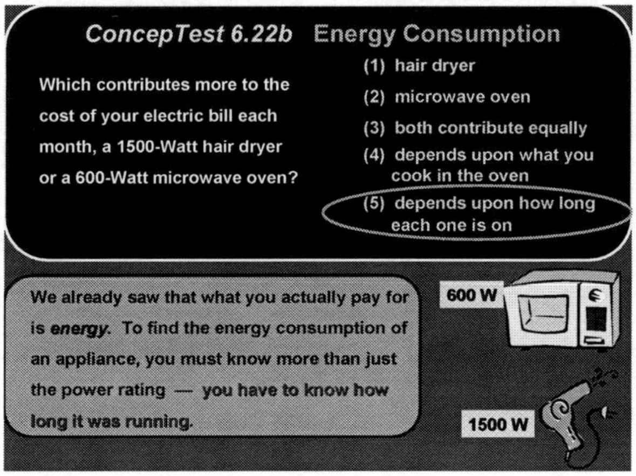

Conceptual Module 7.1

ConcepTest 7.1: Rolling in the Rain
Difficulty Level: 2

This is a case of momentum conservation in its purest form. As the mass of the cart increases (due to the rain collecting), the velocity must decrease to maintain the same value of momentum. Note that the rain falling vertically does not exert any horizontal force on the rolling cart.

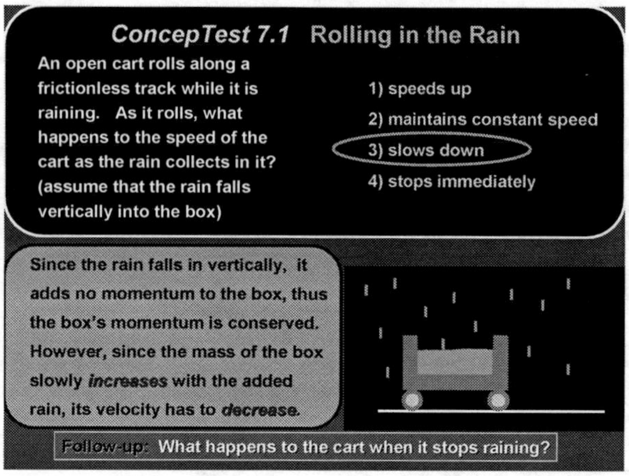

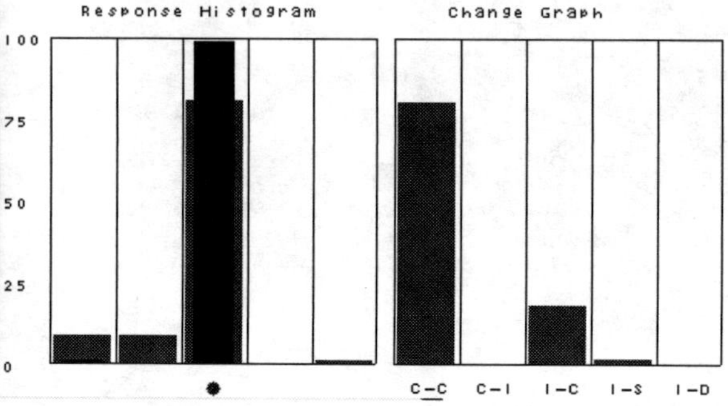

Conceptual Module 7.2 (three parts)

ConcepTest 7.2a: Momentum and KE I
Difficulty Level: 2

In the following questions, the students are asked to think about a system of particles and to draw conclusions about the energy and momentum. In this first case, the only way the kinetic energy (a scalar quantity) can be zero is for all velocities to be zero.

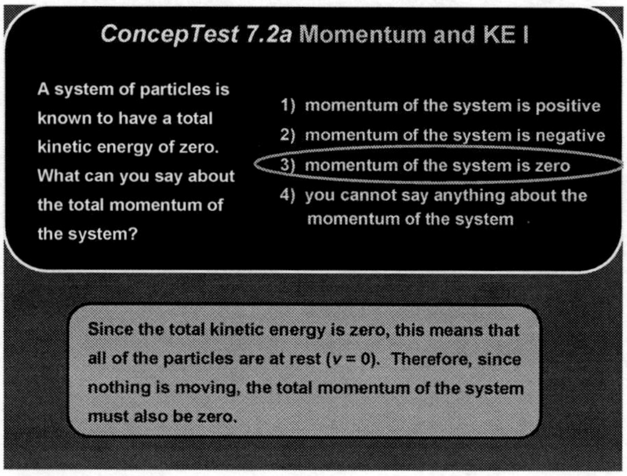

ConcepTest 7.2b: Momentum and KE II
Difficulty Level: 3

In this question, the total momentum can be zero even for finite velocities. The vector nature of momentum means that terms with equal and opposite momenta can cancel.

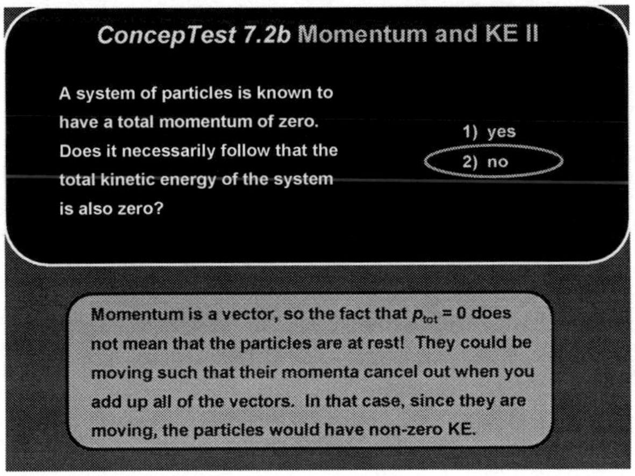

ConcepTest 7.2c: Momentum and KE III
Difficulty Level: 3

While both kinetic energy and momentum depend on mass and velocity, the dependence on these variables is different in each case. So equal momenta do not imply equal values of the mass and velocity, which is what would be necessary for the kinetic energies to be equal as well.

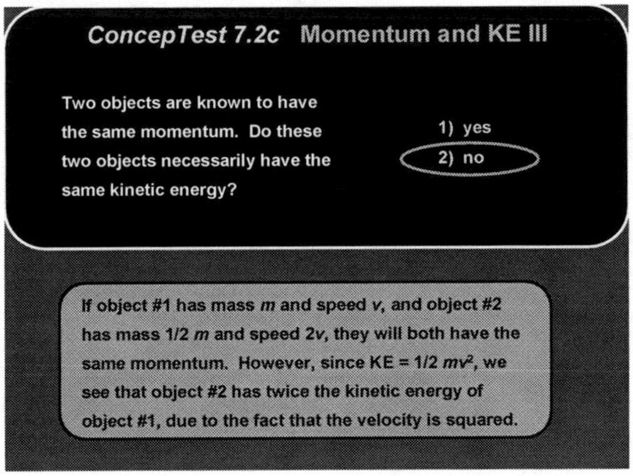

Conceptual Module 7.3 (two parts)

ConcepTest 7.3a: Momentum and Force
Difficulty Level: 2

Force is related to the rate of change of momentum. This question is emphasizing the equivalence of these two quantities. This is essentially a restatement of Newton's Second Law.

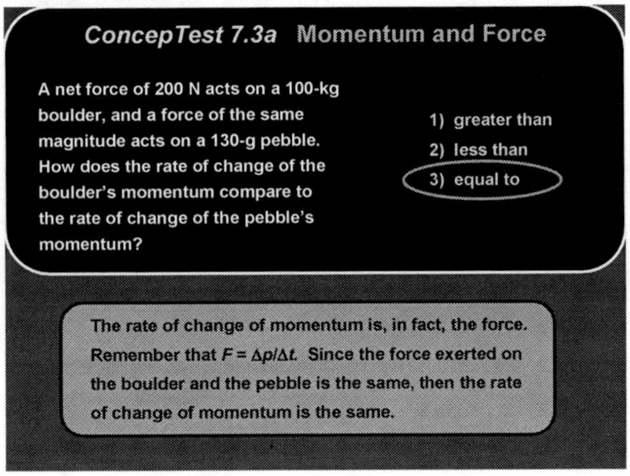

ConcepTest 7.3b: Velocity and Force
Difficulty Level: 3

Using the relationship examined in the previous question, the students must now realize that momentum is related to mass and velocity in order to address the question of the rate of change of velocity (which is the acceleration).

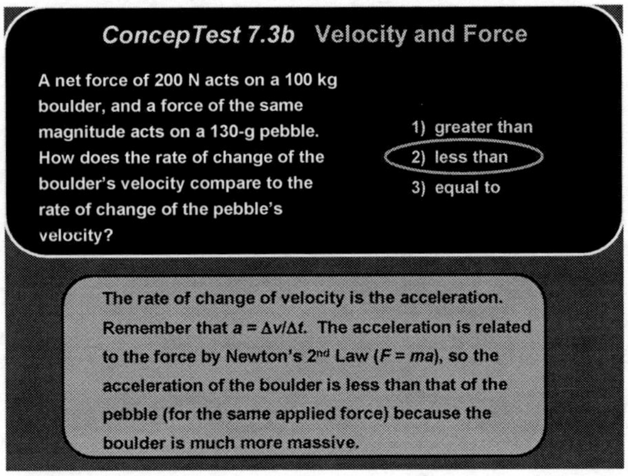

Conceptual Module 7.4

ConcepTest 7.4: Collision Course
Difficulty Level: 2
Suggested follow-up conceptual question:

- **Which one feels the larger acceleration?**

This question examines the relationship between momentum and force as it pertains to internal forces in a closed system. This also underscores the importance of Newton's Third Law in the discussion of momentum conservation.

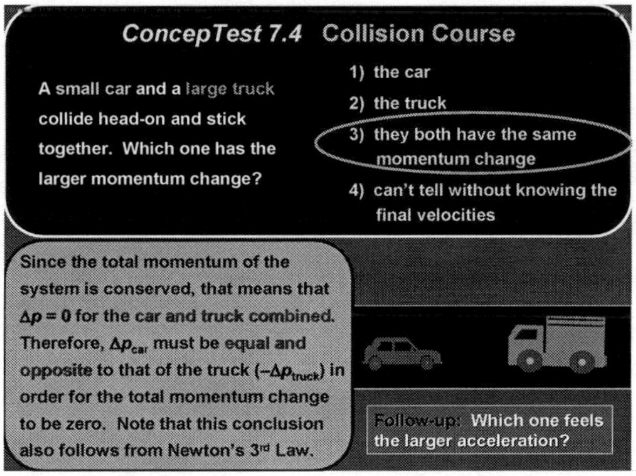

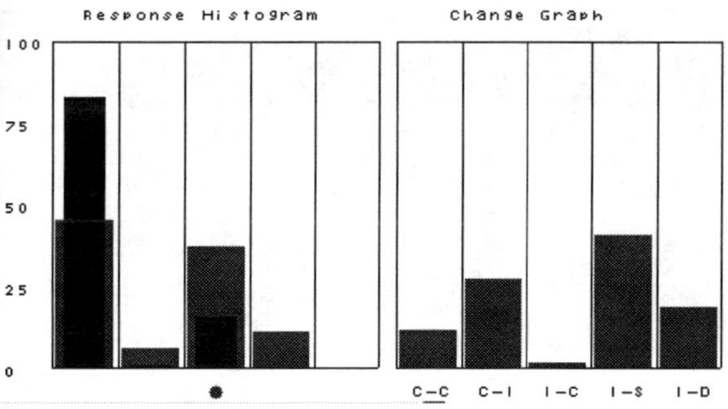

Conceptual Module 7.5 (two parts)

ConcepTest 7.5a: Two Boxes I
Difficulty Level: 2

Impulse is directly related to the change in momentum. In this case, the same force acting over the same time interval in two different situations (regardless of the mass) will give rise to the same momentum change, by the definition of impulse ($F\,\Delta t$).

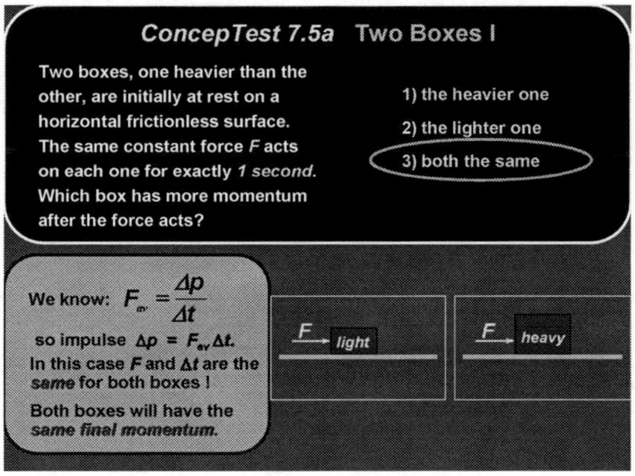

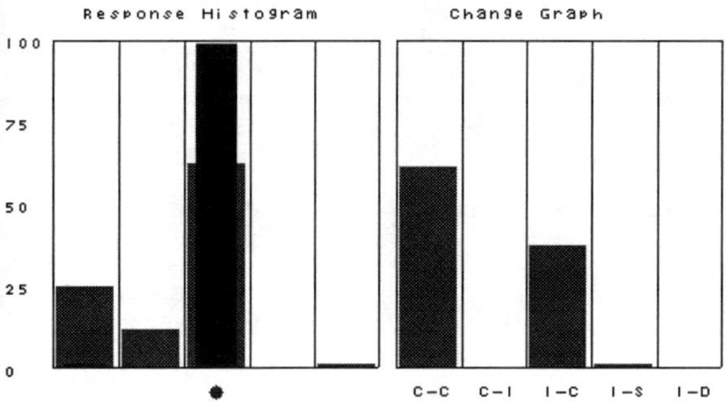

ConcepTest 7.5b: Two Boxes II
Difficulty Level: 3
Suggested follow-up conceptual questions:

- **Which box has gone a larger distance after the force acts?**
- **Which box has gained more KE after the force acts?**

Here the students must break down the momentum into its relationship to mass and velocity in order to determine which box achieves the larger velocity. This result may be more intuitive to the students than the previous result.

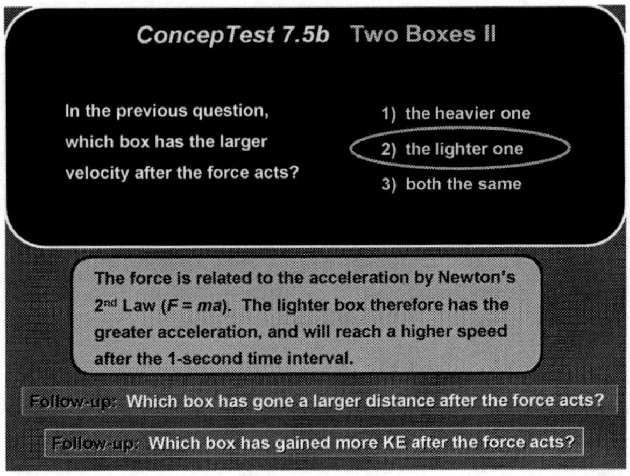

Conceptual Module 7.6

ConcepTest 7.6: Watch Out!
Difficulty Level: 3

This question may again lead to a non-intuitive result, but the fact remains that the momentum change is the same in both cases. The car will come to a dead stop in both cases, and the force can be related to the impulse (momentum change).

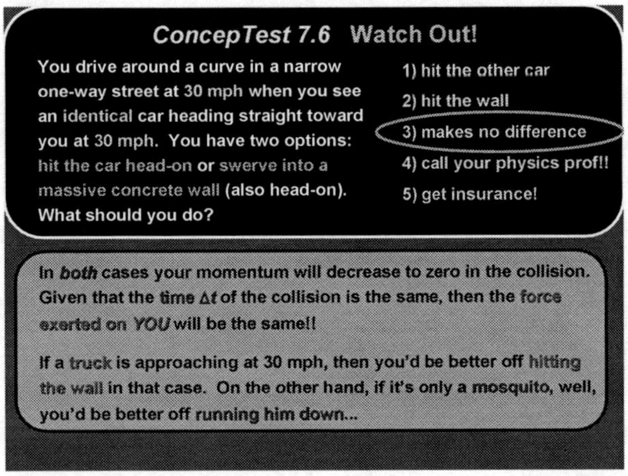

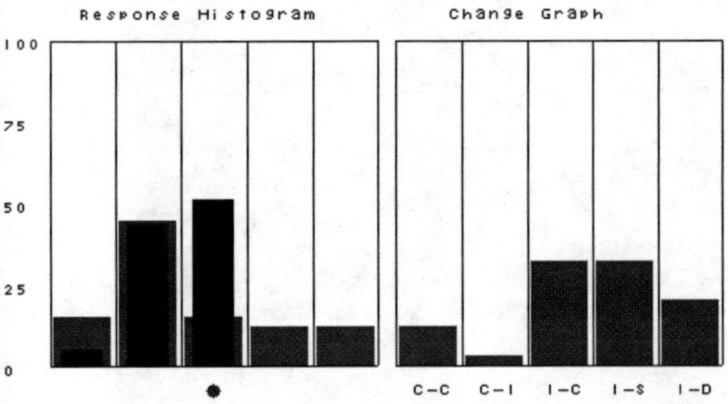

Conceptual Module 7.7

ConcepTest 7.7: Impulse
Difficulty Level: 4
Suggested follow-up conceptual question:

- **Which one imparts the larger force to the floor?**

The impulse is related to the change in momentum. Students must understand that the case where an object rebounds (reverses momentum direction) has a large momentum change, whereas the case where an object comes to rest has a smaller momentum change.

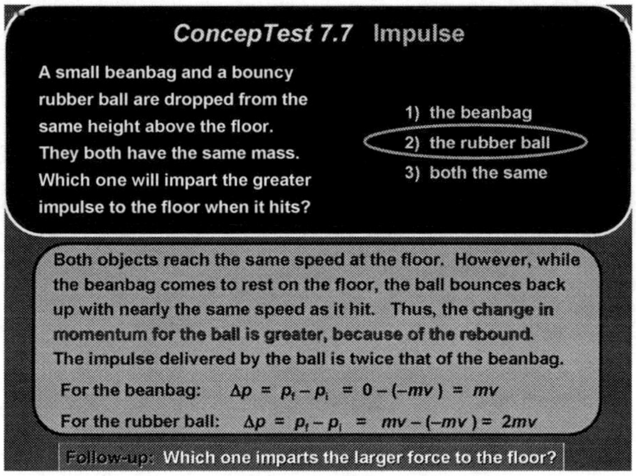

Conceptual Module 7.8

ConcepTest 7.8: Singing in the Rain
Difficulty Level: 4

This question examines the previous issue in more detail. Here the issue of force must be considered, but since the force is a part of the impulse, the extension is not so great. The fact that the hail "bounces off" and the rain does not is what leads to a larger force imparted by the hail.

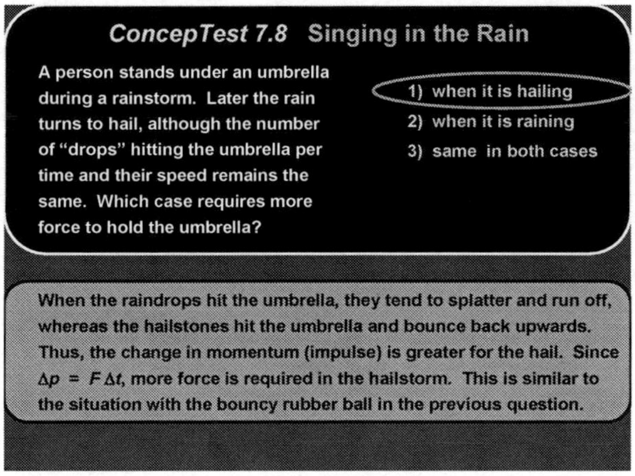

Conceptual Module 7.9 (two parts)
ConcepTest 7.9a: Going Bowling I
Difficulty Level: 3

This question examines the connection between momentum and impulse. Here the students are asked to break down the impulse into its dependence on the force and the time interval over which the force acts.

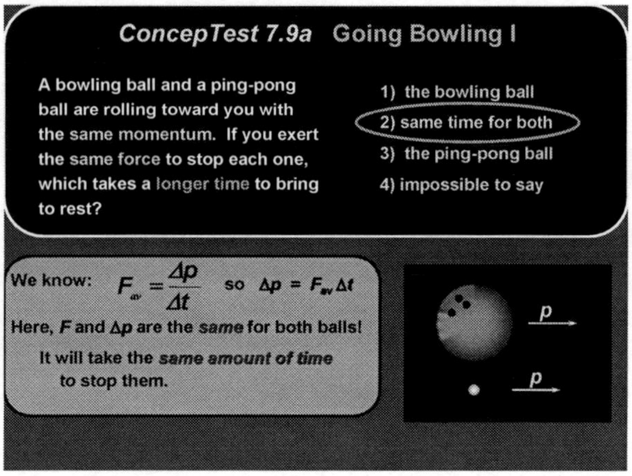

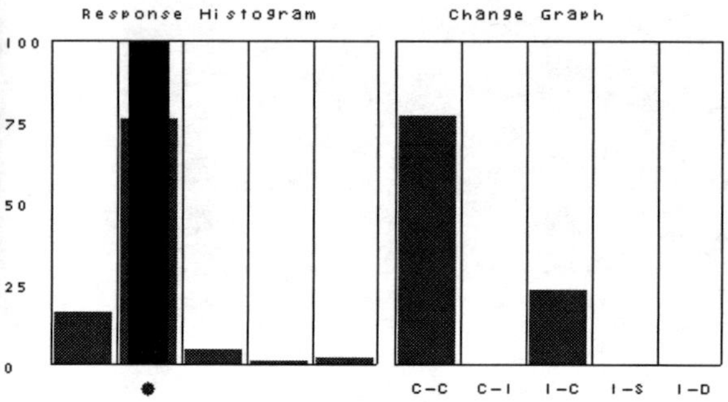

ConcepTest 7.9b: Going Bowling II
Difficulty Level: 5

This more complicated extension requires making the connection between kinetic energy and work. The kinetic energy must be inferred from the momentum, in this case, in order to draw a conclusion about the applied force and the distance (not the time, in this case!) over which it acts.

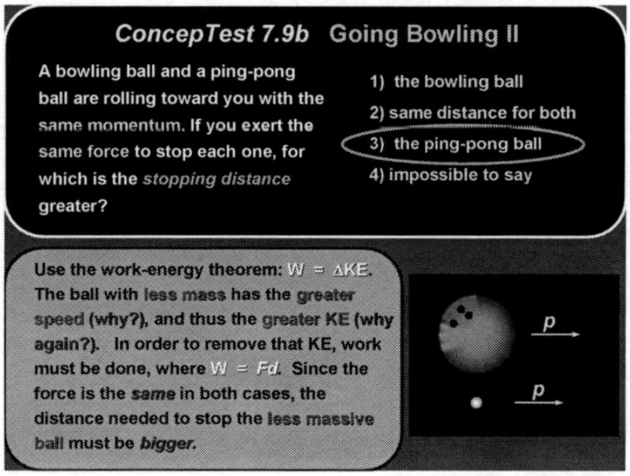

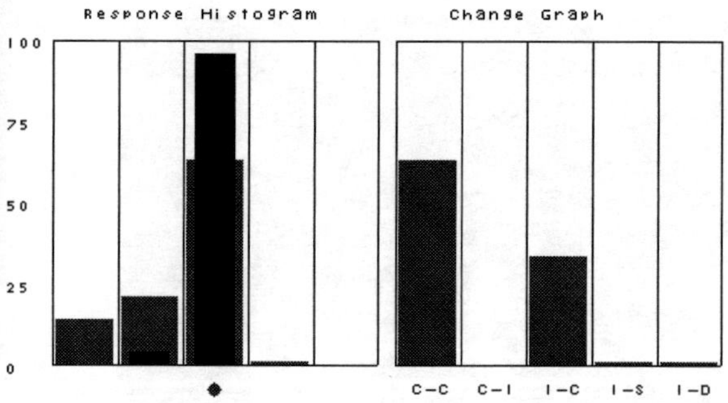

Conceptual Module 7.10 (two parts)

ConcepTest 7.10a: Elastic Collisions I
Difficulty Level: 4

In an elastic collision, the speed of approach before the collision equals the speed of recession after the collision. The students must compare the approach (same in each case) and the recession to determine the answer. Some reasonable assumptions have to be made about the final velocities of the two objects, which gives students a chance to make some independent judgements in the case of $m << M$.

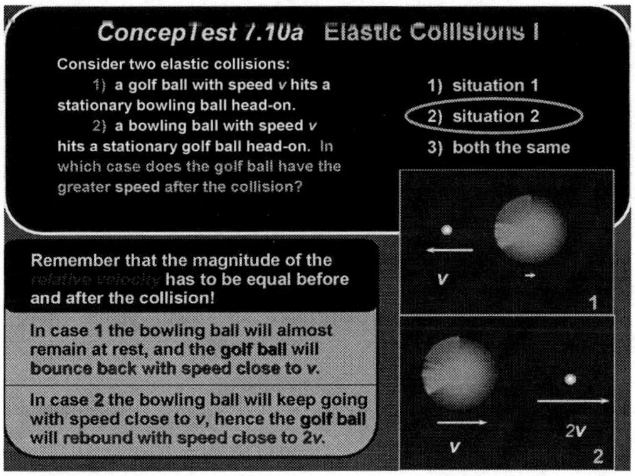

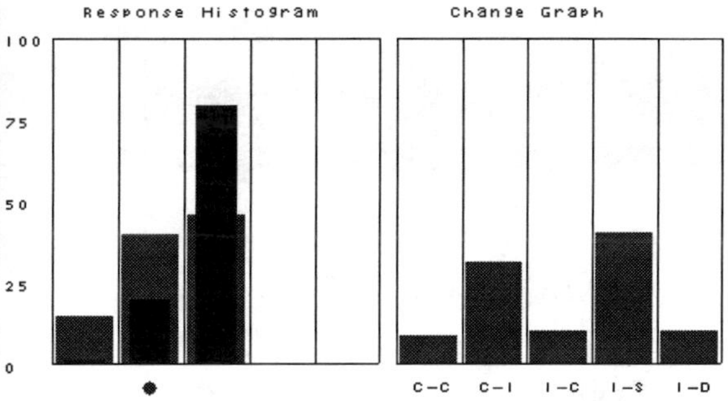

ConcepTest 7.10b: Elastic Collisions II
Difficulty Level: 5
Suggested follow-up conceptual question:

- **With initial drop height *h*, how high does the small rubber ball bounce up?**

This more difficult question relates to a popular classroom demonstration. Here again, the relative velocities must be considered in an elastic collision and some reasonable assumptions must be applied.

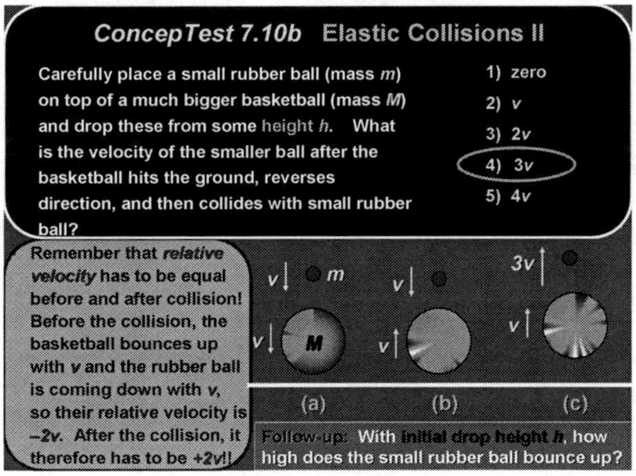

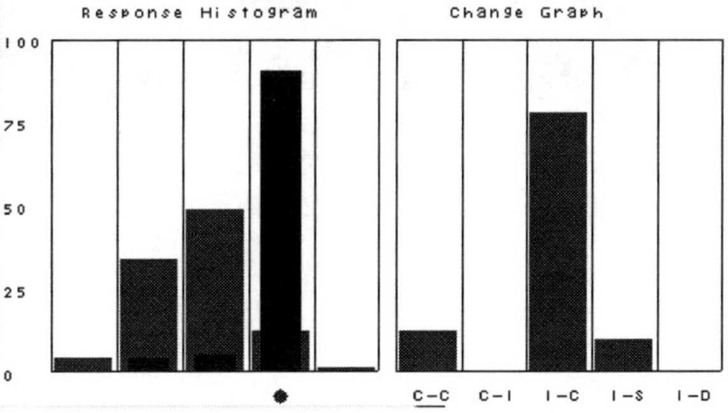

Conceptual Module 7.11

ConcepTest 7.11: Golf Anyone?
Difficulty Level: 3

This revisits the question of the relative speeds in an elastic collision.

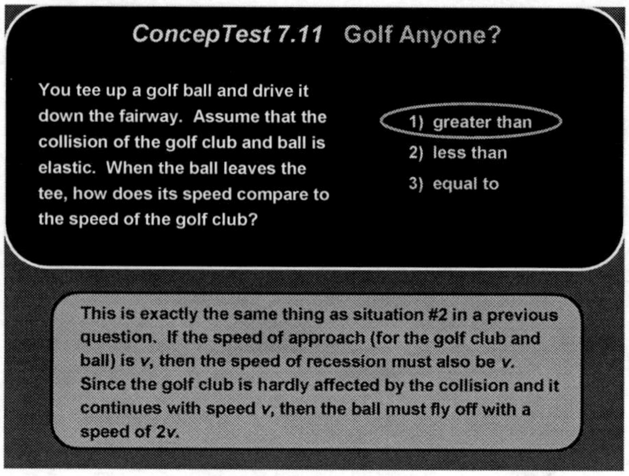

Conceptual Module 7.12 (two parts)

ConcepTest 7.12a: Inelastic Collisions I
Difficulty Level: 3

This is a straightforward application of momentum conservation in the case of an inelastic collision.

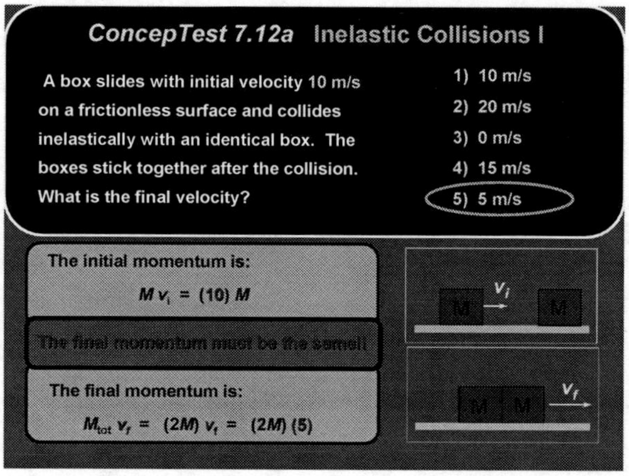

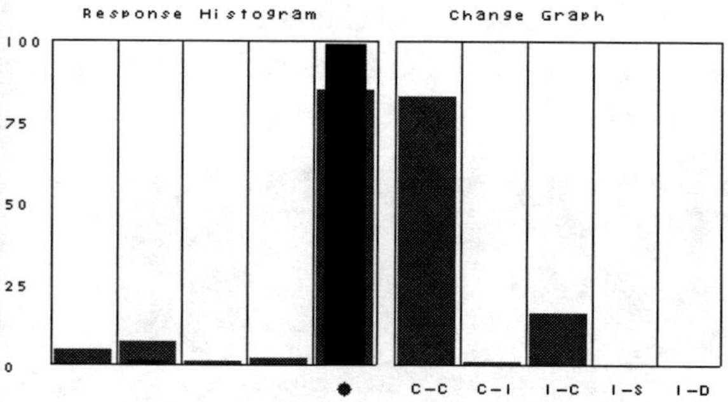

ConcepTest 7.12b: Inelastic Collisions II
Difficulty Level: 4

In this case, the students must extrapolate what they can deduce about the momentum (or actually, the velocity) to the issue of kinetic energy. This requires them to find the final velocity and then use it to compare the initial and final KE values.

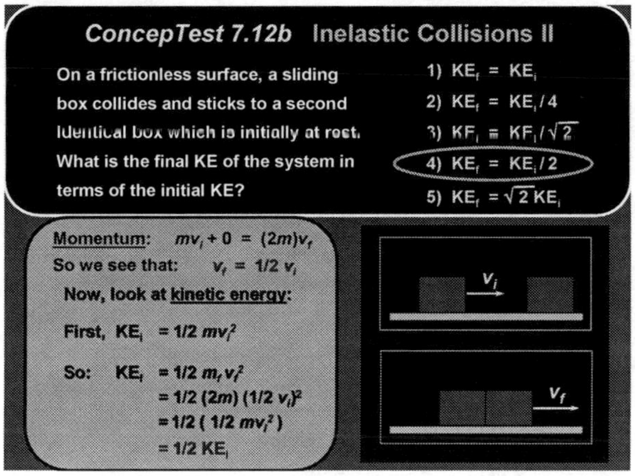

Conceptual Module 7.13 (two parts)
ConcepTest 7.13a: Nuclear Fission I
Difficulty Level: 2

This illustrates momentum conservation in an explosion. This first part establishes that both fragments have the same magnitude of momentum (equal and opposite), since the total momentum of the system must be zero.

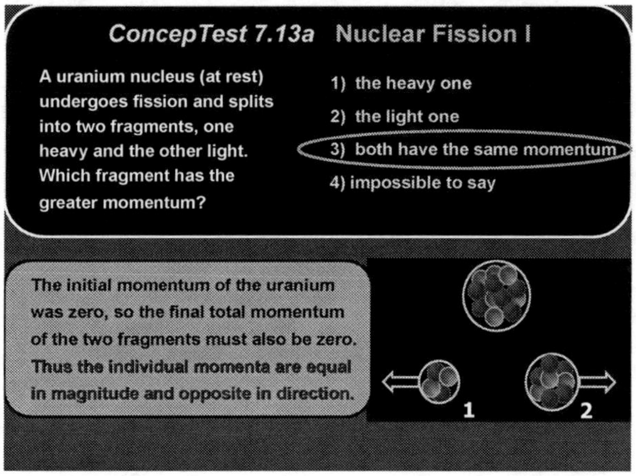

ConcepTest 7.13b: Nuclear Fission II
Difficulty Level: 3

In this part, the students must use what they have learned about the momenta of the fragments to deduce their respective velocities.

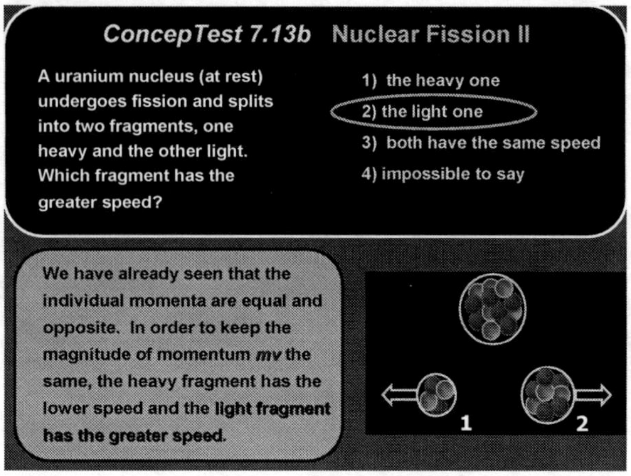

Conceptual Module 7.14 (two parts)

ConcepTest 7.14a: Recoil Speed I
Difficulty Level: 3

This question again examines momentum conservation in an explosion. In the previous ConcepTest, this was treated qualitatively. This case asks for a quantitative analysis.

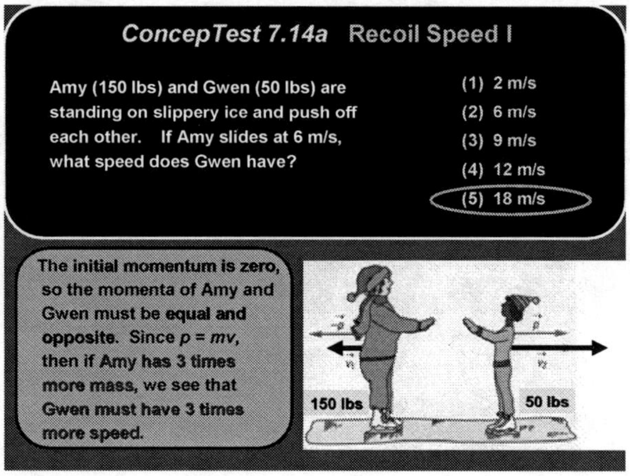

ConcepTest 7.14b: Recoil Speed II
Difficulty Level: 3

This is another quantitative question about momentum conservation in an explosion.

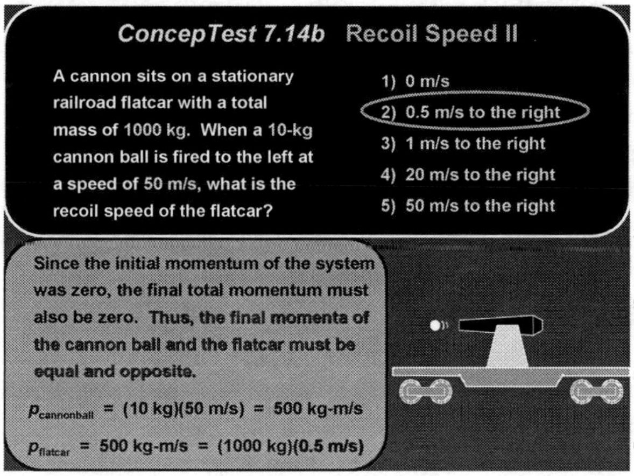

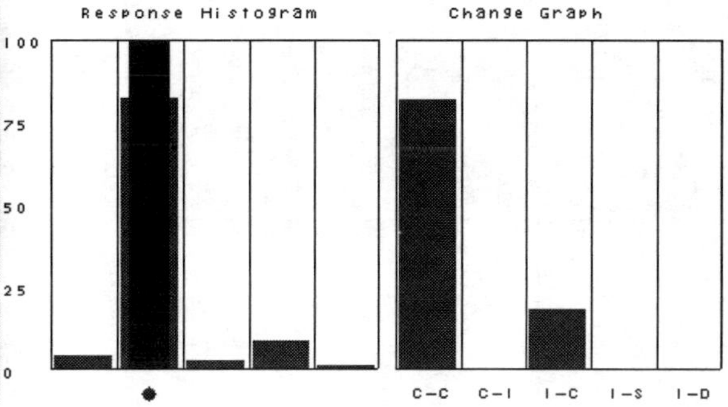

Conceptual Module 7.15

ConcepTest 7.15: Gun Control
Difficulty Level: 3

This question illustrates that the potential for damage is related to the kinetic energy, not specifically the momentum. The recoil of a gun will have the same momentum as the bullet that was fired from the gun, but the kinetic energies can be vastly different due to the fact that KE involves the square of a velocity.

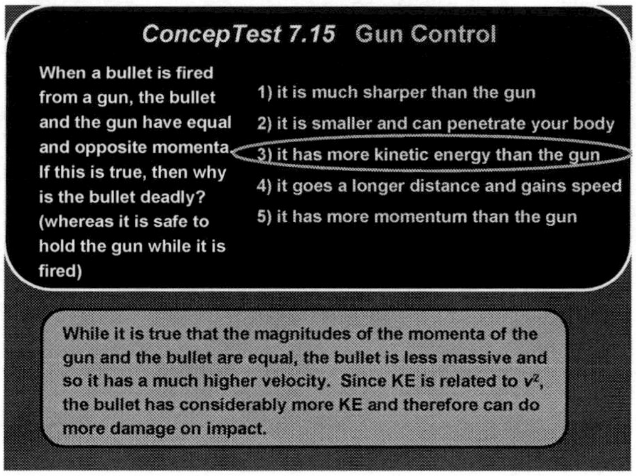

Conceptual Module 7.16 (two parts)

ConcepTest 7.16a: Crash Cars I
Difficulty Level: 3

This is a question of momentum conservation that must be answered by comparing the momenta of the oppositely directed cars.

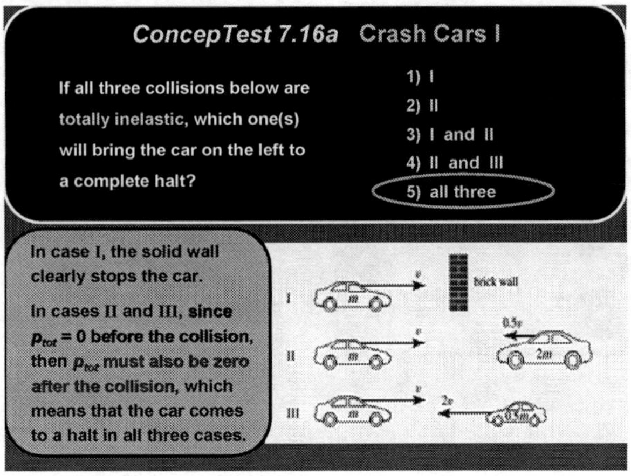

ConcepTest 7.16b: Crash Cars II
Difficulty Level: 4

Having established in the previous ConcepTest that the final velocities are zero in all cases, this next question investigates the damage done in the collisions. In this case, the relevant quantity is the loss of kinetic energy, which will depend heavily on the velocity, due to the fact that KE involves the square of the velocity.

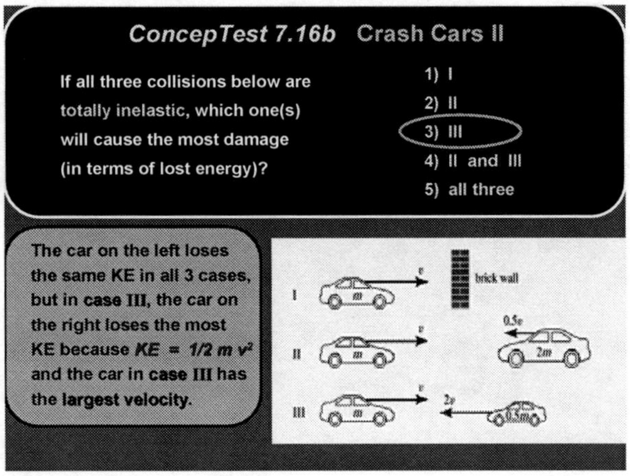

Conceptual Module 7.17
ConcepTest 7.17: Shut the Door!
Difficulty Level: 4

The issue of effectiveness relates to the one that will impart the greatest impulse (momentum change) to the door. As discussed in an earlier ConcepTest, the case where the ball rebounds will involve the greater momentum change, as opposed to the case where the ball simply comes to rest.

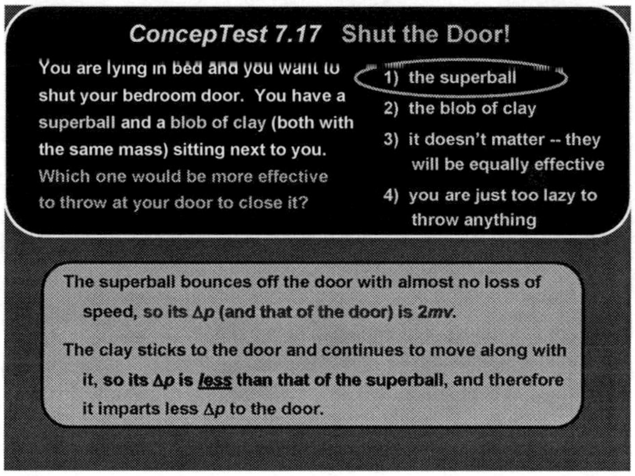

Conceptual Module 7.18
ConcepTest 7.18: Baseball Bat
Difficulty Level: 3

The center of mass must be determined by considering the distribution of mass in an extended object. Here, with more mass concentrated in the thick part of the bat, the center of mass will be closer to that end (as opposed to being at the midpoint).

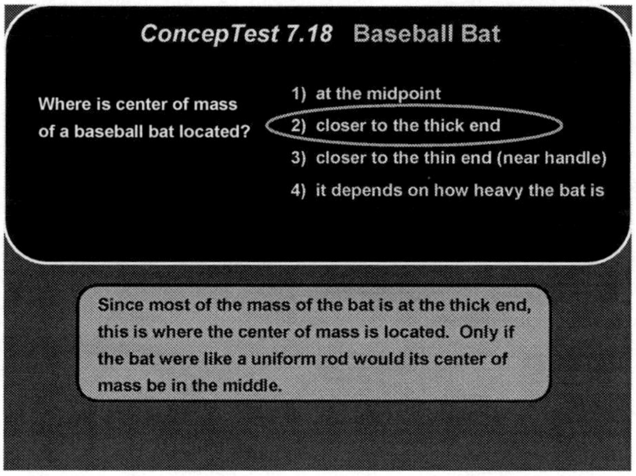

Conceptual Module 7.19
ConcepTest 7.19: Motion of CM
Difficulty Level: 4

This is a question about the center of mass of a two-particle system, where one of the particles is moving. The location of the center of mass must be deduced as a function of time.

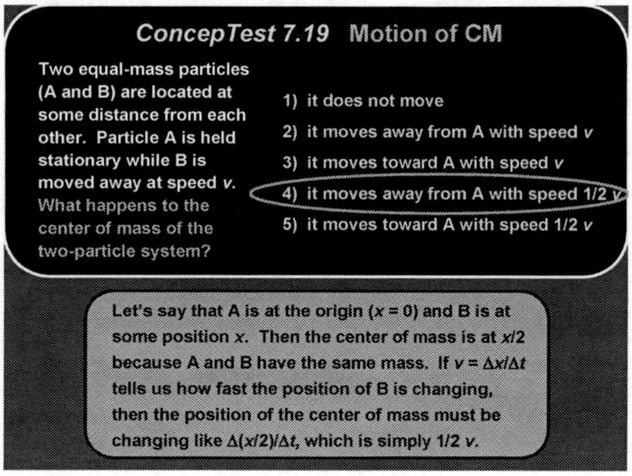

Conceptual Module 7.20
ConcepTest 7.20: Center of Mass
Difficulty Level: 4

This question requires the center of mass of two separate extended objects to be identified, and then the overall center of mass to be determined. Since flipping the two semi-disks alters the mass distribution in the resulting shape (compared to the original disk), the center of mass will change.

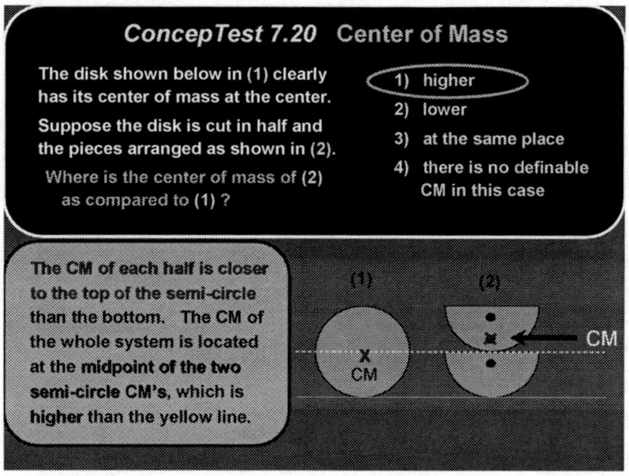

Conceptual Module 8.1 (two parts)

ConcepTest 8.1a: Bonnie and Klyde I
Difficulty Level: 1

Students must distinguish between angular velocity and linear velocity. For two objects on a rigid body that is rotating, the angular velocities are the same (even if the objects are at different radii).

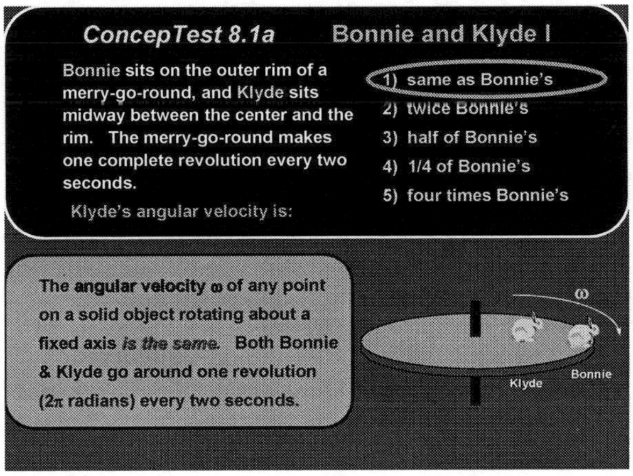

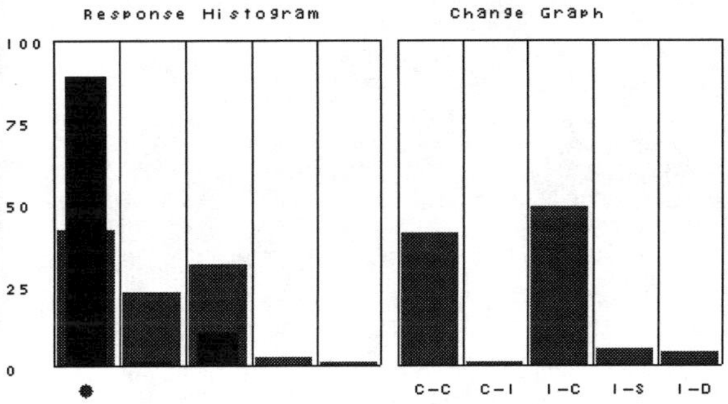

ConcepTest 8.1b: Bonnie and Klyde II
Difficulty Level: 2
Suggested follow-up conceptual question:

- **Who has the larger centripetal acceleration?**

In this case, the linear velocity is examined. Here the position of the object relative to the center will matter. Objects at larger radii will move faster. Students should be able to relate this to their own experience with a rotating turntable (or merry-go-round) usually found in a playground.

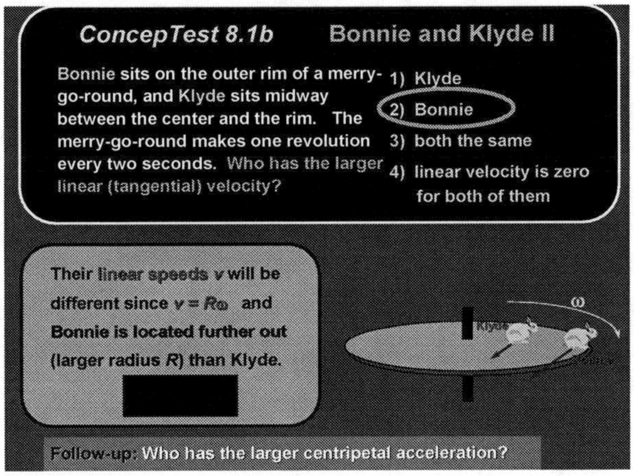

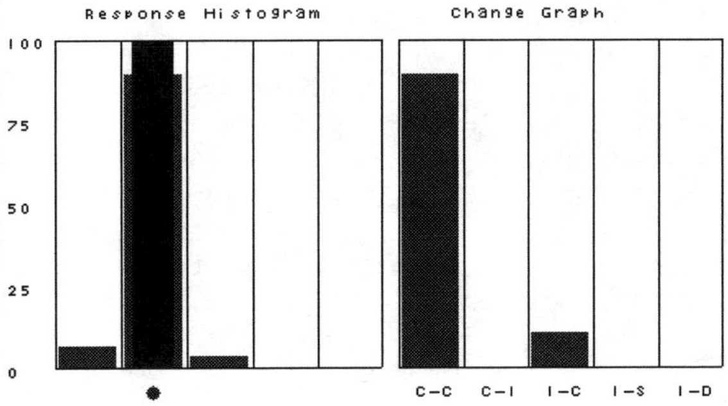

Conceptual Module 8.2

ConcepTest 8.2: Truck Speedometer
Difficulty Level: 4

This question asks students to relate angular velocity to linear velocity. In this case, the speedometer actually measures angular velocity and then converts to linear velocity, assuming a particular tire radius. Students must realize that the larger tires will allow the truck to move faster (linearly) for a given angular velocity of the tires. Thus the speedometer will read too low.

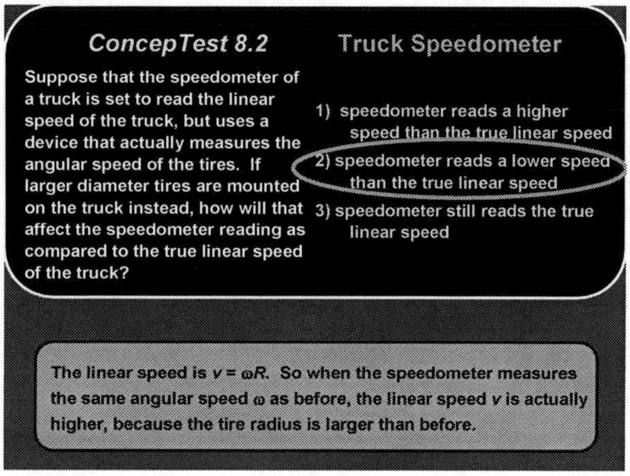

Conceptual Module 8.3 (two parts)

ConcepTest 8.3a: Angular Displacement I
Difficulty Level: 3

This question addresses rotational kinematics, which is quite analogous to linear kinematics. The angular displacement is related to the angular acceleration by the square of the time, so a time period that is half the original time will lead to an angular displacement that is a quarter of the original value.

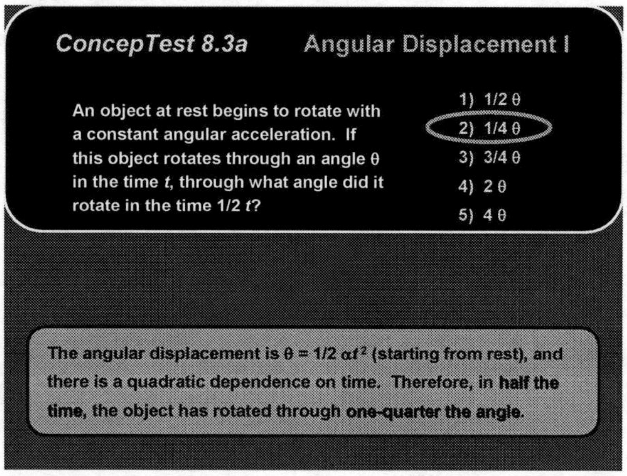

ConcepTest 8.3b: Angular Displacement II
Difficulty Level: 3

This question continues to address rotational kinematics. The angular velocity is linearly related to the angular acceleration, so a time period that is half the original time will lead to an angular velocity that is also half the original value.

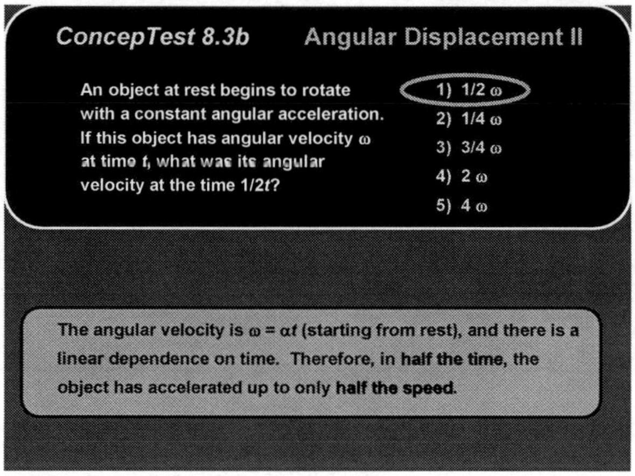

Conceptual Module 8.4

ConcepTest 8.4: Using a Wrench
Difficulty Level: 2
Suggested follow-up conceptual question:

- **What is the difference between arrangement 1 and 4?**

Students often have a great deal of trouble with torque. In this question, the force is assumed fixed, so the issue here is the finding the largest lever arm. This is accomplished in the case where the force is applied at the greatest perpendicular distance from the axis of rotation. Cases #1 and #4 are, in fact, identical in this regard – the extension used in case #4 essentially gains nothing.

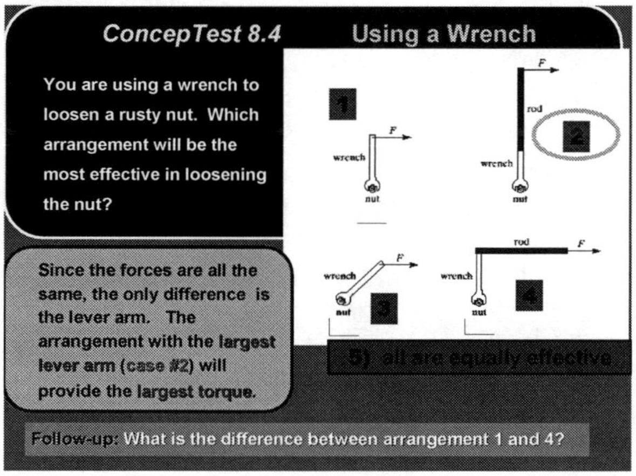

Conceptual Module 8.5

ConcepTest 8.5: Two Forces
Difficulty Level: 2
Suggested follow-up conceptual question:

- **If two torques are identical, does that mean their forces are identical as well?**

Torque is not the same as force. Clearly, force is involved, but the lever arm must also be taken into account. Thus, two equivalent torques do not necessarily imply that the forces are equal.

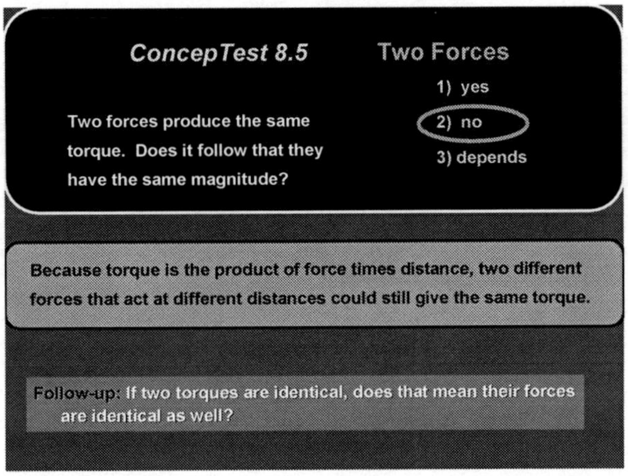

Conceptual Module 8.6

ConcepTest 8.6: Closing a Door
Difficulty Level: 2
Suggested follow-up conceptual question:

- **How large would the force have to be for F_4?**

Students often have difficulty identifying the lever arm when the force is not applied at right angles to the displacement vector from the axis of rotation. This question gives some practice for forces applied at the same position in space (but with different lever arms).

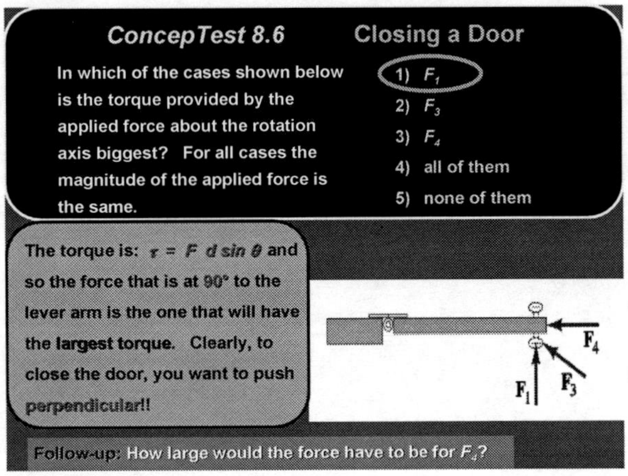

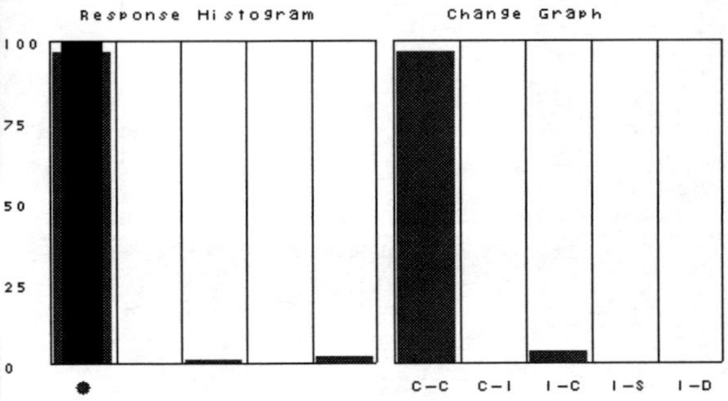

Conceptual Module 8.7

ConcepTest 8.7: Cassette Player

Difficulty Level: 2

In this question, the tension force in a rotating spool of tape is assumed constant, and the students are asked to examine the variation in torque as the effective lever arm changes. The students must recognize that the lever arm is decreasing as the spool empties, so the torque also decreases.

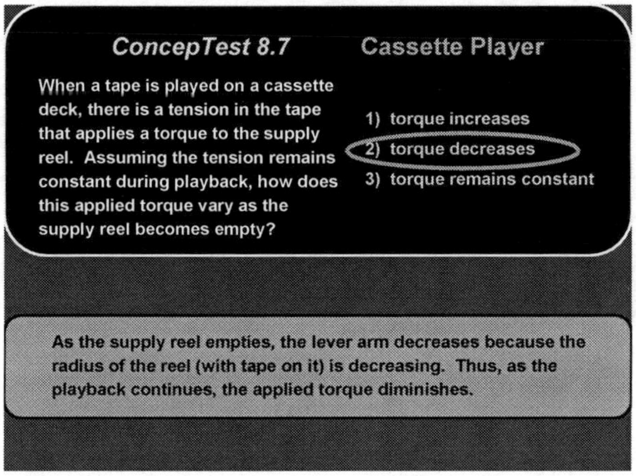

Conceptual Module 8.8 (two parts)

ConcepTest 8.8a: Dumbbell I

Difficulty Level: 2

This question relates linear motion to rotational motion for a given applied force. The impulse ($F\ \Delta t$) to the system is the same in both cases, so the motion of the center of mass will be the same also.

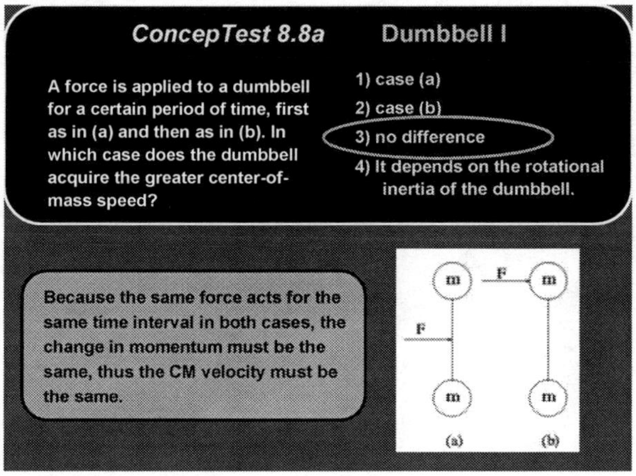

ConcepTest 8.8b: Dumbbell II
Difficulty Level: 3

In this follow-up question, students now have to account for the rotation of the dumbbell. Since the linear motions are the same in both cases, the translational energies will also be the same. The rotational motion of the dumbbell contributes additional energy to the system in case (b).

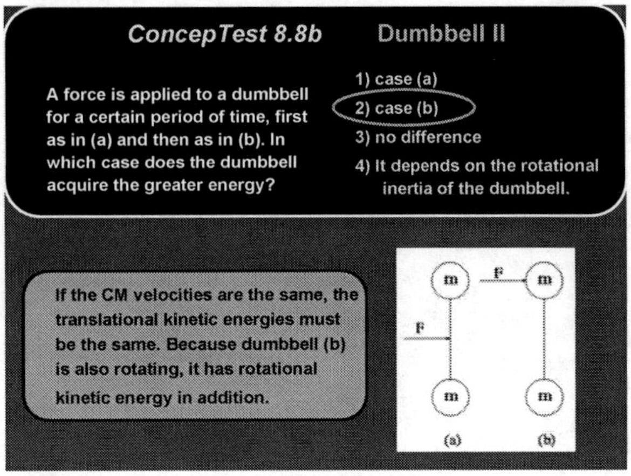

Conceptual Module 8.9

ConcepTest 8.9: Moment of Inertia
Difficulty Level: 3

The moment of inertia depends upon the distribution of mass around an axis of rotation. Considering how this distribution affects the situation is often a challenge for students. The case in which more of the mass is located at a larger distance (the gold sphere) will have the larger moment of inertia.

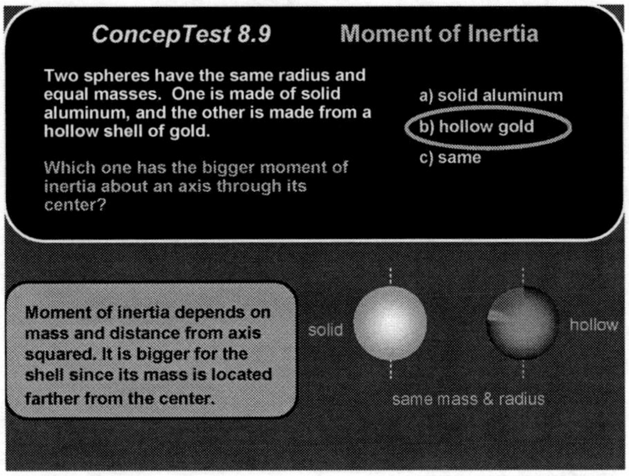

Conceptual Module 8.10

ConcepTest 8.10: Figure Skater

Difficulty Level: 4

Suggested follow-up conceptual question:

- **Where does the extra energy come from?**

The students must make the distinction between angular momentum and rotational energy. By angular momentum conservation, the skater speeds up her rotation by decreasing her moment of inertia. But the rotation KE depends upon the square of the angular speed, so this fact leads to an increase in rotational KE (the decrease in moment of inertia only enters as a linear term).

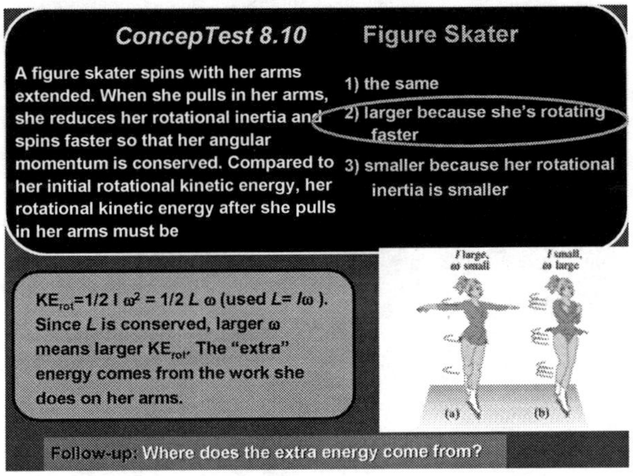

Conceptual Module 8.11

ConcepTest 8.11: Two Disks

Difficulty Level: 4

This question also asks students to examine the moment of inertia and angular velocity in terms of the angular momentum and the rotational KE.

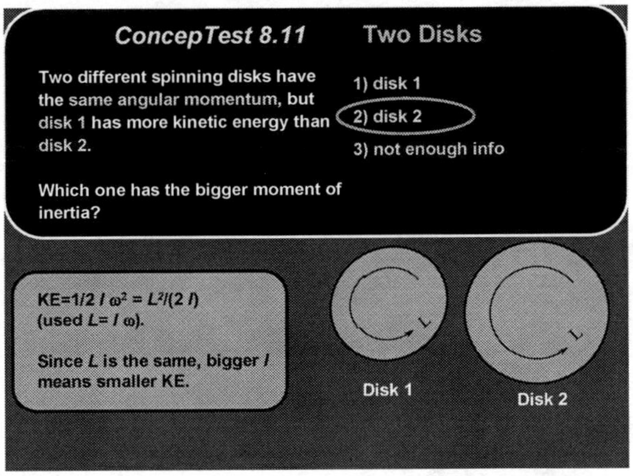

Conceptual Module 8.12

ConcepTest 8.12: Spinning Bicycle Wheel
Difficulty Level: 2

This is a classic demonstration. In order to keep the angular momentum constant, when you flip the spinning bicycle wheel, the turntable must then rotate in the same direction in order to compensate for the "loss" of angular momentum. Students must realize that there are two contributions to the angular momentum in this question – the bicycle wheel and the turntable itself.

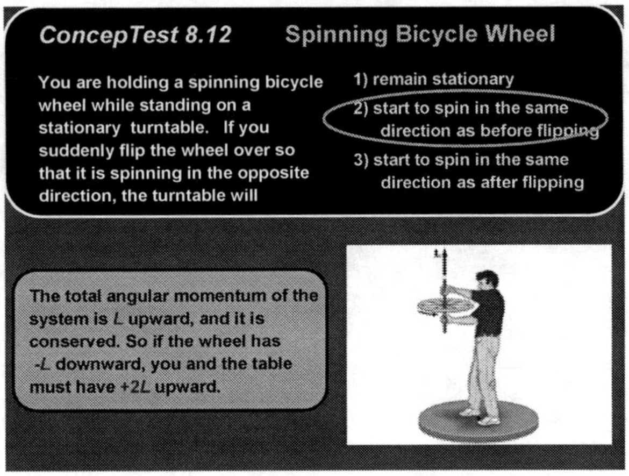

Conceptual Module 9.1
ConcepTest 9.1: Balancing Rod
Difficulty Level: 4

This is a simple balancing problem, but one of the objects is an extended object (the rod). The students must realize that the rod can effectively be "replaced" by a point mass at the rod's center of mass, and then the answer is apparent.

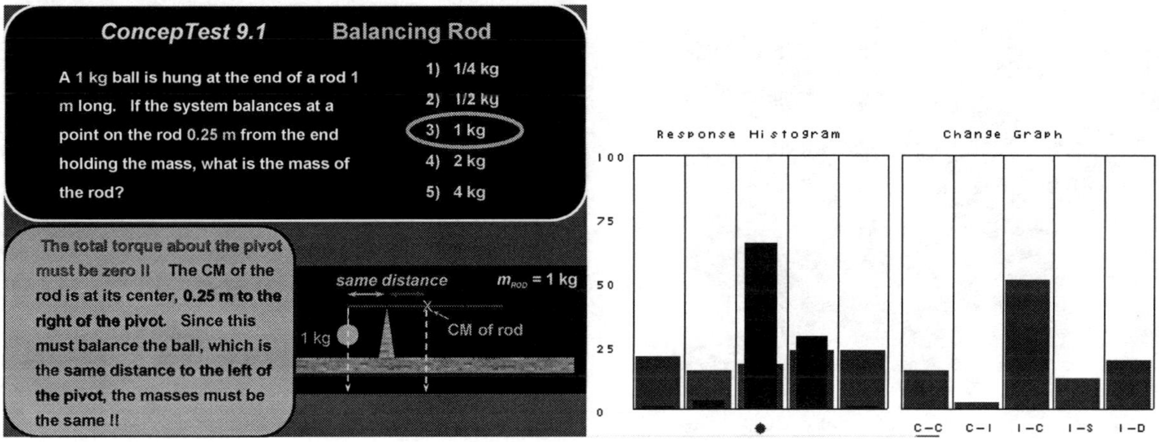

Conceptual Module 9.2
ConcepTest 9.2: Mobile
Difficulty Level: 4

This is another balancing problem, but with multiple steps. As long as the students proceed systematically through the logical steps, starting with the bottom rod, they should be able to work out the values of the missing masses in order to get the total mass.

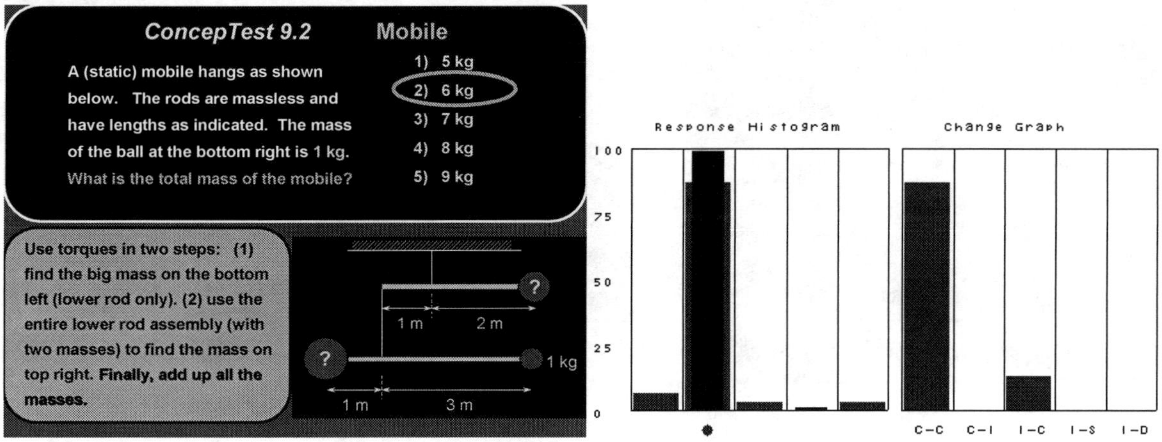

Conceptual Module 9.3 (two parts)

ConcepTest 9.3a: Tipping Over I
Difficulty Level: 2

The stability of an object depends on the position of the center of mass relative to the axis of rotation (for gravitational torques). If the object can rotate in order to lower its center of mass (and hence, reduce its gravitational PE), it will do so. This example also gives an illustration of the problems associated with a top-heavy large object.

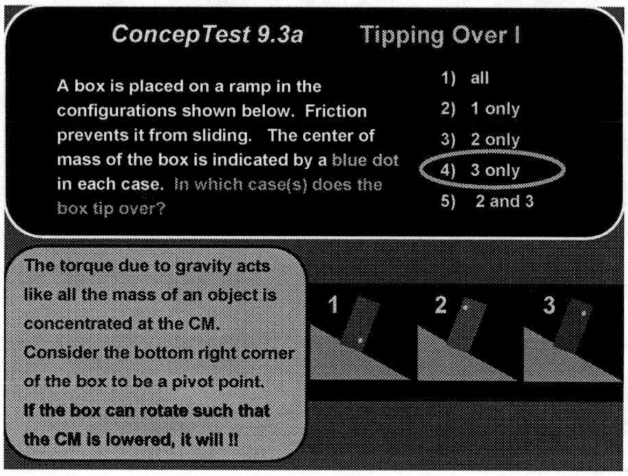

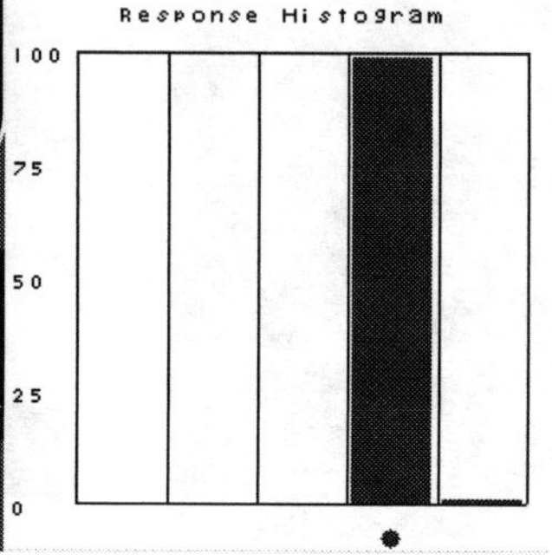

ConcepTest 9.3b: Tipping Over II
Difficulty Level: 4

Here again, the books will topple over if the center of mass of the entire system is out over the edge of the table. To deduce this position, the separate center of mass positions for each book must be considered and then used to find the overall center of mass.

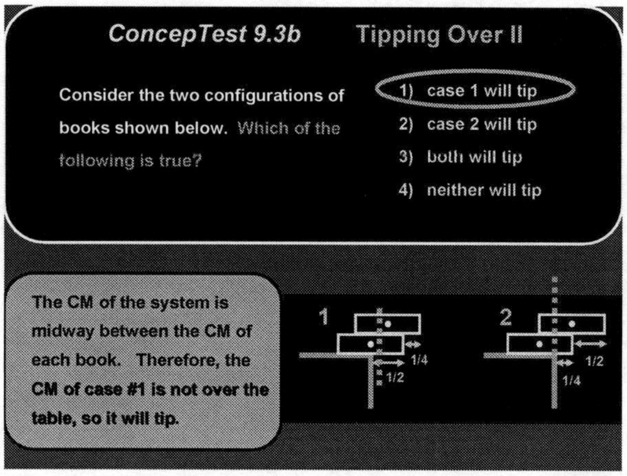

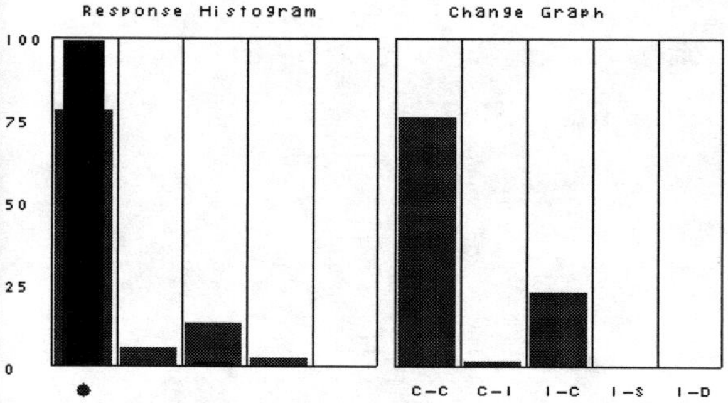

Conceptual Module 10.1
ConcepTest 10.1: Density
Difficulty Level: 1

Density is mass/volume, and so a statement about the density does not necessarily reflect on the mass of the molecules. It could also be referring to the packing of those molecules – that is, more molecules can be packed in a given volume to achieve a higher density.

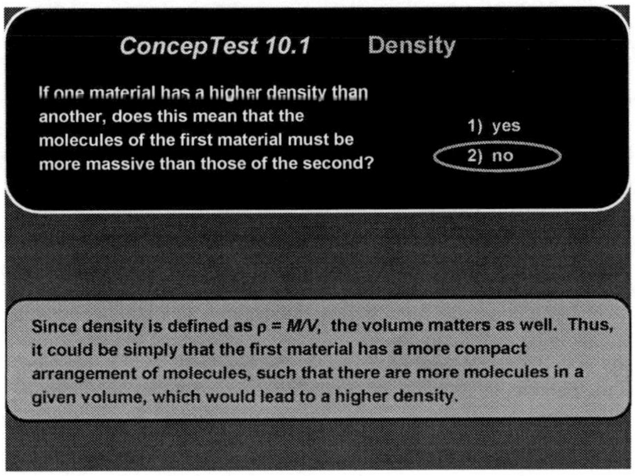

Conceptual Module 10.2
ConcepTest 10.2: Too Much Pressure
Difficulty Level: 2

Pressure is force/area, which is why, for a given force, the pin is more likely to puncture your skin than the blunt end of a pen. The force of the pin is concentrated over a smaller area, making the pressure much greater at the point of contact.

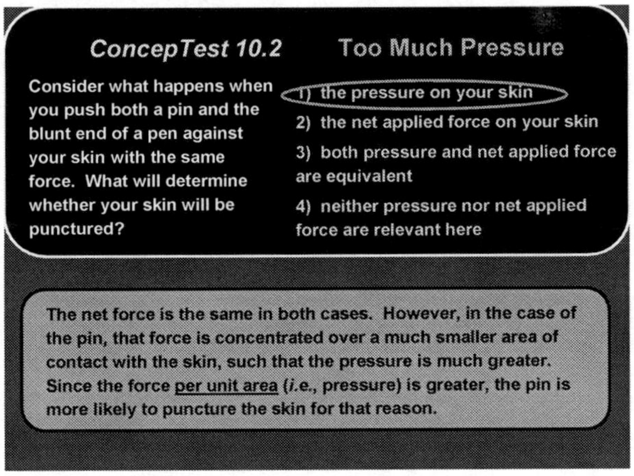

Conceptual Module 10.3

ConcepTest 10.3: On a Frozen Lake
Difficulty Level: 2

This question again addresses the concept of pressure. In this case, the force (your weight) is fixed, so the best thing that you can do is to spread that force over a larger area, which leads to a lower pressure on the ice at any one point.

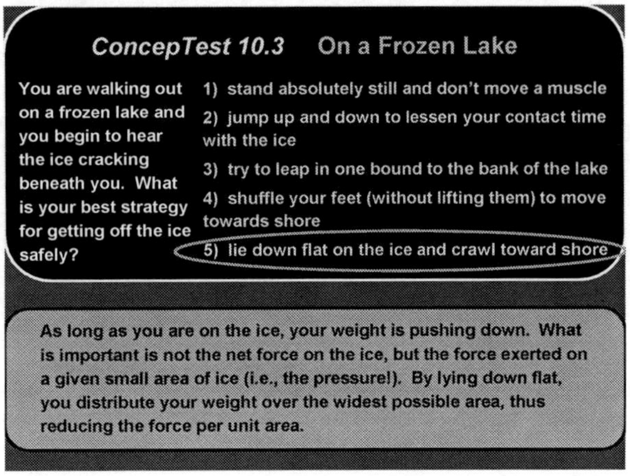

Conceptual Module 10.4

ConcepTest 10.4: Bubbling Up
Difficulty Level: 3

The students must recognize that the water pressure increases with depth, so as the bubble rises, the water pressure outside the bubble is diminishing. This allows the fixed volume of air in the bubble to expand. This also previews the ideal gas law.

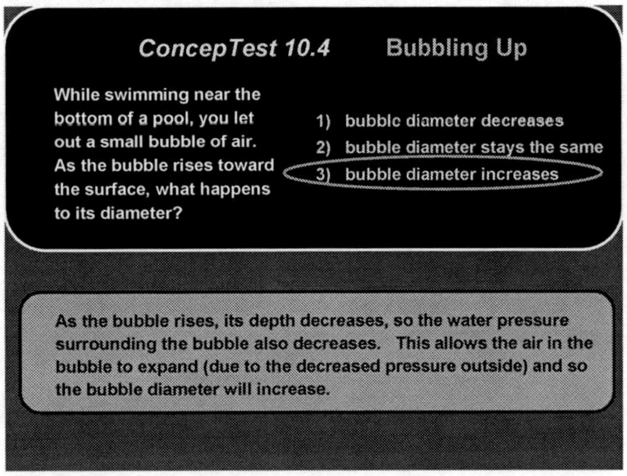

Conceptual Module 10.5
ConcepTest 10.5: Three Containers
Difficulty Level: 1

The pressure depends only on the height of the container and not its shape. This issue also relates to the case where a variety of differently shaped vessels are connected together – the fluid level remains the same in all the vessels, regardless of their rather different shapes.

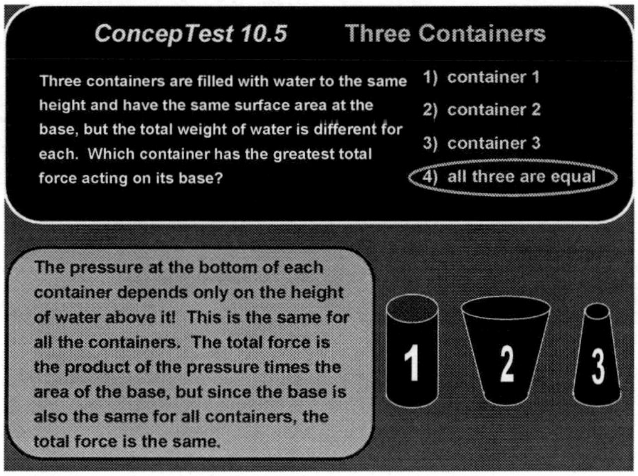

Conceptual Module 10.6
ConcepTest 10.6: The Falling Bucket
Difficulty Level: 3

The students must recognize that when the can is in free fall, the fluid inside is "effectively weightless" and therefore cannot exert any pressure at the bottom of the can. Thus, the fluid flow will cease, due to the lack of water pressure in the can.

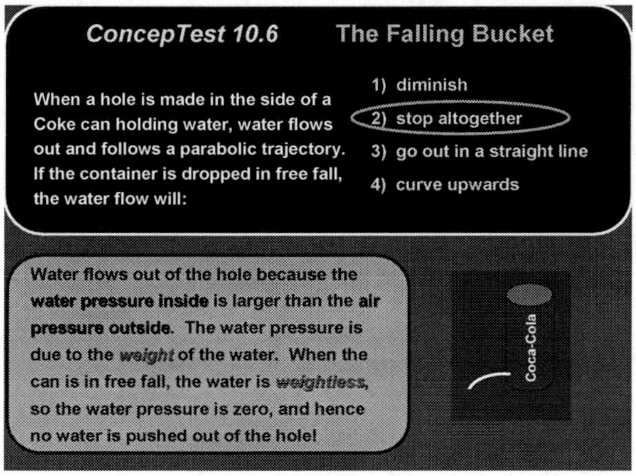

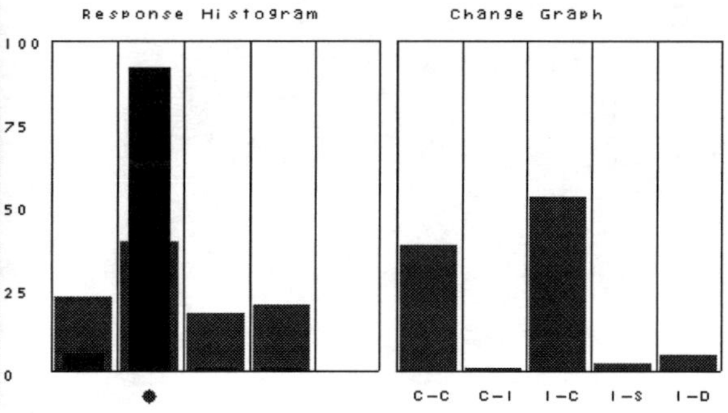

Conceptual Module 10.7 (three parts)

ConcepTest 10.7a: The Straw I
Difficulty Level: 2
Suggested follow-up conceptual question:

- **Is it possible to sip liquid through a straw on the Moon?**

Students often think that "sucking on a straw" involves "pulling" the liquid up the straw, while in reality it is atmospheric pressure "pushing" the liquid from below. The drinker simply reduces the pressure at the top of the straw, and the atmosphere does the rest!

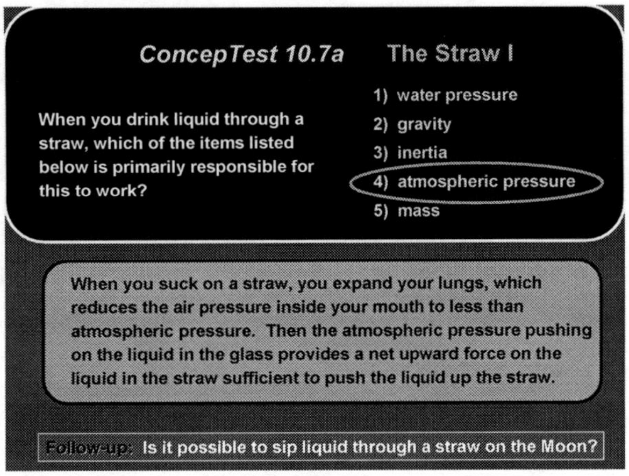

ConcepTest 10.7b: The Straw II
Difficulty Level: 4

This question examines the pressure difference at the top and bottom of a liquid inside a straw that is capped at the top. The students should be able to draw the free-body diagram of the liquid in order to realize that atmospheric pressure is counteracting both the water pressure and the pressure at the top.

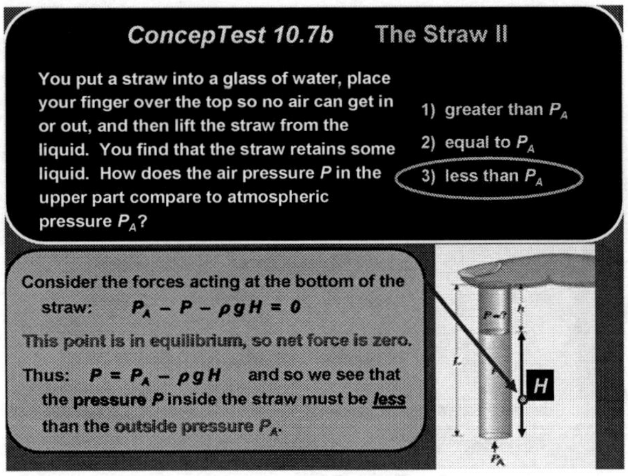

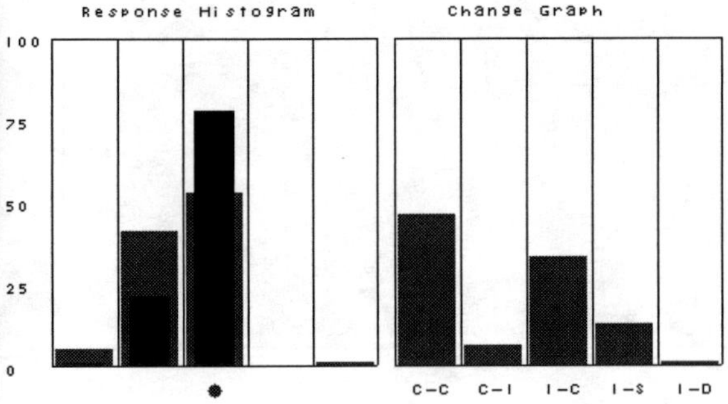

ConcepTest 10.7c: The Straw III
Difficulty Level: 3

Students often have a hard time distinguishing between force and pressure. In this question, the larger barometer does indeed require a larger force to raise the mercury column, but there is a larger area over which the atmospheric pressure can act. Thus, nothing changes, which may be a surprise to students!

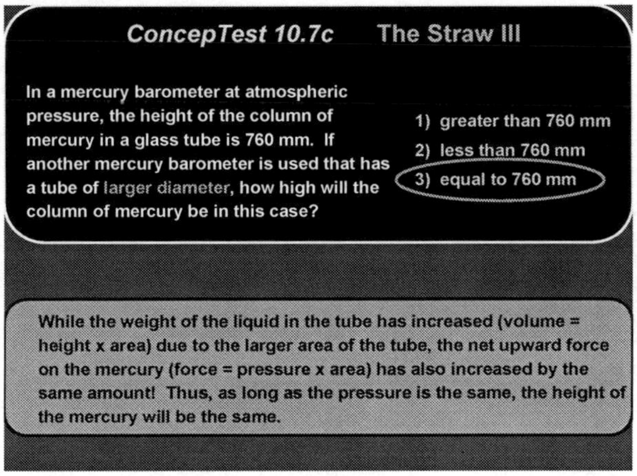

Conceptual Module 10.8
ConcepTest 10.8: Thermometers
Difficulty Level: 2

Any liquid can serve as a barometer, but there is a practical limitation. A water barometer would require 10 m of height! In this question, this realization, coupled with the knowledge that the density of mercury is very high, leads to the practical solution of the everyday mercury barometer.

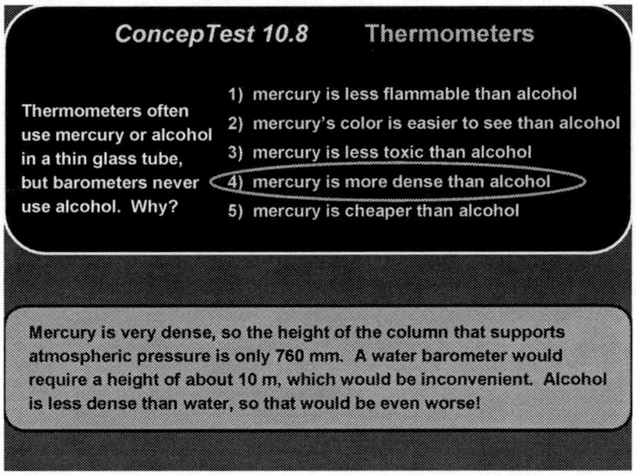

Conceptual Module 10.9
ConcepTest 10.9: Two Bricks
Difficulty Level: 2

The notion that buoyancy is related to the volume of displaced fluid essentially derives from the pressure difference at the top and bottom of a submerged object. While it is true that brick #2 is in a region of higher overall pressure, the *pressure difference* is still the same between the top and bottom of the brick, regardless of its position in the tank! This realization helps demystify the concept of buoyancy for most students.

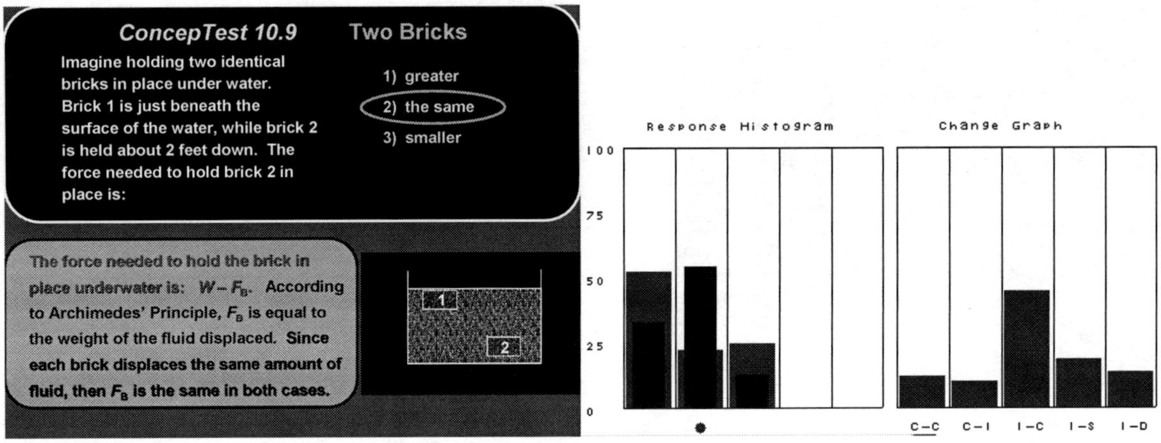

Conceptual Module 10.10 (two parts)
ConcepTest 10.10a: Cylinder and Pail I
Difficulty Level: 3

This question emphasizes Archimedes' Principle (buoyant force equals weight of displaced fluid). By submerging the aluminum cylinder, its "effective weight" is reduced, but the reduction is exactly equal to the water weight displaced, as demonstrated by pouring the displaced water back into the pail.

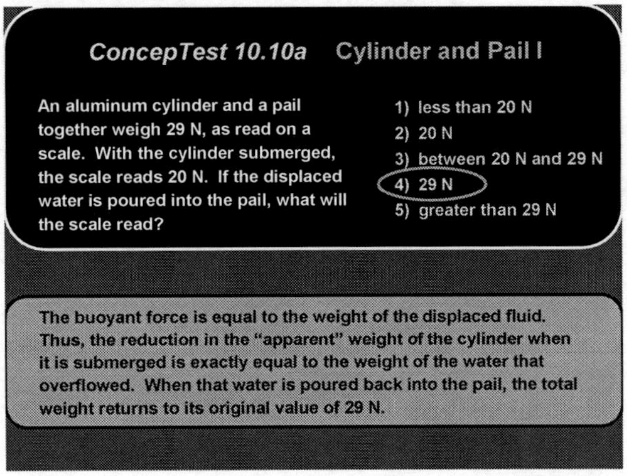

ConcepTest 10.10b: Cylinder and Pail II
Difficulty Level: 5

This is an interesting question involving buoyancy and Newton's Third Law. The buoyant force on the cylinder reduces its "effective weight," as described in the previous question. But this buoyant force must have a reaction force, by Newton's Third Law, and it is this reaction force that is registered by the scale reading for the water.

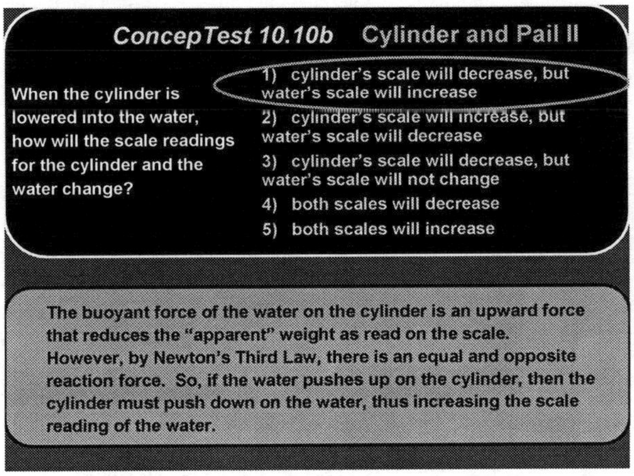

Conceptual Module 10.11

ConcepTest 10.11: On Golden Pond
Difficulty Level: 4

This is actually a situation that most students have already experienced, but probably they have not given it much thought in the past. The question examines the difference between a floating object that displaces its own weight and a sinking object that displaces its own volume. For a massive object, this difference can be substantial.

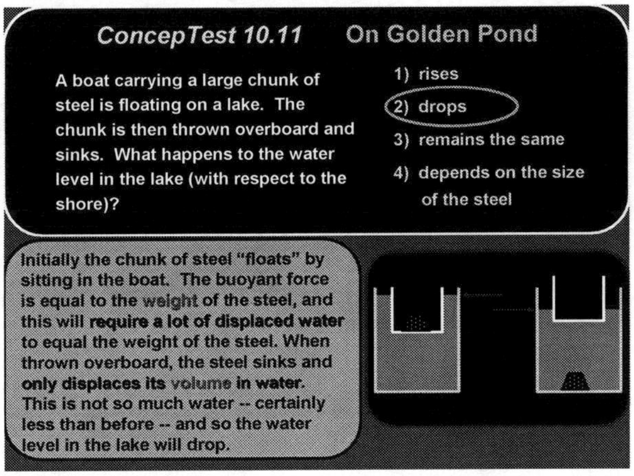

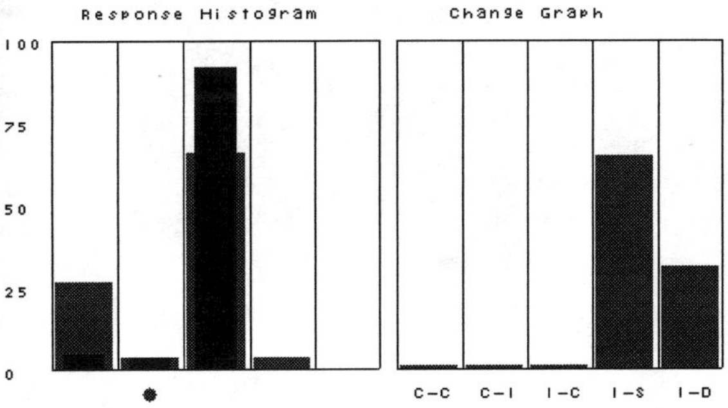

Conceptual Module 10.12 (four parts)
ConcepTest 10.12a: Archimedes I
Difficulty Level: 2

Students should already know that objects float in a fluid of higher density than the object. But there is a specific relation between the fraction of the object that is submerged and the density ratio. It should be intuitive, at least, that the smaller the density ratio, the less material that is actually submerged.

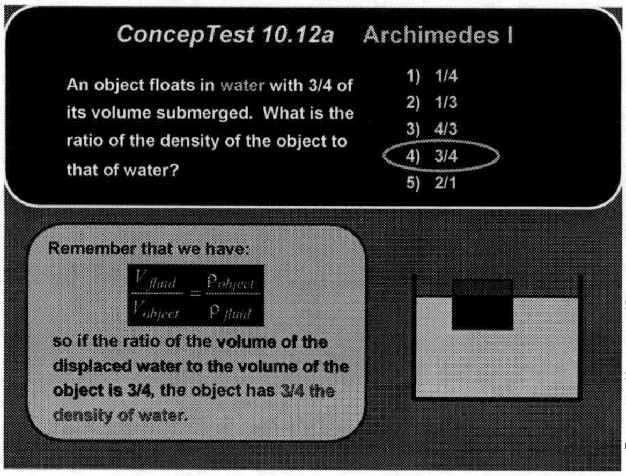

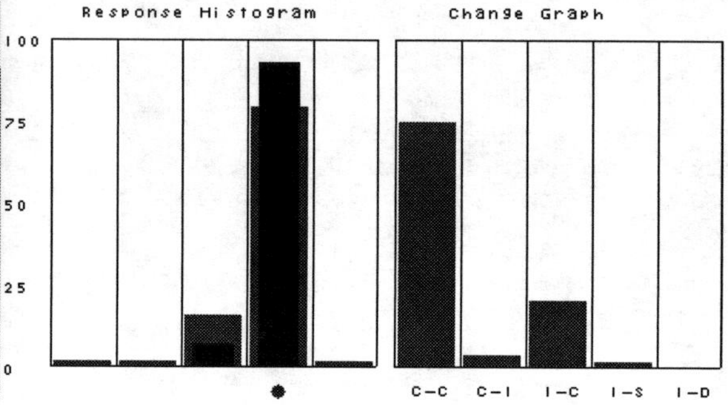

ConcepTest 10.12b: Archimedes II
Difficulty Level: 2

Using the information from the previous ConcepTest, students should realize that in this case, the object is actually denser than the fluid (oil) and will therefore sink.

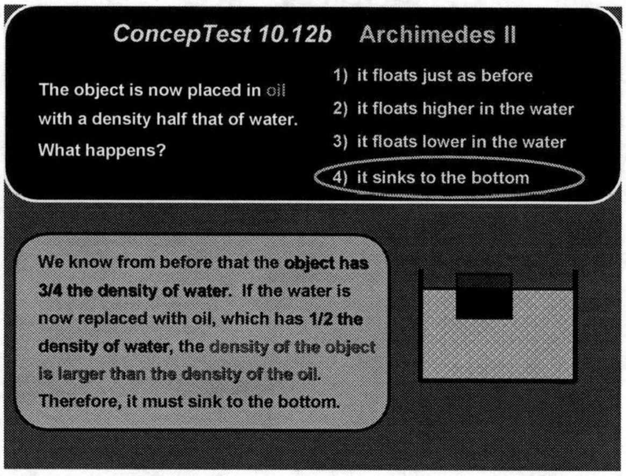

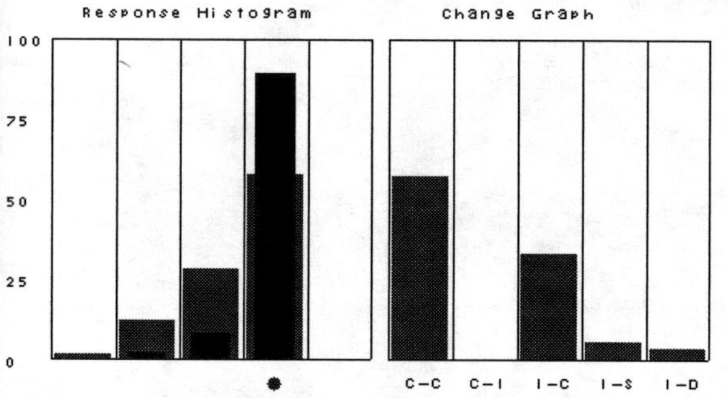

ConcepTest 10.12c: Archimedes III
Difficulty Level: 3

This question may seem rather obvious to students, since most of them will recognize that when covered with water, the block will still rise to the top and float. However, this is a setup for the following question, which leads to some surprising answers!

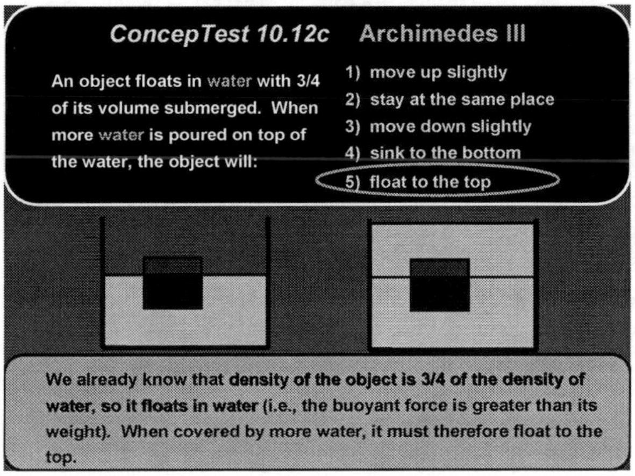

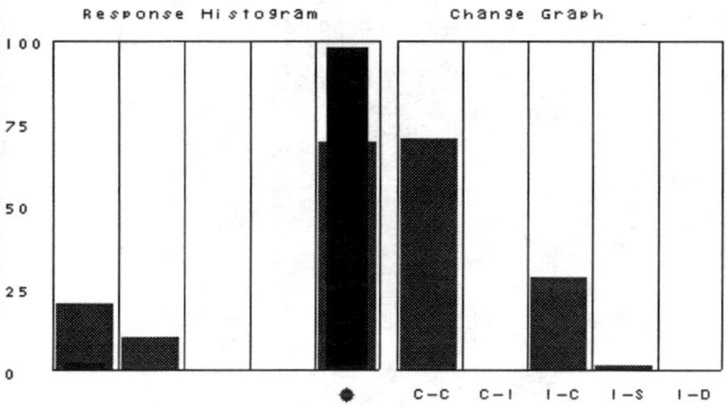

ConcepTest 10.12d: Archimedes IV
Difficulty Level: 4

This question is actually very similar to the previous one, but often elicits the opposite answer! Students tend to think that when the oil is on top of the block, it exerts a downward force, pushing the block lower (or having no effect). However, they fail to recognize that this logic directly contradicts their answer to the previous question (where water, which is more dense, was poured on top!).

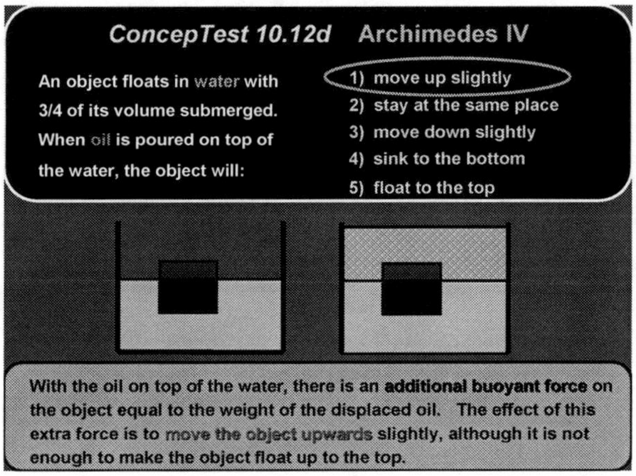

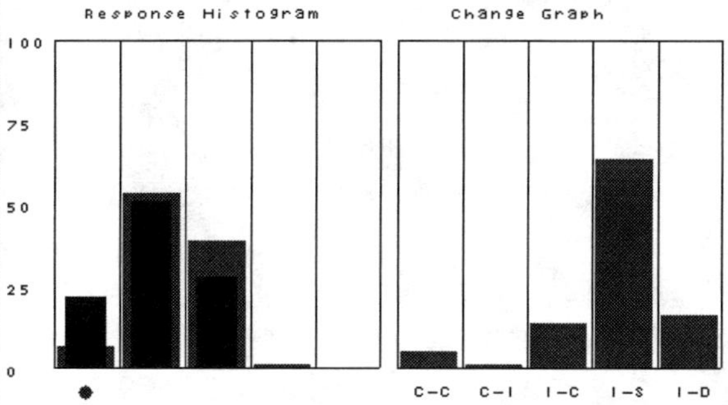

Conceptual Module 10.13 (two parts)

ConcepTest 10.13a: Helium Balloon I

Difficulty Level: 3

A helium balloon in helium gas will do the same thing as an air-filled balloon in air – it will sink! To determine the conditions of buoyancy, it is necessary to consider both the density of the object in question and the density of the surrounding fluid.

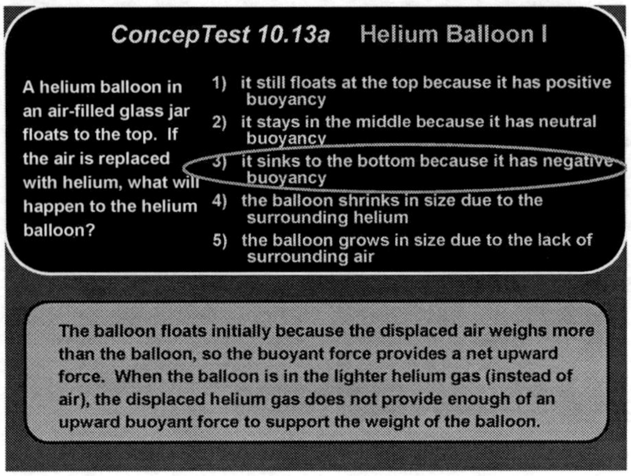

ConcepTest 10.13b: Helium Balloon II

Difficulty Level: 4

Again, the relative densities of objects and fluids are being examined in this follow-up question. In this case, by lifting the jar, the bottom of the jar has an air-helium interface, on the surface of which the helium balloon will float. The helium balloon is still "lighter than air" so it will not sink into the air that is below the bottom of the jar (and the lighter helium, of course, stays in the jar!).

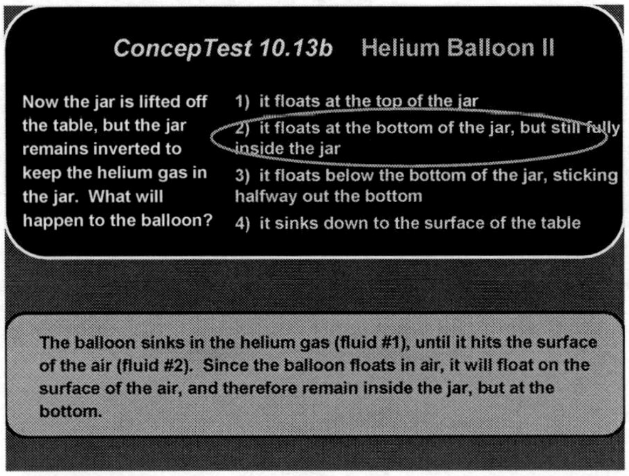

Conceptual Module 10.14 (two parts)

ConcepTest 10.14a: Wood in Water I
Difficulty Level: 4

This question examines the relation between buoyant force and weight of displaced fluid. By adding the wood, an amount of fluid equal to the weight of the wood is displaced (and spills out). So the overall scale reading will remain unchanged, since equal weights were simply exchanged.

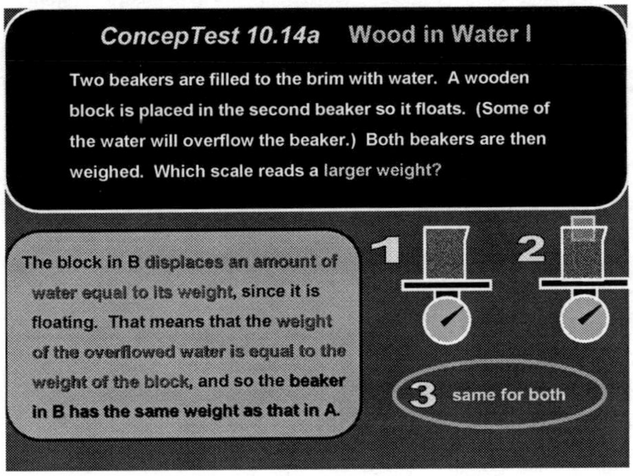

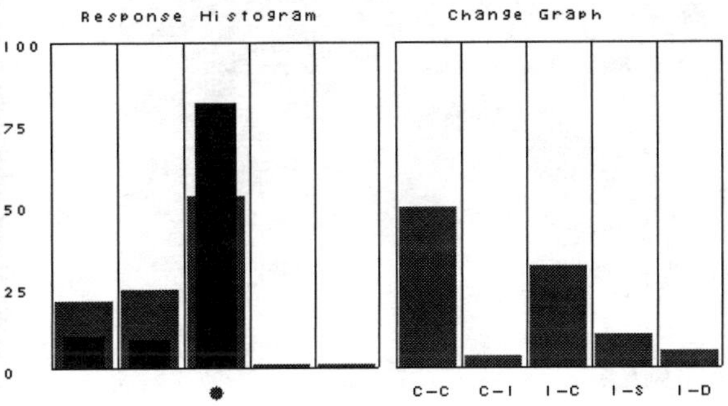

ConcepTest 10.14b: Wood in Water II
Difficulty Level: 2

This question focuses on the issue of weight. On the Moon, while both the weights of the block and the displaced water will be lower, they still will be equal, just as on Earth. So nothing has changed!

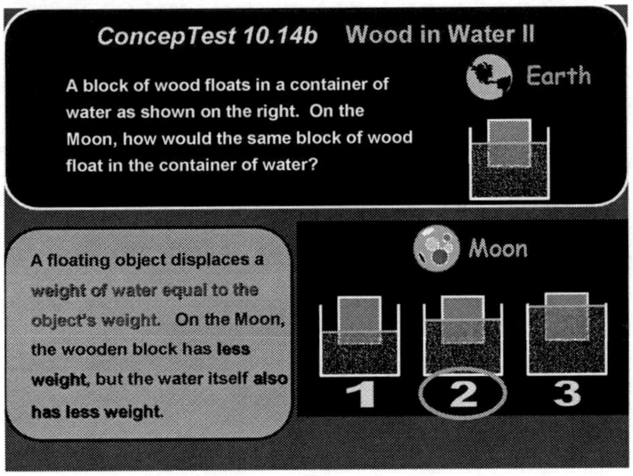

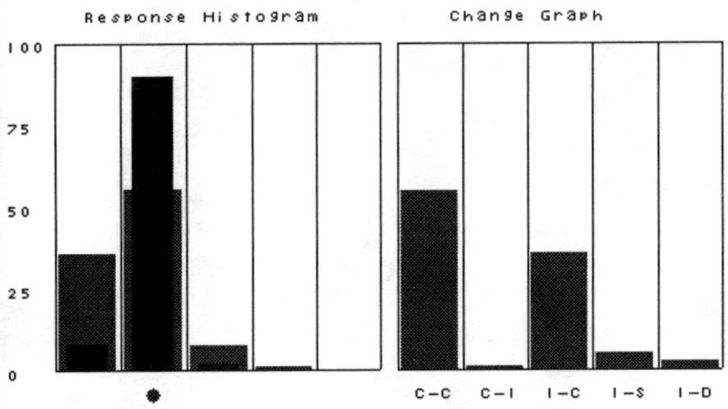

Conceptual Module 10.15 (three parts)

ConcepTest 10.15a: Fluid Flow
Difficulty Level: 3

This question examines the continuity equation, which relates the cross-sectional area and the velocity of fluid flow. In this case, students must realize that the area will be dependent on the square of the radius of the pipe.

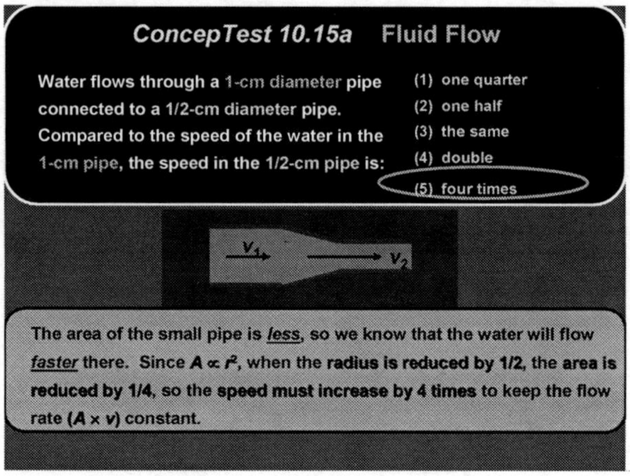

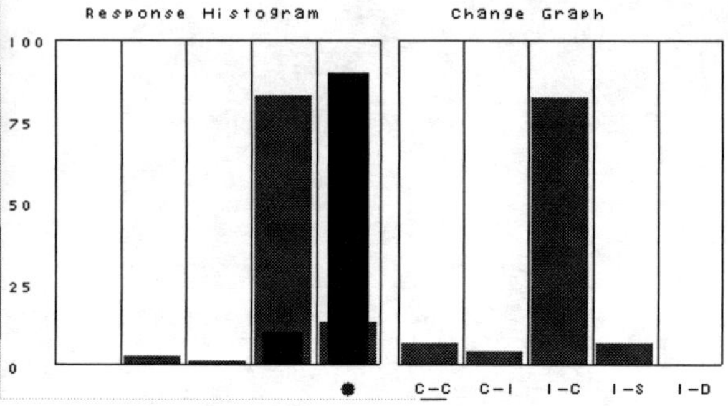

ConcepTest 10.15b: Blood Pressure I
Difficulty Level: 4

This question links the continuity equation with Bernoulli's equation. To determine the pressure (by Bernoulli), the fluid velocity must first be deduced based on the size of the pipe (by continuity).

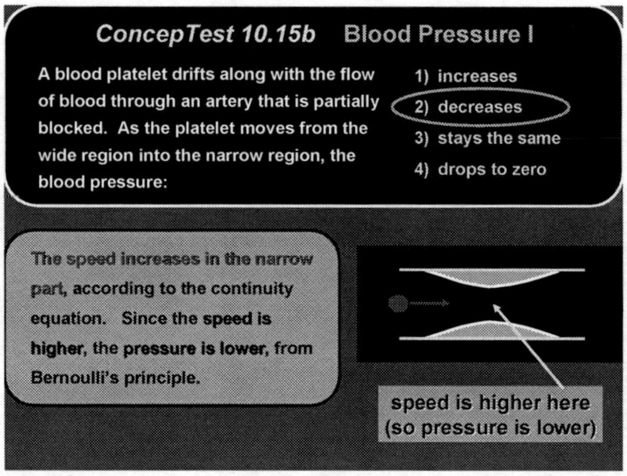

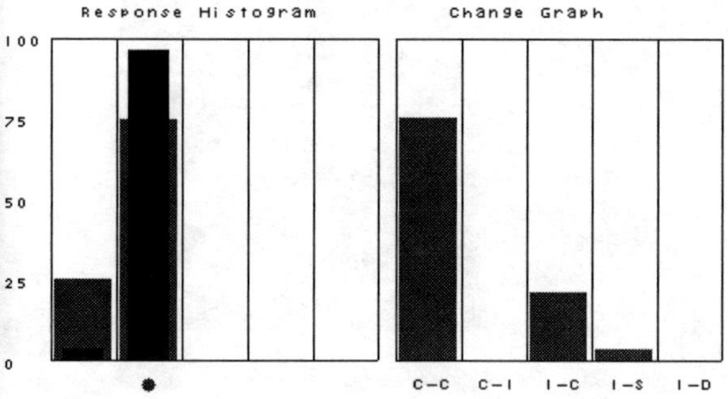

ConcepTest 10.15c: Blood Pressure II
Difficulty Level: 4

This question examines other aspects of Bernoulli's equation, namely the dependence on gravitational potential energy. Since this relation is based on conservation of energy, both kinetic energy (assumed to be constant in this case) and gravitational PE come into play in this case.

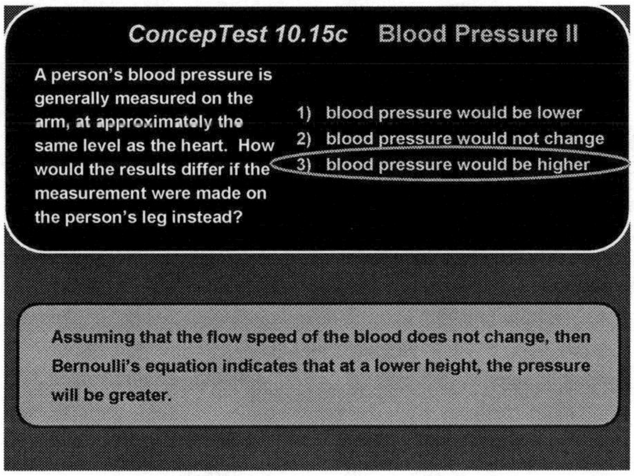

Conceptual Module 10.16
ConcepTest 10.16: The Chimney
Difficulty Level: 3

There are many practical applications of Bernoulli's equation. This case of wind at the top of a chimney is similar to the case of an atomizer, where a rapidly flowing fluid creates a region of low pressure, which allows a fluid (or gas) to be drawn up a shaft from a region of higher pressure.

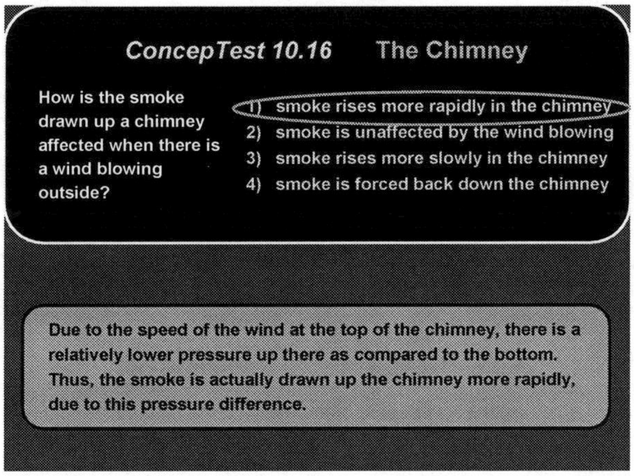

Conceptual Module 11.1 (three parts)

ConcepTest 11.1a: Harmonic Motion I
Difficulty Level: 2

This question is intended to make sure students understand the meaning of a full cycle in simple harmonic motion, which is what is achieved in the time interval T. For example, a half cycle (where the mass returns to the equilibrium position for the first time) might be a common misconception for a full cycle.

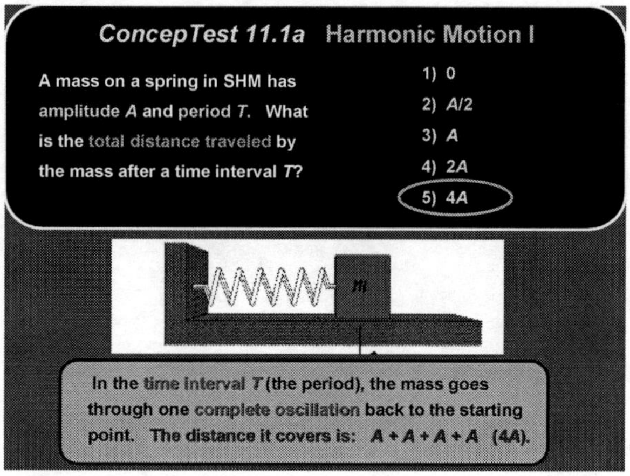

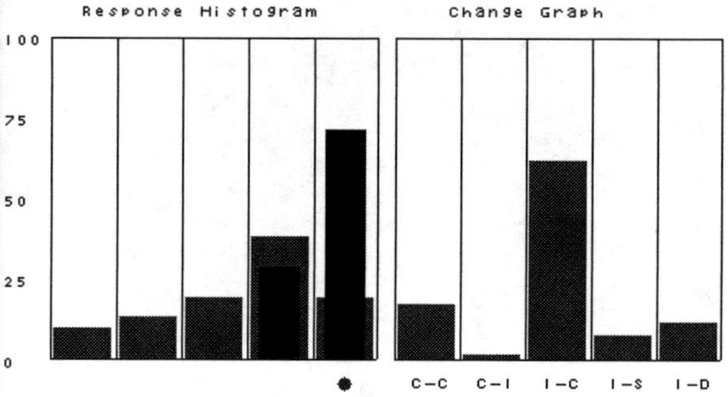

ConcepTest 11.1b: Harmonic Motion II
Difficulty Level: 2
Suggested follow-up conceptual question:

- **What is the net displacement after a half of a period?**

This question addresses a similar issue, but is asking about the displacement (not the distance traveled). This underscores the central idea that a full cycle requires the mass to return to its original position.

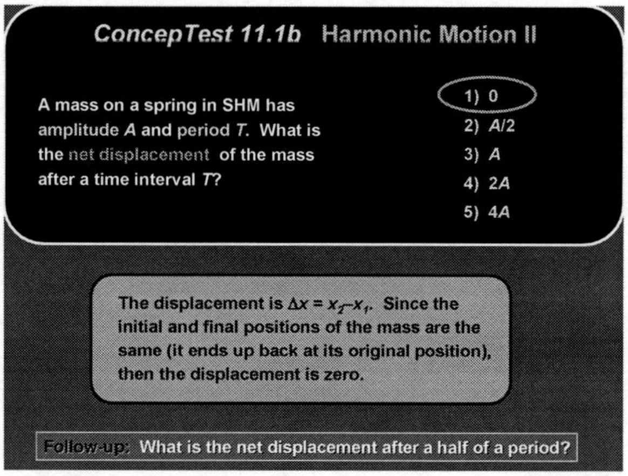

ConcepTest 11.1c: Harmonic Motion III
Difficulty Level: 3
Suggested follow-up conceptual question:
- **What is the net displacement at this particular time?**

By recognizing what is meant by periodic motion and the idea of a complete cycle, the students should be able to extend what they have learned from ConcepTest 11.1a (above) to determine how long (in units of period *T*) it takes to cover a specific total distance.

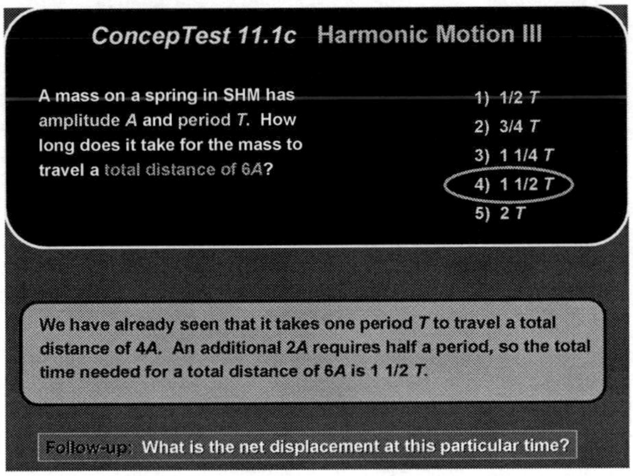

Conceptual Module 11.2

ConcepTest 11.2: Speed and Acceleration
Difficulty Level: 4
Suggested follow-up conceptual question:
- **Where is acceleration a maximum?**

The students need to be familiar with the facts that the acceleration is maximum at the endpoints (where velocity is zero) and minimum at the equilibrium position (where velocity is a maximum). This question requires them to recognize these facts, such as there is no position where both acceleration and velocity are simultaneously zero.

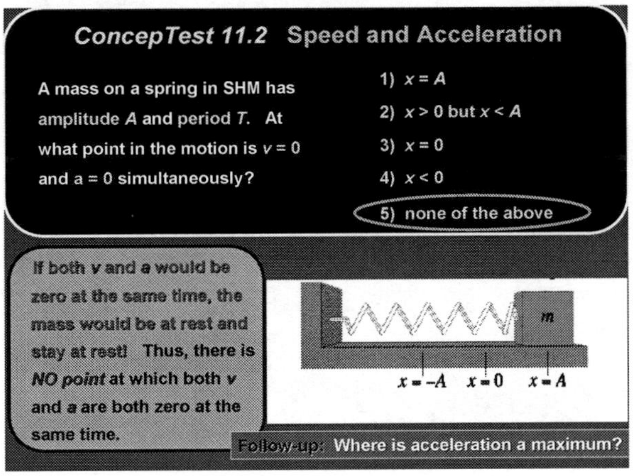

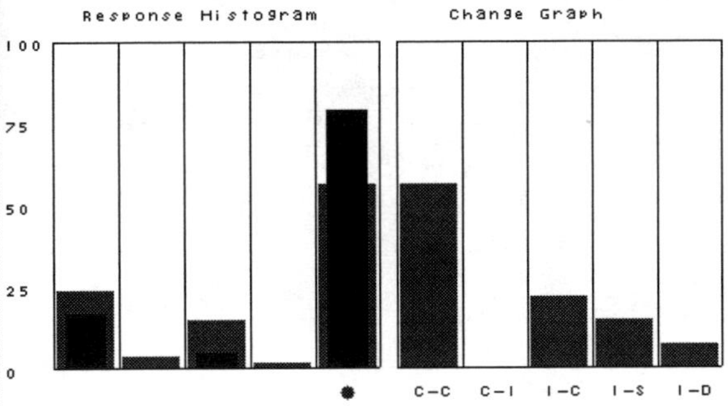

Conceptual Module 11.3 (two parts)

ConcepTest 11.3a: Spring Combination I
Difficulty Level: 4

The idea of two springs in parallel is a more difficult concept and requires students to recognize that each spring will exert the same force at a given displacement. Thus, the forces from the two springs should add, and the total force will be the sum of the two.

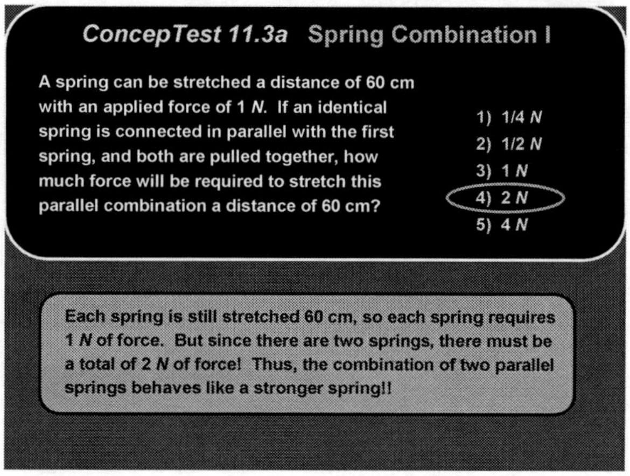

ConcepTest 11.3b: Spring Combination II
Difficulty Level: 5

The idea of two springs in series is also tricky and requires a more detailed analysis of the situation. Here each spring is only stretched half the total distance, so the force on each spring is half the original force (by Hooke's Law). However, since the springs are in series, the force exerted by the first spring is subsequently exerted on the second, so no additional force is needed beyond half the original force.

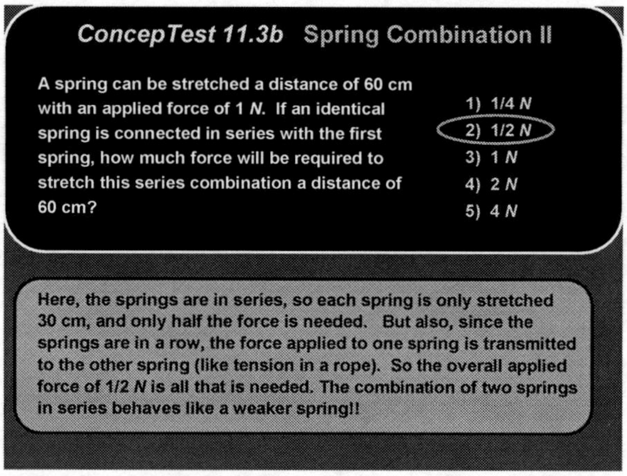

Conceptual Module 11.4

ConcepTest 11.4: To the Center of the Earth
Difficulty Level: 3
Suggested follow-up conceptual question:

- **Where is your acceleration zero?**

There are many examples of simple harmonic motion. This one is rather novel and is amusing to contemplate. The students should be able to reason that there will be zero force at the center (and a maximum velocity) and then a maximum force at the other end (where the person slows to a stop) – which makes this motion sound a lot like SHM.

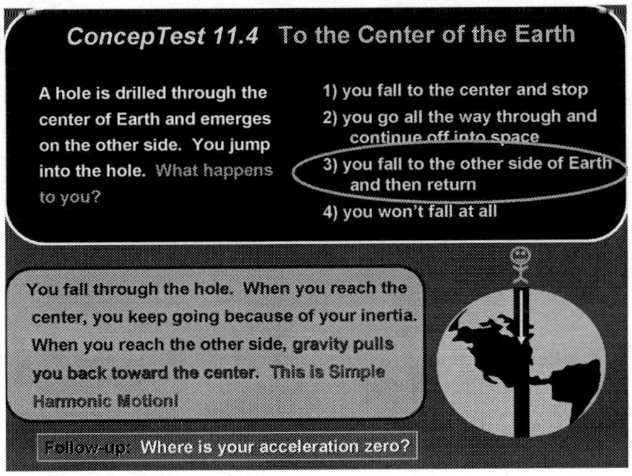

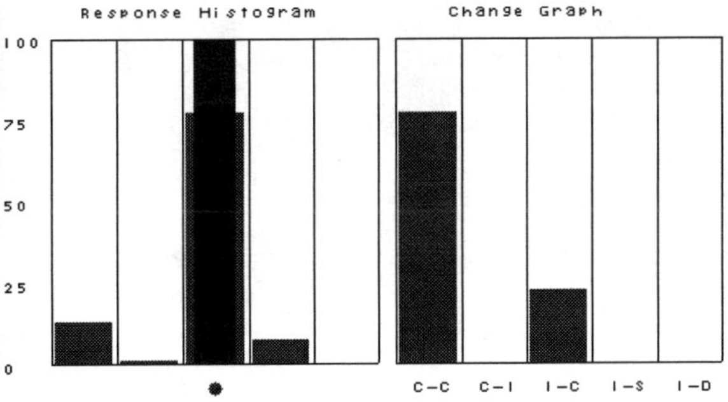

Conceptual Module 11.5 (two parts)

ConcepTest 11.5a: Energy in SHM I
Difficulty Level: 2
Suggested follow-up conceptual question:

- **What happens if you double the amplitude?**

This question examines the dependence of the energy of a mass-spring system on specific variables, such as amplitude (yes) or mass (no).

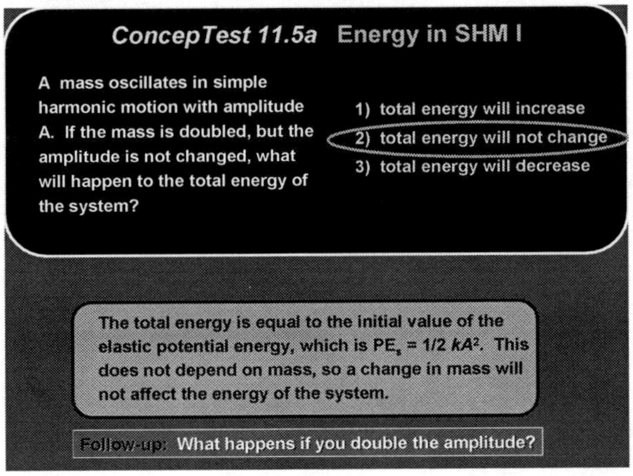

ConcepTest 11.5b: Energy in SHM II
Difficulty Level: 2
Suggested follow-up conceptual question:

- **Why do maximum speed and acceleration double?**

This question examines the dependence of various SHM quantities on the amplitude. As it turns out, all but one of the listed quantities depend on mass, and the only one that depends on amplitude is the answer.

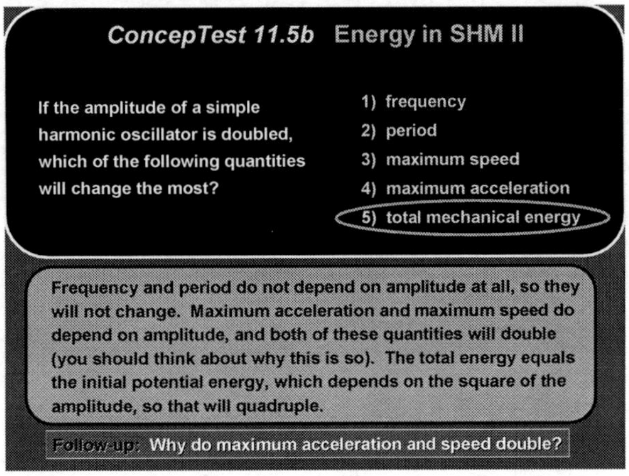

Conceptual Module 11.6 (two parts)

ConcepTest 11.6a: Period of a Spring I
Difficulty Level: 2
Suggested follow-up conceptual question:

- **What happens if the amplitude is doubled?**

This question examines the mass dependence of the period in a mass-spring system.

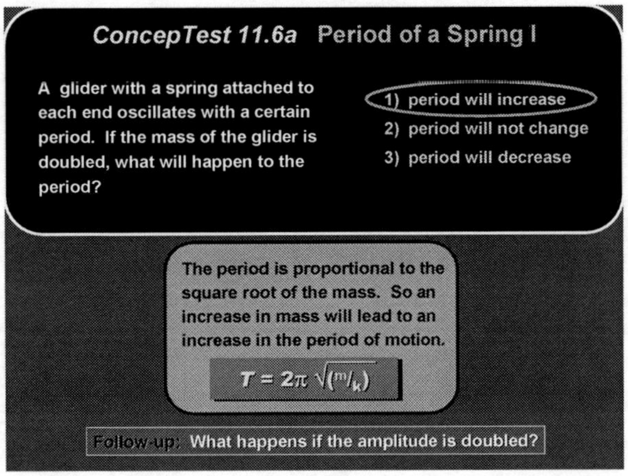

ConcepTest 11.6b: Period of a Spring II
Difficulty Level: 4

This question relates to an earlier one on parallel springs. We saw in the earlier case that the effective spring constant *k* increased. So in this case, the dependence of the period on the spring constant must be known in order to see the effect of the increased value of *k*.

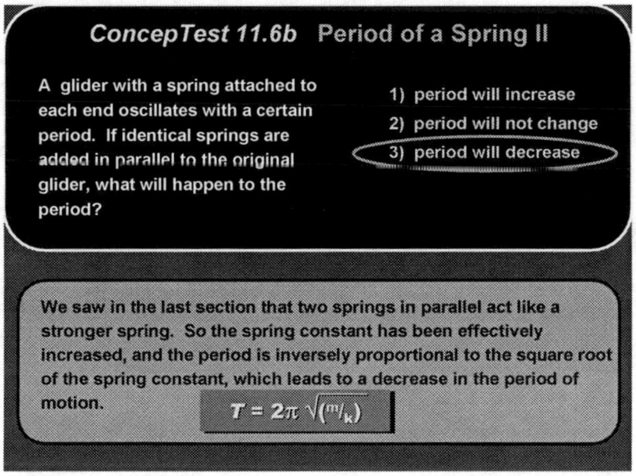

Conceptual Module 11.7 (three parts)

ConcepTest 11.7a: Spring in an Elevator I
Difficulty Level: 3

The vertical motion of the elevator could give rise to an "apparent" value of *g* that is different from usual. However, since *g* does not enter into the period for a mass-spring system, this is irrelevant and will not alter the period. However, it is worth pointing out that the constant vertical motion of the elevator will not alter the expected value of *g*.

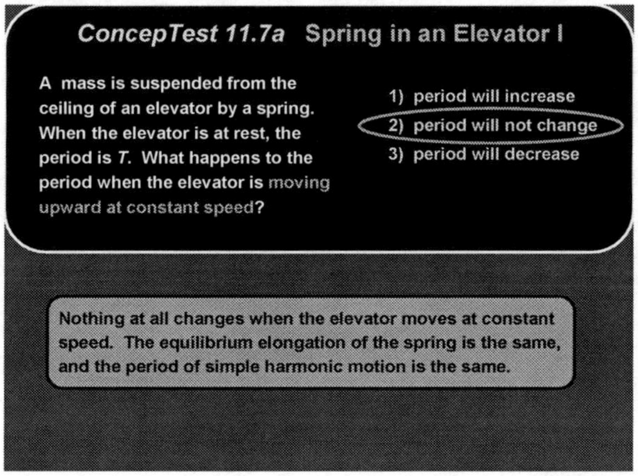

ConcepTest 11.7b: Spring in an Elevator II
Difficulty Level: 3

In this case, the vertical motion of the elevator will give rise to an "apparent" value of *g* that is different from usual. However, since *g* does not enter into the period for a mass-spring system, this is irrelevant and will also not alter the period in this case.

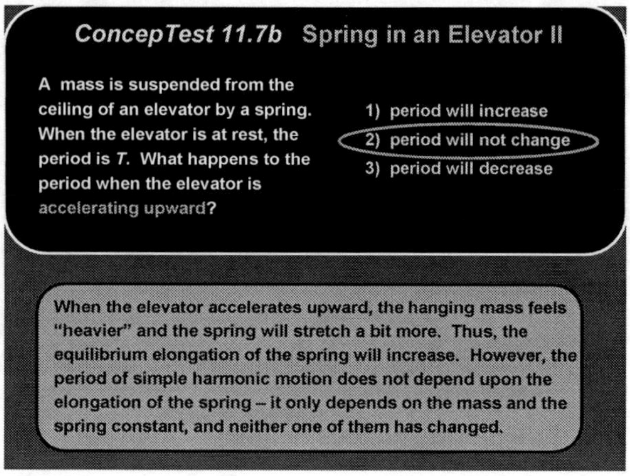

ConcepTest 11.7c: Spring on the Moon
Difficulty Level: 3
Suggested follow-up conceptual question:
- **Will the period be the same on any planet?**

In this case, the value of *g* is changed by going to the Moon. Again, since there is no dependence on *g* for the period of a mass-spring system, the period will not change.

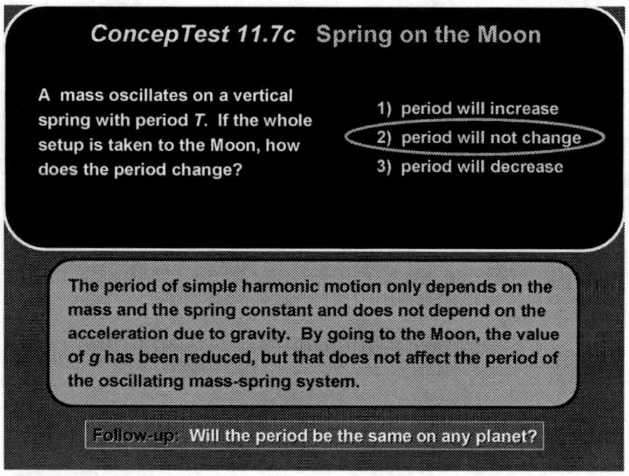

Conceptual Module 11.8 (two parts)

ConcepTest 11.8a: Period of a Pendulum I
Difficulty Level: 2
Suggested follow-up conceptual question:

- **What happens if the amplitude is doubled?**

This question addresses the dependence (or lack thereof, in this case) of the period of a pendulum on the mass of the pendulum bob.

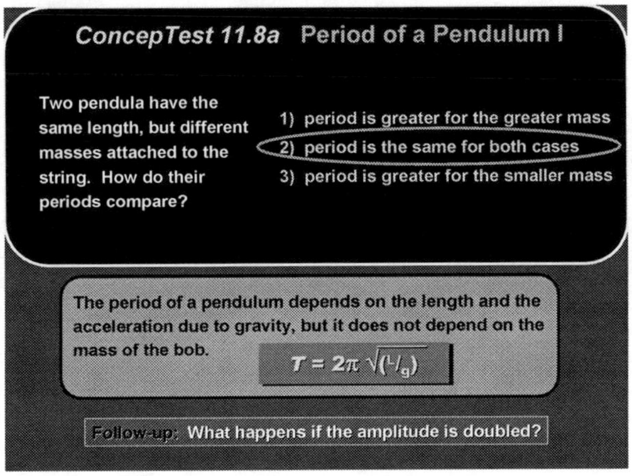

ConcepTest 11.8b: Period of a Pendulum II
Difficulty Level: 3

This question examines the dependence of the period of a pendulum on the length. The period scales as the square root of the length.

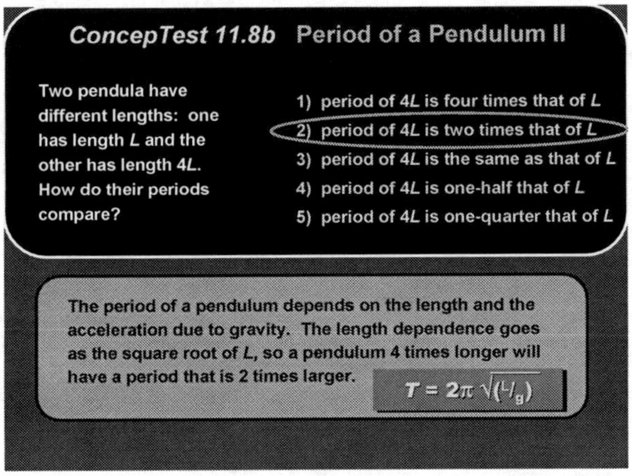

Conceptual Module 11.9

ConcepTest 11.9: Grandfather Clock
Difficulty Level: 4

Since the length of a pendulum affects the period, a grandfather clock can be adjusted by changing the length of the swinging pendulum. In this question, the direction of the needed correction must be recognized, and then this must be applied to the expression for the period (in terms of length) in order to perform the appropriate length adjustment.

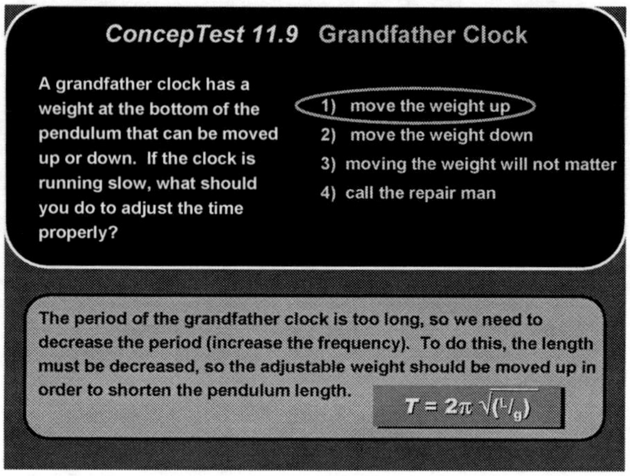

Conceptual Module 11.10 (three parts)

ConcepTest 11.10a: Pendulum in Elevator I
Difficulty Level: 3

The vertical motion of the elevator could give rise to an "apparent" value of *g* that is different from usual, and since the period of a pendulum does depend on *g*, this is entirely relevant in this case. However, the constant vertical motion of the elevator will not alter the expected value of *g*, so the period will not change.

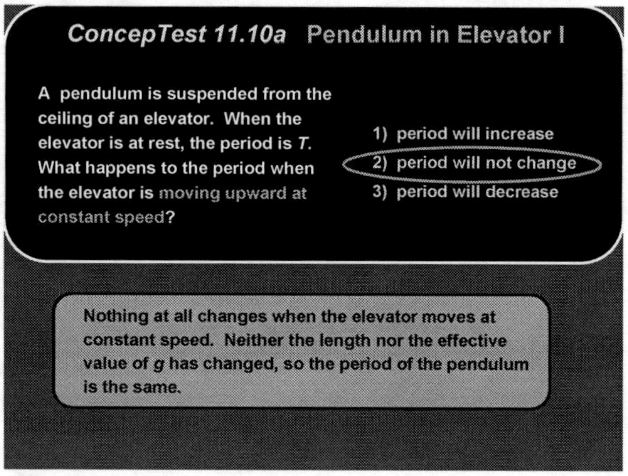

ConcepTest 11.10b: Pendulum in Elevator II
Difficulty Level: 4

The upward acceleration of the elevator gives rise to an "apparent" value of *g* that is greater than usual. Since the period of a pendulum does depend on *g*, the period will be altered by this "increased" value of *g*. The sense of the effect on the period must be deduced from the expression for the pendulum period.

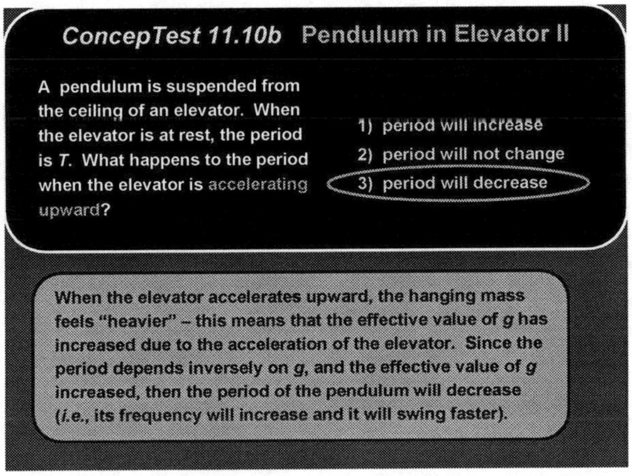

ConcepTest 11.10c: Pendulum in Elevator III
Difficulty Level: 4
Suggested follow-up conceptual question:
- **What can you do to return the pendulum to its original period?**

This is similar to the previous question, except that the value of *g* has been altered by going to the Moon. The only other difference is that in this case, *g* has decreased, while in the previous question, *g* was effectively increased.

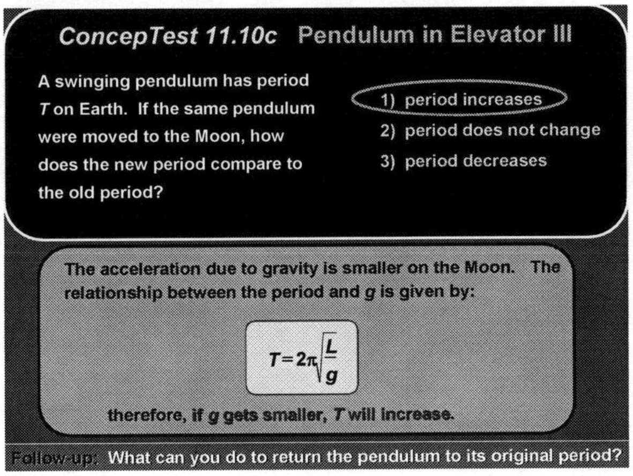

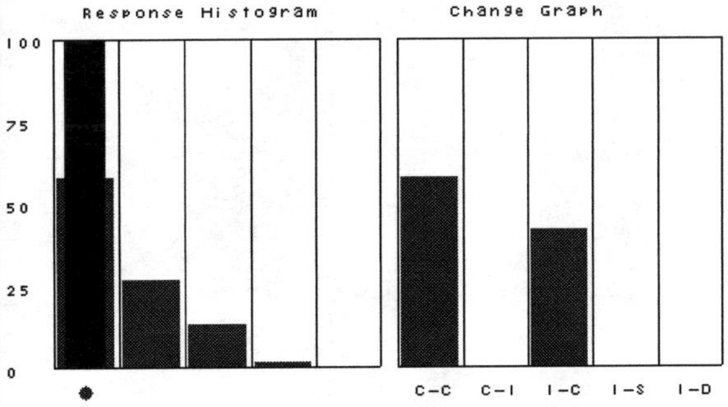

Conceptual Module 11.11
ConcepTest 11.11: Damped Pendulum
Difficulty Level: 3
Suggested follow-up conceptual question:

- **What is happening to the energy of the pendulum?**

The period of a pendulum does not depend on the amplitude, so even though the amplitude is diminishing in this question, the period is not affected.

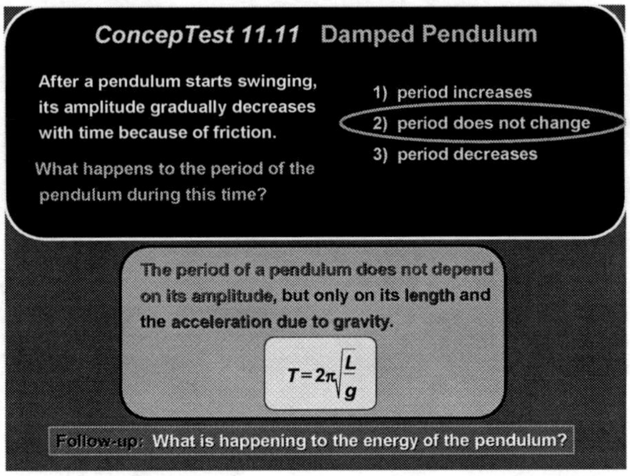

Conceptual Module 11.12
ConcepTest 11.12: Swinging in the Rain
Difficulty Level: 3

Students must realize that when standing, you are effectively reducing the length of the pendulum constituted by you and the swing. Thus, a decrease in length will result in a subsequent decrease in period, as determined by the expression for the pendulum period.

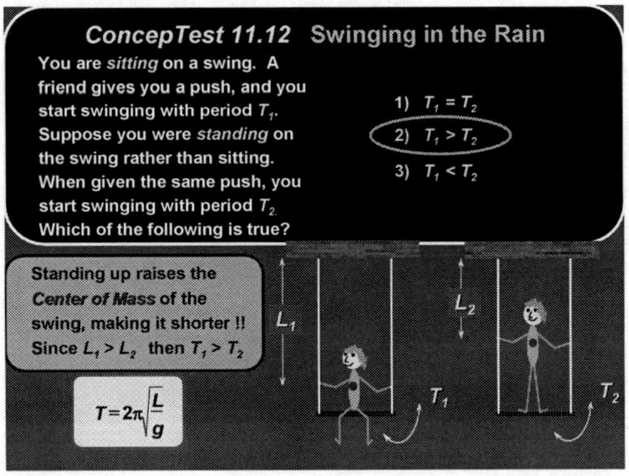

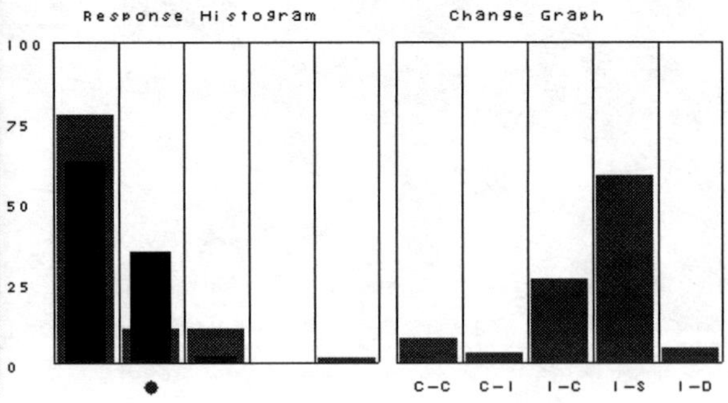

Conceptual Module 11.13

ConcepTest 11.13: Sound it Out
Difficulty Level: 2

Students may tend to think of waves mostly as transverse disturbances, and it is easy to visualize the amplitude in such cases. However, longitudinal waves are just as valid and also have an amplitude – this is a standard characteristic of any wave.

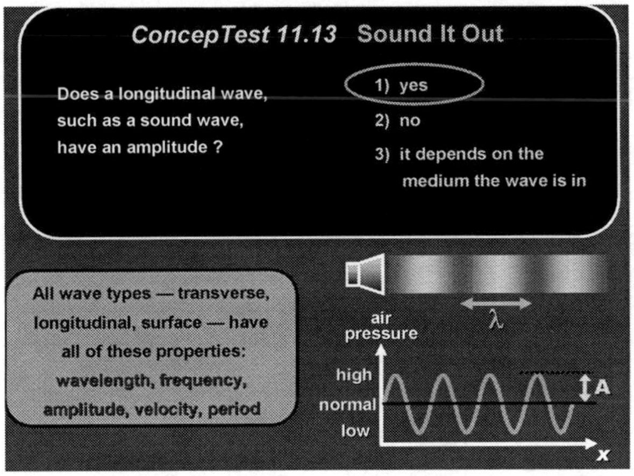

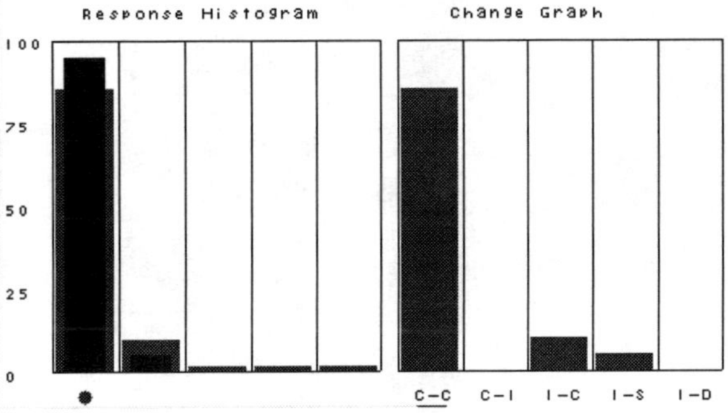

Conceptual Module 11.14

ConcepTest 11.14: The Wave
Difficulty Level: 2
Suggested follow-up conceptual question:

- **What type of wave occurs when you toss a pebble in a pond?**

This is just another way of illustrating the nature of transverse waves. A transverse wave is characterized by a disturbance that is perpendicular to the actual direction of wave motion. In the case of the "wave" at a sporting event, the disturbance is vertical (people moving up and down) while the wave motion is horizontal.

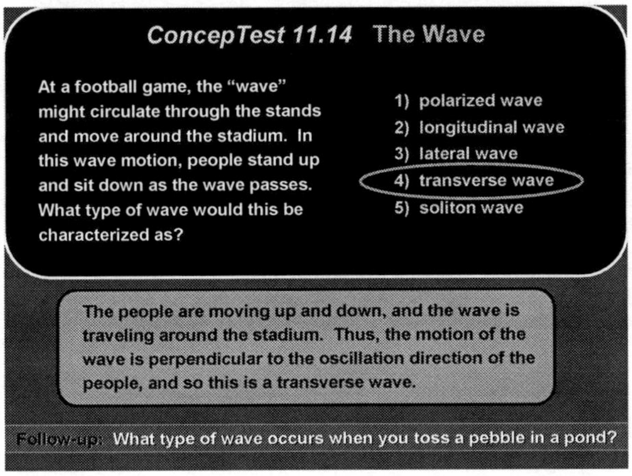

Conceptual Module 11.15 (two parts)

ConcepTest 11.15a: Wave Motion I
Difficulty Level: 4
Suggested follow-up conceptual question:

- **What is the acceleration of the particle at point A?**

This question examines the detailed nature of a transverse wave. Each particle on the string undergoes simple harmonic motion in the vertical direction. At the endpoints of that motion (such as point A), the particle is at rest momentarily.

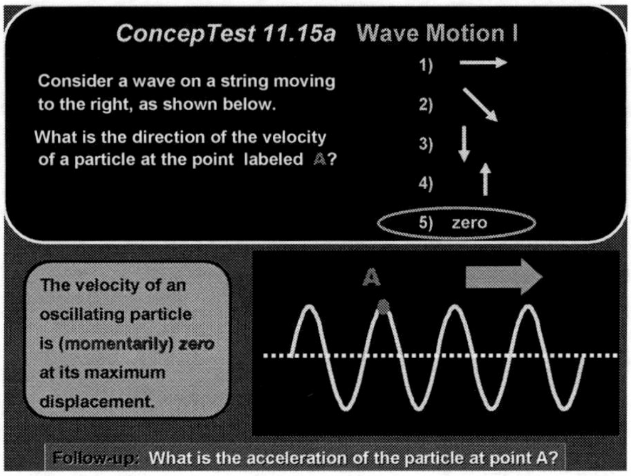

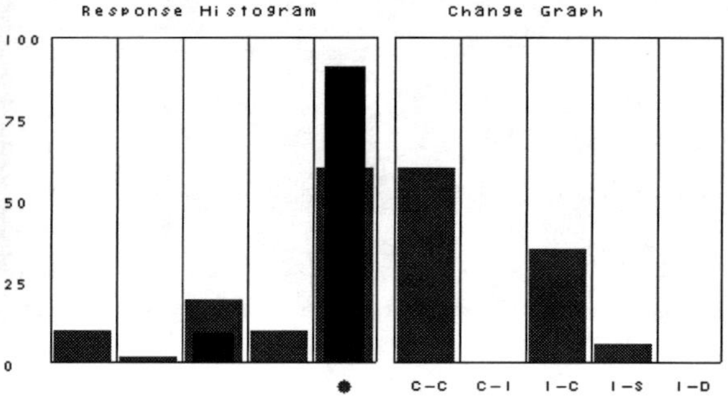

ConcepTest 11.15b: Wave Motion II
Difficulty Level: 5
Suggested follow-up conceptual question:

- **What is the acceleration of the particle at point B?**

In this case, the particle on the string (at B) is in motion, and the question is to recognize the direction of that motion. To accomplish this, the direction of wave motion must be taken into account as well.

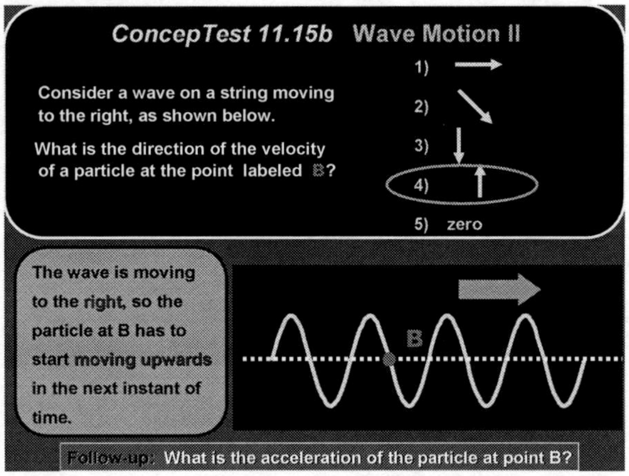

Conceptual Module 11.16
ConcepTest 11.16: Out to Sea
Difficulty Level: 3

The question describes motion that corresponds to half a period, which the students must realize. What remains is for the students to compute the period based on the wave speed and the wavelength.

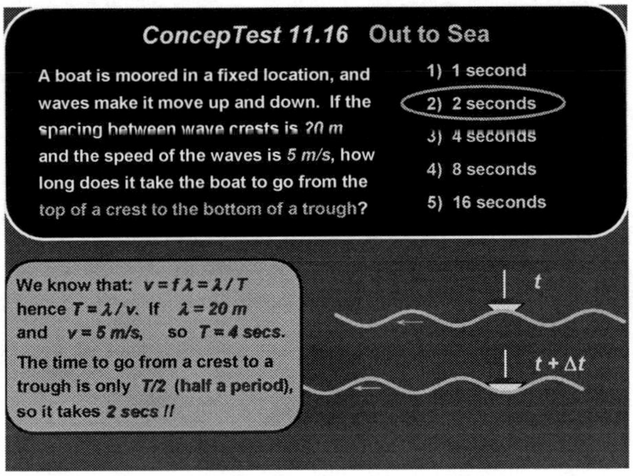

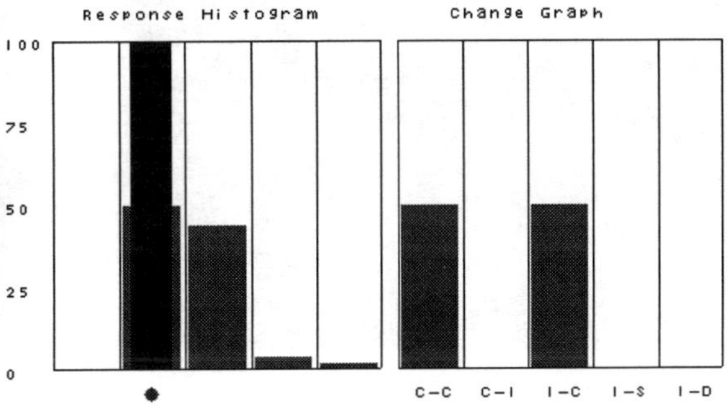

Conceptual Module 11.17
ConcepTest 11.17: Lunch Time
Difficulty Level: 3

This is a direct calculation of wavelength based on the wave speed and the frequency. The numbers are sufficiently simple that this can be done without a calculator. This "real life" example will reveal to the students the wavelength of standing waves in their microwave ovens (which can be tested).

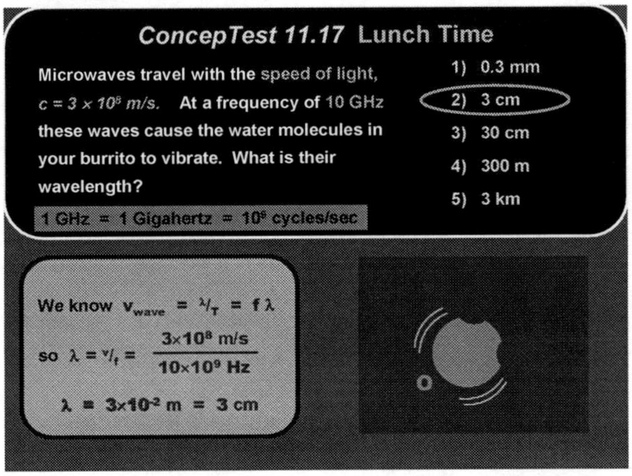

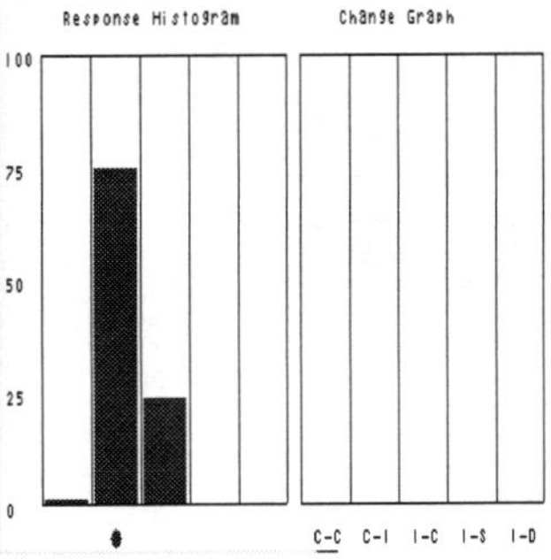

Conceptual Module 11.18 (three parts)
ConcepTest 11.18a: Wave Speed I
Difficulty Level: 2

This question examines the relationship of the wave speed to the tension. The greater the tension, the more rapidly a wave can move on the rope.

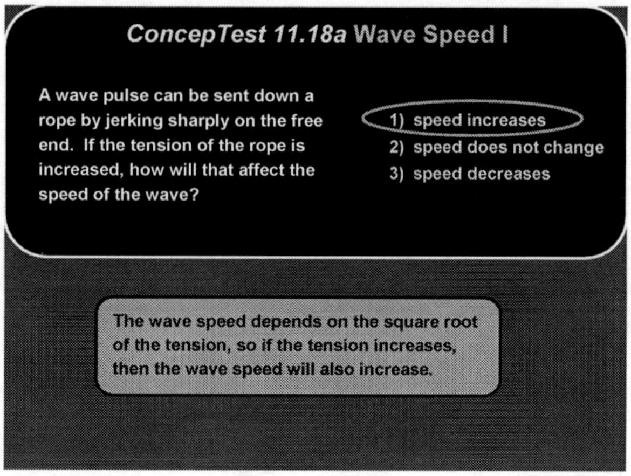

ConcepTest 11.18b: Wave Speed II
Difficulty Level: 3

This question examines the relationship of the wave speed to the mass density. The greater the mass density, the more slowly a wave will move on the rope.

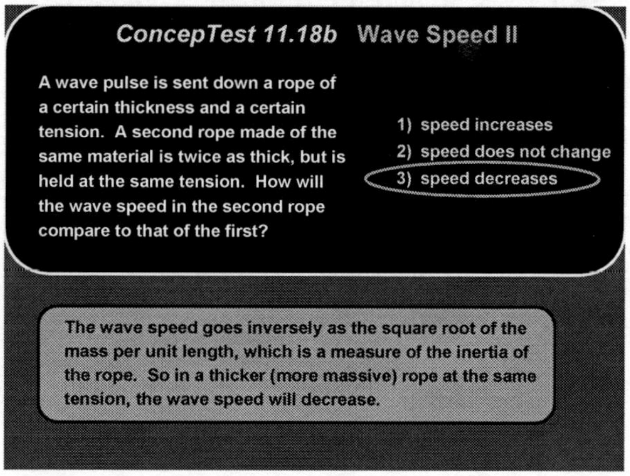

ConcepTest 11.18c: Wave Speed III
Difficulty Level: 5

This final question in the series also deals with the tension, but in a more realistic situation. Students must grasp a new concept, that is, that the tension in a hanging massive rope depends on the weight of rope below a given point. So the tension is increasing in the rope as the pulse rises. This increasing tension affects the wave speed as the pulse moves up.

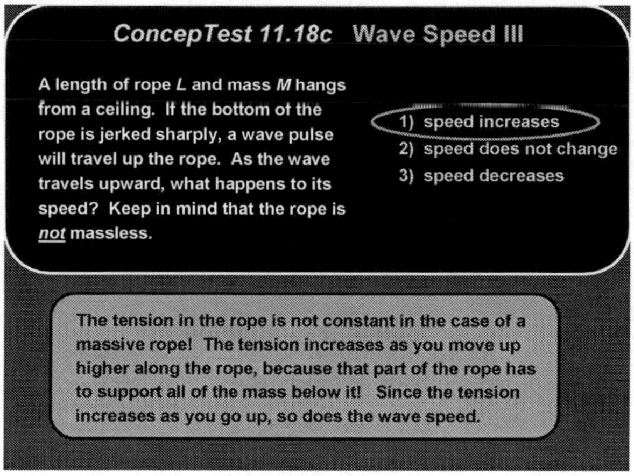

Conceptual Module 11.19 (two parts)
ConcepTest 11.19a: Standing Waves I
Difficulty Level: 4

This question relates a standing wave to the simple harmonic motion of the individual particles in the string. When the particles are at the endpoints, the velocity is zero (and the acceleration is maximum).

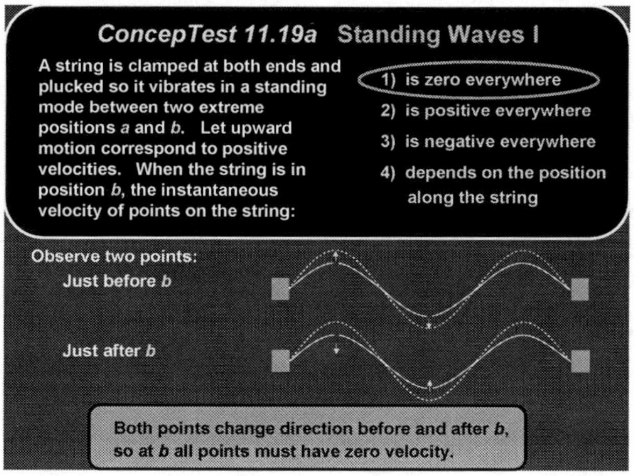

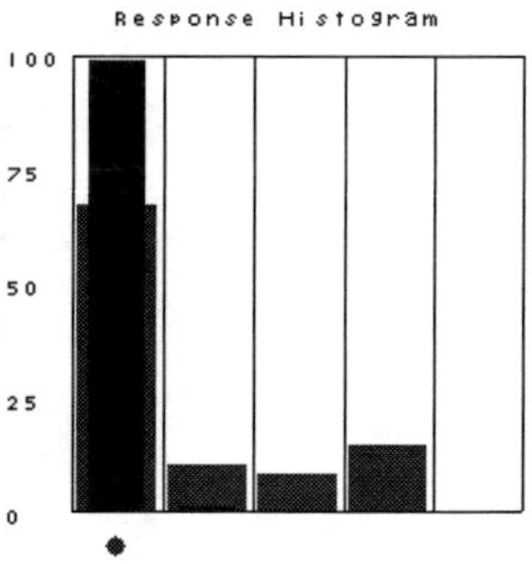

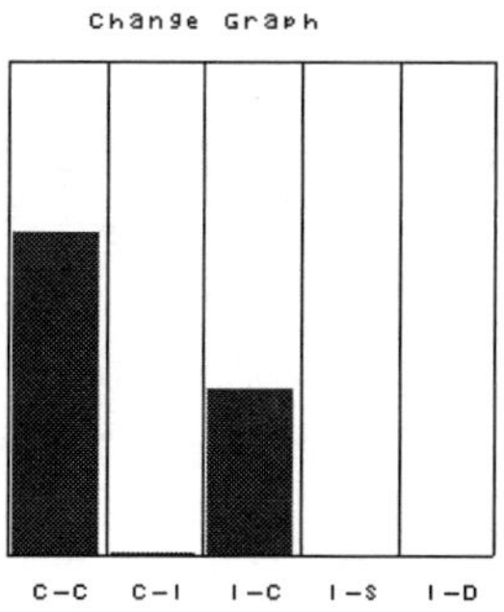

ConcepTest 11.19b: Standing Waves II
Difficulty Level: 4

In this case, the string is in the equilibrium position at all locations. That means that some particles are moving up and others are moving down, so the velocity of a particle does indeed depend upon its position along the string.

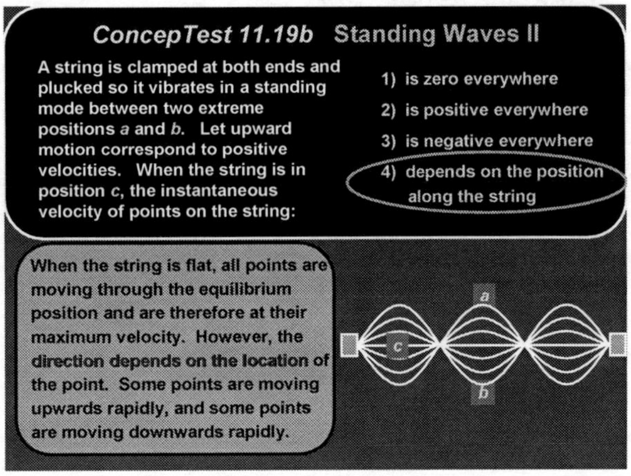

Conceptual Module 12.1 (two parts)

ConcepTest 12.1a: Sound Bite I
Difficulty Level: 2
Suggested follow-up conceptual question:

- **Does the wave speed increase or decrease in water?**

The wave speed depends entirely on the medium. The frequency, on the other hand, is determined by the actual vibrations of the source. The wavelength is a parameter that can change from one medium to another, depending upon changes in the wave speed.

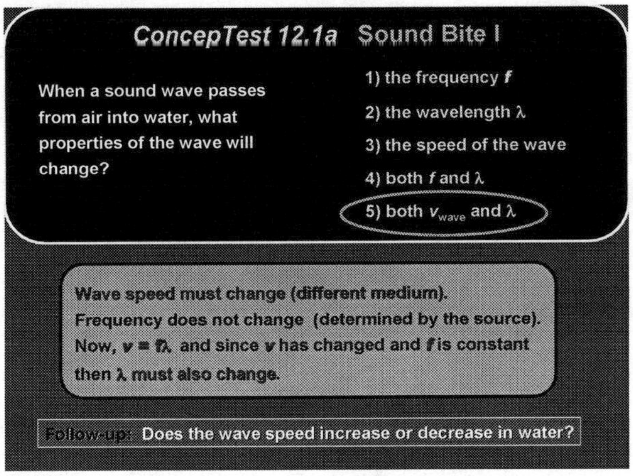

ConcepTest 12.1b: Sound Bite II
Difficulty Level: 3

This follow-up question is asking students to determine the sense of the wavelength change. By knowing how the wave speed will change, the effect on the wavelength can be deduced.

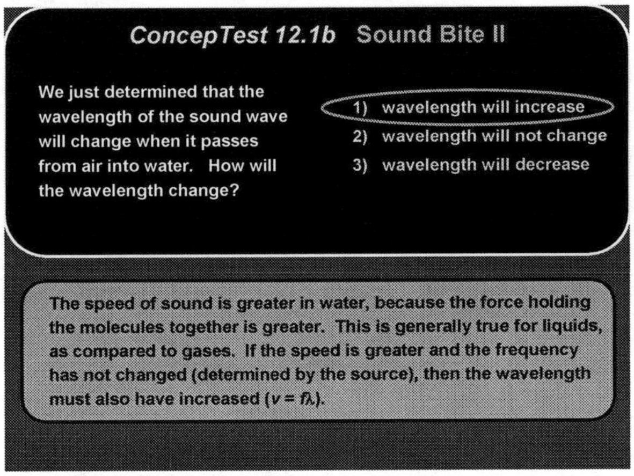

Conceptual Module 12.2 (three parts)
ConcepTest 12.2a: Speed of Sound I
Difficulty Level: 1

This question addresses the effect of the medium on the speed of sound. The solid medium will have the greater restoring force between particles in the medium, leading to a faster wave speed. This characteristic is akin to the tension in a rope, for the case of a transverse wave.

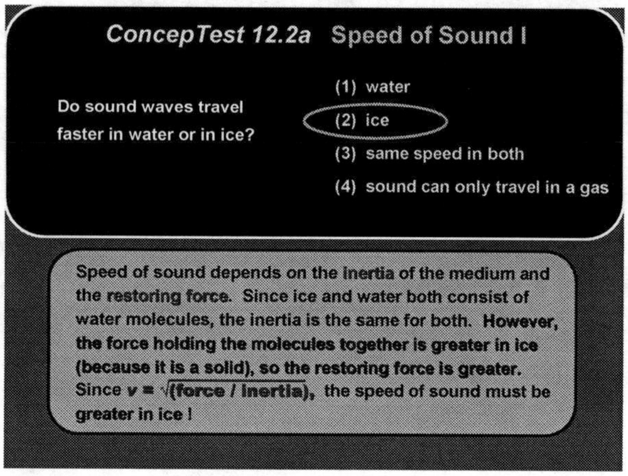

ConcepTest 12.2b: Speed of Sound II
Difficulty Level: 2

The temperature dependence of the speed of sound is also related to the ease with which particles in the medium can "communicate" with each other. At higher temperatures, the air molecules are moving more quickly, so they are able to respond more rapidly to oscillations in the medium.

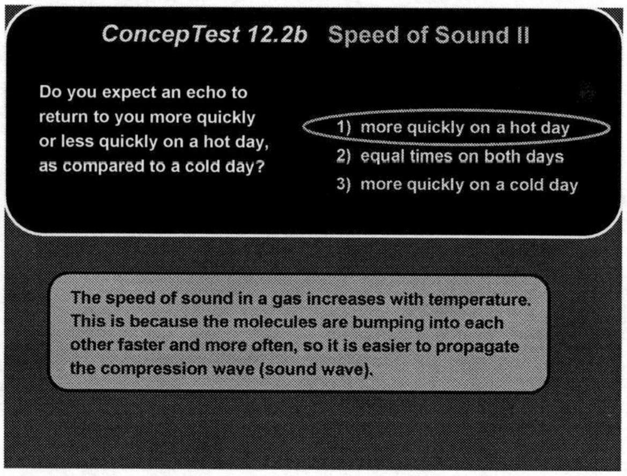

ConcepTest 12.2c: Speed of Sound III
Difficulty Level: 4
Suggested follow-up conceptual question:

- **Why is the speed of sound greater in helium than in air?**

In this case, the wavelength is held fixed by the actual length of the vocal cords. The empirical observation in this problem is that the sound frequency is higher. That must therefore imply that the speed of sound is also higher, given the other conditions stated above.

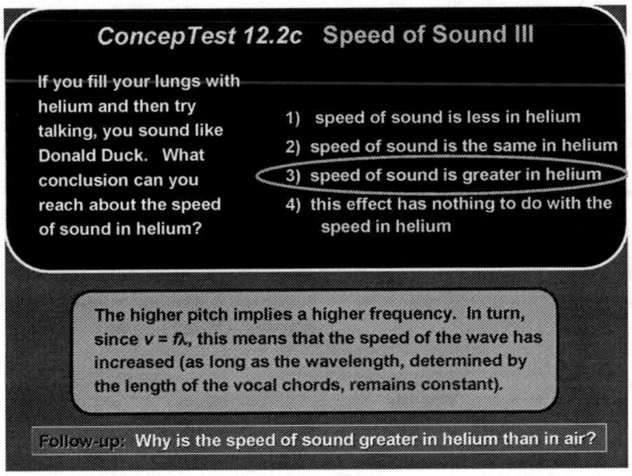

Conceptual Module 12.3

ConcepTest 12.3: Wishing Well
Difficulty Level: 4
Suggested follow-up conceptual question:

- **How long does the sound take to travel the depth of the well?**

This is more a question of logic (or kinematics) than of sound speed. Assuming the transit time of the sound wave to be negligible compared to the fall time of the rock, the students are then simply comparing the fall times in the two cases. In the second case, the rock will cover the second half of the well much more quickly than the first half, so the time will be less than double the 1.5 seconds.

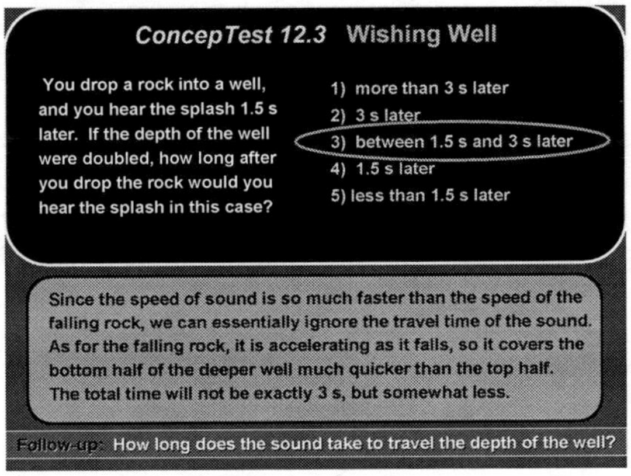

Conceptual Module 12.4 (two parts)

ConcepTest 12.4a: Sound Intensity I
Difficulty Level: 2
Suggested follow-up conceptual question:

- **What distance would reduce the intensity by a factor of 100?**

This question emphasizes the inverse square relation of the intensity to the distance from the source.

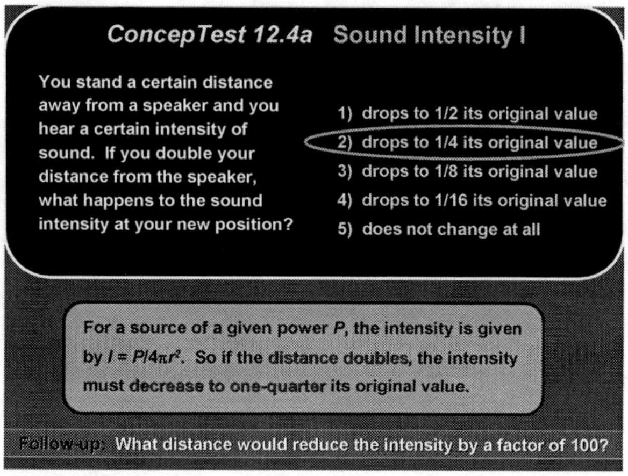

ConcepTest 12.4b: Sound Intensity II
Difficulty Level: 2

This question also relates to distance and intensity. In this case, the students must make some approximation for the square root of 10.

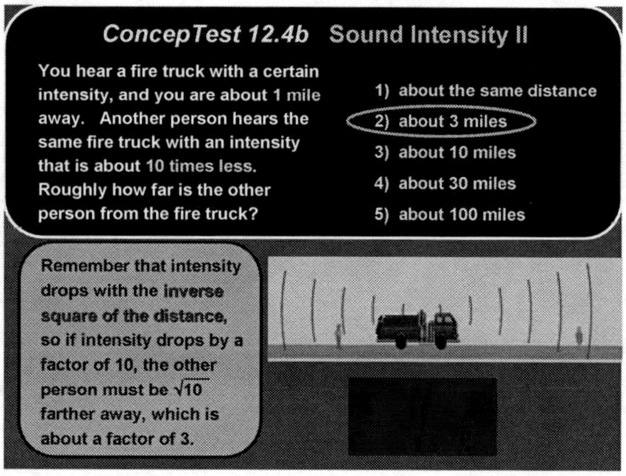

Conceptual Module 12.5 (three parts)

ConcepTest 12.5a: Decibel Level I
Difficulty Level: 4

This question deals with the addition of intensities and the relation of intensity to decibel level. While the intensities simply add, the decibel level involves a logarithm of the doubled intensity, so it will only change by a small amount.

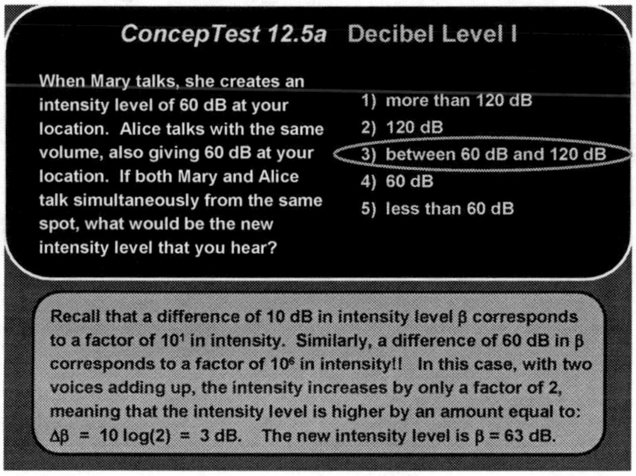

ConcepTest 12.5b: Decibel Level II
Difficulty Level: 4
Suggested follow-up conceptual question:

- **What decibel level gives an intensity a million times greater?**

In this question, the decibel level is given first, and the intensity needs to be determined. This exercise emphasizes the connection between *factors of 10* in the intensity and *increments of 10* in the decibel level. This provides a guideline for students to quickly (and approximately) relate the two quantities.

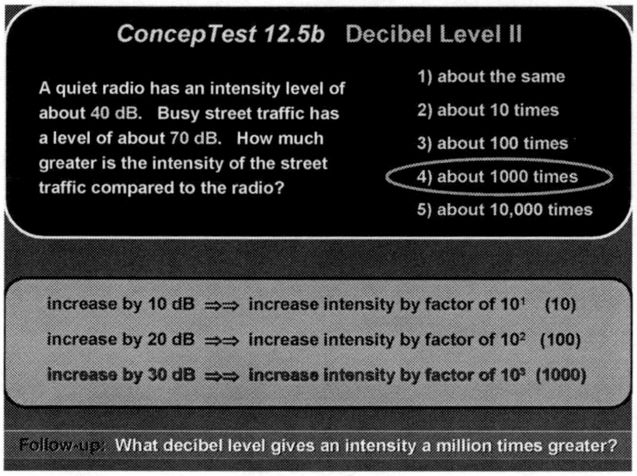

ConcepTest 12.5c: Decibel Level III
Difficulty Level: 2

This question addresses the meaning of the logarithm in the decibel level. The value of β can be zero only when $I/I_0 = 1$, which uniquely determines the value of the intensity at the point $\beta = 0$.

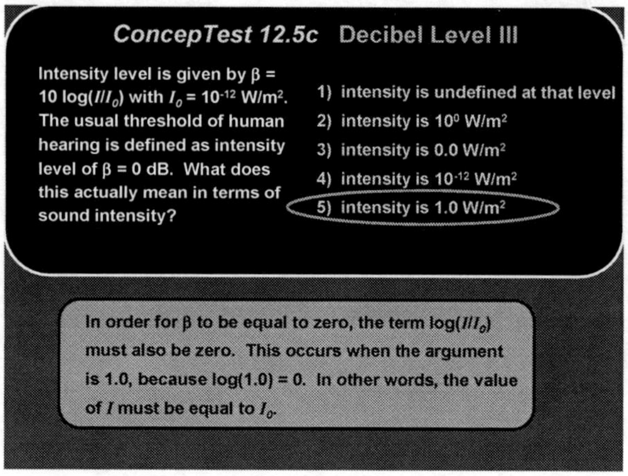

Conceptual Module 12.16 (three parts)

ConcepTest 12.6a: Pied Piper I
Difficulty Level: 2

The students must understand the relationship between the frequency (or wavelength) of a sound wave and the length of the pipe generating that sound. The longer pipe has the longer wavelength, but that must be translated into terms of frequency by the wave equation.

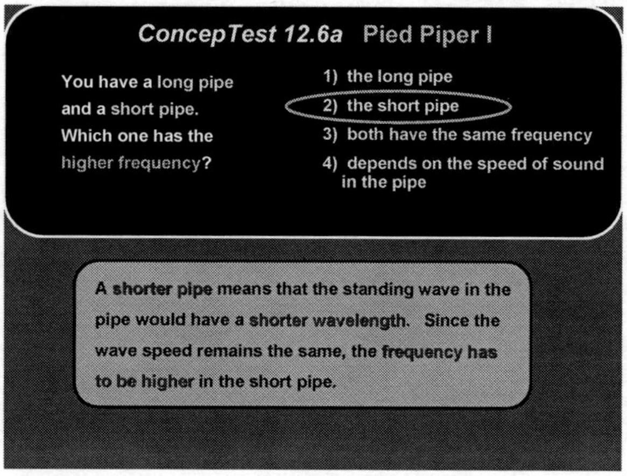

ConcepTest 12.6b: Pied Piper II
Difficulty Level: 2

This question also relates to frequency of sound and length of pipe. As described in this question, the situation can be demonstrated with a slide whistle, which allows a continuous change in length.

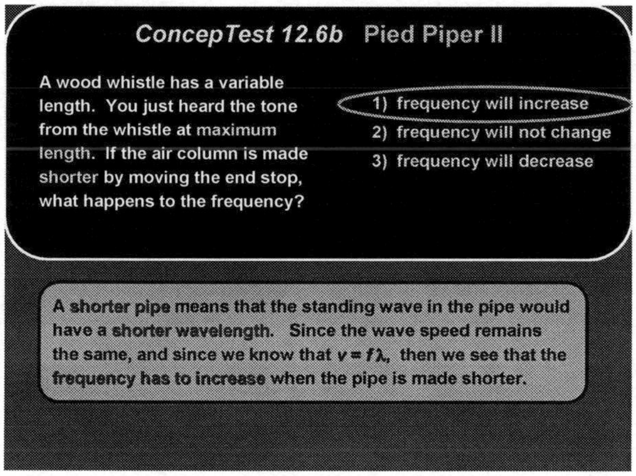

ConcepTest 12.6c: Pied Piper III
Difficulty Level: 2
Suggested follow-up conceptual question:

- **Why doesn't the wave speed change?**

In this case, the effective length of the pipe is changed by removing liquid from a soda bottle. Here the students must realize that the pipe is the column of air bounded by the top of the bottle and the surface of the liquid. Many students probably have some experience with this method of generating sound.

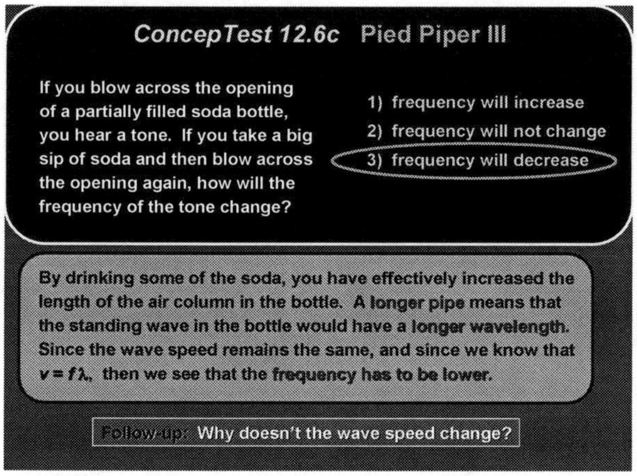

Conceptual Module 12.7

ConcepTest 12.7: Open and Closed Pipes
Difficulty Level: 2
Suggested follow-up conceptual question:

- **What would you have to do to the pipe to increase the frequency?**

This question examines the comparison between an open and closed pipe of fixed length. The formation of standing waves in these pipes is central to understanding the differences between these two cases (which leads to the variations in frequency).

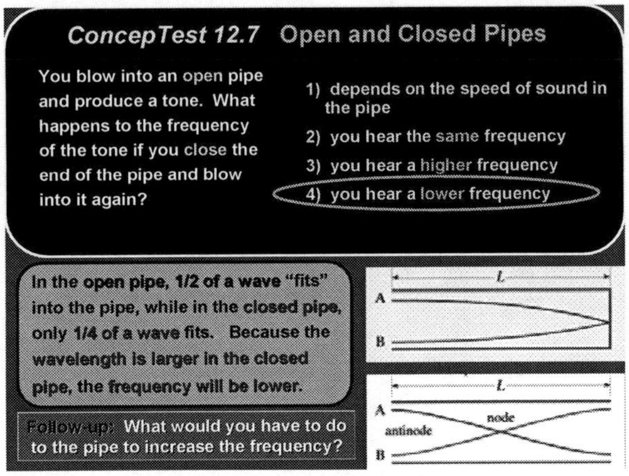

Conceptual Module 12.8

ConcepTest 12.8: Out of Tune
Difficulty Level: 2
Suggested follow-up conceptual question:

- **To increase frequency, do you tighten or loosen the strings?**

This "real life" question asks students to recall the relation between frequency and wave speed, as well as the connection between wave speed and tension. Both of these relations are needed to address the issue of guitar tuning.

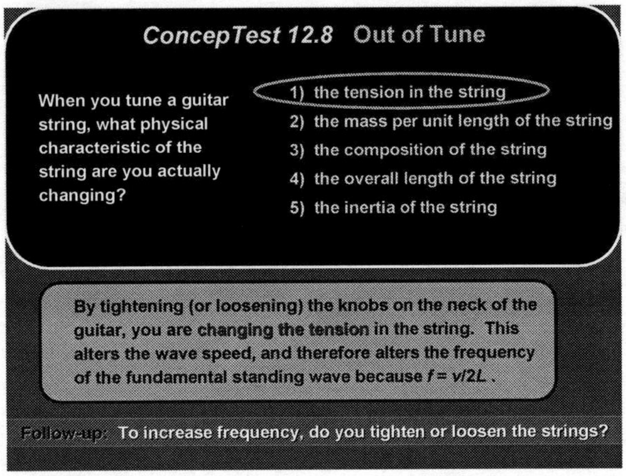

Conceptual Module 12.9

ConcepTest 12.9: Interference
Difficulty Level: 3
Suggested follow-up conceptual question:

- **What if you move back by four meters?**

The concept of interference is embodied in the notion of waves being in phase or out of phase. This ConcepTest reveals this basic premise in the simple situation of two sources and an observer all located along a line.

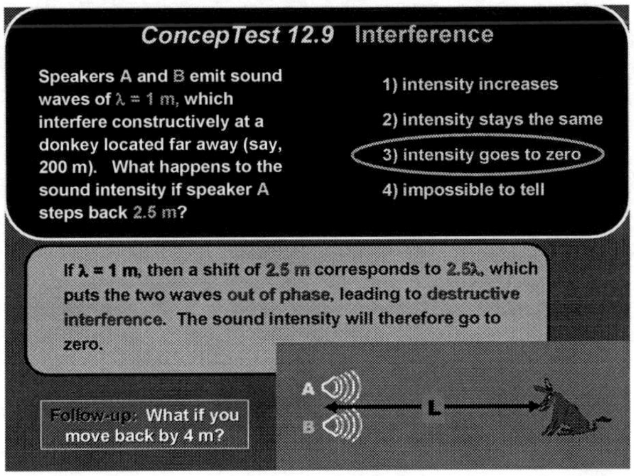

Conceptual Module 12.10

ConcepTest 12.10: Beats
Difficulty Level: 3

The beat frequency is the difference between the frequencies of the two interfering sounds. Students must be able to identify the trace that represents the higher beat frequency, which corresponds to the case of the largest difference between the two interfering sound waves.

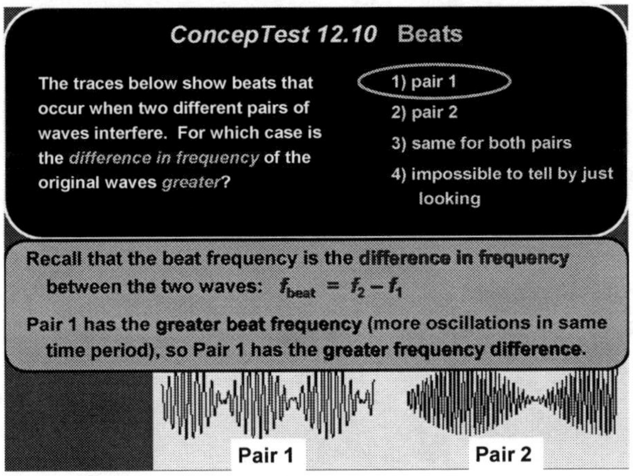

Conceptual Module 12.11 (three parts)

ConcepTest 12.11a: Doppler Effect I

Difficulty Level: 1

Suggested follow-up conceptual question:

- **Where is the frequency lowest?**

In this question, the students must understand that higher frequencies correspond to more closely packed wavefronts (for a given wave speed). This question leads into the concept of Doppler Effect.

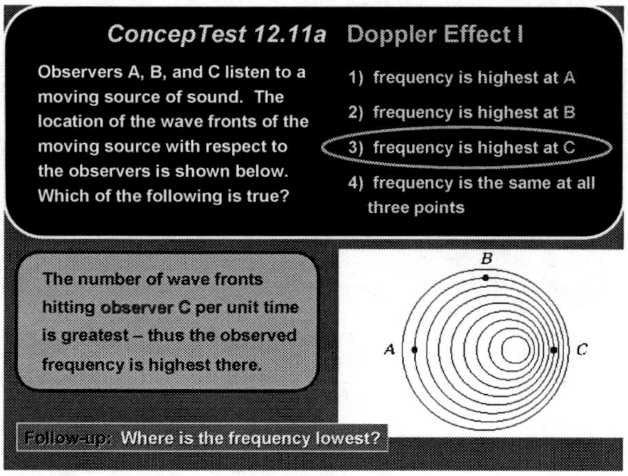

ConcepTest 12.11b: Doppler Effect II

Difficulty Level: 2

In this case, the Doppler Effect will play a role in determining the frequency due to the approaching sound source. The students should be able to relate this situation to the figure shown in the previous ConcepTest, which shows explicitly the bunching of wavefronts in front of a moving source.

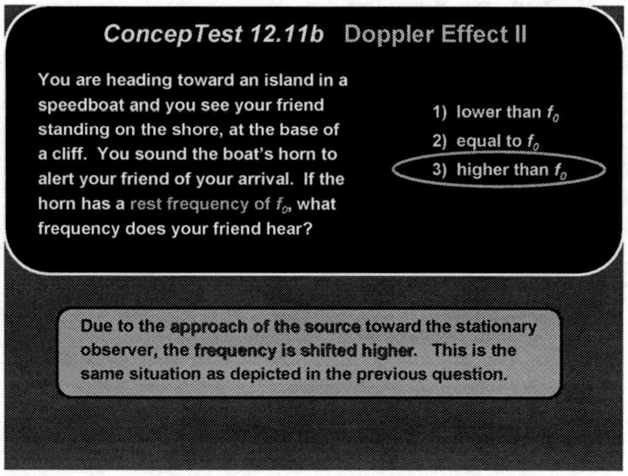

ConcepTest 12.11c: Doppler Effect III
Difficulty Level: 3

This question goes one step further by asking about the sound heard by an approaching observer. In this case, the source is fixed (sound bouncing off a stationary cliff) and it is the observer who is moving toward the source, which leads to another increase in frequency.

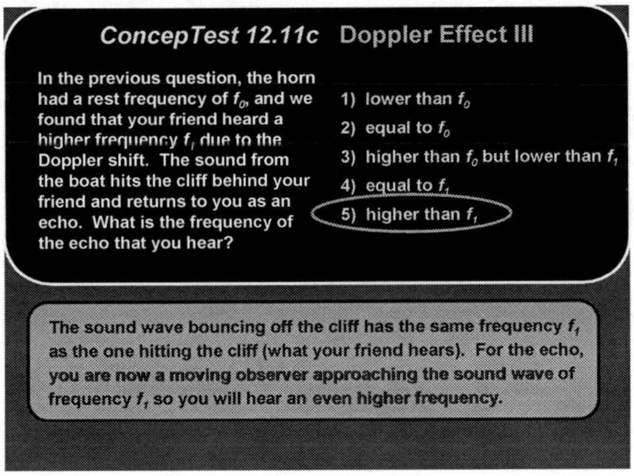

Conceptual Module 13.1
ConcepTest 13.1: Degrees
Difficulty Level: 1

While the three principal temperature scales have different calibration points, they also have different size degree units. This question emphasizes the difference between the Fahrenheit and Celsius degrees, and the fact that Celsius and Kelvin scales only differ by an offset.

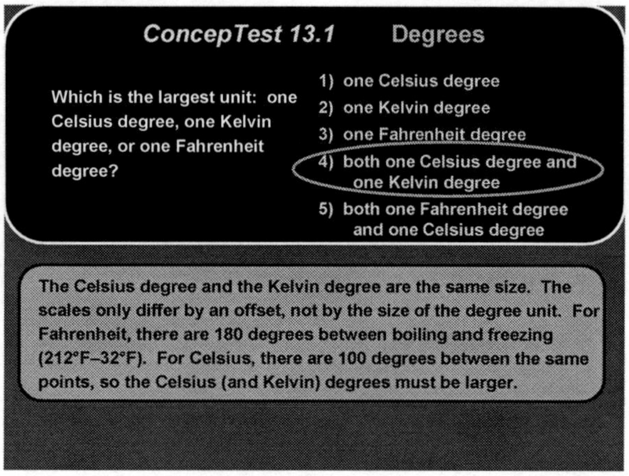

Conceptual Module 13.2
ConcepTest 13.2: Freezing Cold
Difficulty Level: 2

The students should recognize that the Celsius and Kelvin scales only differ by any offset. Thus, there can never be a temperature at which the two scales give the same reading.

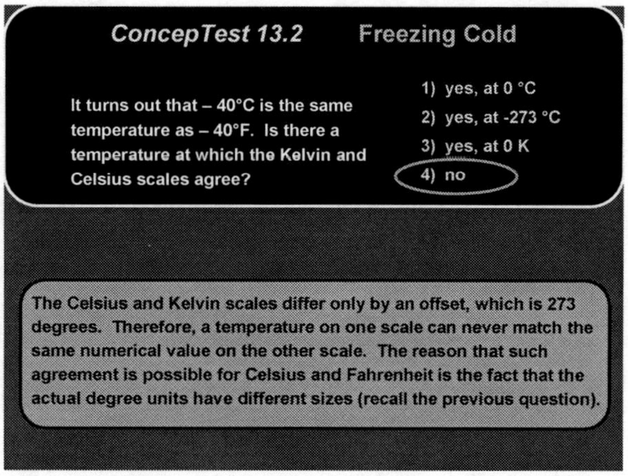

Conceptual Module 13.3

ConcepTest 13.3: Thermometers
Difficulty Level: 2
Suggested follow-up conceptual question:

- **Is it possible to have the mercury first rise and later drop?**

This question addresses the concept of thermal expansion. While the glass, and then the mercury, expand in a certain time sequence, the fact that the thermal expansion coefficient of mercury is greater than glass means that it will expand by a greater amount for a given temperature change.

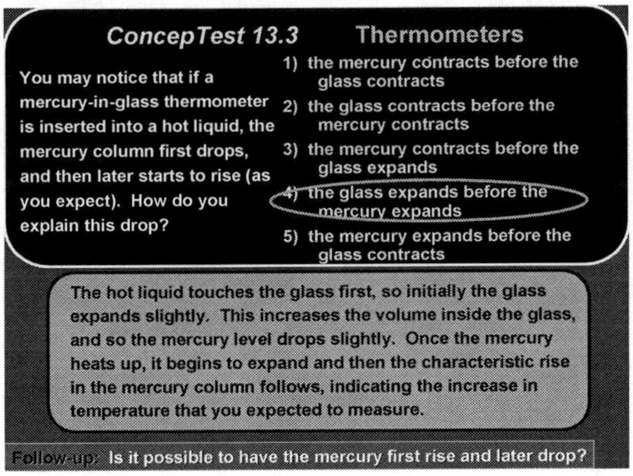

Conceptual Module 13.4

ConcepTest 13.4: Glasses
Difficulty Level: 2

This is a "trick" that most of the students will know from practical experience in the kitchen. This question will give an indication of whether or not they understand why this "trick" works.

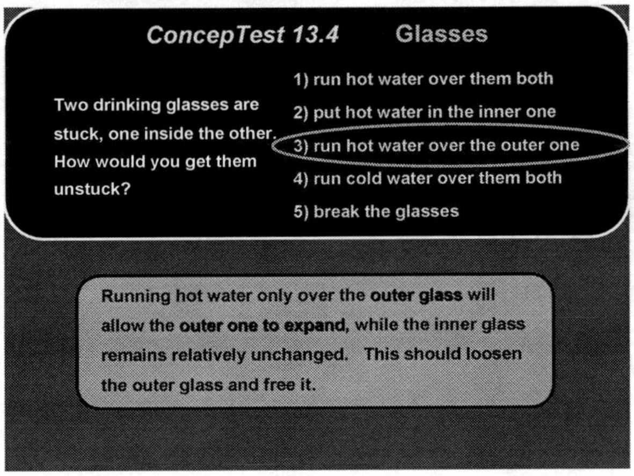

Conceptual Module 13.5 (two parts)

ConcepTest 13.5a: Steel Expansion I
Difficulty Level: 3

In this case, students must make the extrapolation from thermal expansion to an actual measurement. Here they must apply the expanded tape measure to a specific length in order to infer that the resulting measurements will, in fact, be too small.

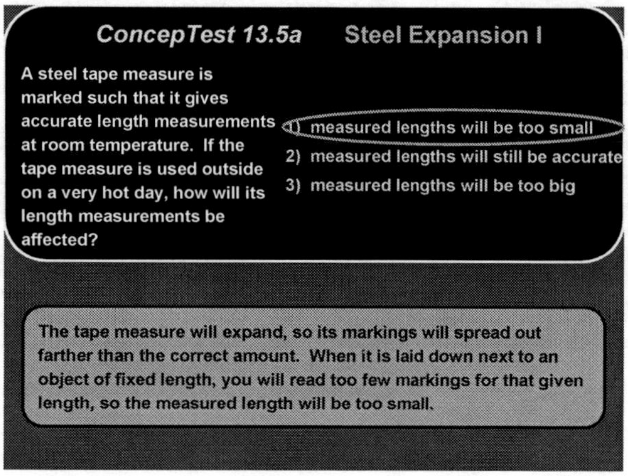

ConcepTest 13.5b: Steel Expansion II
Difficulty Level: 3

This is often a difficult question for students. The tendency is to believe that the metal "expands" into the hole, thus making it smaller. The analogy given, that of a circle sketched on a solid plate, generally convinces the students that the hole will, in fact, expand along with the rest of the plate.

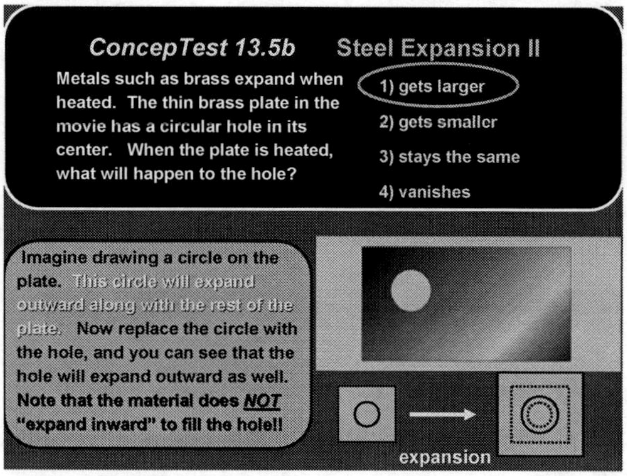

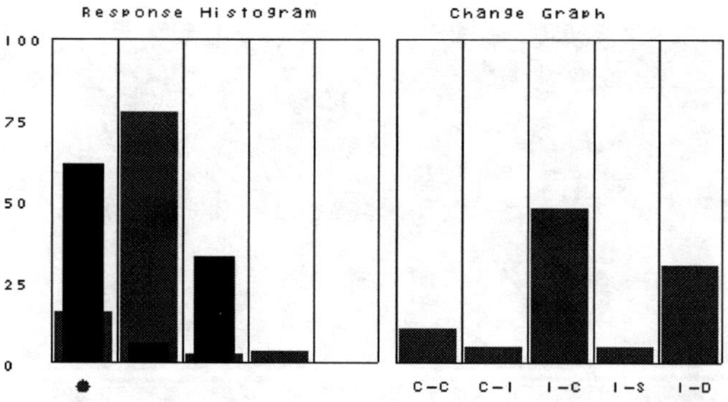

Conceptual Module 13.6 (two parts)

ConcepTest 13.6a: Steel Ring I
Difficulty Level: 4

In this case, the thermal expansion coefficients for different materials must be compared in order to arrive at the answer. The rod must be able to expand more than the ring (for a given temperature change) in order for the rod to touch the top of the ring, so the rod must have a higher coefficient.

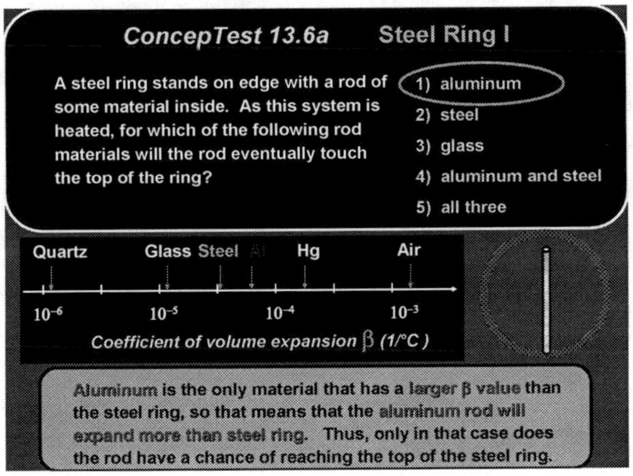

ConcepTest 13.6b: Steel Ring II
Difficulty Level: 4

This is the same type of comparative question. In this case, the two materials are given, and the action (heating or cooling) must be deduced. Students must realize that either the steel screws must shrink more or the aluminum parts must expand more (for a given temperature change) in order to loosen the stuck screws.

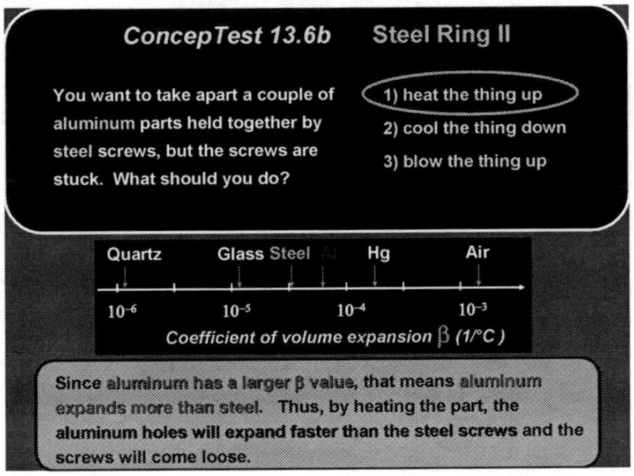

Conceptual Module 13.7

ConcepTest 13.7: Grandfather Clock
Difficulty Level: 4
Suggested follow-up conceptual question:

- **Roughly by how much will it run slower?**

This question links thermal expansion with motion of a pendulum. Once students realize that the pendulum length will increase, the effect of this change on the period must be deduced.

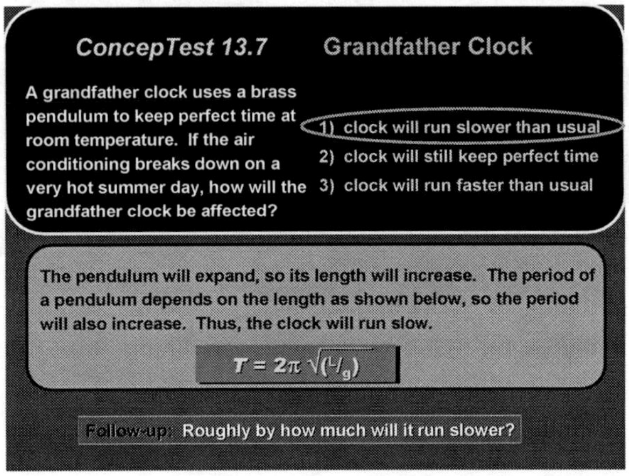

Conceptual Module 13.8 (two parts)

ConcepTest 13.8a: Nitrogen and Oxygen I
Difficulty Level: 1

The definition of a mole is a fixed quantity of particles (Avogadro's number). This does not depend on the substance involved, so a mole of any quantity contains the same number of constituents.

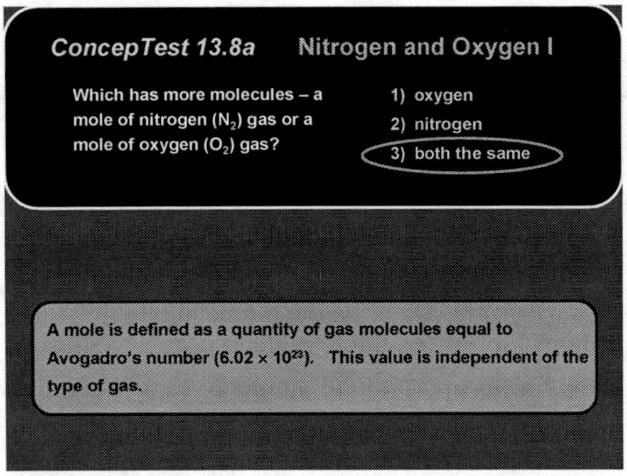

ConcepTest 13.8b: Nitrogen and Oxygen II
Difficulty Level: 2
Suggested follow-up conceptual question:

- **Which one will take up more space?**

This follow-up question now examines the mass of a mole of a gas. Since the number of particles is the same for each gas, the mass of the gas depends entirely on the masses of the individual molecules, which is larger for oxygen.

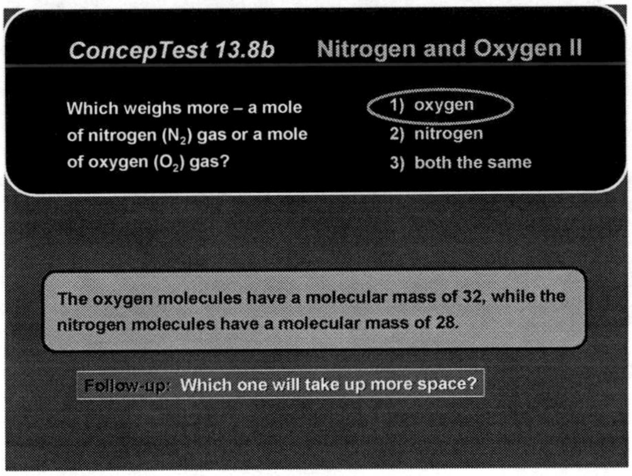

Conceptual Module 13.9 (three parts)

ConcepTest 13.9a: Ideal Gas Law I
Difficulty Level: 3

This question examines the ideal gas law. Students must learn to manipulate the parameters of the ideal gas law to simplify their problem-solving methods. In this case, the volume and temperature are held fixed, leaving a direct comparison between pressure (P) and moles (n) of gas.

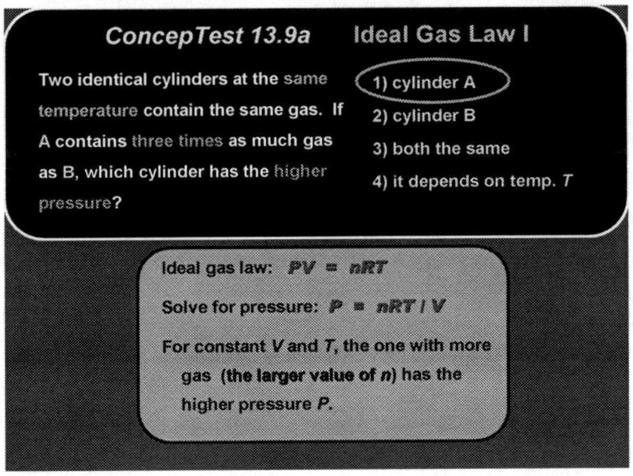

ConcepTest 13.9b: Ideal Gas Law II
Difficulty Level: 3

This is the same type of question for the ideal gas law, except that in this case, the volume and temperature are held fixed, leaving a direct comparison between temperature (T) and moles (n) of gas.

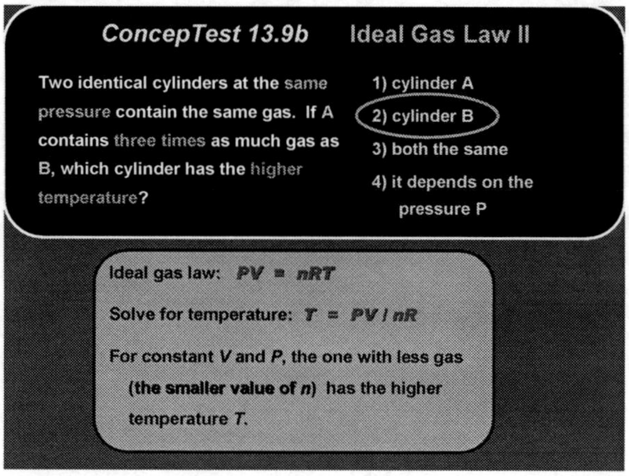

ConcepTest 13.9c: Ideal Gas Law III
Difficulty Level: 4

This is a slightly more complicated application of the ideal gas law, but it still follows the same logic. In this case, the question is quantitative and students must keep track of the volume and the number of moles of gas in order to deduce the pressure change.

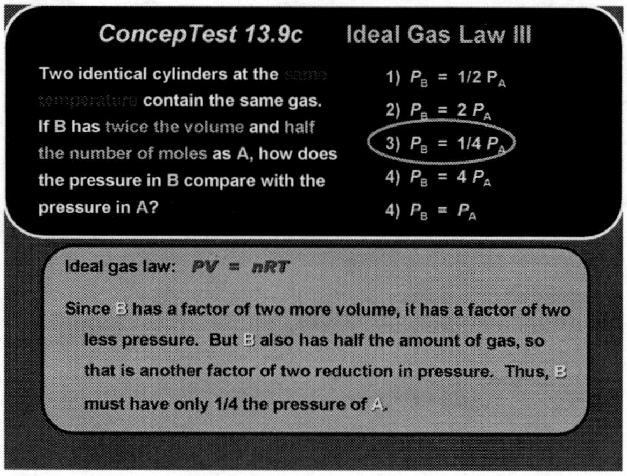

Conceptual Module 13.10
ConcepTest 13.10: Soda Bottle
Difficulty Level: 3

This "real life" example has probably happened to many of the students. Here again, while they may have noticed a partially crushed plastic bottle in the refrigerator, they may not have given much thought to what might have caused this effect.

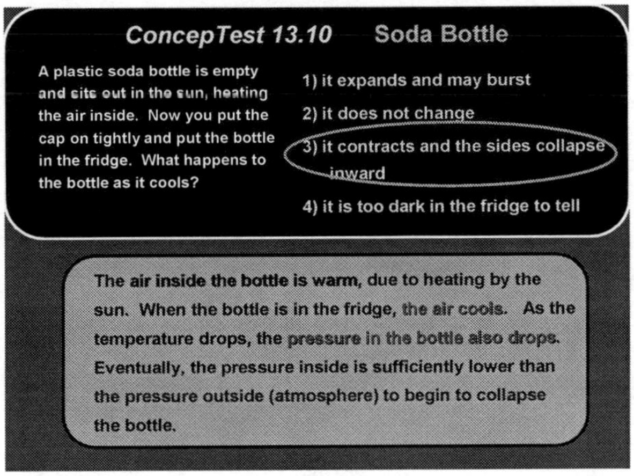

Conceptual Module 13.11
ConcepTest 13.11: Balloon in Freezer
Difficulty Level: 3
Suggested follow-up conceptual question:

- **What happens to the volume as the balloon rises in the air?**

This is the same type of question as the previous one. Using the ideal gas law, the students should deduce that the volume will decrease as the temperature decreases (at constant pressure).

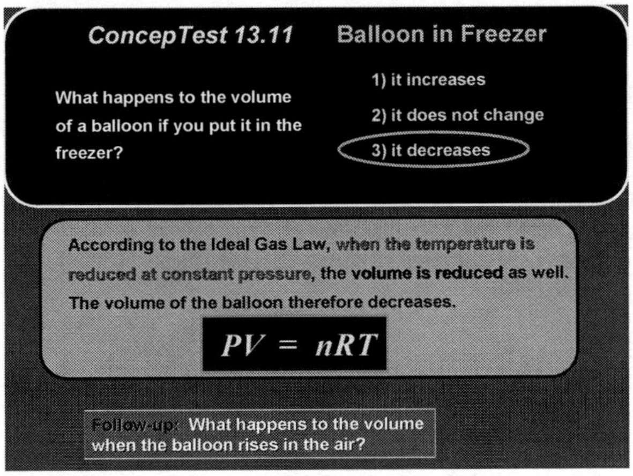

Conceptual Module 14.1 (two parts)

ConcepTest 14.1a: Thermal Contact I
Difficulty Level: 2

The temperature change ΔT of each object as the combined system approaches thermal equilibrium will depend upon the mass m of the objects and their specific heats c. Since they are composed of the same material, only the masses matter in this case, and the less massive object will have the greater temperature change (based on $\Delta Q = m\, c\, \Delta T$).

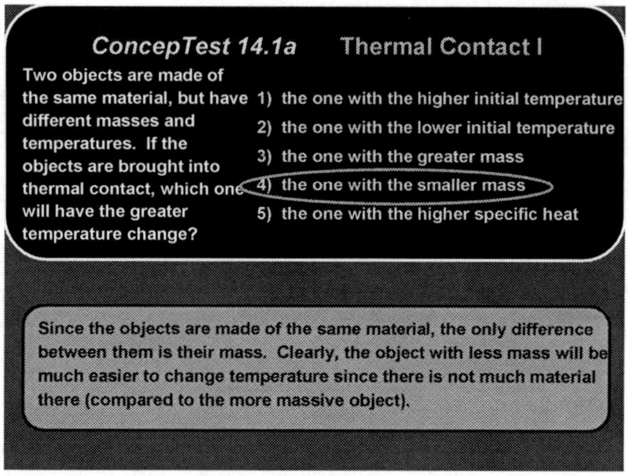

ConcepTest 14.1b: Thermal Contact II
Difficulty Level: 1

As seen from the relationship $\Delta Q = m\, c\, \Delta T$, the temperature change depends on the mass and specific heat of an object, but it does *not* depend upon the initial temperature. The same amount of heat added to an object at 20°C or 80°C will still result in the same temperature change.

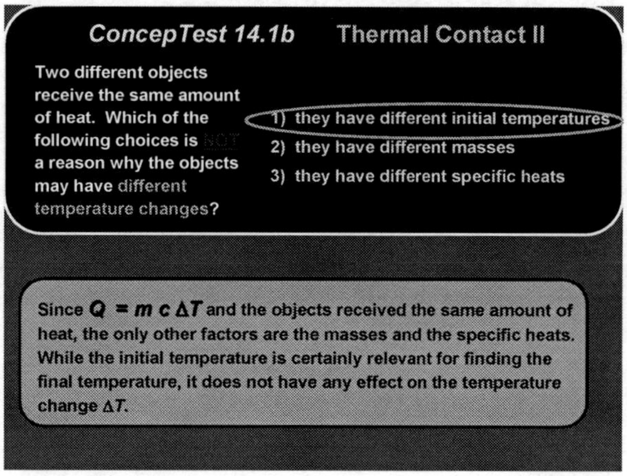

Conceptual Module 14.2
ConcepTest 14.2: Two Liquids
Difficulty Level: 2

Based on the equal times of heating, students must infer that the same amount of heat was added to each liquid. Since the liquids have the same mass, the only variable in the problem is the specific heat. We can think of specific heat as a "thermal inertia"—the liquid with the higher specific heat will have a harder time changing temperature and will therefore have a lower ΔT.

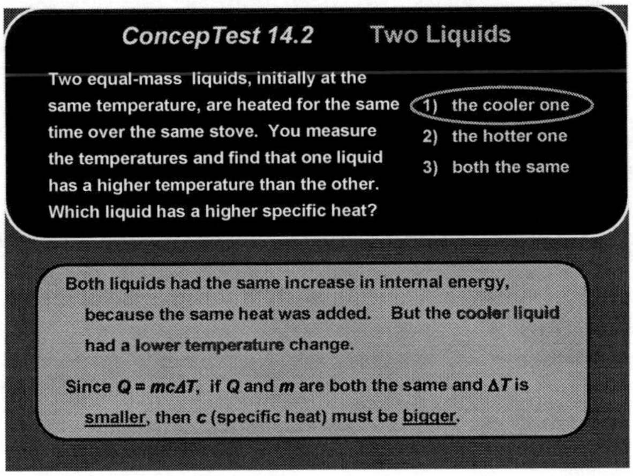

Conceptual Module 14.3 (two parts)
ConcepTest 14.3a: Night on the Field
Difficulty Level: 2

This again refers to specific heat and the fact that it is a type of "thermal inertia." The material with the lower specific heat (less "thermal inertia") will cool more rapidly; in this case, the soil.

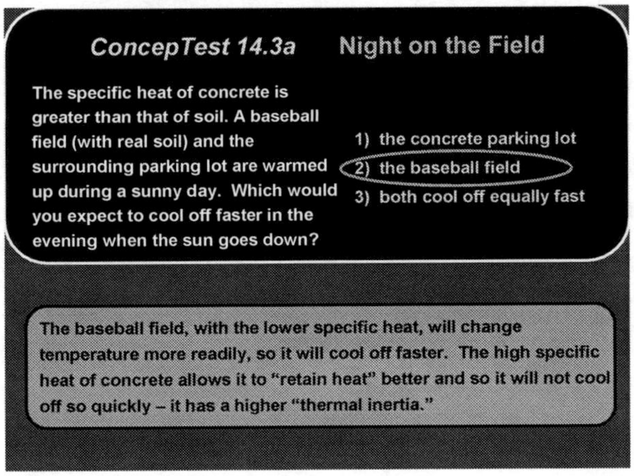

ConcepTest 14.3b: Night on the Beach
Difficulty Level: 4

This follow-up question introduces the concept of convection. The notion of specific heat dictates which material gets hotter during the day or cooler at night. The flow of air then must be considered (keeping in mind that hot air rises) in order to deduce the direction of the convection currents during the day or the night.

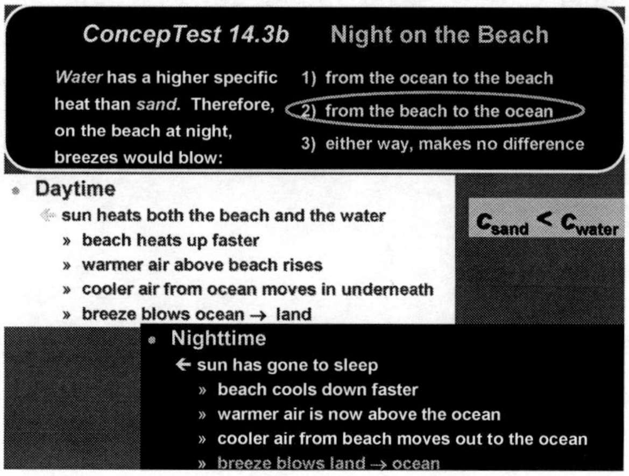

Conceptual Module 14.4

ConcepTest 14.4: Calorimetry
Difficulty Level: 3

Calorimetry is based on the concept of energy conservation (heat loss = heat gain). In this case, with the two materials being the same, the temperature change of each will depend entirely upon the masses. Since equal masses would result in an equilibrium temperature right in the middle (50°C), the students must realize that the larger mass will have the smaller ΔT, so the final temperature must be closer to (but not equal to) the initial temperature of the larger mass.

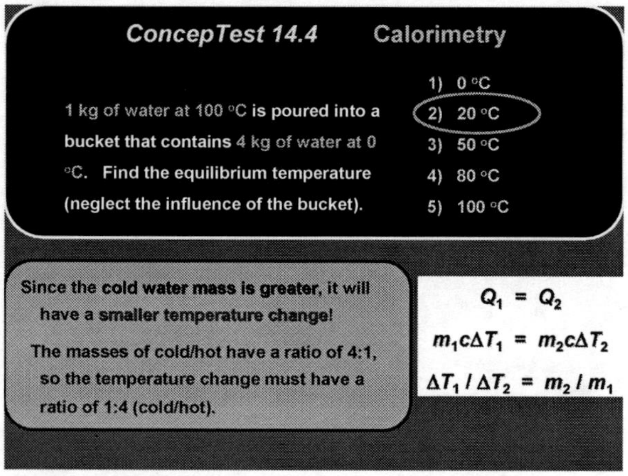

Conceptual Module 14.5

ConcepTest 14.5: More Calorimetry
Difficulty Level: 3

In this case, it is not the materials that are the same, but, rather, the masses. So the ΔT will depend on the different specific heats. The water has the higher specific heat, which means that it is more difficult to change its temperature. Therefore the final temperature will be closer to that of the water.

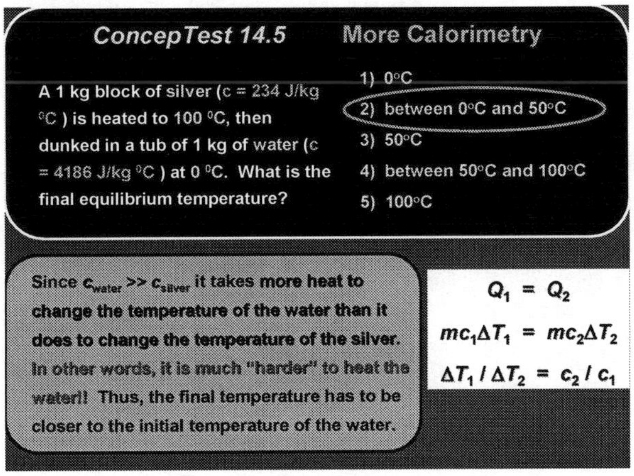

Conceptual Module 14.6

ConcepTest 14.6: Adding Heat
Difficulty Level: 1
Suggested follow-up conceptual question:

- **Does that depend on the substance?**

This may be regarded as a "trick" question, but it is certainly representative of an important physical process. The notion of a phase change is specifically related to changing the "binding" of a substance and not changing the kinetic energy (related to temperature) of its constituent particles. This is a crucial distinction for students to make in order to fully understand what a phase change is.

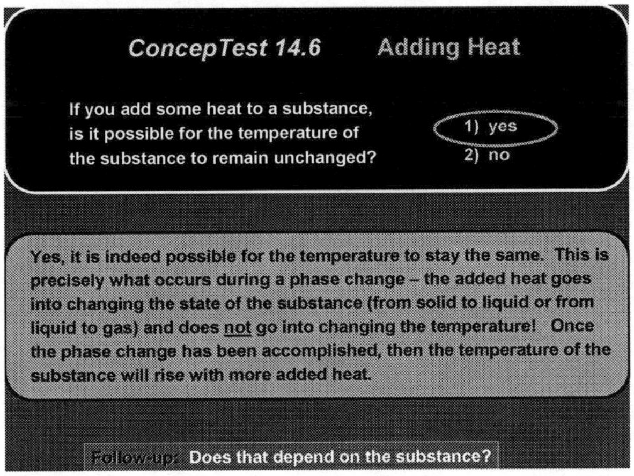

Conceptual Module 14.7

ConcepTest 14.7: Hot Potato
Difficulty Level: 2
Suggested follow-up conceptual question:

- **How can you cook the potatoes faster?**

This question (again) relates to phase changes. When the water is boiling, a phase change is taking place, so the temperature of the water is *not increasing* as heat is added. Adding more heat will only make the phase change occur more rapidly (*i.e.*, faster boiling), but will still not alter the temperature of the water as long as the phase change is taking place. Thus, the potatoes will *not* cook faster.

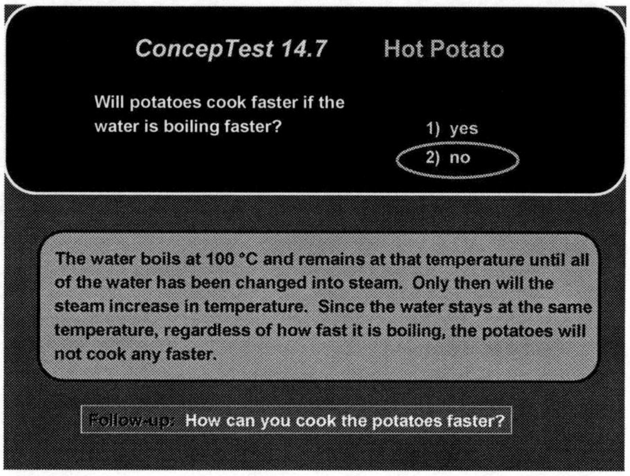

Conceptual Module 14.8

ConcepTest 14.8: Water and Ice
Difficulty Level: 3
Suggested follow-up conceptual question:

- **How much more water at 50°C water would you need?**

Both temperature changes and phase changes must be taken into account here. The heat released by the 50°C water as it cools down to 0°C goes into melting the ice (which does not completely melt).

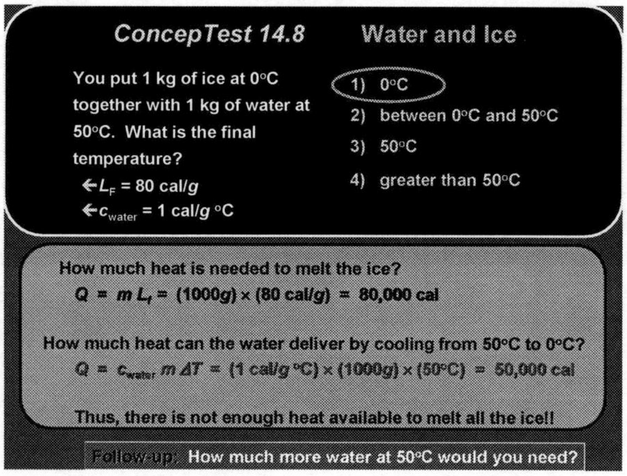

Conceptual Module 14.9

ConcepTest 14.9: Ice and Steam
Difficulty Level: 4
Suggested follow-up conceptual question:

- **How much more ice would you need?**

This follow-up question is similar to the previous one in the sense that temperature changes and phase changes need to be considered. In this case, due to the very high value of the heat of vaporization of water (540 cal/g), there is actually more than enough steam to not only melt all of the ice, but also to raise the temperature of the melted ice (0°C water) up to 100°C and still have steam to spare!

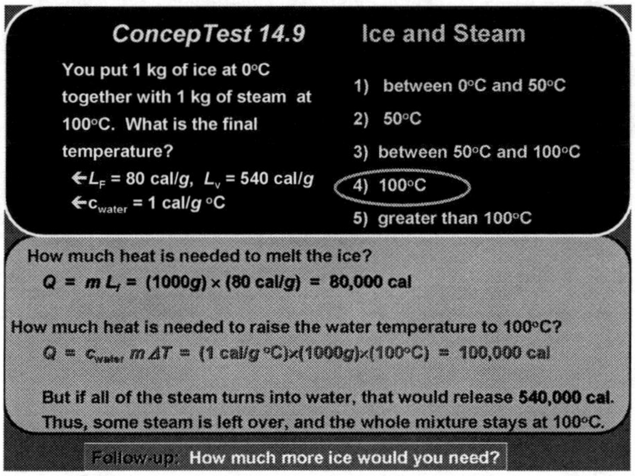

Conceptual Module 14.10

ConcepTest 14.10: You're in Hot Water!
Difficulty Level: 2

At first glance, the fact that both the water and the steam are at the same temperature suggests that they will both burn equally. However, the key point here is that when the steam condenses into water, it releases the heat of vaporization, which is a very large quantity of heat. So, bearing that in mind, the steam is therefore much more likely to cause severe burns than the hot water.

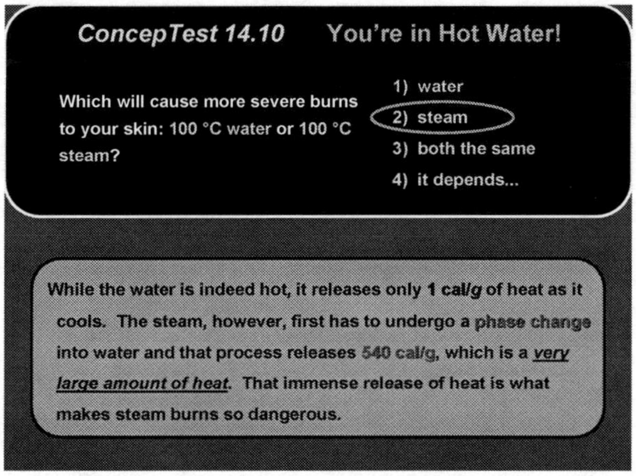

Conceptual Module 14.11

ConcepTest 14.11: Spring Break
Difficulty Level: 2

Evaporation is essentially a cooling process, which is why we sweat. In order to vaporize the water on your skin, the heat of vaporization must be absorbed from your body, which is why you tend to cool off. This evaporation is much more likely to take place in a dry climate (like Phoenix) where the air is not already saturated with moisture. So you will tend to feel cooler in the drier environment, where a more effective evaporation process can take place.

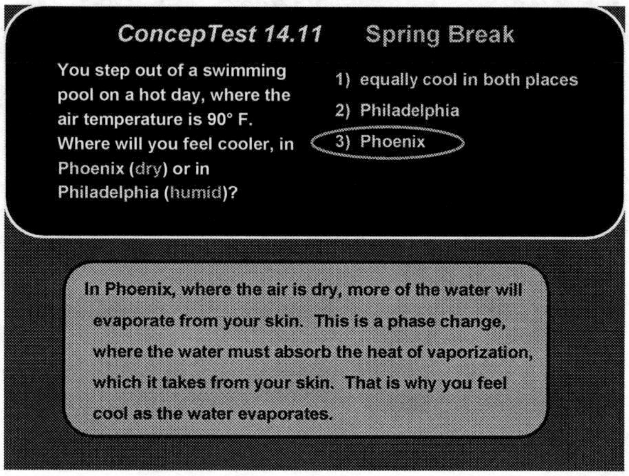

Conceptual Module 14.12

ConcepTest 14.12: Heat Conduction
Difficulty Level: 2

The higher the thermal conductivity of a surface, the faster it can draw (conduct) heat away from your body. The metal surface will always feel cooler (compared to bare concrete or a rug); this is a direct result of its higher thermal conductivity.

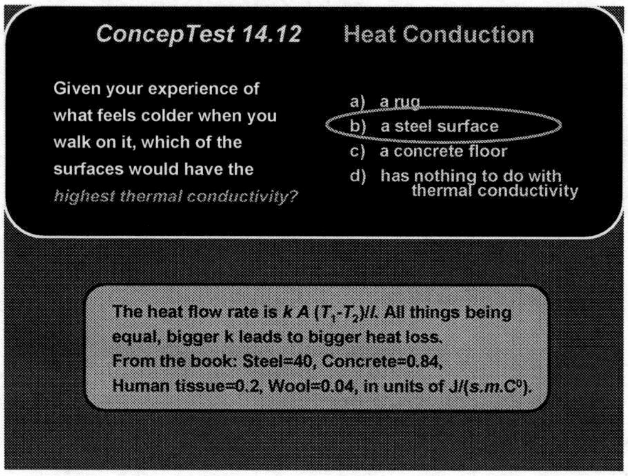

Conceptual Module 14.13

ConcepTest 14.13: Radiation

Difficulty Level: 3

This is a direct application of the Stefan-Boltzmann Law. According to that relationship, the radiated energy depends on T^4, so if the temperature drops by a factor of 2, the energy will correspondingly drop by a factor of 16.

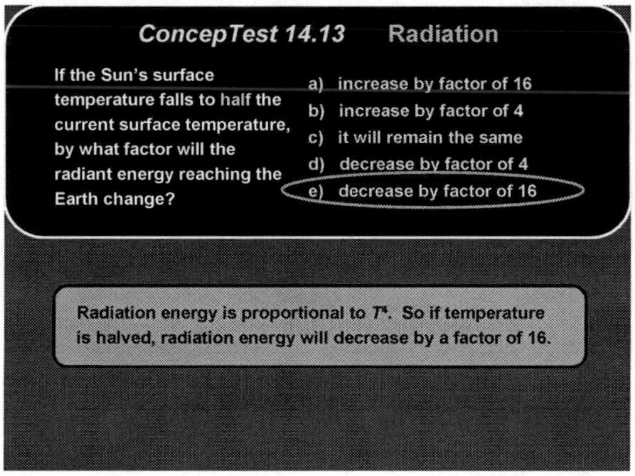

Conceptual Module 15.1

ConcepTest 15.1: Free Expansion
Difficulty Level: 2

This question examines the First Law of Thermodynamics. In this case, with no work done and no heat added, the total energy of the system must remain constant (this is essentially a restatement of conservation of energy). Thus, if the energy of the system does not change, the temperature will not change.

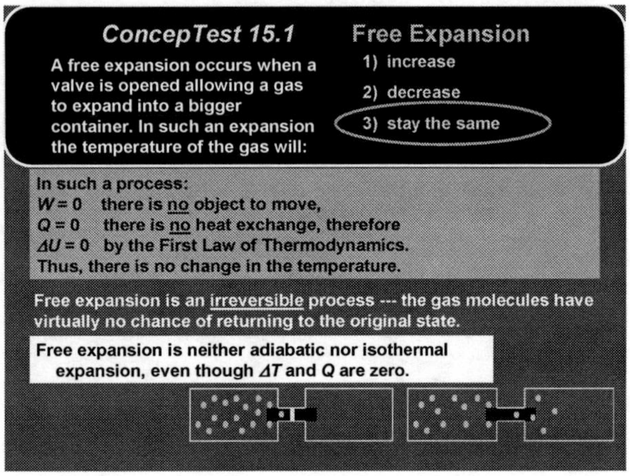

Conceptual Module 15.2

ConcepTest 15.2: Work
Difficulty Level: 2

This question is designed to familiarize students with a P-V diagram. The area under the P-V curve represents the work done by a gas. Since the top curve (expansion at high pressure) corresponds to positive work and the lower curve (compression at low pressure) corresponds to negative work, it is clear that the net work is positive.

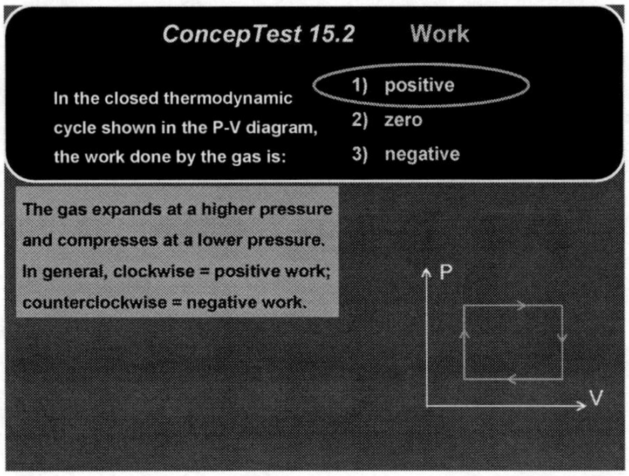

Conceptual Module 15.3

ConcepTest 15.3: Heat Engine
Difficulty Level: 3
Suggested follow-up conceptual question:

- **What would you need to change to make it a Carnot engine?**

Students must get a sense of idealized and real processes and understand the efficiency that pertains to each. The Carnot efficiency denotes an ideal efficiency corresponding to a reversible process. In this case, the actual efficiency is less than the Carnot efficiency, so this situation describes a real process, which is by definition irreversible.

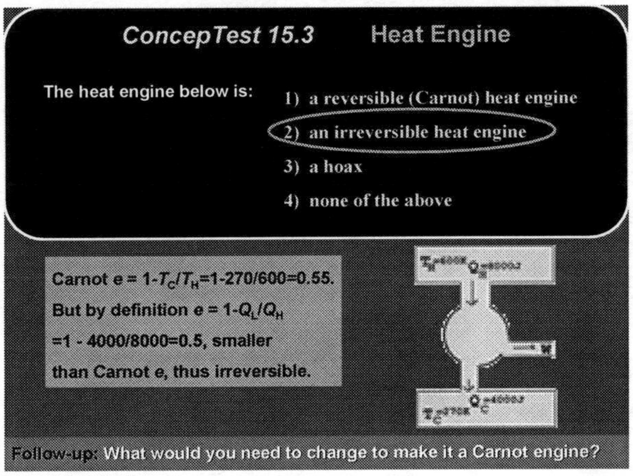

Conceptual Module 16.1 (two parts)

ConcepTest 16.1a: Electric Charge I
Difficulty Level: 1
Suggested follow-up conceptual questions:

- **What does the picture look like if the two balls are oppositely charged?**
- **What if both balls are neutral?**

The repulsion of the two charged balls indicates only that they have the same type of charge, but it is impossible to distinguish between two positive or two negative balls. Either case will result in a repulsion as shown in the figure.

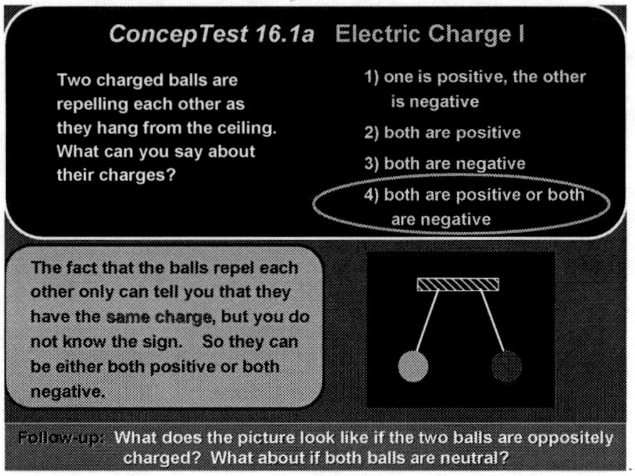

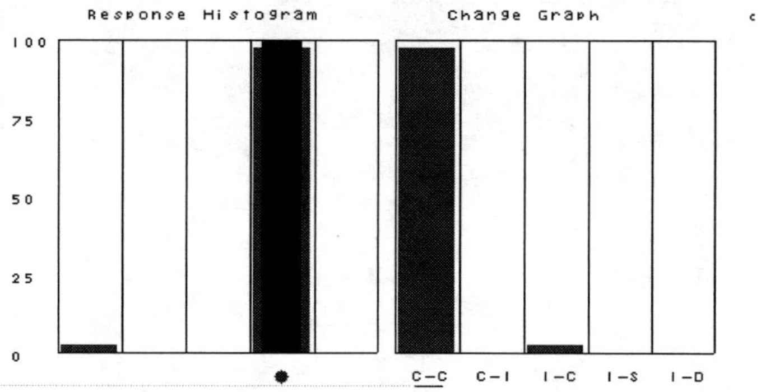

ConcepTest 16.1b: Electric Charge II
Difficulty Level: 2

This is an exercise in logic, as it pertains to like and unlike charges. Students must deduce from the first figure that the green and pink balls have the same charge and from the second figure that the green and yellow balls also have the same charge. These deductions should lead them to the final conclusion that all the balls have the same charge.

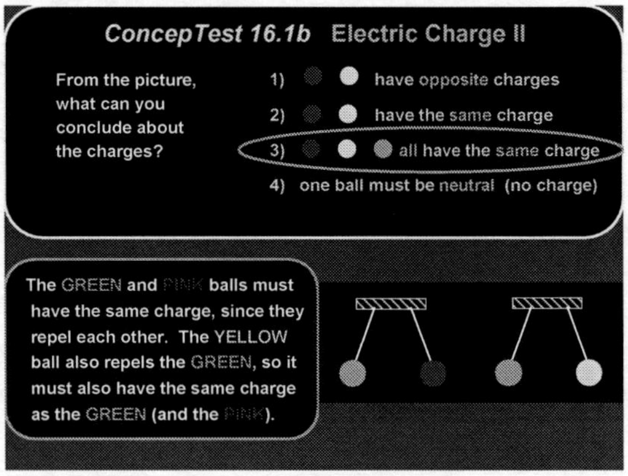

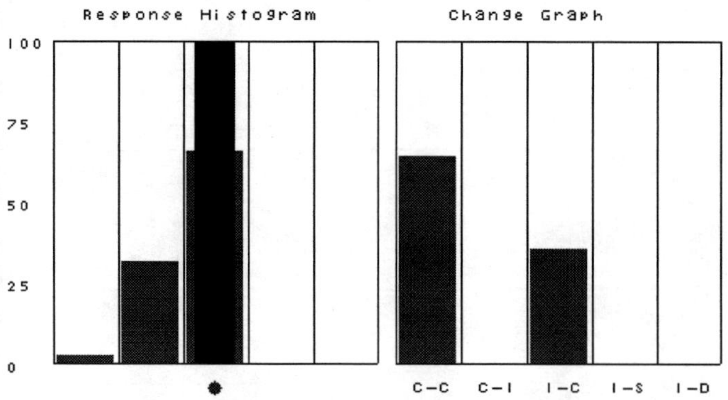

Conceptual Module 16.2 (two parts)

ConcepTest 16.2a: Conductors I
Difficulty Level: 2
Suggested follow-up conceptual question:

- **What happens if the metal ball is replaced by a plastic ball?**

This question addresses the concept of induction. Students must realize that the metal ball need not be charged in order to be attracted to the charged rod. The separation of mobile charges (induction) in the metal conductor will also give rise to a net attraction between the two objects.

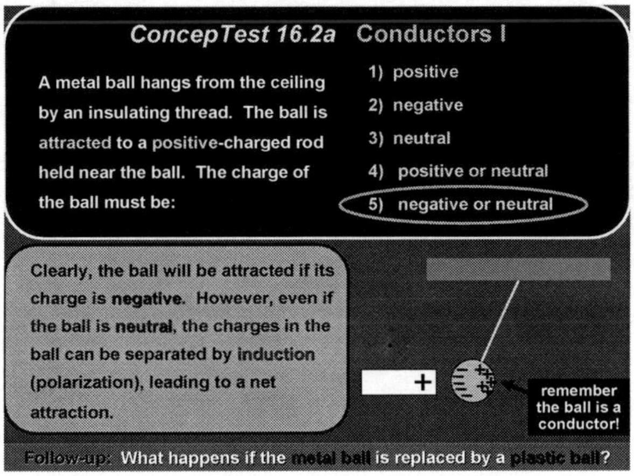

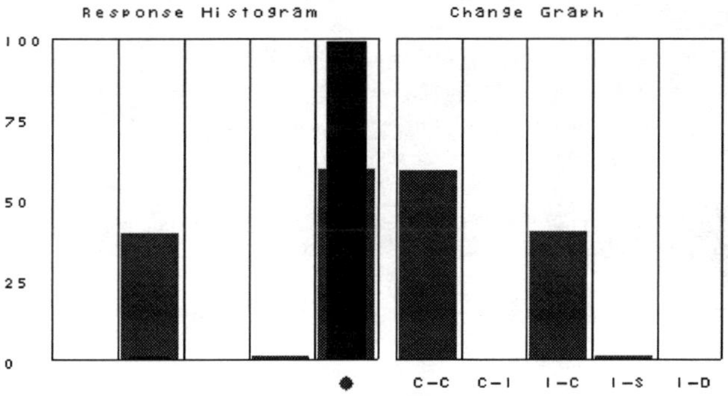

ConcepTest 16.2b: Conductors II
Difficulty Level: 2
Suggested follow-up conceptual question:

- **What will happen when the conductors are reconnected with a wire?**

This question also relates to induction. The separation of charges that is achieved when the conducting balls are linked together can be "locked" in place by cutting the connection between the balls. The students must understand that the near ball will have an opposite charge to the rod, since these charges are attracted when the rod is brought close.

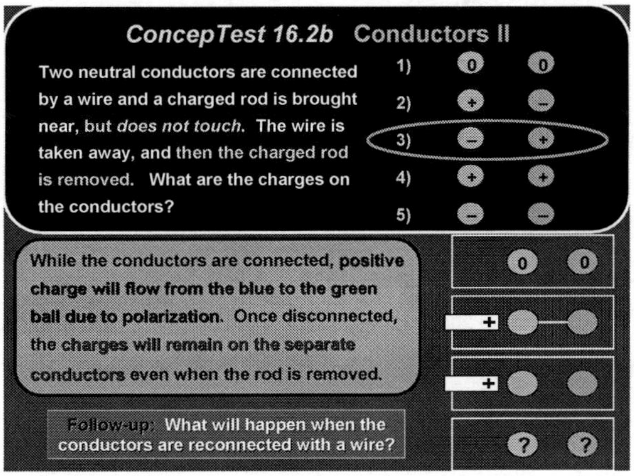

Conceptual Module 16.3 (three parts)
ConcepTest 16.3a: Coulomb's Law I
Difficulty Level: 2

This question is intended to begin familiarizing students with Coulomb's Law, which is symmetric in the two charges that are interacting. The answer to this question can also be deduced from Newton's Third Law, since the forces must be equal and opposite.

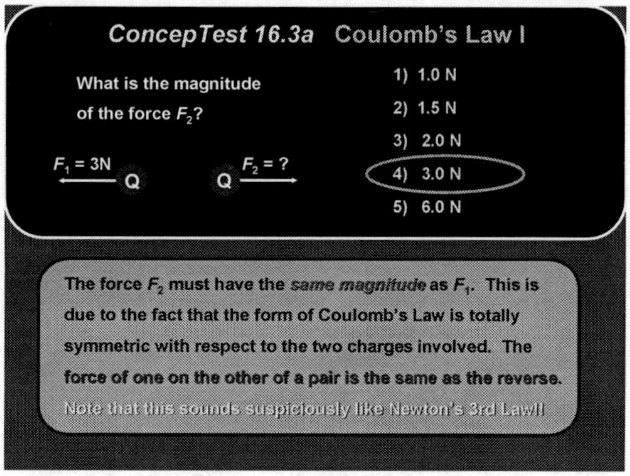

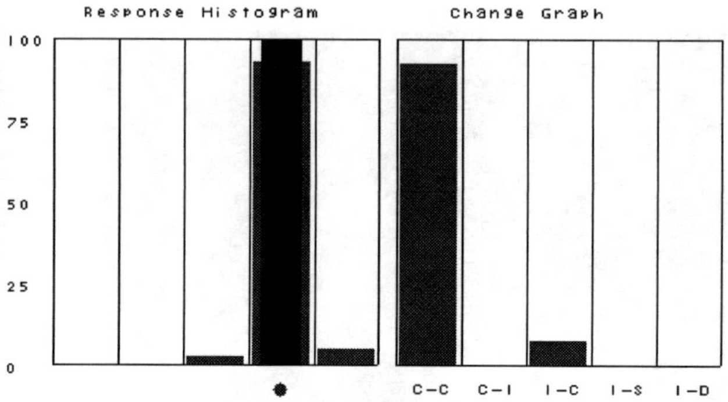

ConcepTest 16.3b: Coulomb's Law II
Difficulty Level: 3
Suggested follow-up conceptual question:
- **Now what is the magnitude of F_2?**

Continuing with Coulomb's Law, the students should be able to determine the new force on the charges when one of them is increased. The force simply scales as the product of the charges, so changing one charge by a certain factor will change the force by that same factor.

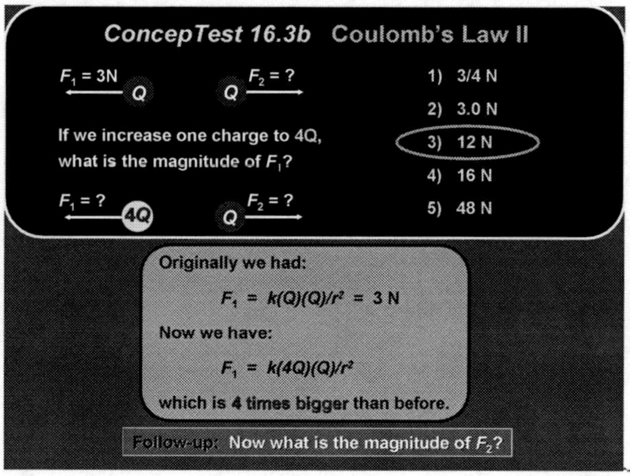

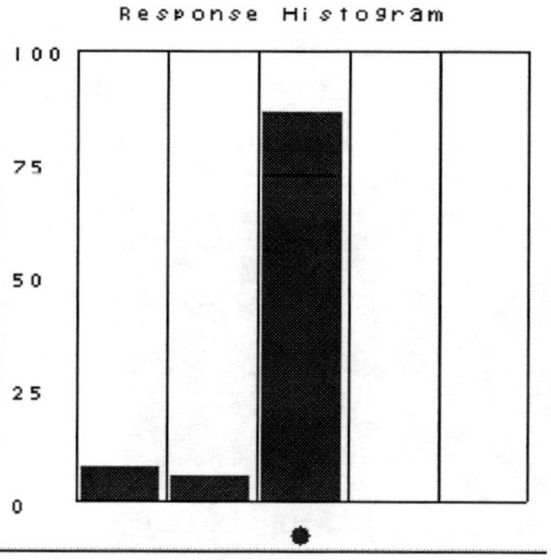

ConcepTest 16.3c: Coulomb's Law III
Difficulty Level: 3
Suggested follow-up conceptual question:

- **What is the force if the original distance is halved?**

This question examines the distance dependence of Coulomb's Law. Using the inverse square law, an increase in distance by a factor of 3 will lead to a decrease in force by a factor of 9.

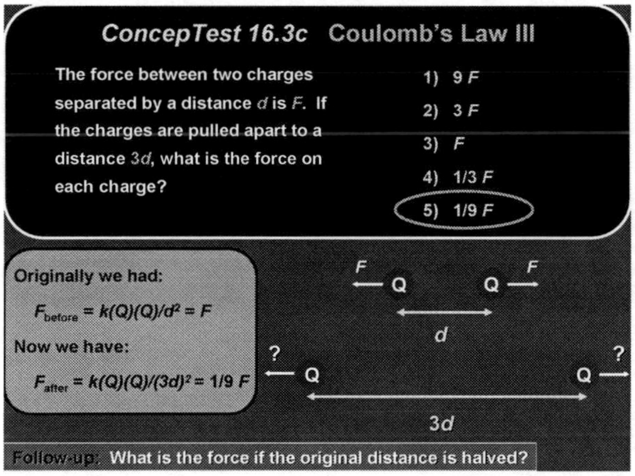

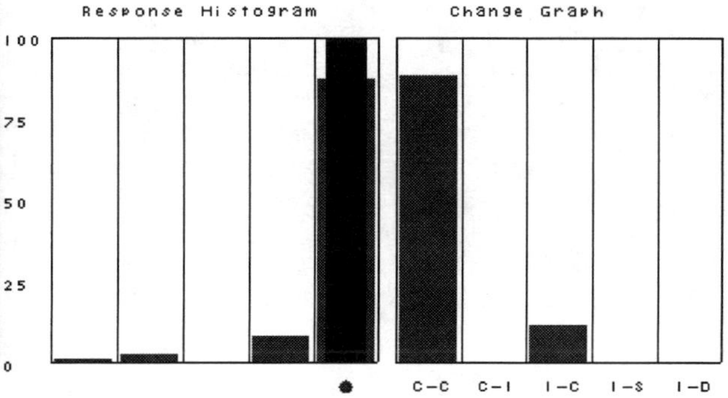

Conceptual Module 16.4 (three parts)

ConcepTest 16.4a: Electric Force I
Difficulty Level: 3
Suggested follow-up conceptual questions:

- **What happens if both charges are +*Q*? Where would the *F* = 0 point be in this case?**

Superposition of electric fields and forces is a common difficulty for students. This question requires the students to consider the vector addition of two forces for a "virtual" charge that can be either positive or negative. In the end, it does not matter, but this is a realization that the students need to make for themselves. In essence, they are actually finding the point where the net electric field is zero.

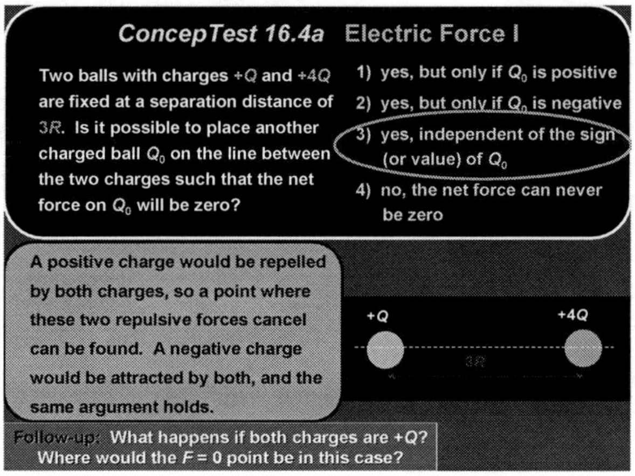

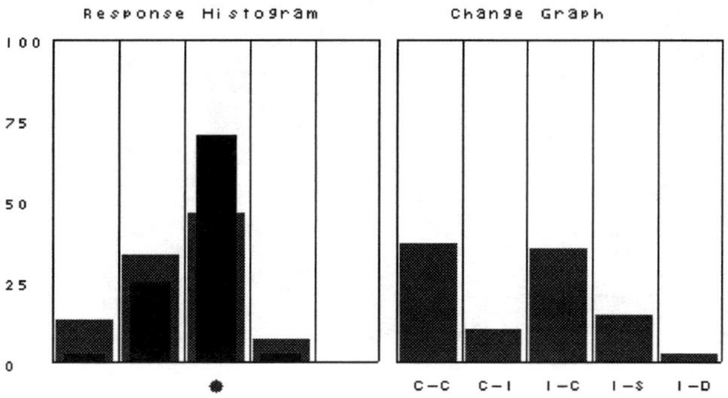

ConcepTest 16.4b: Electric Force II
Difficulty Level: 4

This follow-up question asks students to address the previous case quantitatively. Now they have to determine the actual position where the net force is zero, taking into account the relative sizes and positions of the two charges. Students should easily be able to see that the point must be closer to the smaller charge, but then it is a matter of evaluating the specific position.

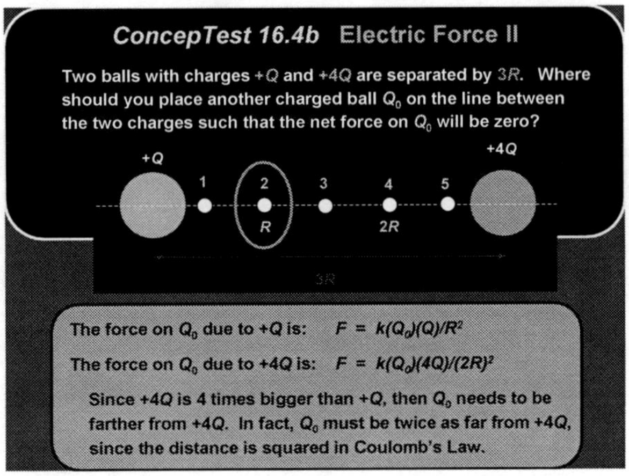

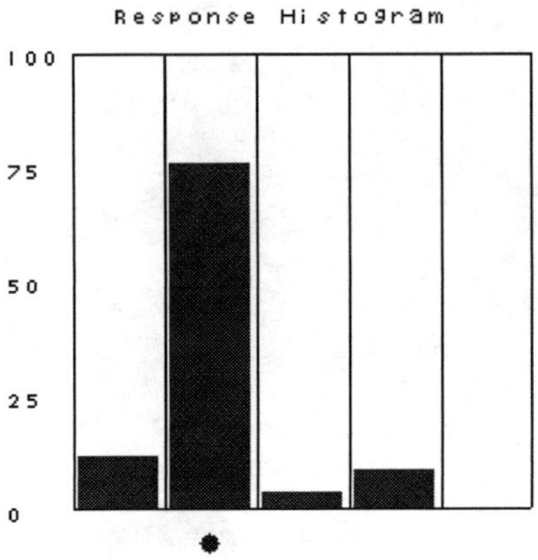

ConcepTest 16.4c: Electric Force III
Difficulty Level: 3

Suggested follow-up conceptual question:

- **What happens if one charge is +*Q* and the other is –*Q*?**

This examines an alternate case to that of ConcepTest 16.4a where the charges now have opposite signs. This means that the position of zero force cannot be in between the charges and in fact must be off tó one side or the other. Again, this question previews the upcoming introduction of electric field.

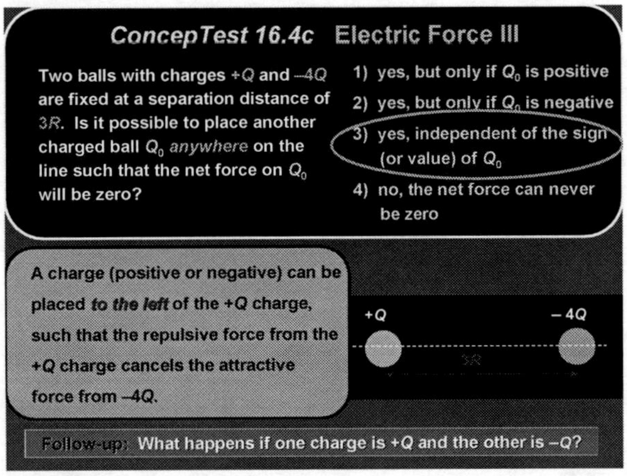

Conceptual Module 16.5 (three parts)

ConcepTest 16.5a: Proton and Electron I
Difficulty Level: 2
Suggested follow-up conceptual question:

- **Which particle feels the larger force at any one moment?**

This is a question relating to the distance dependence of Coulomb's Law. Due to the inverse square law, as the separation between the charges decreases, the force between them will increase. While this result may seem obvious, students do not often have the experience of a non-constant force acting.

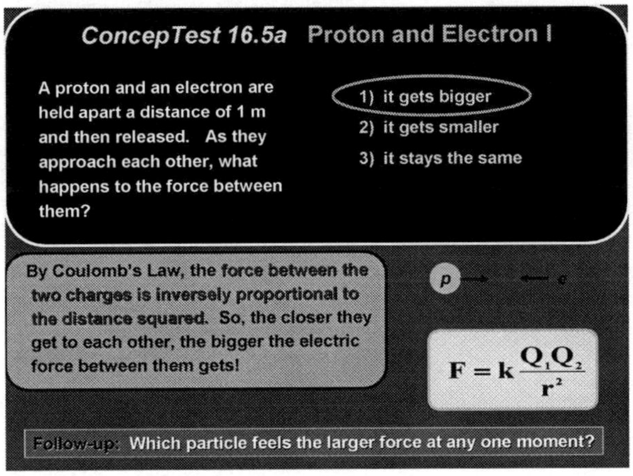

ConcepTest 16.5b: Proton and Electron II
Difficulty Level: 3

This is a reminder of earlier work with Newton's Second Law. At any one moment, the force on each particle is the same, but the one with the smaller mass will have the larger acceleration ($F = ma$).

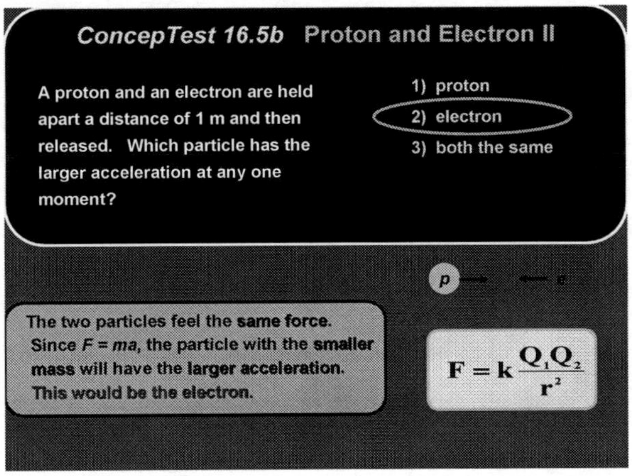

ConcepTest 16.5c: Proton and Electron III
Difficulty Level: 3
Suggested follow-up conceptual question:

- **Which particle will be moving faster when they meet?**

For this final question, the students must utilize their answer for acceleration in the previous question to recognize that if the electron accelerates faster, it will cover more distance more rapidly, and so the two particles will meet much closer to the original proton position (and *not* in the middle!).

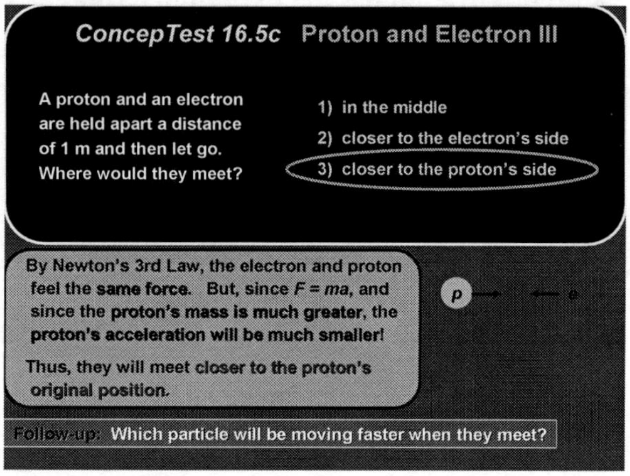

Conceptual Module 16.6

ConcepTest 16.6: Forces in 2D
Difficulty Level: 4
Suggested follow-up conceptual question:

- **What happens if the yellow charge would be +3*Q*?**

This is a problem relating to Coulomb's Law in two dimensions. Here the students have to perform a vector addition with two forces that they must first determine from Coulomb's Law. In this case, the distances are the same, so the variation in the two forces is related to the magnitude of the charges.

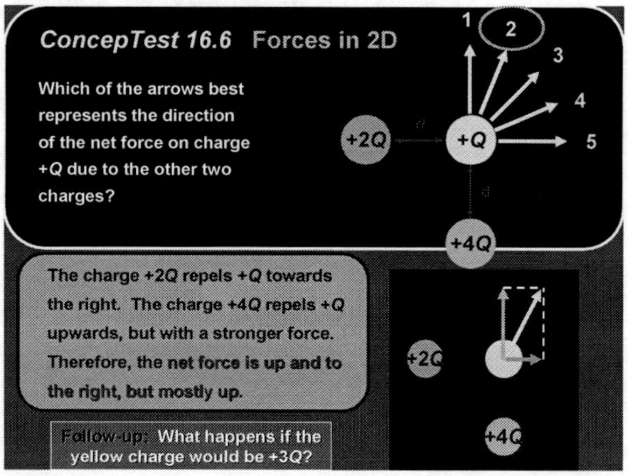

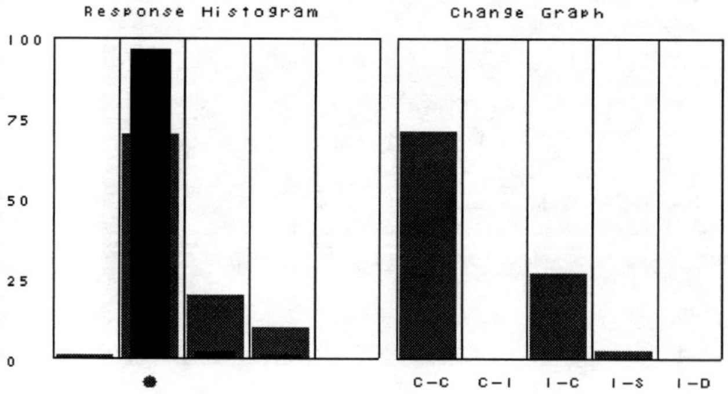

Conceptual Module 16.7

ConcepTest 16.7: Electric Field

Difficulty Level: 3

Suggested follow-up conceptual question:

- **If your distance is doubled, what must you do to the charge to maintain the same *E* field at your new position?**

This question examines the expression for electric field strength in terms of the charge and distance. Related to Coulomb's Law, this also involves an inverse square law for the distance, which must be taken into account in this problem to obtain the correct answer.

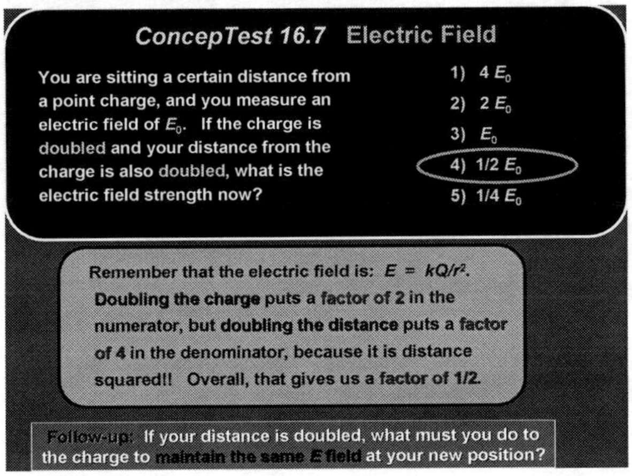

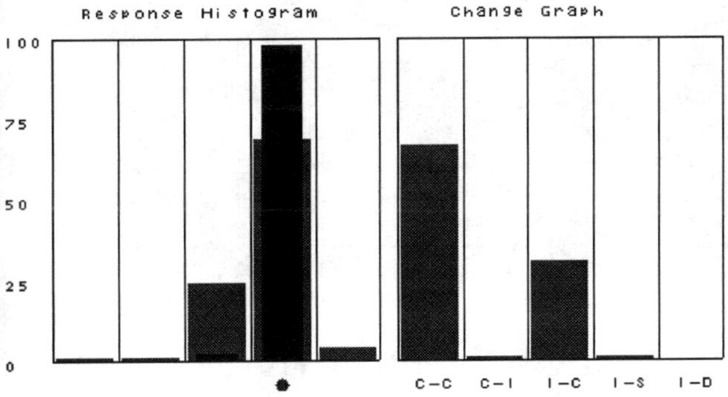

Conceptual Module 16.8 (two parts)

ConcepTest 16.8a: Field and Force I

Difficulty Level: 2

Students have tremendous difficulty understanding that there can be an electric field at an empty point in space. There is a tendency to think that a charge must be at the field location. To emphasize that the electric field only has to do with the source charge, this question places two different charges (which are irrelevant) at the field location and asks for the electric field in both cases.

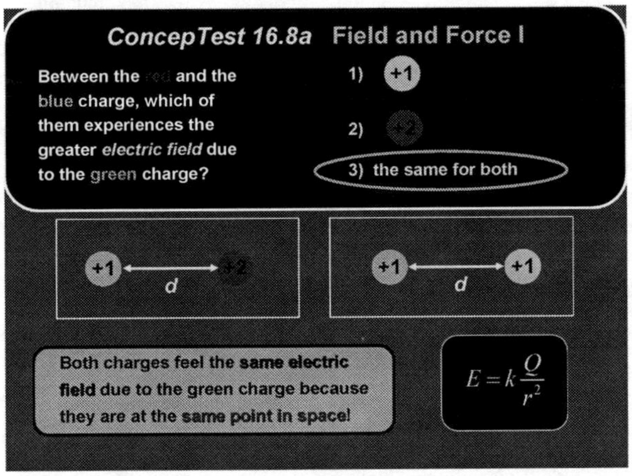

ConcepTest 16.8b: Field and Force II
Difficulty Level: 3

Following on the previous question, this one now asks for the force. In this case, the charge located at the field position is relevant ($F = qE$). So while the field is the same at the position of the red or blue charges, the force on those charges is different, depending on the value of those charges. This series of questions is intended to clarify the difference between electric field and electric force.

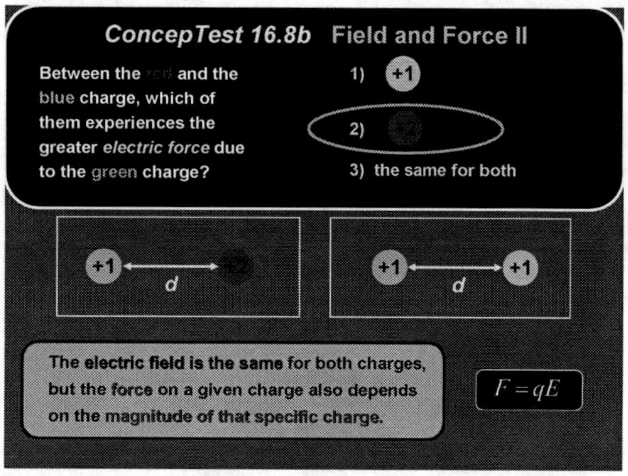

Conceptual Module 16.9 (three parts)
ConcepTest 16.9a: Superposition I
Difficulty Level: 4
Suggested follow-up conceptual questions:
- **What if the lower charge was +2C?**
- **What if both charges were +2C?**

This question examines the superposition of electric field vectors for the simple case of two point charges. Students must recognize that the field points toward the negative charges and that the superposition in this symmetric case leads to a net leftward force.

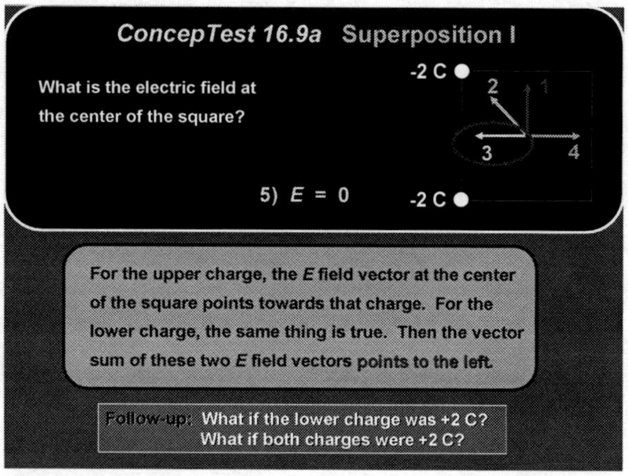

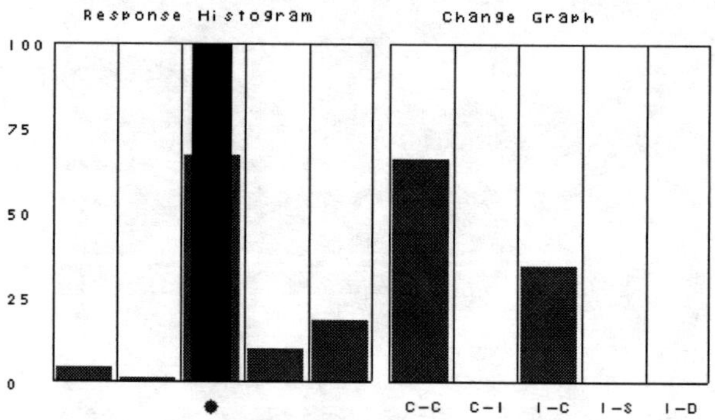

ConcepTest 16.9b: Superposition II
Difficulty Level: 4
Suggested follow-up conceptual questions:

- **What if the upper two charges were +2C?**
- **What if the right-hand charges were +2C?**

In this case, the superposition involves four charges. Still, the situation is symmetric, and so in this case, the net electric field is zero at the center. In fact, the sign of the charges will not matter, as long as they are all the same sign.

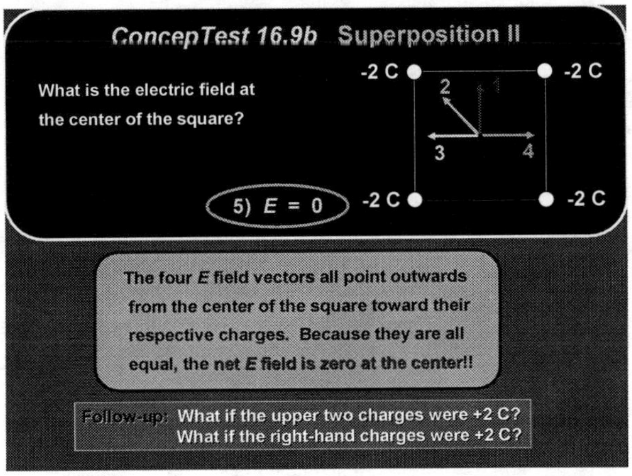

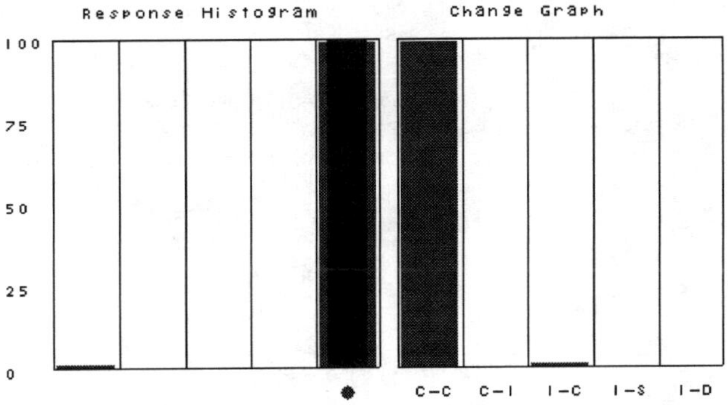

ConcepTest 16.9c: Superposition III
Difficulty Level: 4
Suggested follow-up conceptual question:

- **What if all three charges reversed their signs?**

This last question involves a case of superposition that is not so symmetric. The effect of the two closer positive charges compared to the more distant single negative charge has to be taken into account. Both distances from the field point and directions of the contributions must be considered.

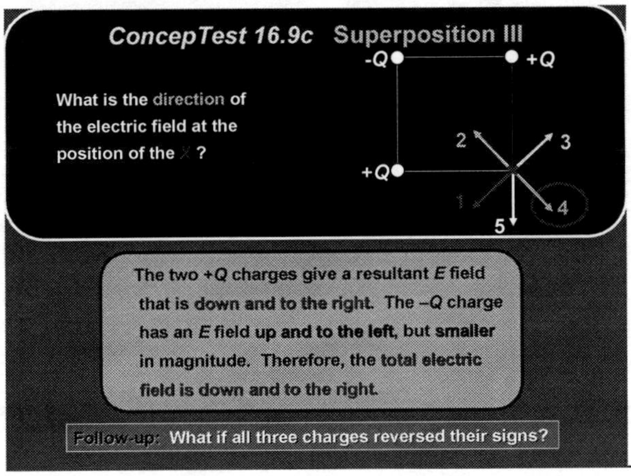

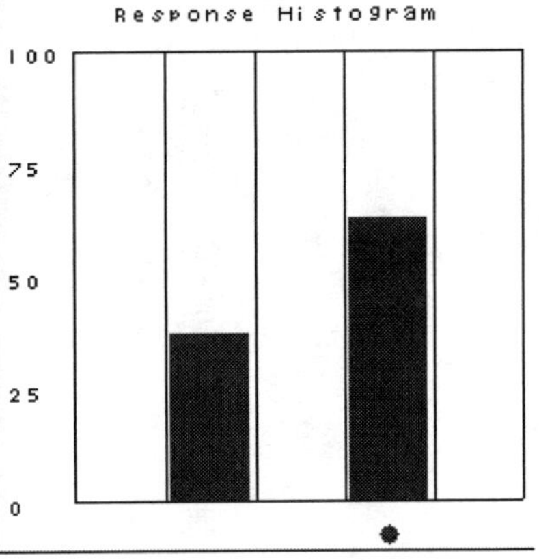

Conceptual Module 16.10
ConcepTest 16.10: Find the Charges
Difficulty Level: 4
Suggested follow-up conceptual question:
- **How would you get the *E* field to point toward the right?**

This question employs a different tactic. Here the electric field is given, and something must be inferred about the charges that have created that field. Using the superposition practice from the previous series of questions, students should be able to reconstruct the nature of the charges.

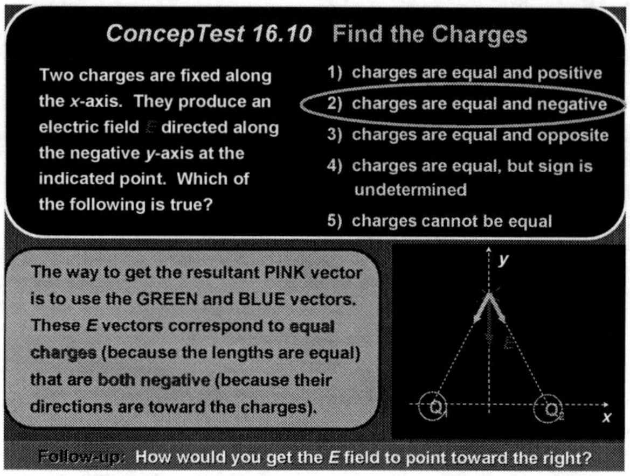

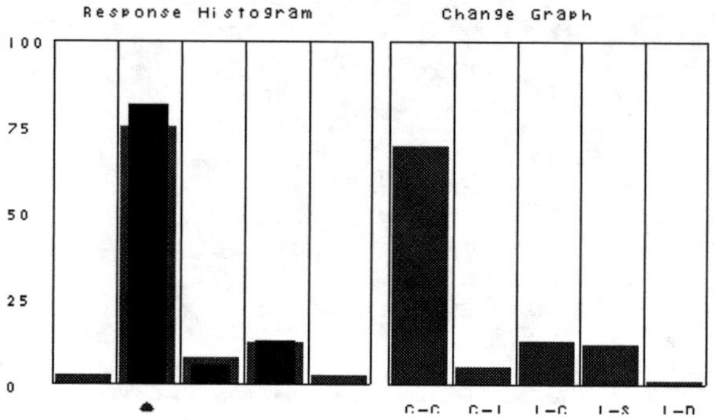

Conceptual Module 16.11
ConcepTest 16.11: Uniform Electric Field
Difficulty Level: 3
Suggested follow-up conceptual question:
- **What if the charge is placed at a different position in the field?**

The force on a charge in a uniform electric field depends on the magnitude of the charge ($F = qE$). So if the charge varies by a certain factor, then the force will change by the same factor.

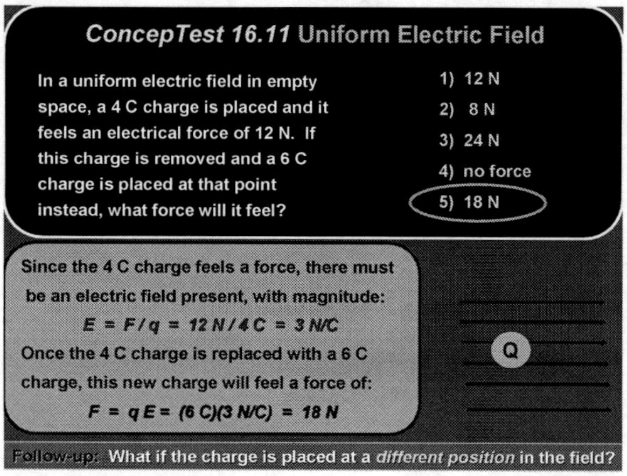

Conceptual Module 16.12 (two parts)
ConcepTest 16.12a: Electric Field Lines I
Difficulty Level: 1

To help with the understanding of electric field lines, the students must interpret a field plot with regard to the charges that created it. In this case, this primarily relates to the sign of the charges, which is determined by viewing the direction of the electric field vectors.

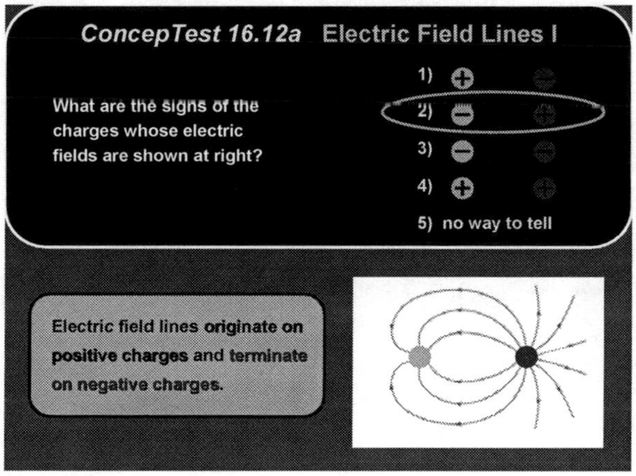

ConcepTest 16.12b: Electric Field Lines II
Difficulty Level: 2
Suggested follow-up conceptual question:
- **What is the red/green ratio of magnitudes for the two charges?**

In this follow-up question, the students must now make a judgement about the relative magnitude of the two charges, based on the density of field lines in the vicinity of each charge.

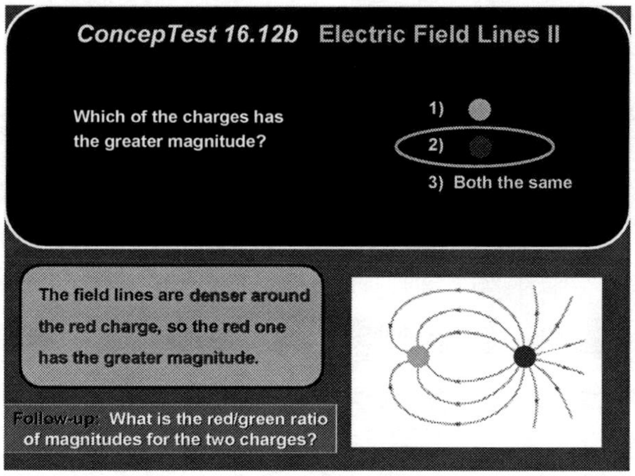

Conceptual Module 17.1 (three parts)

ConcepTest 17.1a: Electric Potential Energy I
Difficulty Level: 1

Students must realize that the electric force depends on the charge of the particle in the uniform field and that the electron and proton have the same magnitude of charge. The forces will have the same magnitude, although the directions will be opposite because the particles have opposite charges.

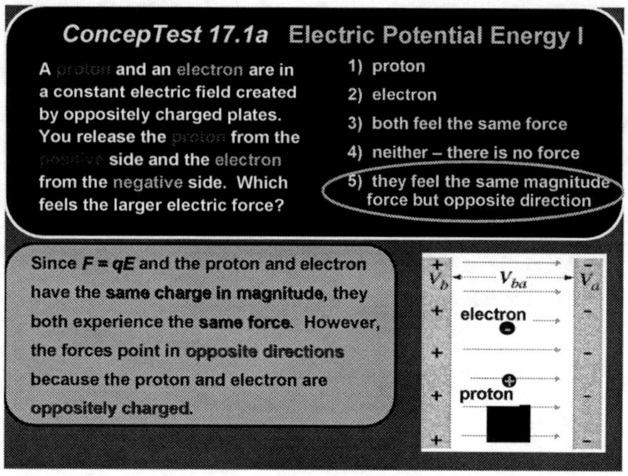

ConcepTest 17.1b: Electric Potential Energy II
Difficulty Level: 2

Again reminding the students of Newton's Second Law: while the forces on the proton and electron may be the same, the accelerations of each particle are quite different, with the less massive particle experiencing the greater acceleration ($F = ma$).

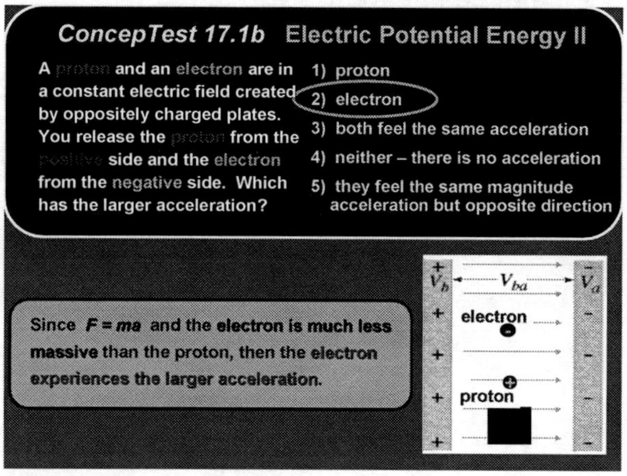

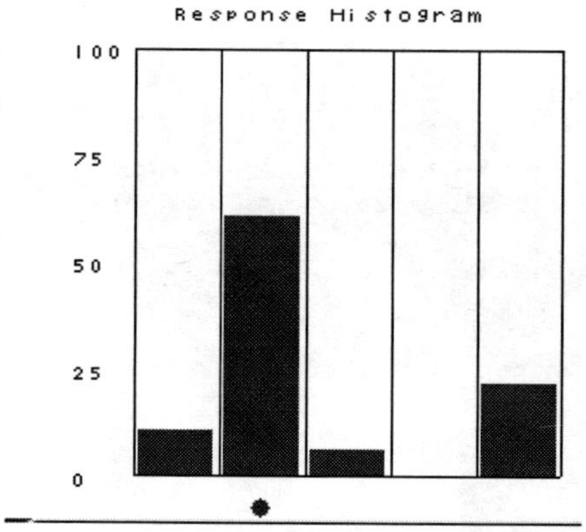

ConcepTest 17.1c: Electric Potential Energy III
Difficulty Level: 3

The initial electrostatic potential energy is converted into kinetic energy as the particle moves. Since the initial PE is the same for both particles ($PE = qV$), the final kinetic energy will also be the same.

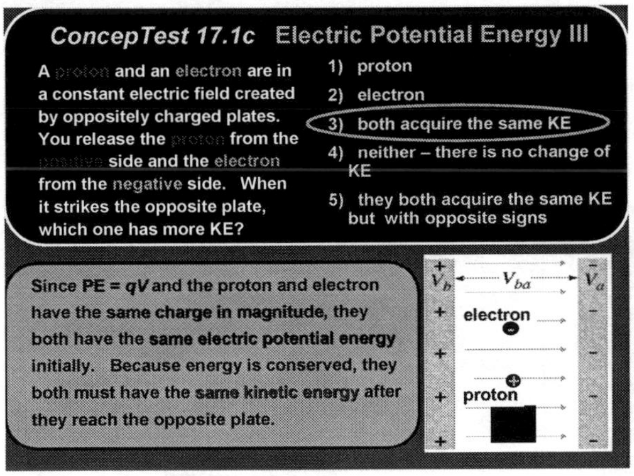

Conceptual Module 17.2

ConcepTest 17.2: Work and Potential Energy
Difficulty Level: 4

The work to assemble a charge distribution is equivalent to the total electrostatic potential energy. The electrostatic PE of a charge configuration must be determined by adding the PE contributions pairwise. For the first case, there is only one contribution (but with a larger charge involved), whereas for the second case, there are three contributions. By treating these contributions quantitatively, it works out that the second case has the higher electrostatic PE.

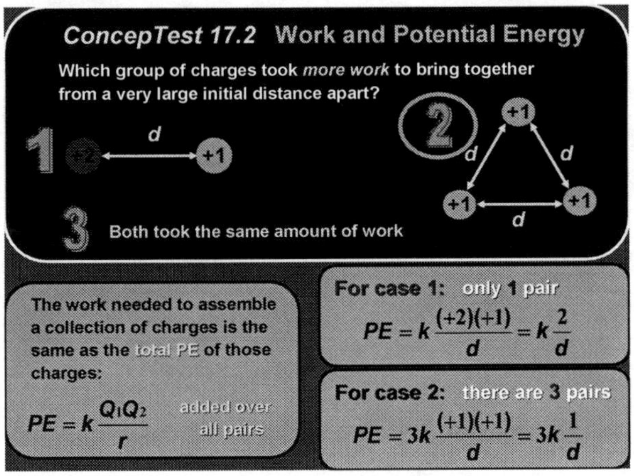

Conceptual Module 17.3 (two parts)

ConcepTest 17.3a: Electric Potential I
Difficulty Level: 3

Unlike the electric field, which is a vector, the electric potential is a scalar quantity. So in this case, the contributions to the potential simply add numerically (not vectorially). In this question, the difference in the relative contributions arises due to the distances, where the positive charge is closer to point A.

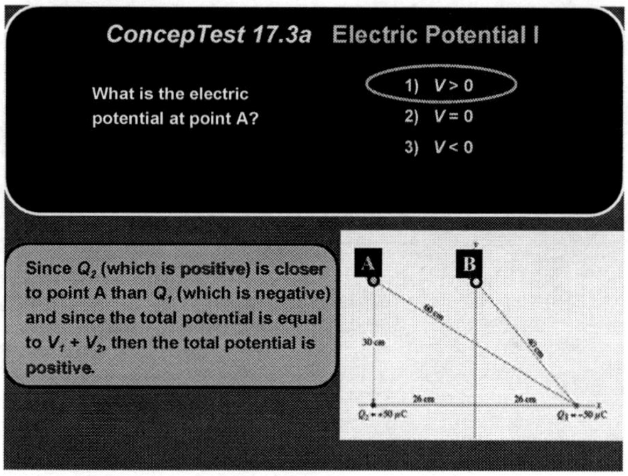

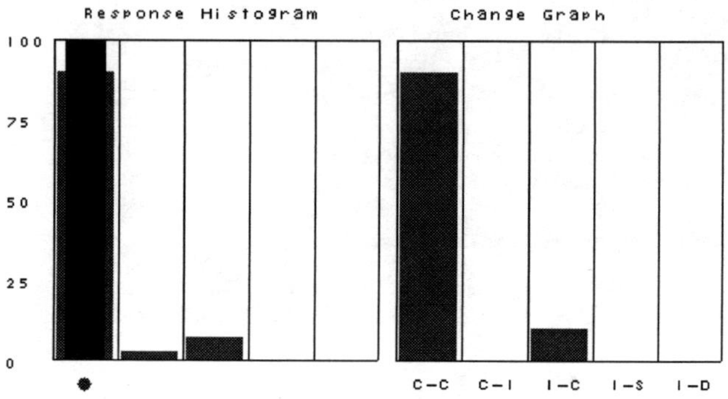

ConcepTest 17.3b: Electric Potential II
Difficulty Level: 3
Suggested follow-up conceptual question:

- **What is the potential at the origin of the *x*-*y* axes?**

In this follow-up question, point B is equidistant between a positive and a negative charge, so the contributions to the potential are equal and opposite (and therefore add up to zero).

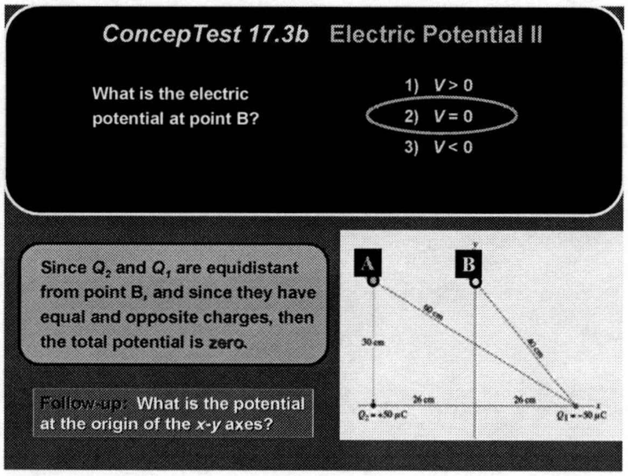

Conceptual Module 17.4

ConcepTest 17.4: Hollywood Square

Difficulty Level: 4

Suggested follow-up conceptual question:

- **What is the direction of the electric field at the center?**

This question allows students to confront the differences between electric field and electric potential. The square configuration of charges is a case they have seen before, but now they are asked about both the field and the potential. In the former, the vector nature leads to a non-zero result, but in the latter, the scalar nature ends up giving zero potential.

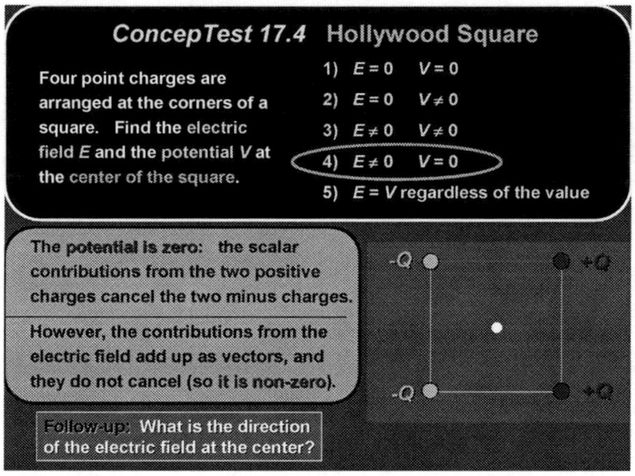

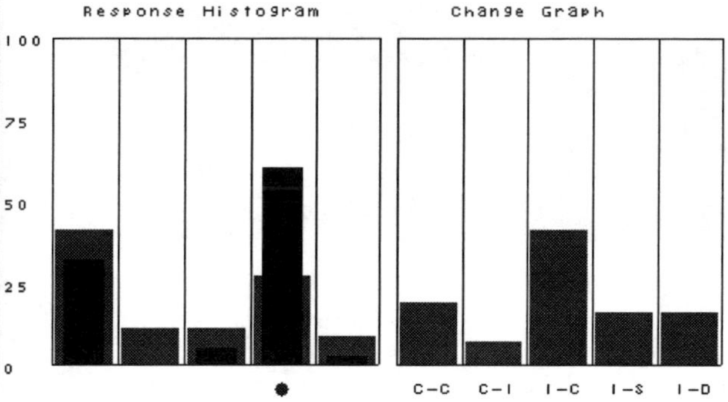

Conceptual Module 17.5 (three parts)

ConcepTest 17.5a: Equipotential Surfaces I

Difficulty Level: 3

Suggested follow-up conceptual question:

- **What is the direction of the electric field at all four points?**

This question serves as a lead-in for the concept of equipotential surfaces. All of the points shown are equidistant from both charges, so the net potential at each point will be zero.

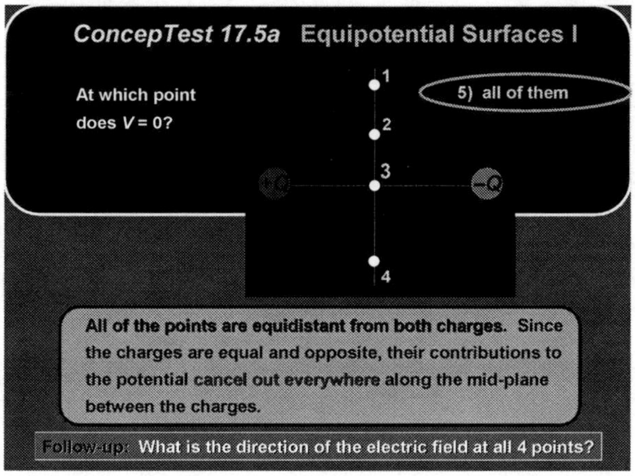

ConcepTest 17.5b: Equipotential Surfaces II
Difficulty Level: 3

This question is examining the concept of equipotential surfaces. As in the previous question, as long as all of the points along the surface are equidistant from the equal and opposite charges, the potential contributions will always cancel out. For an equipotential on the x-axis, this is only true in case #1.

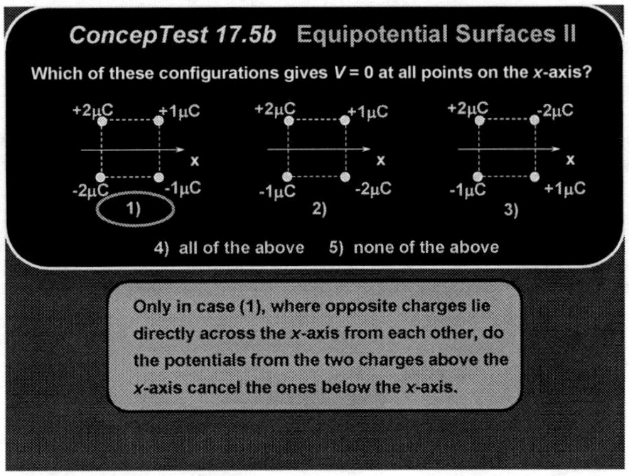

ConcepTest 17.5c: Equipotential Surfaces III
Difficulty Level: 3
Suggested follow-up conceptual question:

- **Where is $V = 0$ for configuration #2?**

This is essentially the same question as the previous one, except that the equipotential surface being sought is perpendicular to the previous case.

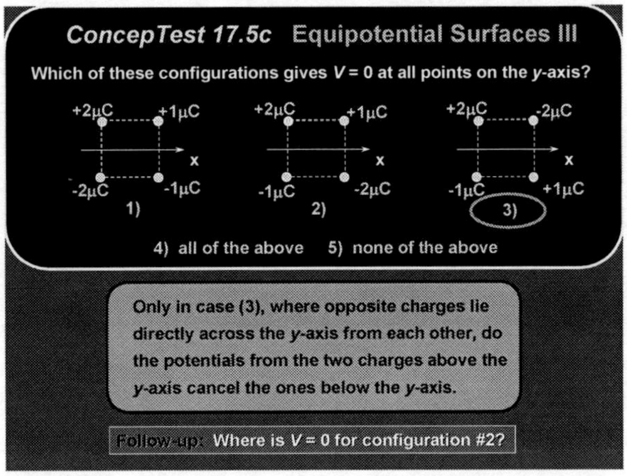

Conceptual Module 17.6

ConcepTest 17.6: Equipotential of Point Charge

Difficulty Level: 2

Suggested follow-up conceptual question:

- **Which point has the smallest potential?**

This question is asking about the equipotential surfaces of a point charge. In this case, such a surface is characterized by its radius, since the situation is spherically symmetric about the point charge.

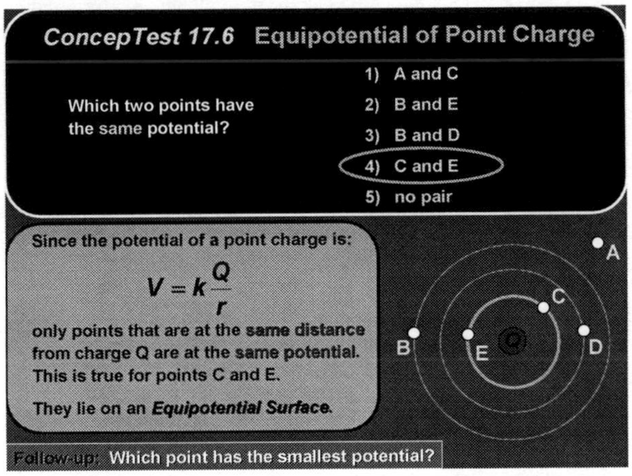

Conceptual Module 17.7 (two parts)

ConcepTest 17.7a: Work and Electric Potential I

Difficulty Level: 2

The concept of work is central to the notion of equipotentials and electric fields. It requires no work to move along an equipotential, while moving along a field line involves positive (or negative) work. In this question, students must be able to distinguish between the cases above and realize that moving a positive charge *against* the electric field will require the most work. This is analogous to the case of lifting a rock vertically upwards, against Earth's gravitational field.

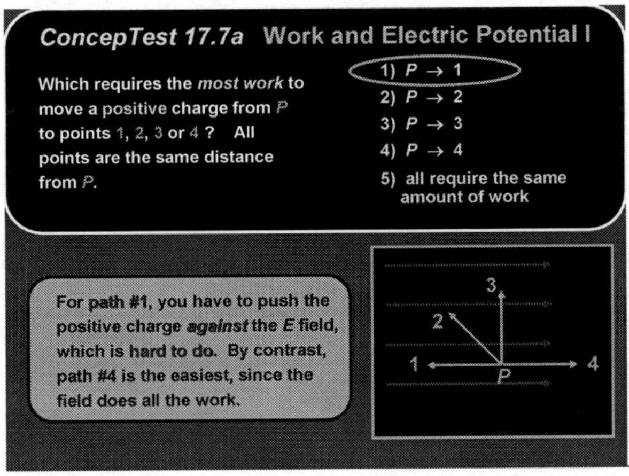

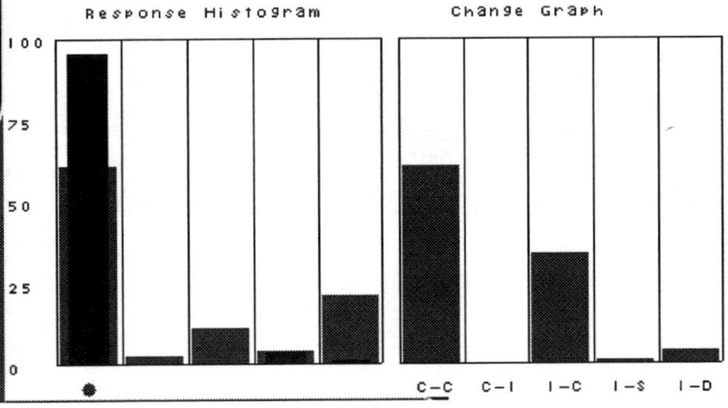

ConcepTest 17.7b: Work and Electric Potential II
Difficulty Level: 2
Suggested follow-up conceptual question:

- **Which path requires the least work?**

As mentioned in the previous question, moving along an equipotential surface requires no work. Thus, the students must be able to identify the equipotentials for a uniform electric field, and then recognize that these surfaces provide the path of zero work.

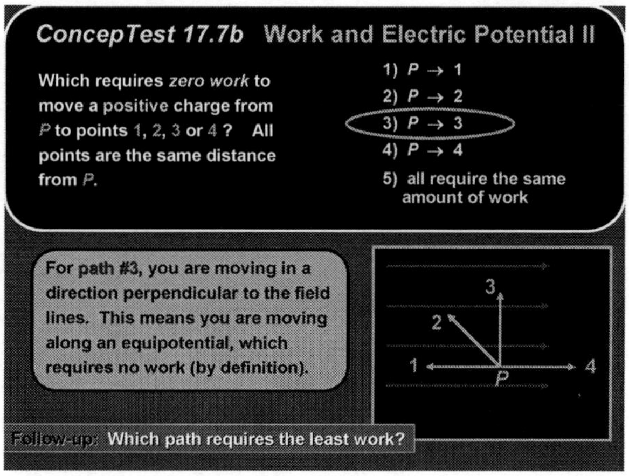

Conceptual Module 17.8

ConcepTest 17.8: Capacitors
Difficulty Level: 2

This is a direct application of $Q = CV$ for capacitors. Since the capacitors are identical, then the only other parameter that the charge depends on is the voltage. A higher voltage will supply more charge.

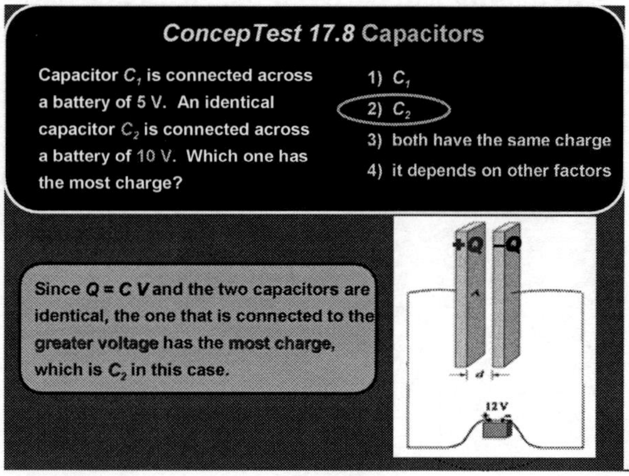

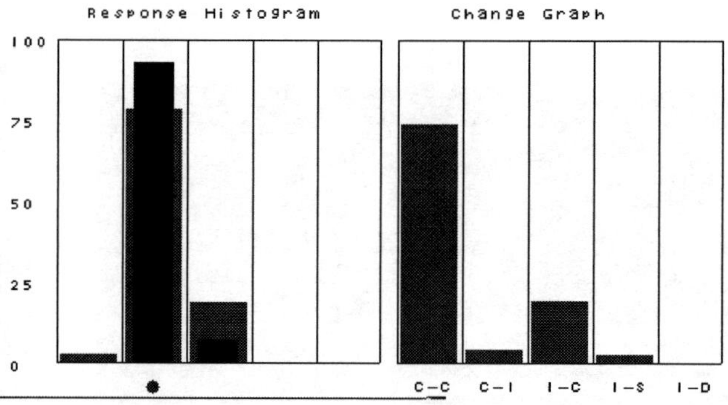

Conceptual Module 17.9 (three parts)
ConcepTest 17.9a: Varying Capacitance I
Difficulty Level: 3

The capacitance depends exclusively on geometrical factors such as plate area and separation. The relationship between these two quantities must be understood in the context of how they affect the capacitance.

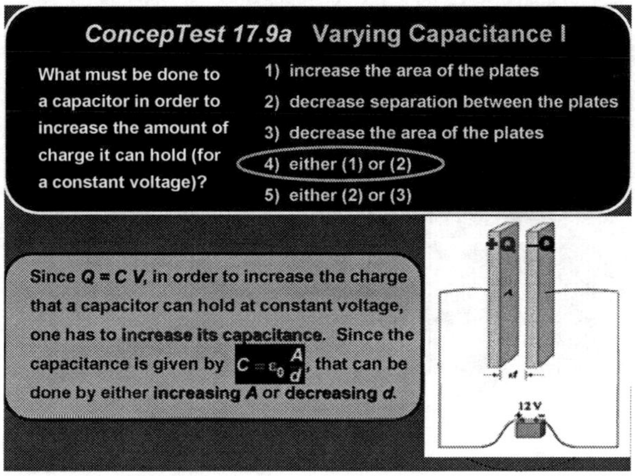

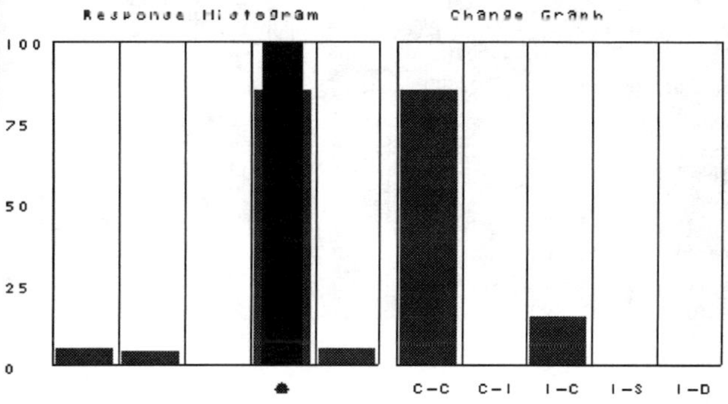

ConcepTest 17.9b: Varying Capacitance II
Difficulty Level: 3
Suggested follow-up conceptual question:
- **How do you increase the charge?**

This question addresses the cause-and-effect relationship between capacitance and charge. The students must first realize that leaving the battery connected means that the voltage remains fixed. So the increase in the plate spacing, which leads to a corresponding decrease in the capacitance, will result in a decrease in charge (using $Q = CV$, which can be rewritten as $V = Q/C$ or $C = Q/V$).

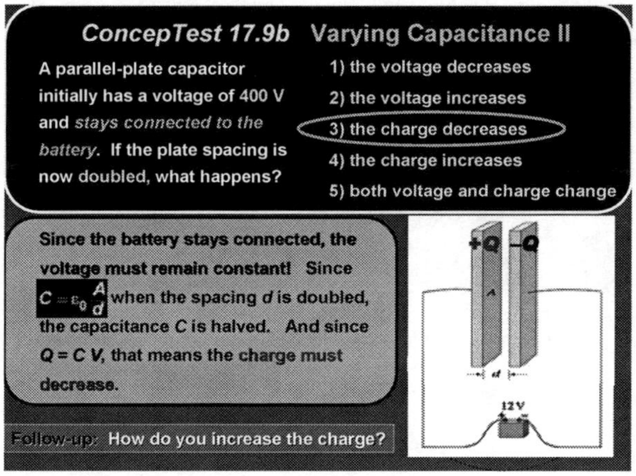

ConcepTest 17.9c: Varying Capacitance III
Difficulty Level: 3

In this follow-up question, with the battery disconnected, the charge must remain fixed. So the increase in the plate spacing, which leads to a corresponding decrease in the capacitance, will result in an increase in voltage (using $Q = CV$, which can be rewritten as $V = Q/C$ or $C = Q/V$).

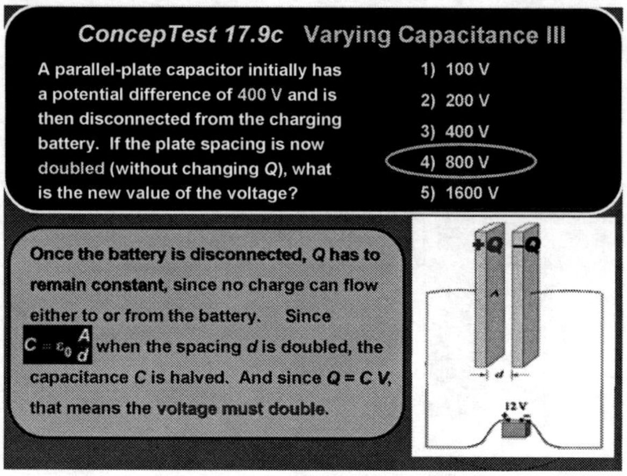

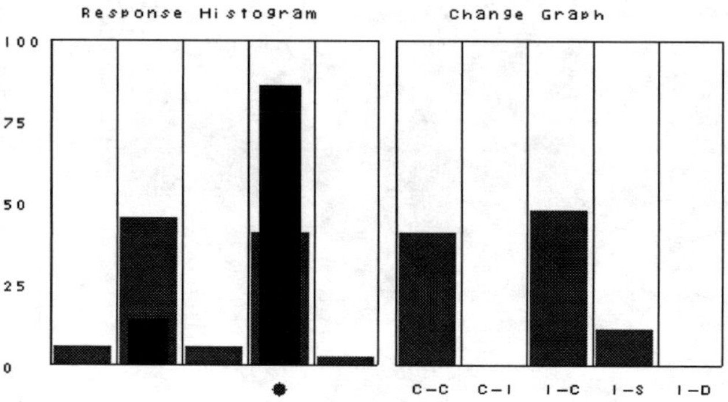

Conceptual Module 18.1

ConcepTest 18.1: Connect the Battery
Difficulty Level: 2

The concept of a complete circuit is very important to emphasize when discussing DC circuits. The notion that current must flow through a complete loop is often missed by students. For example, simply touching the bulb to the battery (case #1) will not light the bulb, but many students may think otherwise. Moreover, case #2 may appear to be a loop, but it is certainly not a complete circuit.

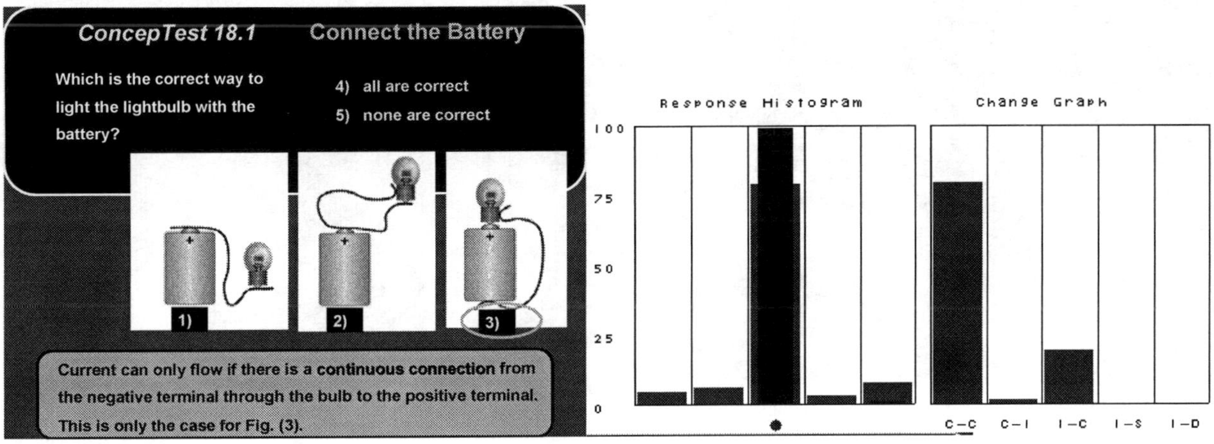

Conceptual Module 18.2

ConcepTest 18.2: Ohm's Law
Difficulty Level: 1
Suggested follow-up conceptual question:

- **Where could this situation occur?**

Ohm's Law is a linear relationship between voltage and current. There are, however, devices that do not follow Ohm's Law, and it is useful to point out this distinction to the students (i.e., that not every device necessarily obeys Ohm's Law).

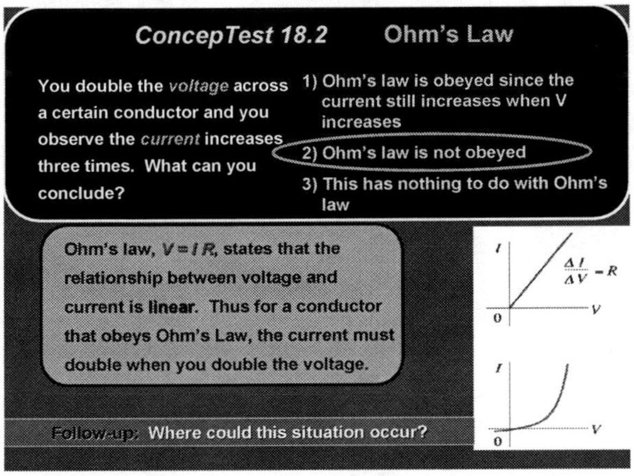

Conceptual Module 18.3 (two parts)

ConcepTest 18.3a: Wires I
Difficulty Level: 3

This question examines the relationship of resistance to the geometry of a wire. Since the resistance depends inversely (but linearly) on the area, the students must realize that the area involves the square of the radius (or diameter). This is why a factor of 4 in resistance relates to a factor of ½ in diameter.

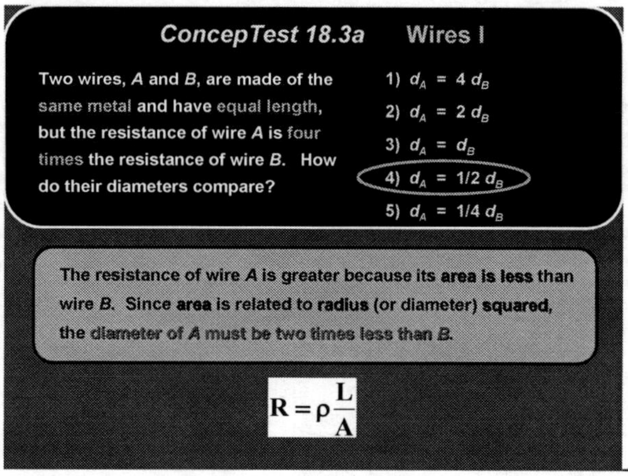

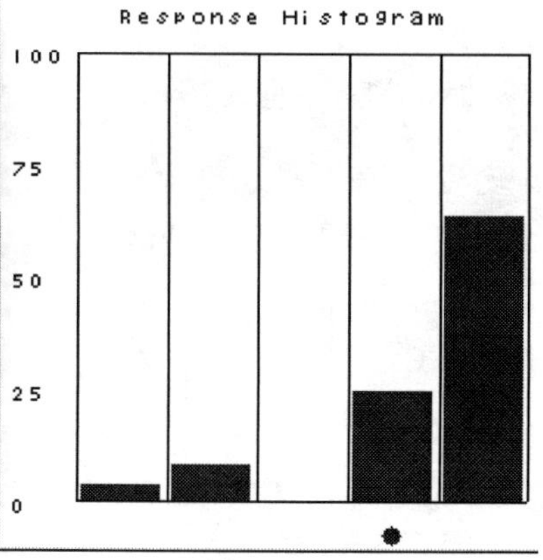

ConcepTest 18.3b: Wires II
Difficulty Level: 4

This is a tricky problem in which students must realize that to preserve the volume of the wire, when the length is doubled, the area is halved. These two factors in the expression for resistance lead to an overall increase in resistance by a factor of 4.

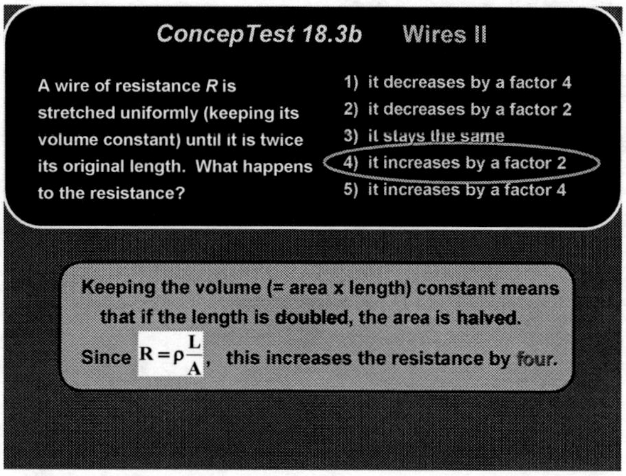

Conceptual Module 18.4

ConcepTest 18.4: Dimmer
Difficulty Level: 3
Suggested follow-up conceptual question:

- **Why does the voltage not change?**

Students must recognize that the voltage is constant for appliances in the house. With this in mind, the current must be changed in order to operate the dimmer. This is achieved by changing the resistance. It is impossible to change only one parameter in Ohm's Law without changing something else.

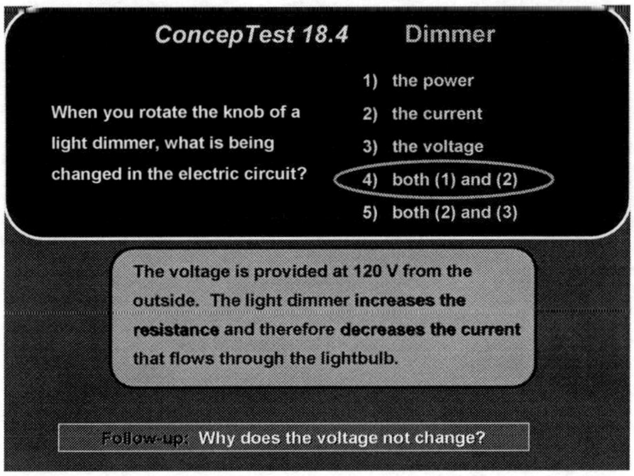

Conceptual Module 18.5 (two parts)

ConcepTest 18.5a: Lightbulbs
Difficulty Level: 3
Suggested follow-up conceptual question:

- **Which one carries the greater current?**

Given that the voltage in a typical household circuit is constant, then the power of a light bulb must be related to the current through it, which is determined by the resistance. Since $P = IV = V^2/R$, and since the voltage is constant, then the lower power comes about from a higher resistance.

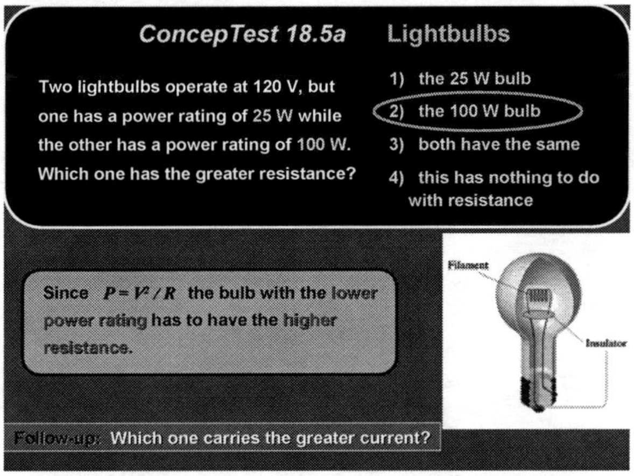

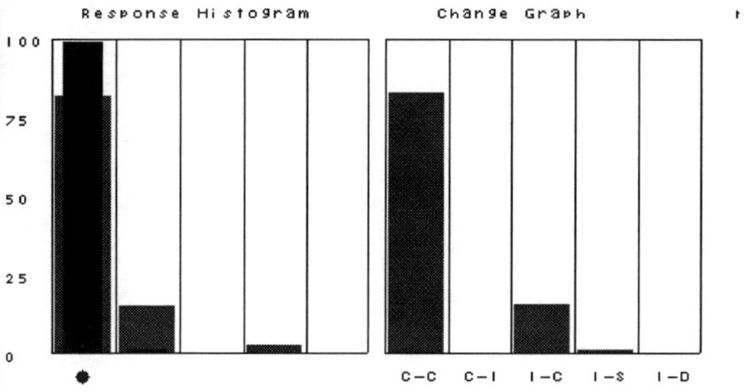

ConcepTest 18.5b: Space Heaters I
Difficulty Level: 3
Suggested follow-up conceptual question:

- **Which one carries the greater current?**

This question relates to the same issue of power and resistance (for constant voltage). Again, using the expression for power $P = V^2/R$, the lower resistance heater will have the higher power (more heat).

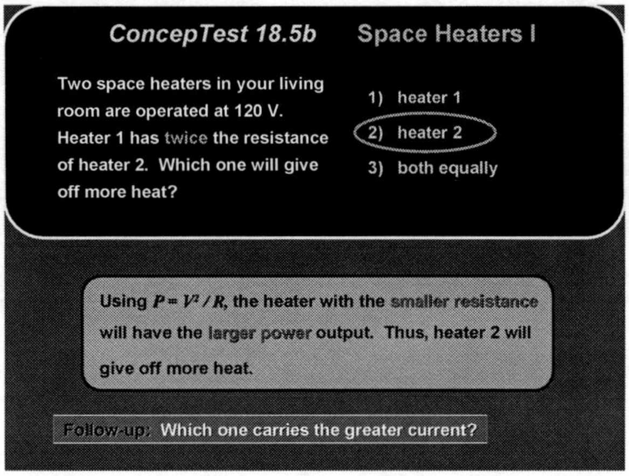

Conceptual Module 19.1 (two parts)

ConcepTest 19.1a: Series Resistors I
Difficulty Level: 3
Suggested follow-up conceptual question:

- **What would be the potential difference if $R = 1\ \Omega, 2\ \Omega, 3\ \Omega$**

For series resistors, the individual voltages add to give the total (battery) voltage. In this case, with three equal resistors, the voltages are simply 1/3 of the total voltage. This gives the students a gentle warm up to the concept of resistors in series and the associated voltages.

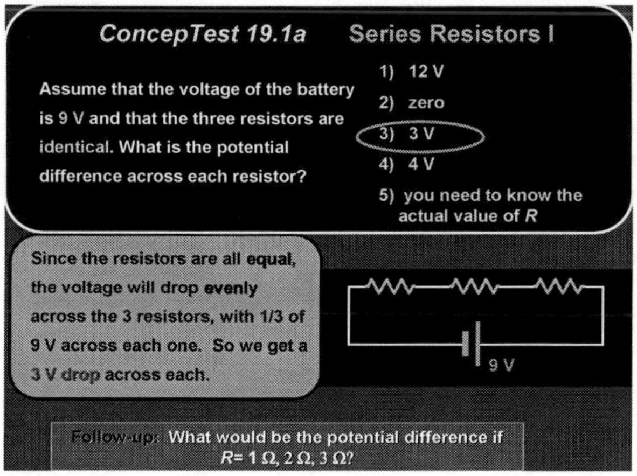

ConcepTest 19.1b: Series Resistors II
Difficulty Level: 3
Suggested follow-up conceptual question:

- **What happens if the voltage is doubled?**

In this follow-up question, the resistors are no longer equal, so the voltage drops will have to scale with the magnitudes of the resistors. A simple way to treat this is to look at the ratio of resistors and realize that the voltages must be in that same ratio.

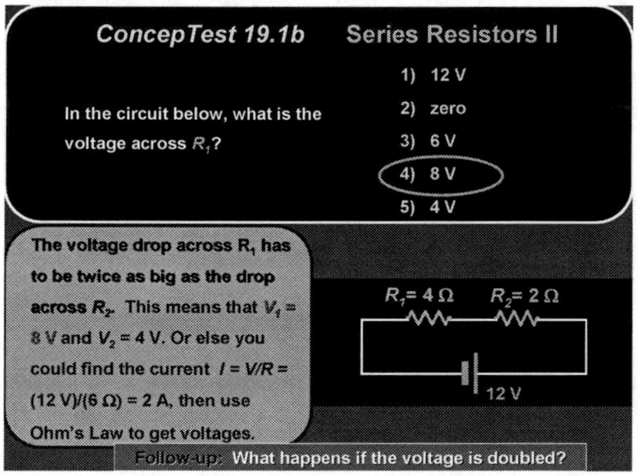

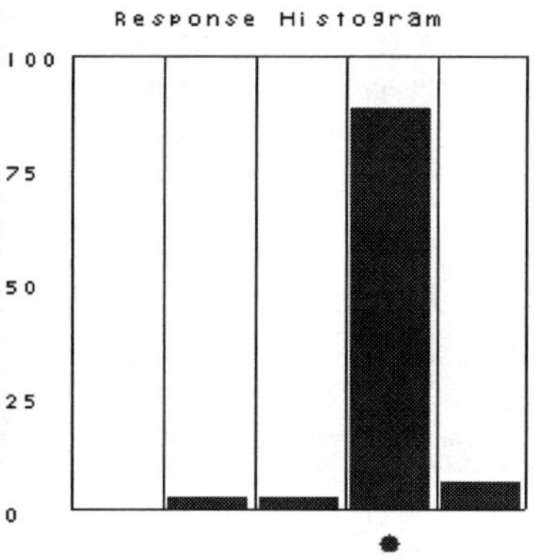

Conceptual Module 19.2 (two parts)

ConcepTest 19.2a: Parallel Resistors I
Difficulty Level: 3
Suggested follow-up conceptual question:

- **What is the total current through the battery?**

This is a simple application of Ohm's Law. Knowing the voltage across R_1 (which is simply the battery voltage) and the value of R_1 gives the current immediately. The parallel combination does not actually enter into this problem at all.

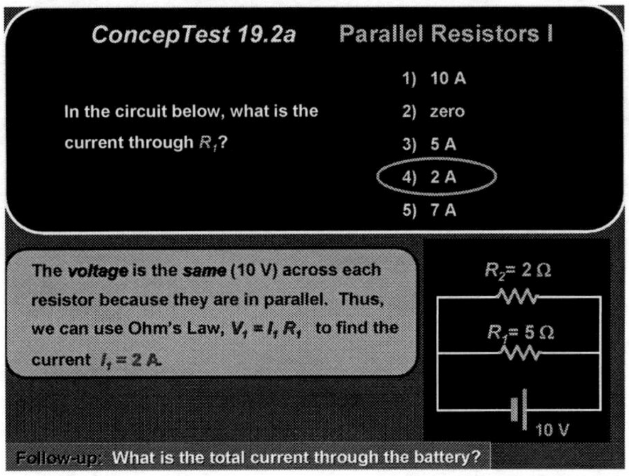

ConcepTest 19.2b: Parallel Resistors II
Difficulty Level: 4
Suggested follow-up conceptual question:

- **What happens to the current through each resistor?**

In parallel circuits, the voltage across each parallel element is the same, and so the current drawn by each element contributes to an ever-increasing total current as more elements are added. The students may have the notion that a battery is a constant current source, which is not true.

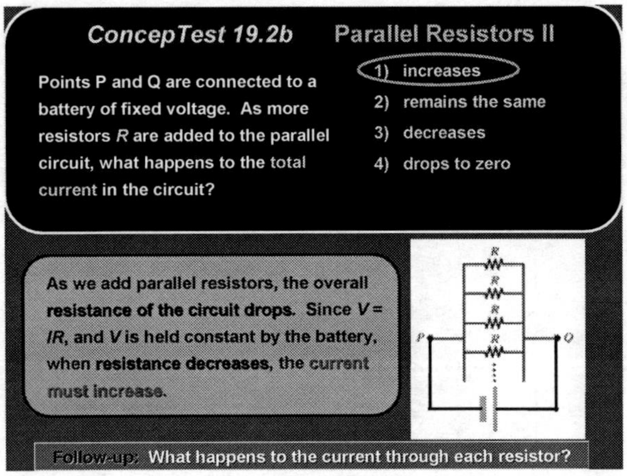

Conceptual Module 19.3 (two parts)

ConcepTest 19.3a: Short Circuit I
Difficulty Level: 3
Suggested follow-up conceptual question:

- **Doesn't the wire have SOME resistance?**

The wire is connected in parallel with the lightbulb, and the current will split based on the resistance of the branches. If the wire has nearly zero resistance, then nearly all of the current will flow through that branch. The current clearly does not split up evenly between the wire and the lightbulb.

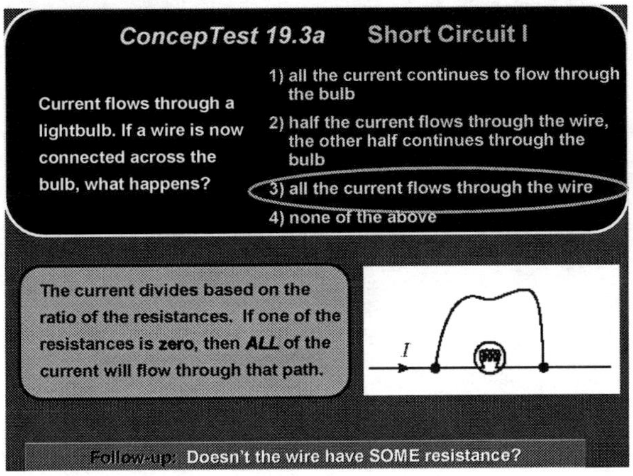

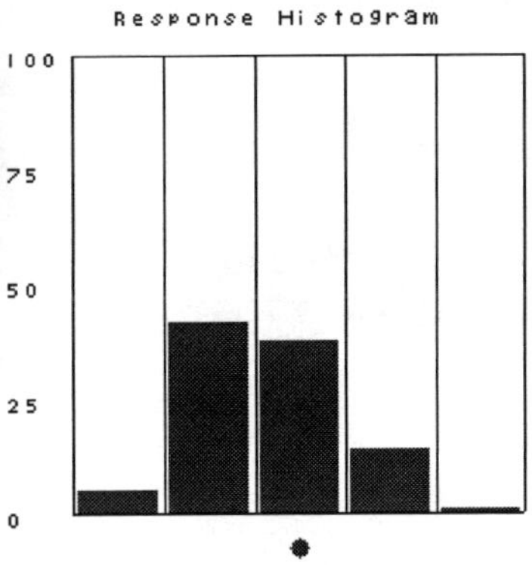

ConcepTest 19.3b: Short Circuit II
Difficulty Level: 4
Suggested follow-up conceptual question:

- **What happens to bulb B?**

By connecting the wire, the overall resistance of the circuit decreases, which in turn increases the total current flowing through the circuit (the total voltage is held fixed by the battery). When the current increases, bulb A will be brighter than before. The modification to the circuit near bulb B has an impact on the entire circuit, not just bulb B.

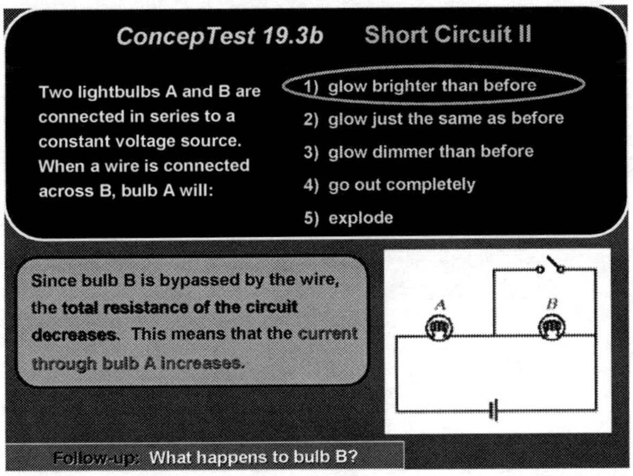

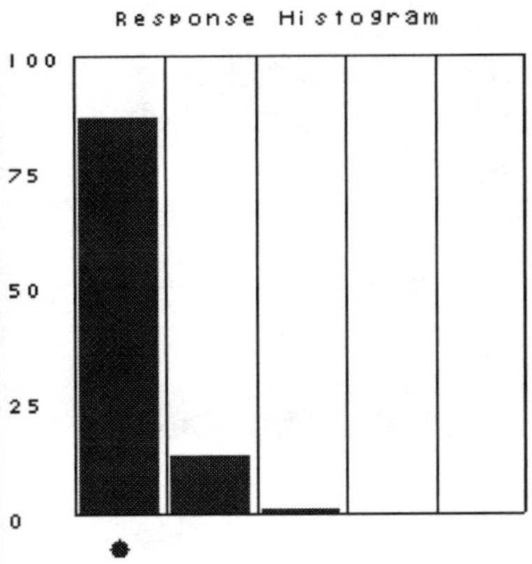

Conceptual Module 19.4 (two parts)
ConcepTest 19.4a: Circuits I
Difficulty Level: 4

This question addresses the total power ($P = IV$) in a circuit. In each case, the total voltage is the same, but the total current is greater in the parallel combination. Thus the power dissipated by the parallel circuit will be greater, which means that the lightbulbs will be brighter.

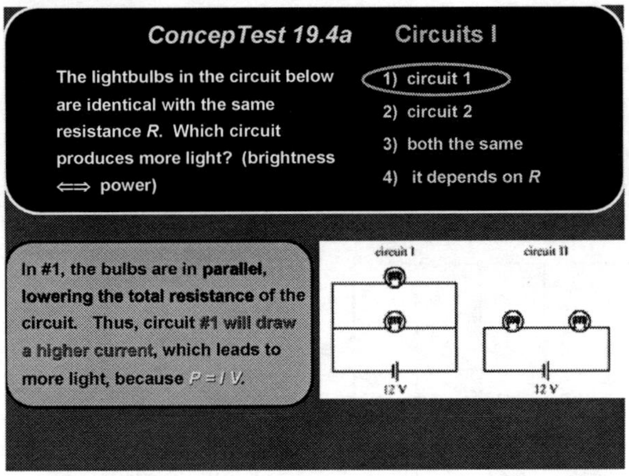

ConcepTest 19.4b: Circuits II
Difficulty Level: 5
Suggested follow-up conceptual question:
- **What is the total current in the circuit?**

This quantitative question examines the power dissipated in specific branches of a circuit. The students must obtain either the voltage or the current for each lightbulb in order to compare the power dissipated by each element. In this case, since $P = I^2R$, then the smaller current (by a factor of 2) in the bottom branch leads to a reduction in power by a factor of 4 (due to the I^2 term).

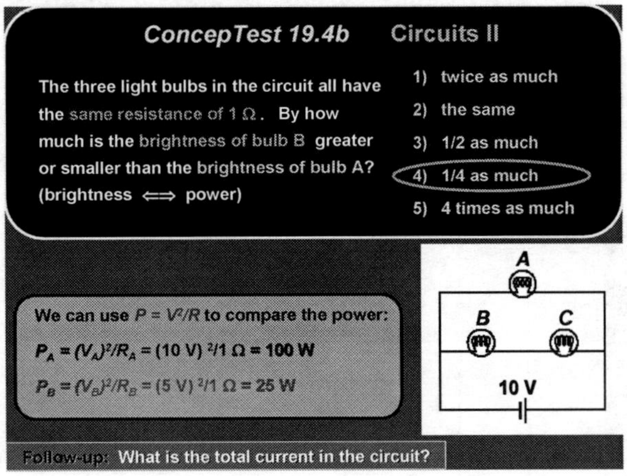

Conceptual Module 19.5 (two parts)

ConcepTest 19.5a: More Circuits I
Difficulty Level: 4
Suggested follow-up conceptual question:

- **What happens to the current through R_3?**

Here again, a change somewhere in the circuit affects the entire circuit. By closing the switch, the equivalent resistance of the parallel combination is changed, affecting the overall current in the circuit. By reducing the parallel resistance, the current in the circuit increases and therefore so does V_1.

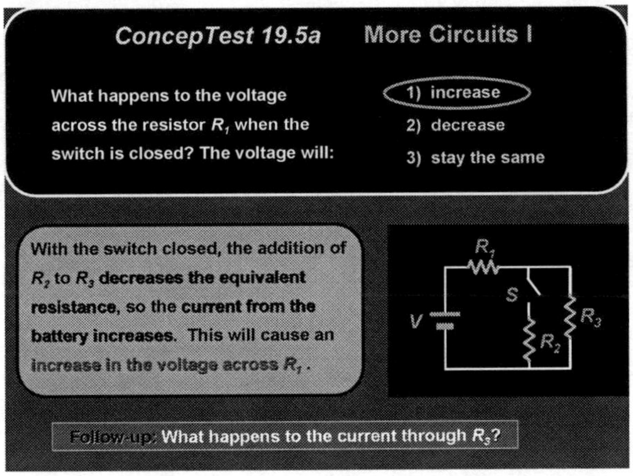

ConcepTest 19.5b: More Circuits II
Difficulty Level: 4
Suggested follow-up conceptual question:

- **What happens to the current through R_4?**

In this question, several steps are necessary, which follow from the previous question. We just saw that when the current in the circuit increases, then V_1 increases – but that means that the voltage across the parallel combination must decrease, since the total voltage is constant (from the battery). The voltage across the combination is also the voltage across any element in the combination.

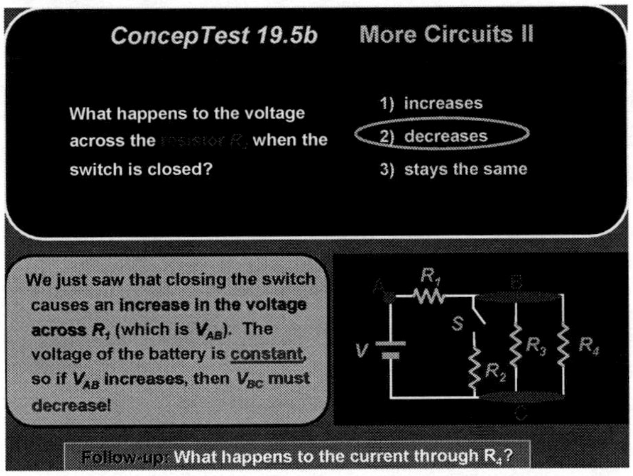

Conceptual Module 19.6

ConcepTest 19.6: Even More Circuits
Difficulty Level: 4
Suggested follow-up conceptual question:

- **Which one has the smallest voltage drop?**

In this question, the students must be able to do some basic qualitative circuit analysis. Realizing that the current is the same through both combinations (because they are in series) leads to an examination of each combination and the way the current will split in each case. The 2:1 ratio of resistors in the second group favors R_5 as the resistor with the greatest current.

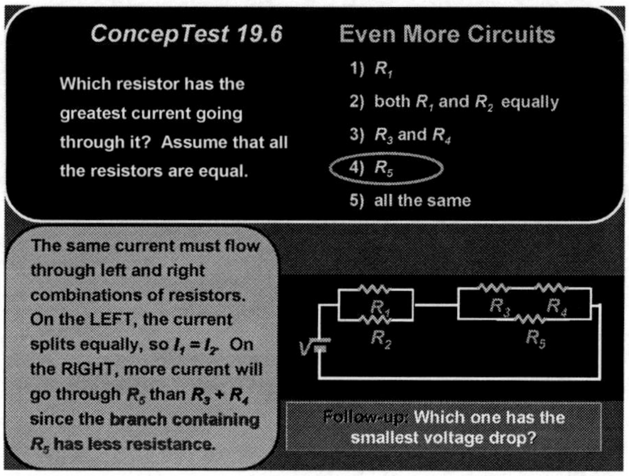

Conceptual Module 19.7

ConcepTest 19.7: Junction Rule
Difficulty Level: 2

This question addresses the junction rule, which states that the total current entering a point must equal the total current leaving that point. By inspecting the currents shown, the students should be able to deduce the value of the missing current.

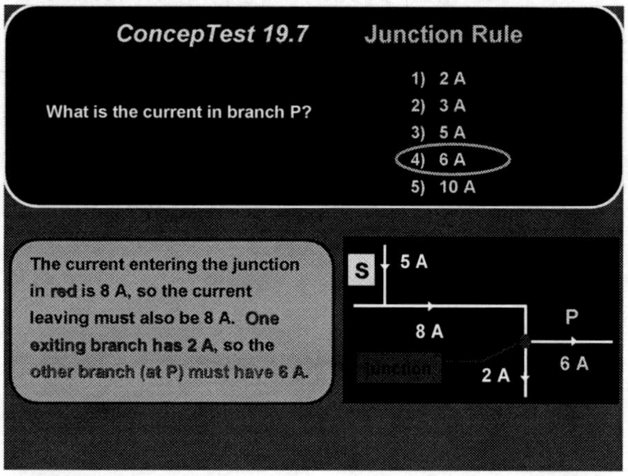

Conceptual Module 19.8

ConcepTest 19.8: Kirchhoff's Rules
Difficulty Level: 5
Suggested follow-up conceptual question:

- **What happens if the bottom battery is replaced by a 24-V battery?**

This question requires a reasonable insight into what is going on in the original circuit. Knowing that the total voltage is 24 V and that this total will drop evenly (12 V each) across the two equal resistors leads to the conclusion that the potential must be 12 V at the point between the two resistors. Since this is the same value of the potential between the batteries, the result is that nothing happens.

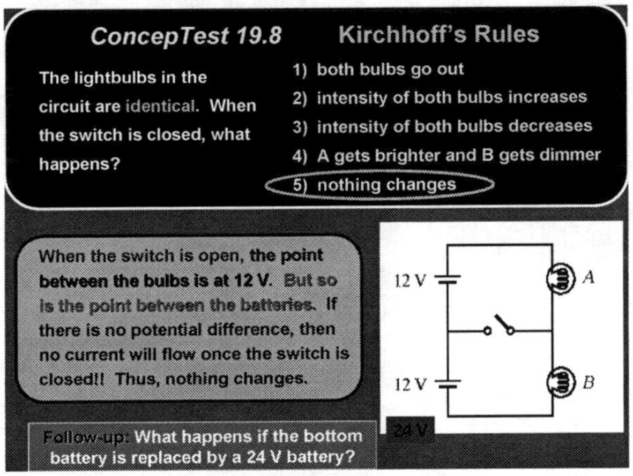

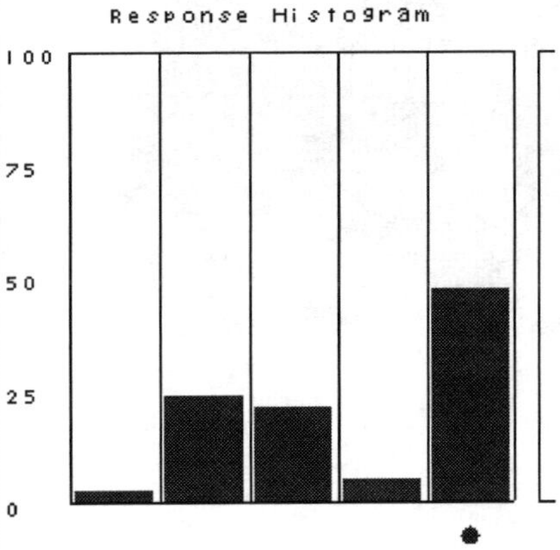

Conceptual Module 19.9

ConcepTest 19.9: Wheatstone Bridge
Difficulty Level: 5

A Wheatstone Bridge operates by adjusting the value of a variable resistor so as to "balance" the bridge and make the current in the ammeter zero. In this case, as before, the students must realize that the potentials at points (a) and (b) are the same, which means that no current will flow. Actually, this equality of potential is due to the ratio of resistors in each branch being the same and does not require the resistors to have the exact same values themselves.

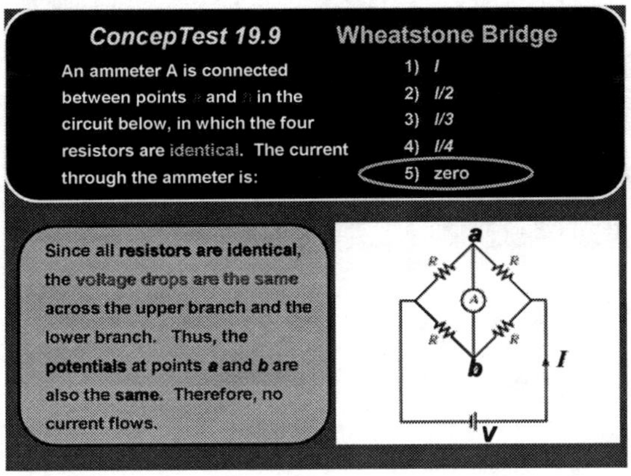

Conceptual Module 19.10
ConcepTest 19.10: More Kirchhoff's Rules
Difficulty Level: 4

In this question, the students must work to set up a valid loop equation for one of the possible loops in the circuit shown. This is an exercise in identifying the proper signs of voltage drops and voltage gains as the current goes through batteries and resistors in the directions shown.

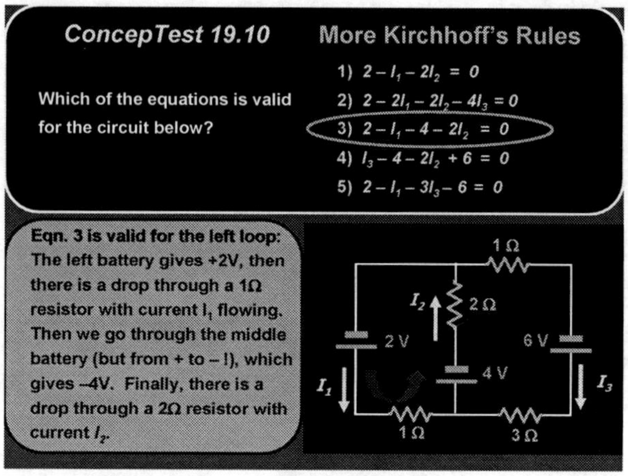

Conceptual Module 19.11 (three parts)
ConcepTest 19.11a: Capacitors I
Difficulty Level: 3

This is a direct application of the rules for equivalent capacitance of combinations. In this case, two series capacitors are combined in parallel with another single capacitor, so there are two steps that the students must account for.

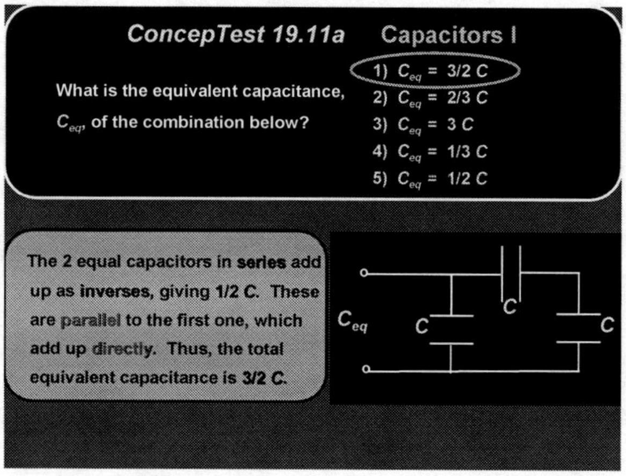

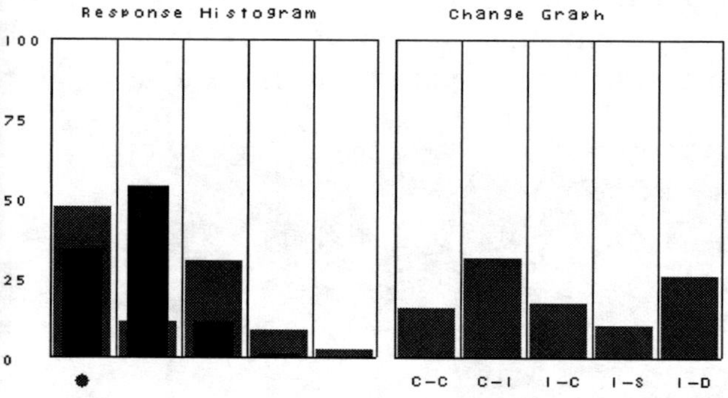

ConcepTest 19.11b: Capacitors II
Difficulty Level: 4
Suggested follow-up conceptual question:
- **What is the current in this circuit?**

This question addresses the issue of voltage in capacitor combinations. Students should recognize immediately that V_1 must be the battery voltage. Then the voltages V_2 and V_3 must add up to the battery voltage, so each of them individually must be less than V_1.

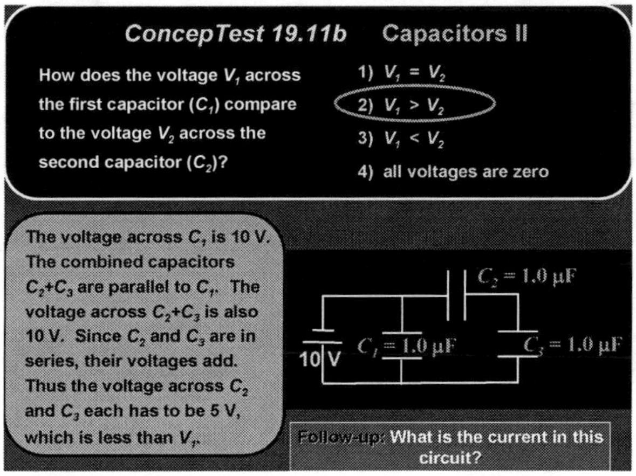

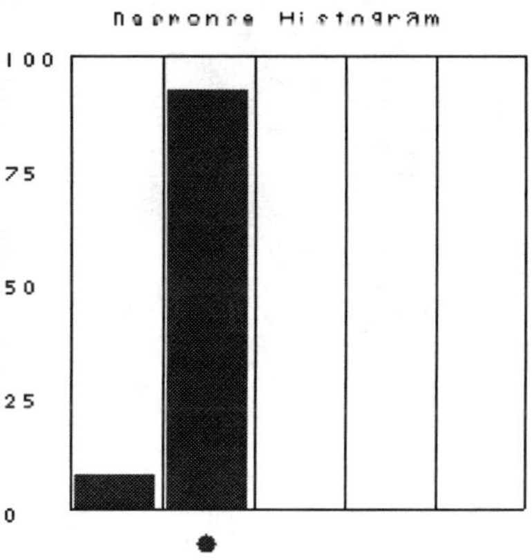

ConcepTest 19.11c: Capacitors III
Difficulty Level: 4

This question addresses the issue of charge in capacitor combinations. Students must utilize what they have learned in the previous question about the voltages to deduce the charges in the present case.

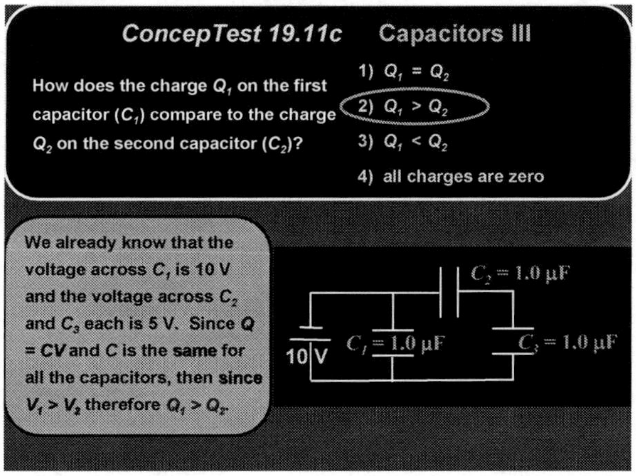

Conceptual Module 20.1 (four parts)

ConcepTest 20.1a: Magnetic Force I
Difficulty Level: 3

This is a direct application of the right-hand rule for determining the direction of the magnetic force on a moving charged particle.

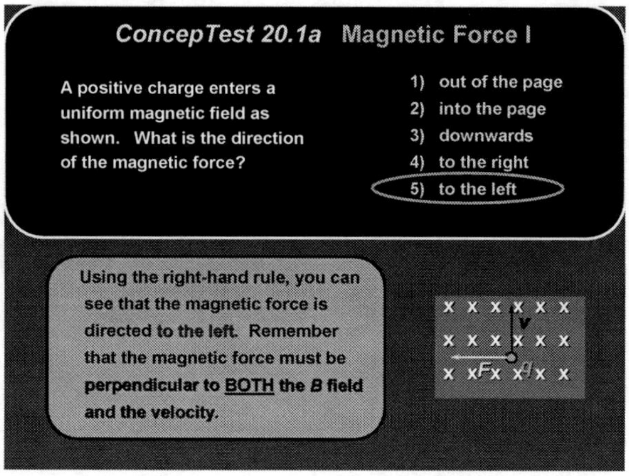

ConcepTest 20.1b: Magnetic Force II
Difficulty Level: 3

This is a direct application of the right-hand rule for determining the direction of the magnetic force on a moving charged particle.

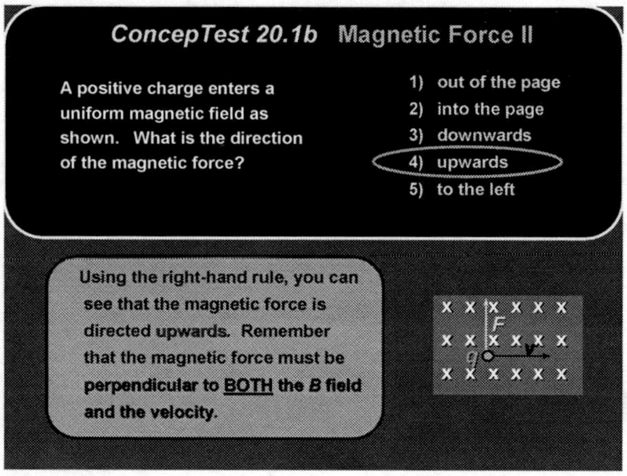

ConcepTest 20.1c: Magnetic Force III
Difficulty Level: 3

This is a direct application of the right-hand rule for determining the direction of the magnetic force on a moving charged particle.

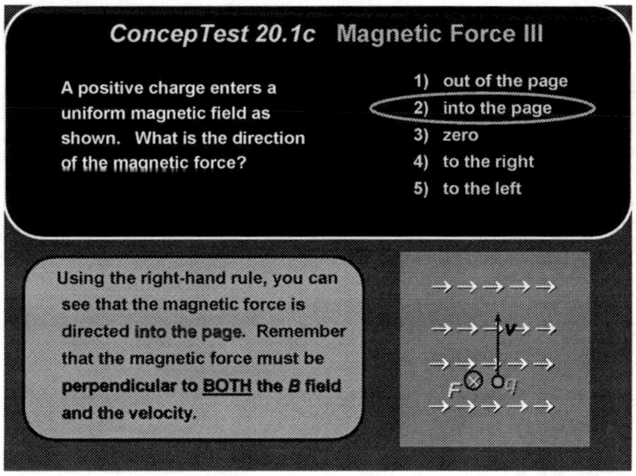

ConcepTest 20.1d: Magnetic Force IV
Difficulty Level: 2

In this final question, the velocity is parallel to the magnetic field, so there will be no magnetic force. This comes directly from the expression $F = qvB(\sin\theta)$ for the magnitude of the magnetic force.

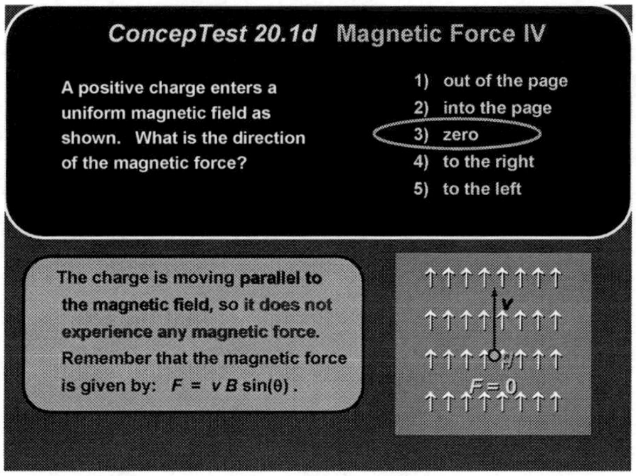

Conceptual Module 20.2

ConcepTest 20.2: Atomic Beams
Difficulty Level: 2
Suggested follow-up conceptual question:

- **What charge would follow path #3?**
- **What charge would follow path #1?**

To some extent, this is a "trick" question. Since atoms are neutral, they will not experience any magnetic force. Thus their paths will be undeflected by the presence of the magnetic field.

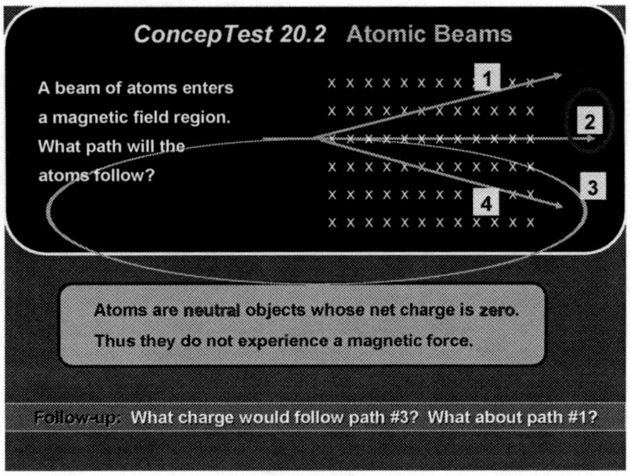

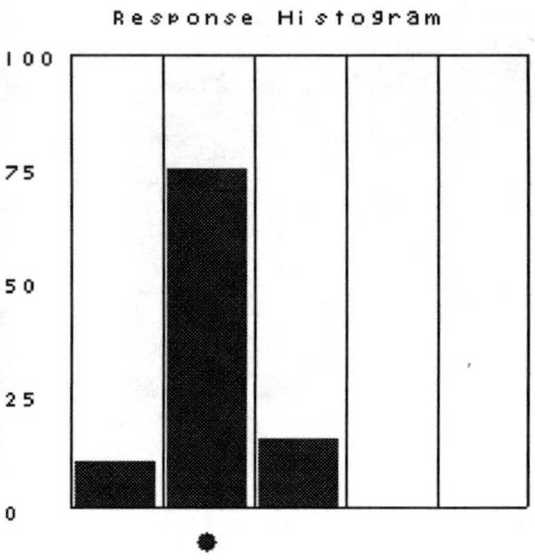

Conceptual Module 20.3

ConcepTest 20.3: Magnetic Field
Difficulty Level: 3
Suggested follow-up conceptual question:

- **What would happen to a beam of atoms?**

This question addresses the right-hand rule, but in reverse. That is, in this case, the force is indicated and the students must work backwards to infer the direction of the magnetic field. By realizing that the field must be perpendicular to both the force and the velocity will reduce the choices down to two.

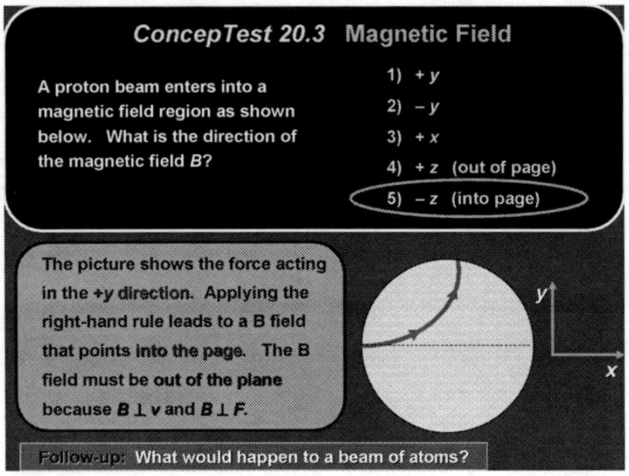

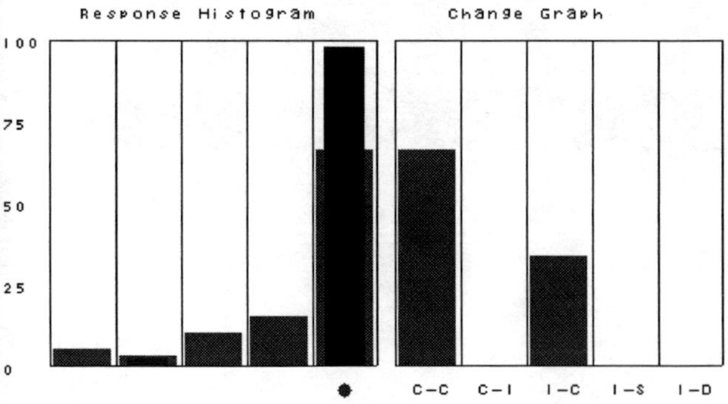

Conceptual Module 20.4 (two parts)
ConcepTest 20.4a: Mass Spectrometer I
Difficulty Level: 3
Suggested follow-up conceptual question:
- **What is the sign of the charges in the picture?**

In addition to using the expression for radius of curvature in a magnetic field, students may also notice that the larger charge will have the larger force (if velocity and B field are fixed) and therefore will also have the larger acceleration (if mass is fixed), which results in the tighter circle as the trajectory.

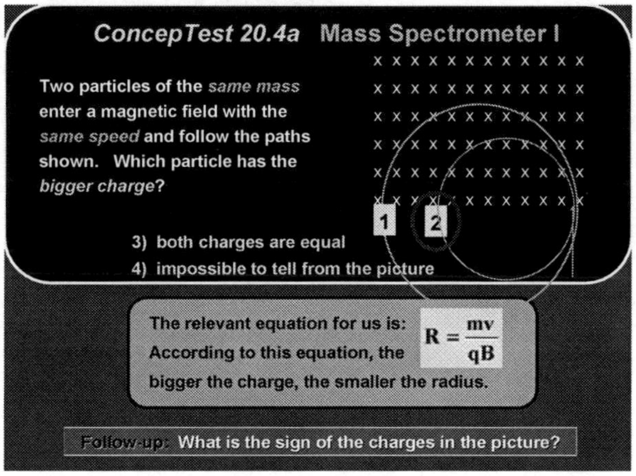

ConcepTest 20.4b: Mass Spectrometer II
Difficulty Level: 2

This question relates to work and kinetic energy. The students have seen this type of question before, in terms of circular motion. When the force acts perpendicular to the direction of motion (as it does for magnetic forces), no work is done and therefore the kinetic energy does not change.

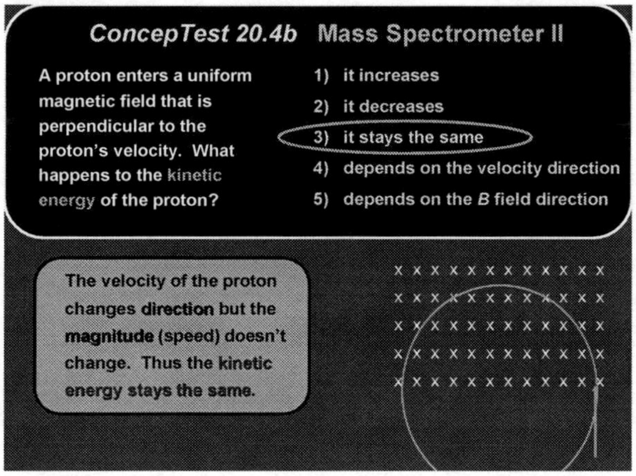

Conceptual Module 20.5

ConcepTest 20.5: Velocity Selector

Difficulty Level: 4

This question links the electric and magnetic forces on a charged particle. First, the electric force must be determined. Then, knowing that the magnetic force must act in the opposite direction, the students must invoke the right-hand rule to infer the direction of the magnetic field that will accomplish this.

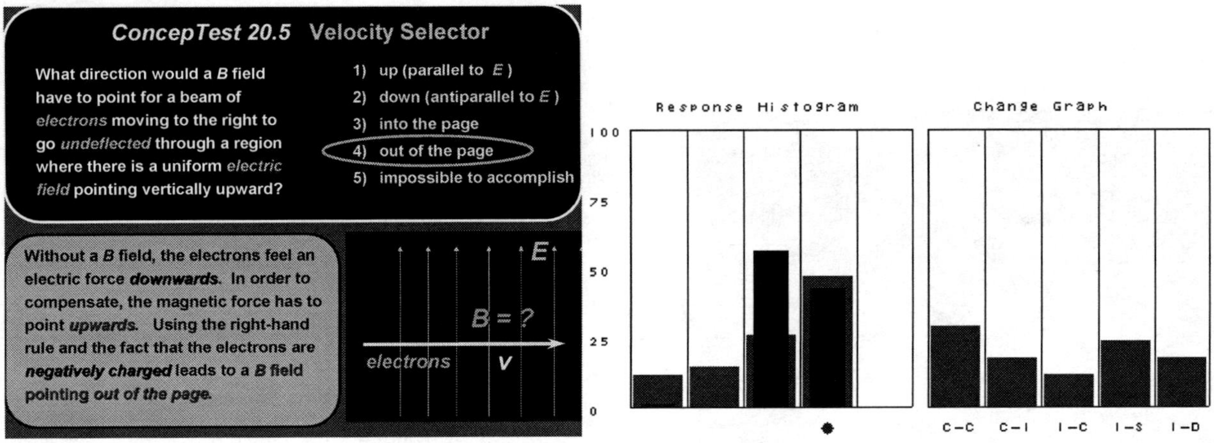

Conceptual Module 20.6 (two parts)

ConcepTest 20.6a: Magnetic Force on a Wire I

Difficulty Level: 3

This is an exercise in the right-hand rule, where in this case the moving charges are a current in a wire. This changes nothing with regard to the direction of the force, but helps students realize that there is no difference between a single moving charge and a current (which consists of many moving charges).

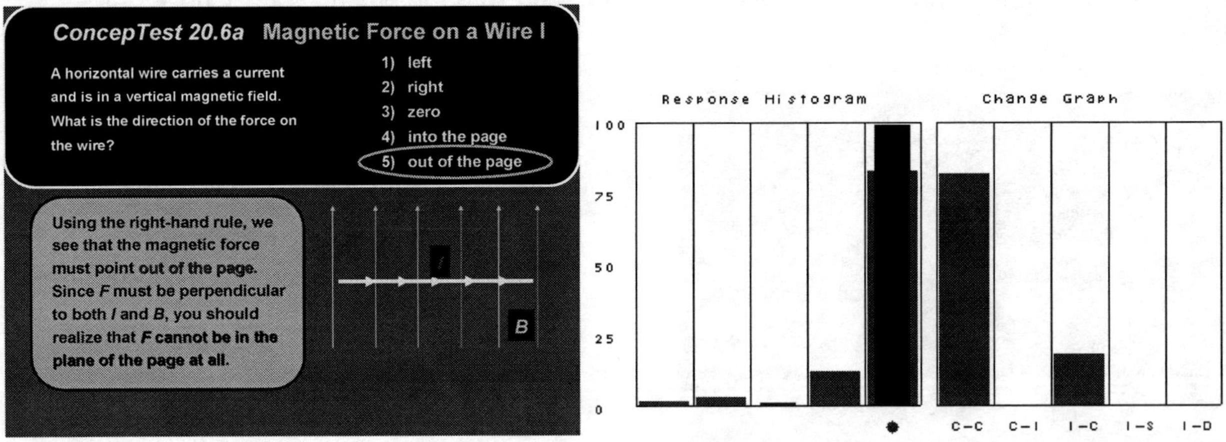

ConcepTest 20.6b: Magnetic Force on a Wire II
Difficulty Level: 2

In this follow-up question, the current is parallel to the direction of the magnetic field, so there is no magnetic force. This comes directly from the expression $F = ILB(\sin\theta)$ for the magnitude of the magnetic force.

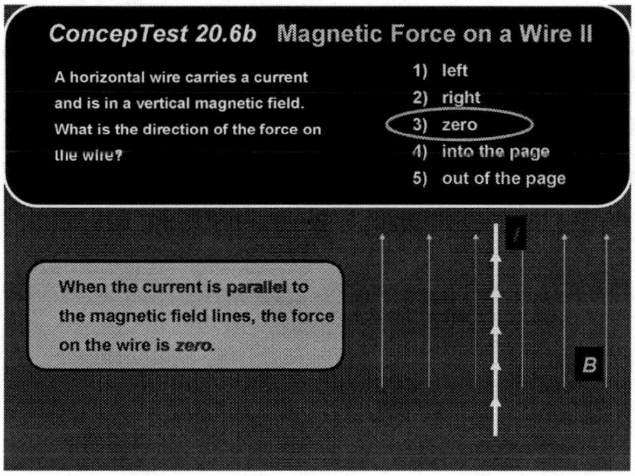

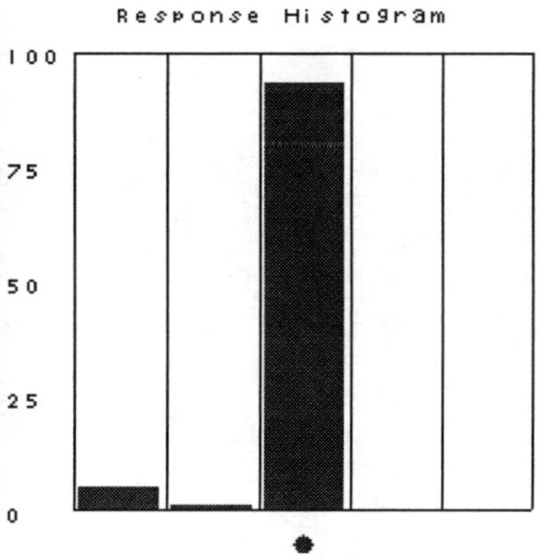

Conceptual Module 20.7

ConcepTest 20.7a: Magnetic Force on a Loop I
Difficulty Level: 4

In this question, the right-hand rule must again be applied. By finding an outward force on one of the legs, students may be able to use symmetry arguments to deduce that each leg will also experience an outward force, such that the vector sum of all the four outward forces will be zero.

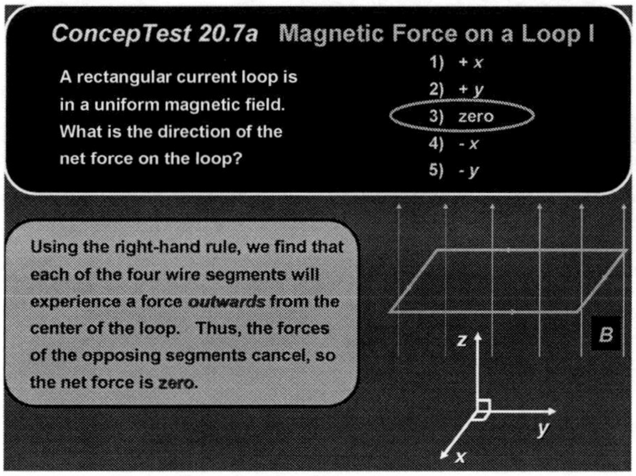

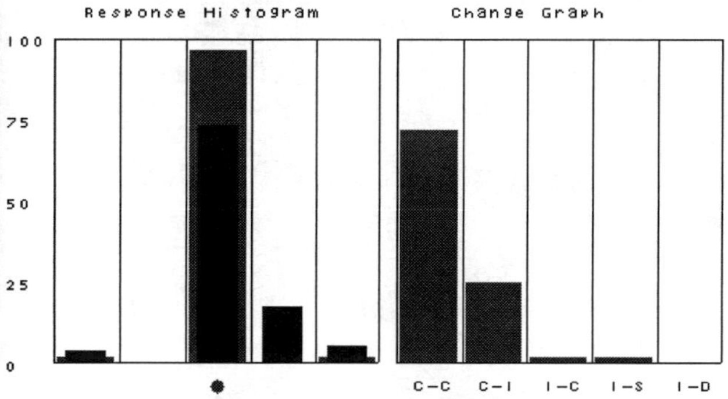

ConcepTest 20.7b: Magnetic Force on a Loop II
Difficulty Level: 4

In this case, the right-hand rule leads to opposite forces on each of the two legs, but the directions of these forces are not in the plane of the loop. Thus, while the loop will not undergo any translational motion, it will in fact rotate. This is a precursor to the introduction of motors, which utilize the same combination of forces in conjunction with an alternating current in the loop.

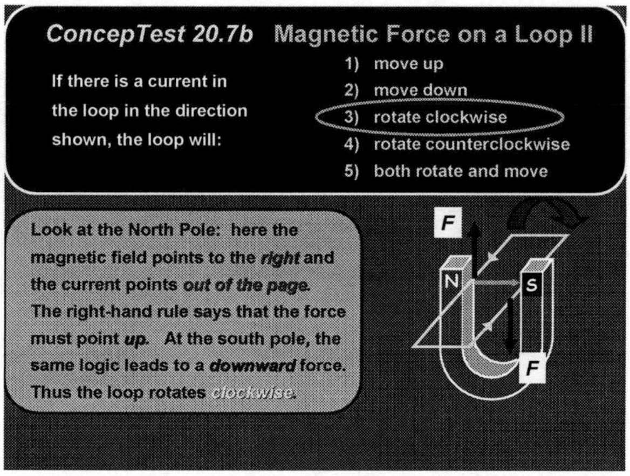

Conceptual Module 20.8 (two parts)

ConcepTest 20.8a: Magnetic Field of a Wire I
Difficulty Level: 4

This question deals with the magnetic field of a straight wire and the superposition of two separate forces. In this case, the students must find the direction of the field from each wire and then perform the vector addition of the two forces to obtain the resultant.

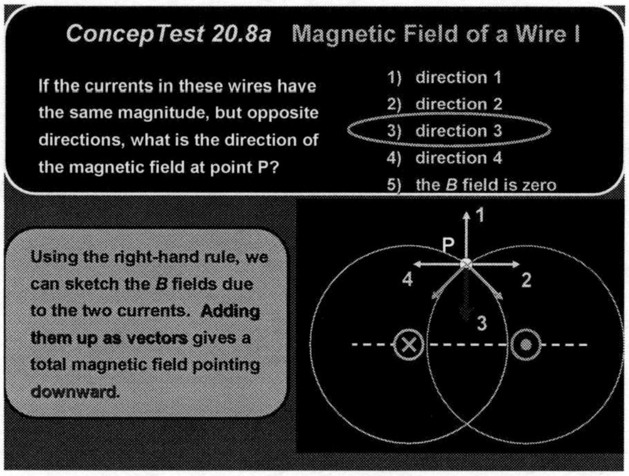

ConcepTest 20.8b: Magnetic Field of a Wire II
Difficulty Level: 4

This follow-up question includes four wires, instead of two, and is asking for the case in which the net magnetic field is largest. In cases #1 and #3, the similar currents at opposite corners of the square will end up canceling, such that these cases give a net field of zero. It is only in case #2 that the field contributions add up to a maximum value.

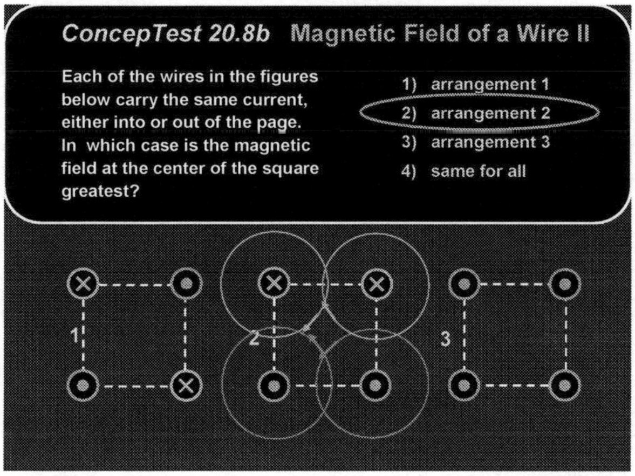

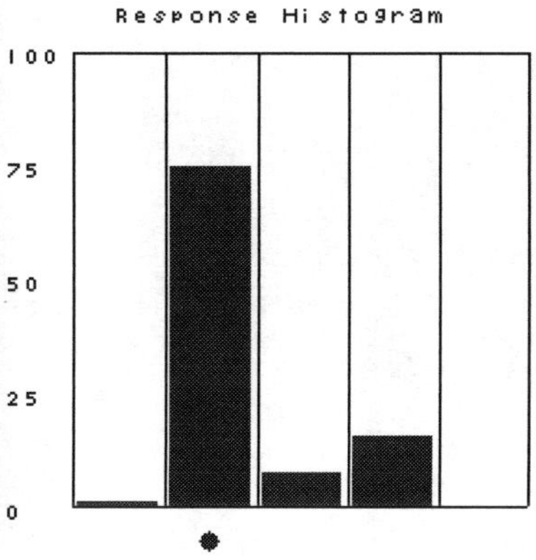

Conceptual Module 20.9 (two parts)

ConcepTest 20.9a: Field and Force I
Difficulty Level: 4

This question involves two separate parts. First, the students must ascertain the direction of the magnetic field due to the wire at the position of the charge. Second, knowing that field and the velocity of the moving charge, the students must then apply the right-hand rule to obtain the direction of the force.

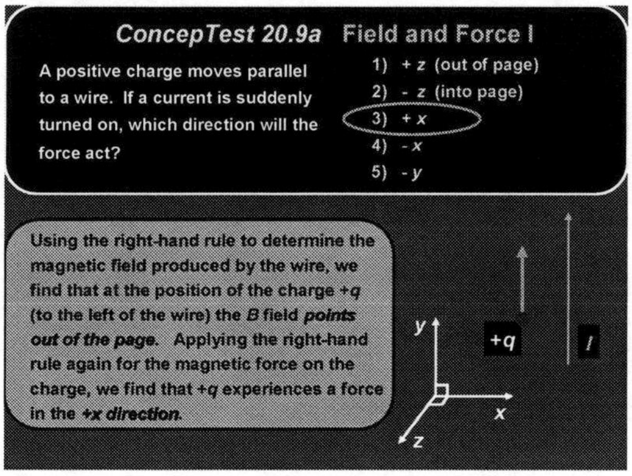

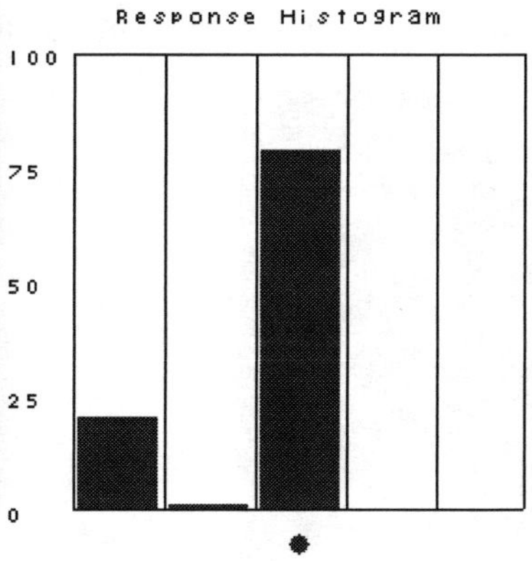

ConcepTest 20.9b: Field and Force II
Difficulty Level: 4
Suggested follow-up conceptual question:

- **What happens when one of the currents is turned off?**

This question is very similar to the previous one, except for the fact that the moving charge has now been replaced by a current in a wire. The same steps must be applied, and the students will arrive at the well-known result that two wires carrying current in the same direction will attract each other.

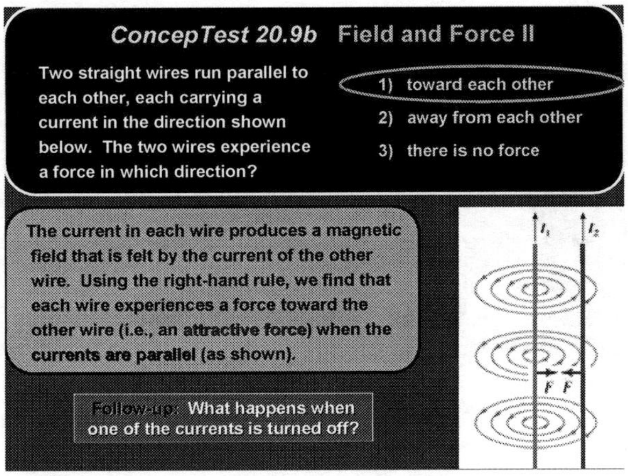

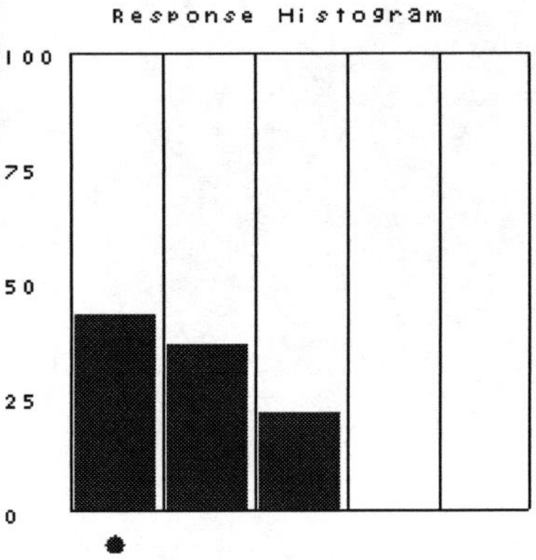

Conceptual Module 20.10

ConcepTest 20.10: Current Loop
Difficulty Level: 4

This is an exercise to get the students to derive for themselves the direction of the magnetic field at the center of a current loop. The simplest case of a current loop is a square loop, which is given in this question. After obtaining this result, a new version of the right-hand rule will soon be introduced to give directly the direction of the magnetic field inside a current loop.

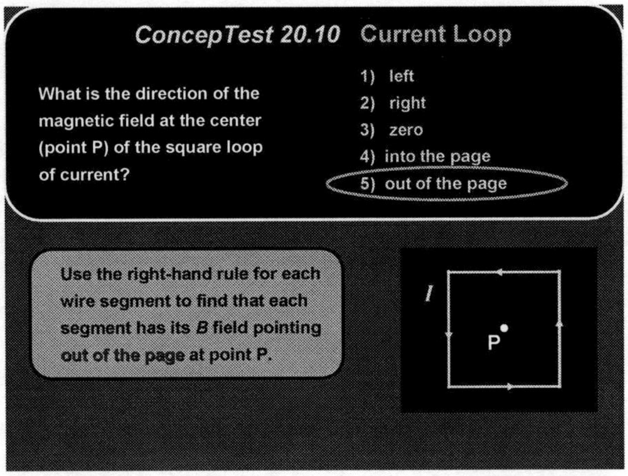

Conceptual Module 21.1 (two parts)

ConcepTest 21.1a: Magnetic Flux I
Difficulty Level: 1

The concept of magnetic flux is troublesome for students, but an understanding of this concept is necessary for the subsequent work on electromagnetic induction. In this question, some of the various ways in which the magnetic flux can be varied are examined.

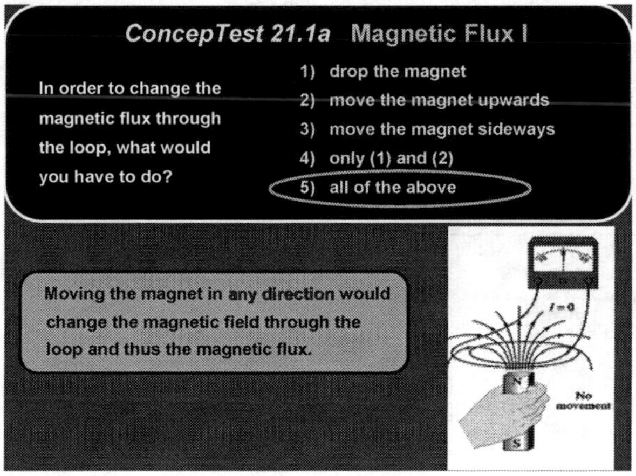

ConcepTest 21.1b: Magnetic Flux II
Difficulty Level: 1

This follow-up question continues to explore the ways in which the magnetic flux can be varied.

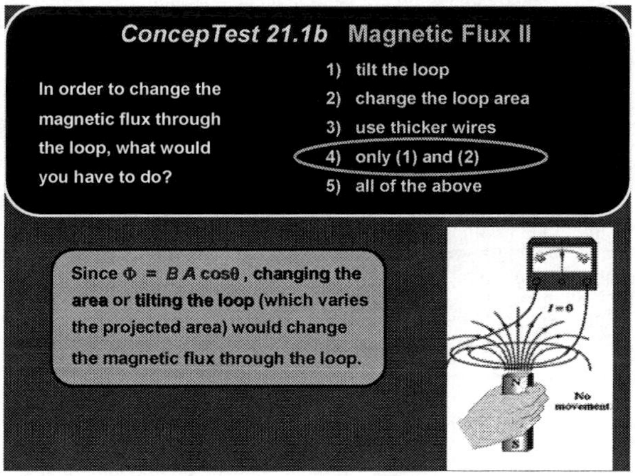

Conceptual Module 21.2 (two parts)

ConcepTest 21.2a: Moving Bar Magnet I
Difficulty Level: 3
Suggested follow-up conceptual question:

- **What happens if the magnet is stationary and the loop moves?**

The next series of questions examines the concept of electromagnetic induction, mainly in terms of the direction of the induced current as given by Lenz's Law. In each situation, the students must identify two specific things: (1) the initial direction of the flux and (2) the sense of the change in flux. These two facts will then dictate the direction of the induced current, which will act to oppose the change.

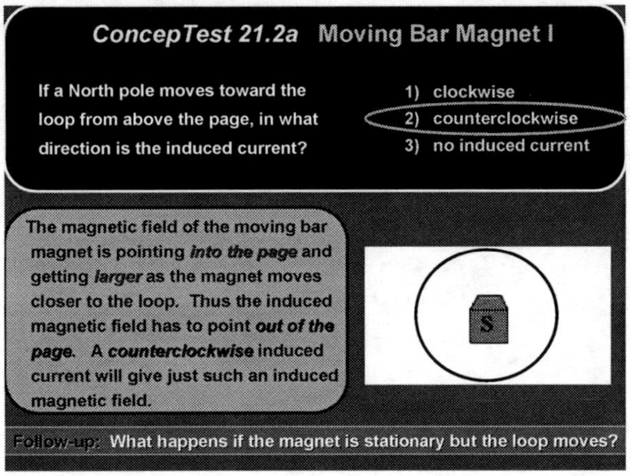

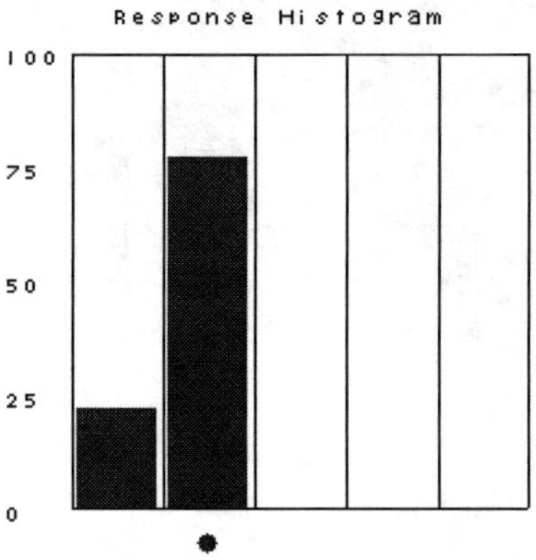

ConcepTest 21.2b: Moving Bar Magnet II
Difficulty Level: 2

In this case, the initial flux through the loop is zero, since it is in the same plane as the magnet. So even though the magnet is moving, the flux through the loop remains zero (i.e., does not change).

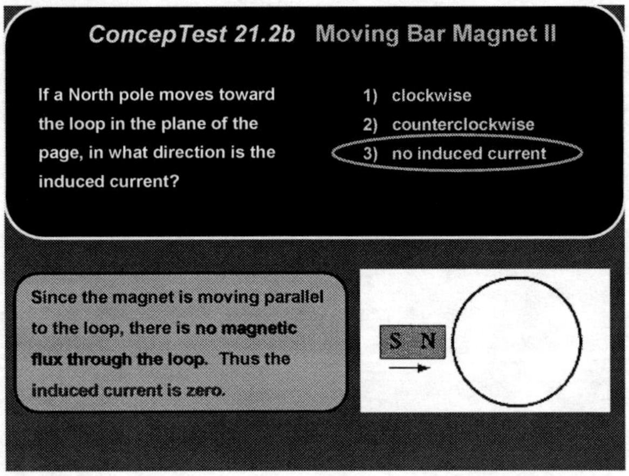

Conceptual Module 21.3 (three parts)

ConcepTest 21.3a: Moving Wire Loop I
Difficulty Level: 2
Suggested follow-up conceptual question:

- **What happens if the loop moves out of the page?**

In this case, even though the loop is moving, the flux through the loop is not changing. Thus, there will be no induced current.

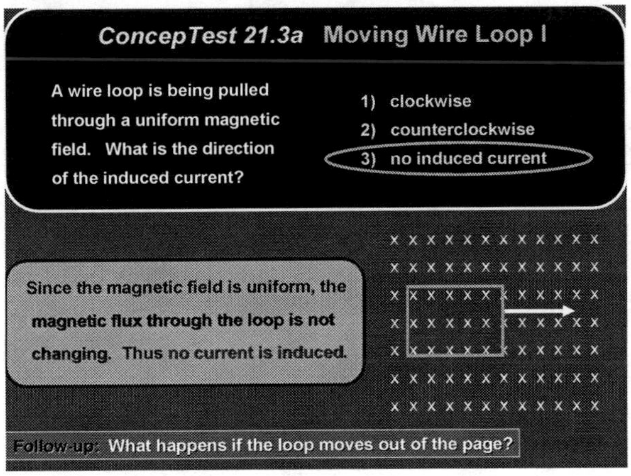

ConcepTest 21.3b: Moving Wire Loop II
Difficulty Level: 3
Suggested follow-up conceptual question:

- **What happens when the loop is completely out of the field?**

In this follow-up question, the loop is emerging from the field region, so the flux is indeed changing. Thus, there will be an induced current, whose direction must be determined by using Lenz's Law.

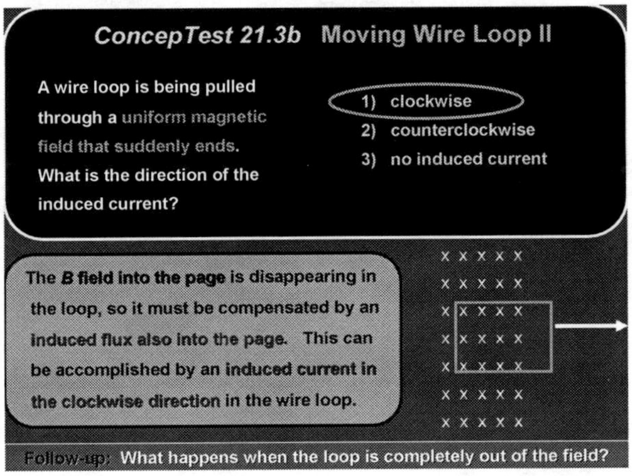

ConcepTest 21.3c: Moving Wire Loop III
Difficulty Level: 3
Suggested follow-up conceptual question:

- **What if the loop stops moving while the field increases?**

In this case, the magnetic flux is also changing, but this is due to an explicit variation in the strength of the magnetic field. Again, Lenz's Law must be used to determine the direction of the induced current.

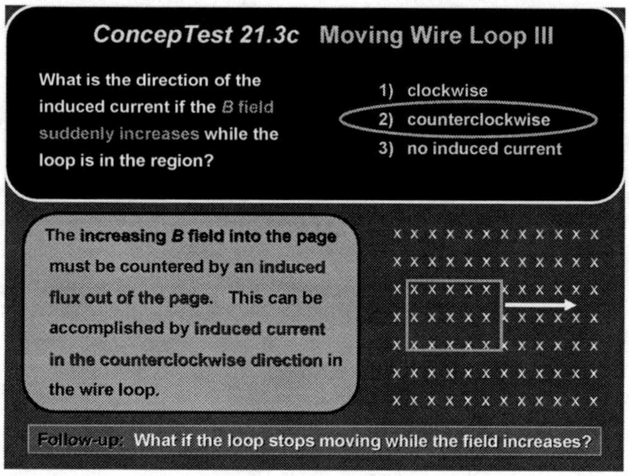

Conceptual Module 21.4

ConcepTest 21.4: Shrinking Wire Loop
Difficulty Level: 3
Suggested follow-up conceptual question:

- **What if the *B* field is oriented at 90° to its present direction?**

In this case, the flux is changing due to a variation in the area of the loop.

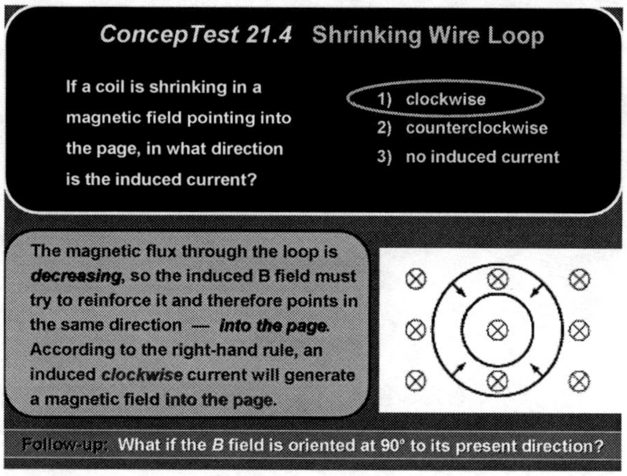

Conceptual Module 21.5

ConcepTest 21.5: Rotating Wire Loop
Difficulty Level: 3

It is also possible to change the flux through the loop by varying its orientation. In this case, the angle between the loop and the magnetic field is changing as the loop rotates, which leads to a corresponding variation in the flux. This is a preview of how a generator works.

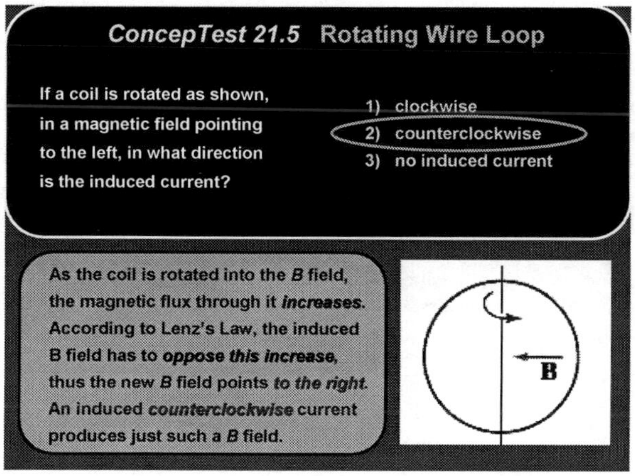

Conceptual Module 21.6 (two parts)

ConcepTest 21.6a: Voltage and Current I
Difficulty Level: 3

The induced voltage is determined by Faraday's Law, which involves the number of turns of wire in the loop. In this case, the two-turn loop will have twice the induced voltage as the single-turn loop.

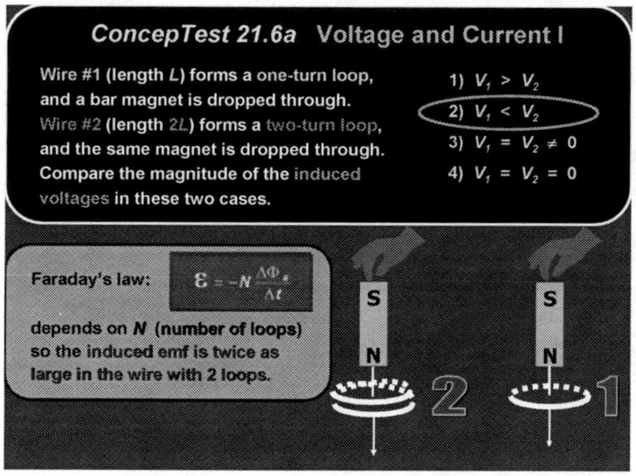

ConcepTest 21.6b: Voltage and Current II
Difficulty Level: 4

This follow-up question now brings in Ohm's Law to relate the current to the voltage. In the previous question, the voltages were different by a factor of 2, due to the number of turns. But the resistances of the loops will also be in that ratio, so in the end, the resulting currents will be the same.

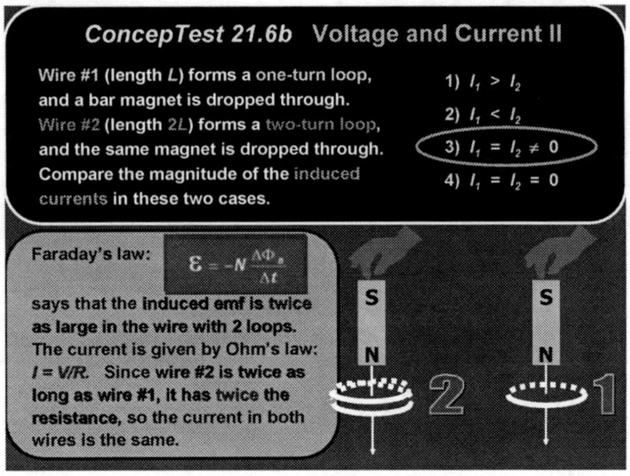

Conceptual Module 21.7 (two parts)

ConcepTest 21.7a: Falling Magnet I
Difficulty Level: 4
Suggested follow-up conceptual question:

- **What happens in case 2 if you flip the magnet so that the South pole is on the bottom as the magnet falls?**

This is a classic demonstration, where a magnet is dropped through a long, hollow aluminum rod. The magnet takes significantly longer to fall through than an unmagnetized piece of metal. This question should be asked before showing the demonstration.

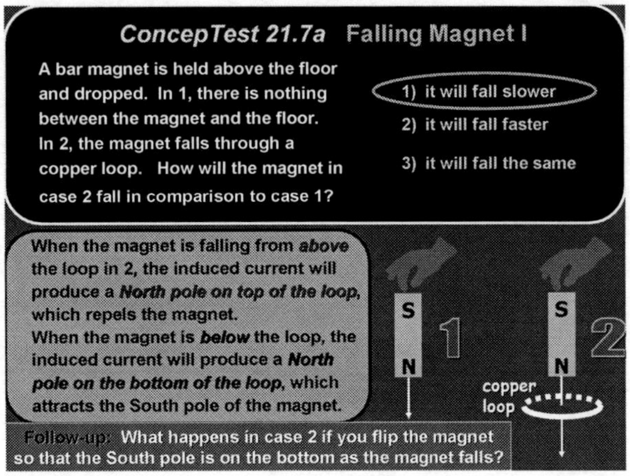

ConcepTest 21.7b: Falling Magnet II
Difficulty Level: 4

In this follow-up question, students must ponder the source of energy to make the induced current. Realizing that the falling magnet moves slower should lead the students to the conclusion that the "lost" kinetic energy is what accounts for the additional energy for the induced current.

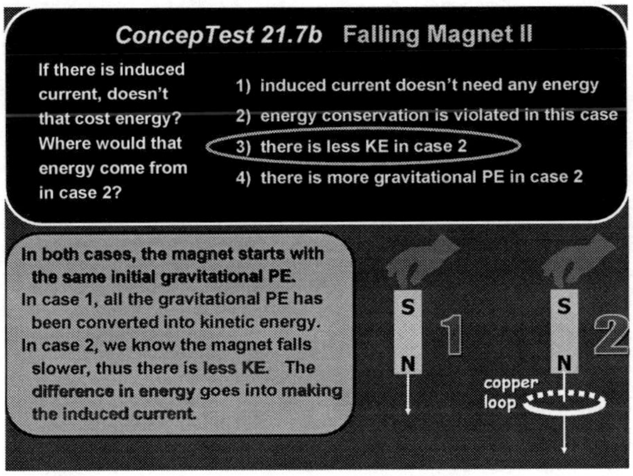

Conceptual Module 21.8 (two parts)
ConcepTest 21.8a: Loop and Wire I
Difficulty Level: 3

In this question, the magnetic field must first be deduced by examining the long straight wire. The fact that the field decreases with distance is what leads to a change in flux as the loop moves away.

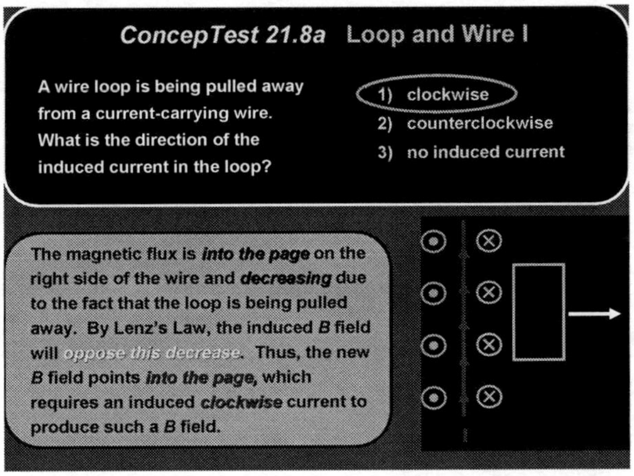

ConcepTest 21.8b: Loop and Wire II
Difficulty Level: 2

In this follow-up question, the direction of motion of the loop is now parallel to the wire. While it is true that the loop is covering a region of non-uniform magnetic field, its motion in that field is such that the flux does not change, so there is no induced current.

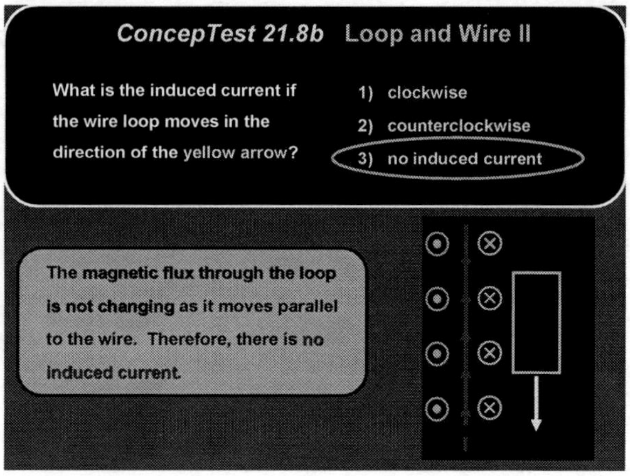

Conceptual Module 21.9

ConcepTest 21.9: Motional EMF
Difficulty Level: 3
Suggested follow-up conceptual question:

- **What direction is the magnetic force on the rod as it moves?**

The concept of motional EMF can be understood in terms of changing flux through a loop, as shown in this question. As the area of the loop effectively changes (as the rod moves), the flux is changing and therefore gives rise to an induced current.

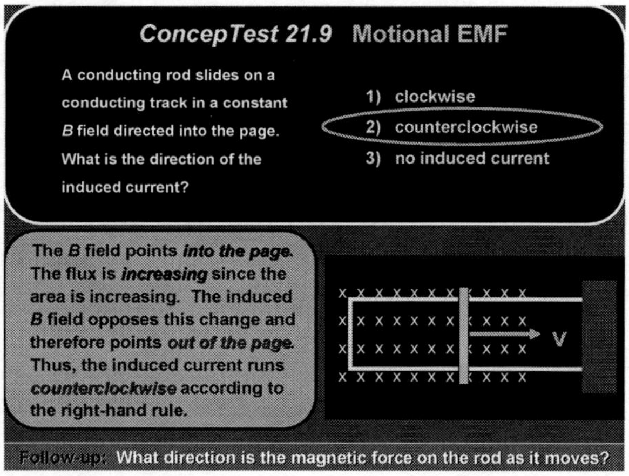

Conceptual Module 21.10

ConcepTest 21.10: Generators
Difficulty Level: 1

This is a direct application of the generator equation, which includes the frequency ω in the amplitude. As the rotation rate increases, the frequency increases but so does the amplitude of the generator.

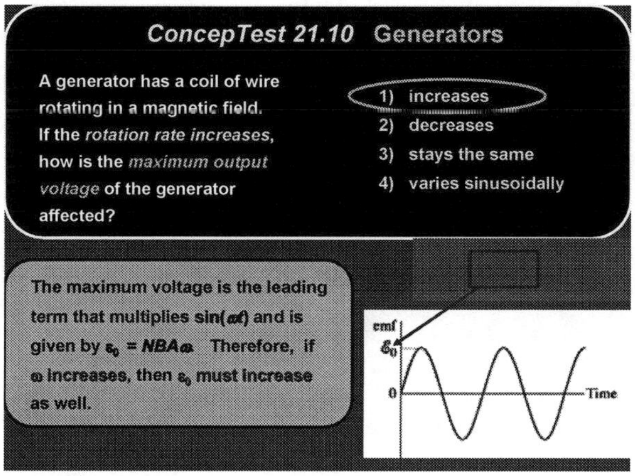

Conceptual Module 21.11

ConcepTest 21.11: Magic Loop
Difficulty Level: 3

This question is intended to introduce the idea of a motor. The direction of current flow in the loop (which is given in this problem) leads to a net torque on the loop. In the case of a motor, the current is provided by an external source and is alternating current (AC), which is synchronized with the rotation of the loop.

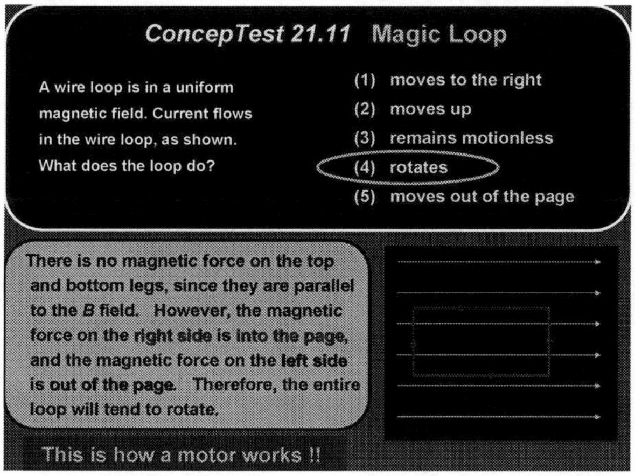

Conceptual Module 21.12 (three parts)
ConcepTest 21.12a: Transformers I
Difficulty Level: 3

The students have to apply the transformer relation twice in order to determine the output voltage of the second transformer. The ratio of primary/secondary voltages is just the same as the ratio of turns in the transformer coils.

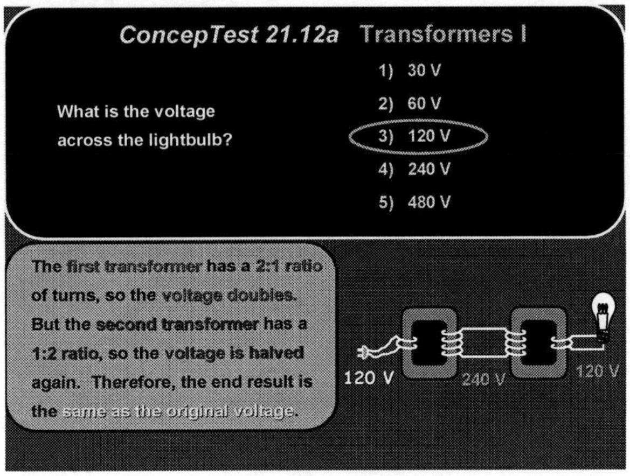

ConcepTest 21.12b: Transformers II
Difficulty Level: 3

While the voltage of a step-down transformer is reduced, the output current is increased by the same factor. This comes from conservation of energy, such that the input power must equal the output power in a transformer. In this case, since the output voltage has decreased by a factor of 2 (given by the ratio of turns of wire), the output current must be doubled.

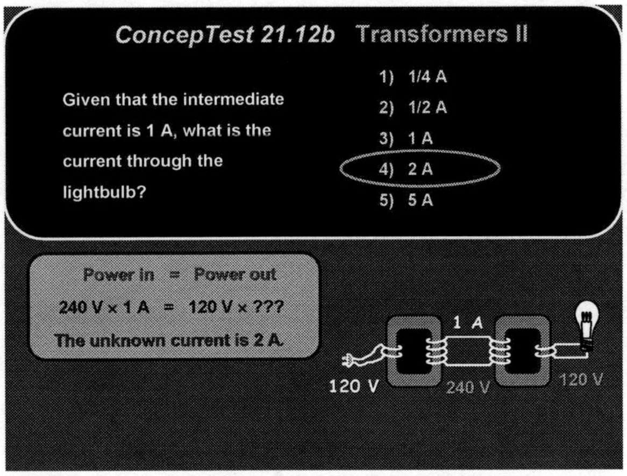

ConcepTest 21.12c: Transformers III
Difficulty Level: 2

This is a "trick" question where the students must realize that a transformer can only work with an AC input current. In this case, a DC battery will not give any changing magnetic flux in the transformer core, so there will be no induced output voltage.

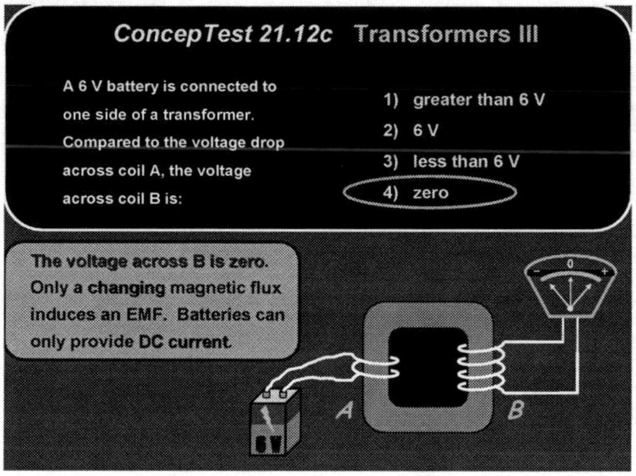

Conceptual Module 22.1 (three parts)
ConcepTest 22.1a: EM Waves I
Difficulty Level: 2

The induced EMF depends entirely on the changing flux through the loop and not on the composition of the loop itself. This is a precursor to introducing the students to the concept of electromagnetic waves, in which an induced alternating voltage can be produced in empty space by an alternating magnetic field (and vice versa).

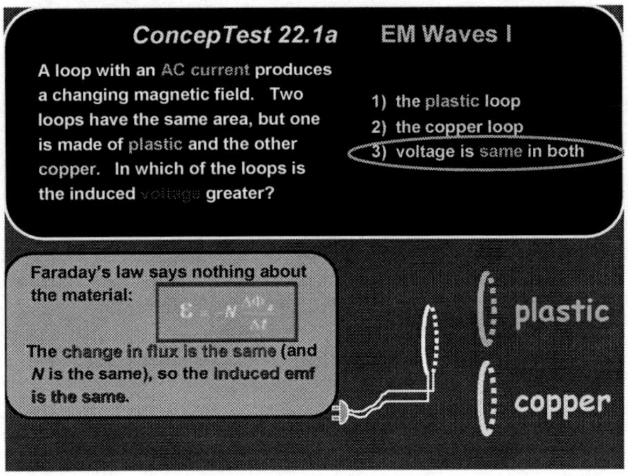

ConcepTest 22.1b: EM Waves II
Difficulty Level: 3

When asking about the current in the loop, now the material becomes relevant in terms of the resistance of the loop itself. The current is determined by Ohm's Law ($V = IR$), and the material with the lower resistance (in this case, copper) will carry the higher current.

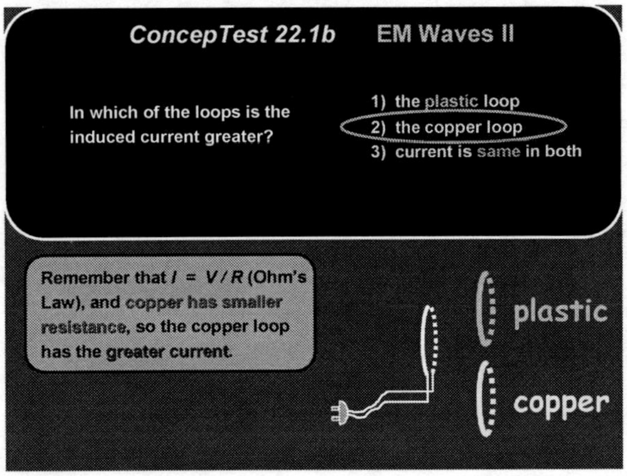

ConcepTest 22.1c: EM Waves III
Difficulty Level: 2

This follow-up question touches on the concept of EM waves. In this case, the plastic loop is replaced by a "loop of air," and the electric field inside both "loops" is seen to be the same. Here, again, the type of material (or even the presence of a material) has no bearing on the induced electric field created by the changing magnetic field.

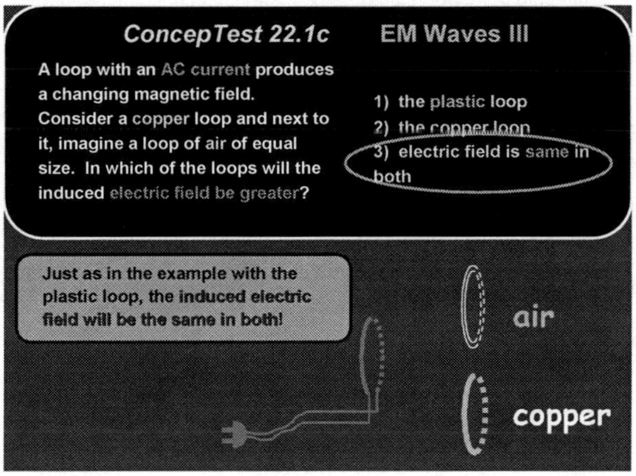

Conceptual Module 22.2

ConcepTest 22.2: Oscillations
Difficulty Level: 2

The electric field, magnetic field and direction of motion of an EM wave are mutually perpendicular. In this case, two of the above are specified (electric field and travel direction), and the students must use this fact about EM waves to infer the direction of the third (magnetic field).

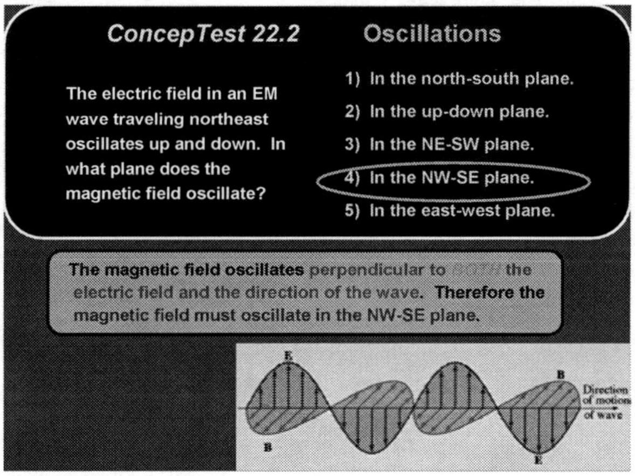

Conceptual Module 22.3
ConcepTest 22.3: TV Antennas
Difficulty Level: 2

Students have probably seen both types of antennas, but have not given much thought to their shapes. The straight antenna is designed to pick up electric oscillations in the vertical direction, since charges can oscillate up and down the antenna. On the other hand, the circular antenna is designed to pick up magnetic oscillations in the horizontal direction, since this changing magnetic field will induce a current in the loop through which the flux is oscillating.

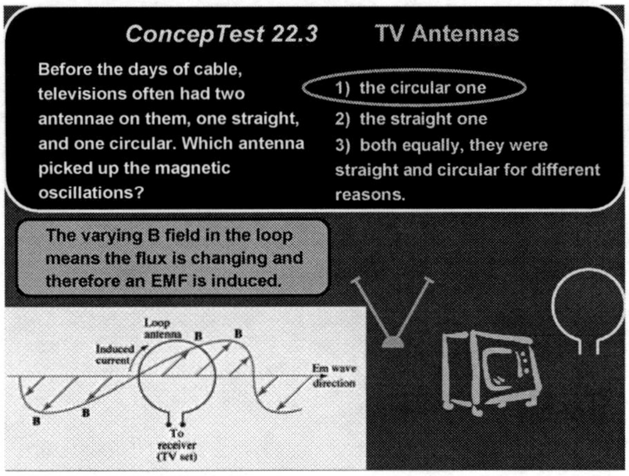

Conceptual Module 22.4
ConcepTest 22.4: Radio Antennas
Difficulty Level: 2

Since the transmitter antenna is vertical, the electric field of the EM wave is also vertical. In order to allow charges to oscillate due to electric forces, the receiving antenna must also be oriented vertically.

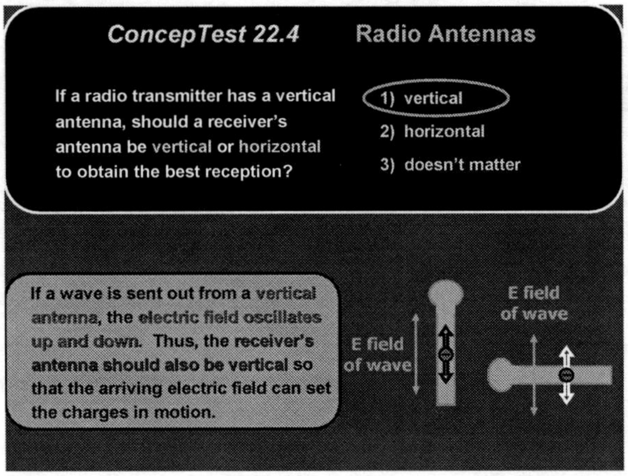

Conceptual Module 22.5
ConcepTest 22.5: Heat Insulation
Difficulty Level: 2

We are accustomed to detecting visible light, but in this case, the students are asked to make the extrapolation to the case of infrared light. Bright spots would be characterized by high intensities of radiation, and since human bodies are warm, they emit infrared radiation. These are the spots that would appear bright in infrared spectrum.

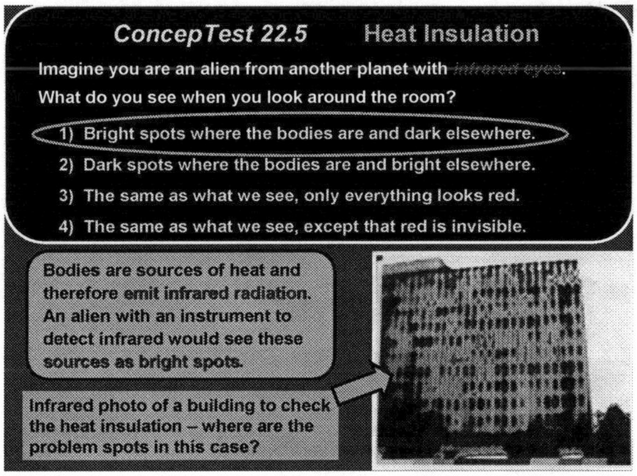

Conceptual Module 22.6
ConcepTest 22.6: Superman
Difficulty Level: 3

This question intends to get students to make the distinction between transmission and reception of EM waves. We see things because our eyes receive visible radiation that is reflected from objects. With no light source present, we cannot see (for example, we cannot see objects in a totally dark room). Thus, unless Lois Lane was irradiated by an external source of x-rays, then Superman would not be able to see her (since there is no reflected x-ray radiation).

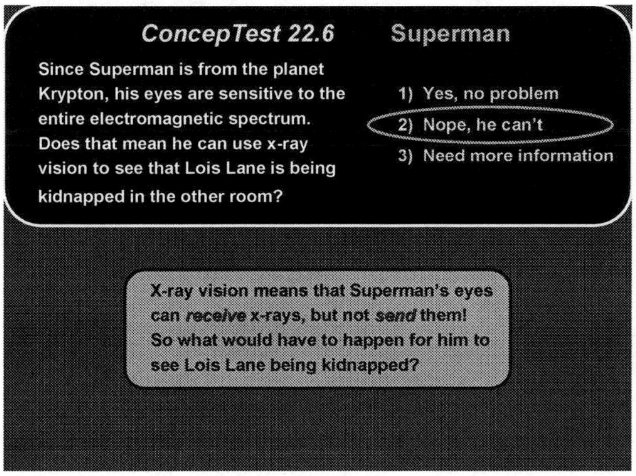

Conceptual Module 23.1

ConcepTest 23.1: Reflection
Difficulty Level: 2
Suggested follow-up conceptual question:

- **Where else does this occur?**

The reflection of light from water is such a common occurrence that we do not give it much thought. However, the difference between that case and the case of reflections from a planar mirror is striking. There are clear reflections from a very calm surface of water, which gives a mirror-like image, but more generally, the reflection is "shimmery" due to the undulation of the reflecting water surface.

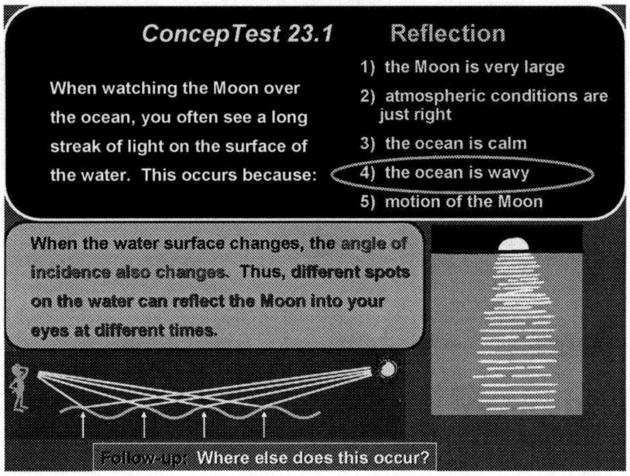

Conceptual Module 23.2 (three parts)

ConcepTest 23.2a: Mirror I
Difficulty Level: 3
Suggested follow-up conceptual question:

- **What happens when the observer starts moving toward the mirror?**

This question gives students a chance to utilize the law of reflection to reconstruct the position of an image from a planar mirror. The ray reaching the observer on the left side appears to emanate directly from the (virtual) image located behind the mirror (right side).

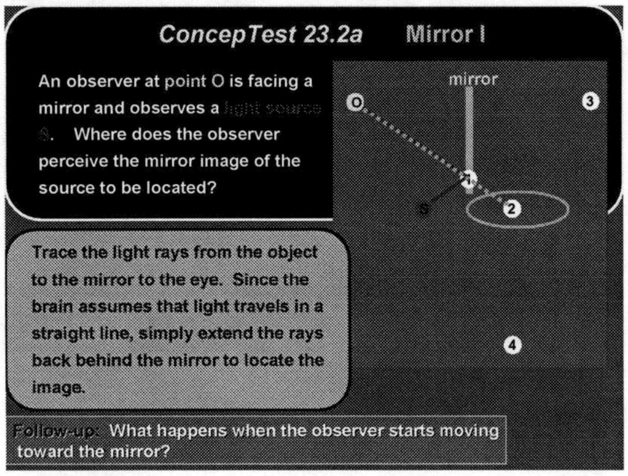

ConcepTest 23.2b: Mirror II
Difficulty Level: 3

Students can use the law of reflection to prove that a "full-length" mirror does not have to be as tall as the viewing object. In fact, students can convince themselves that if the mirror extended lower than the one shown, it would only serve to provide an image of the floor!

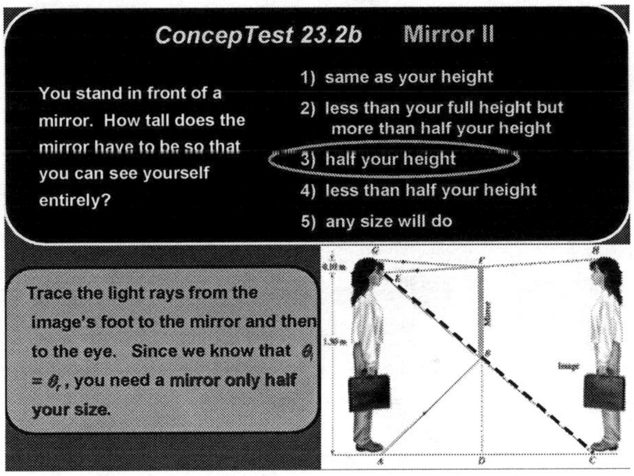

ConcepTest 23.2c: Mirror III
Difficulty Level: 2

The previous result, regarding the minimum size of a "full-length" mirror, is entirely dependent on the law of reflection and has nothing to do with the object distance from the reflecting surface. The bottom ray, which is the one providing an image of the feet, will still be illuminating the same object, regardless of the position of that object.

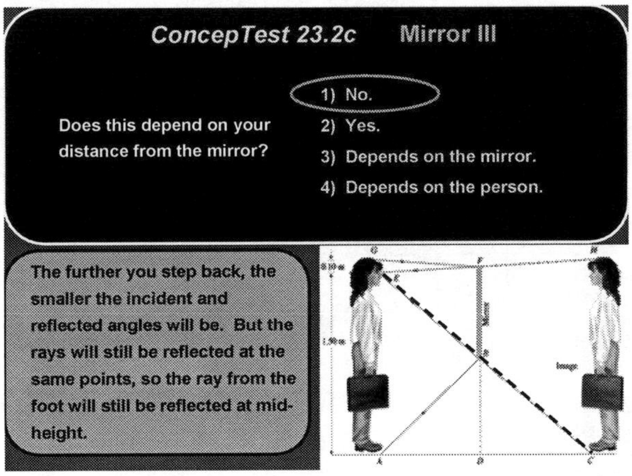

Conceptual Module 23.3

ConcepTest 23.3: All Smoke and Mirrors
Difficulty Level: 4

This is a multi-step problem in reflection that requires the students to understand the nature of object and image. In this case, the object of the full-length mirror is actually the image in the hand mirror. A careful accounting of object and image distance for each reflection stage is needed in order to arrive at the answer.

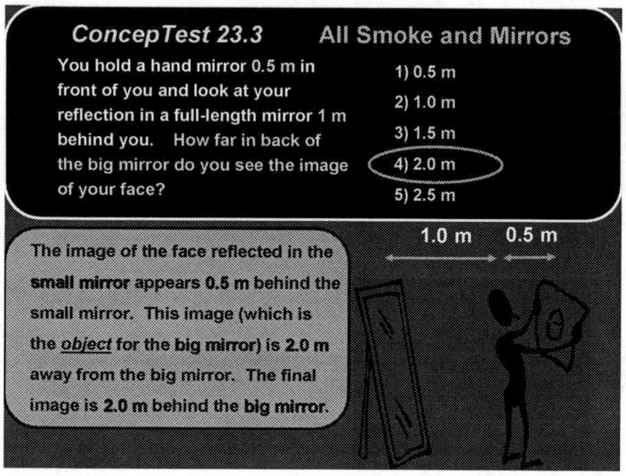

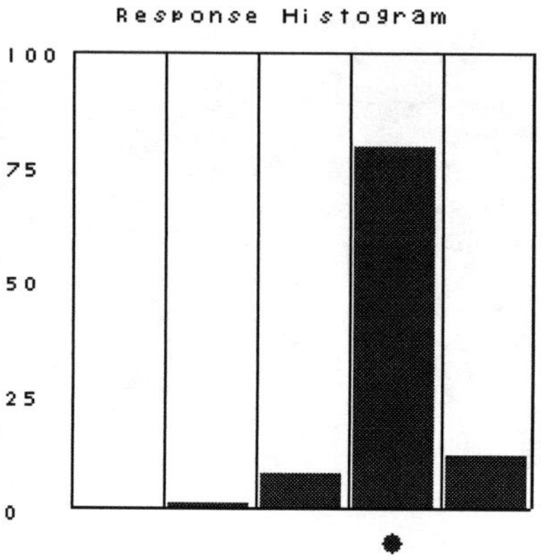

Conceptual Module 23.4 (two parts)

ConcepTest 23.4a: Refraction I
Difficulty Level: 2
Suggested follow-up conceptual question:

- **How does the speed in air compare to that of 1 or 2?**

The angle of refraction is related to the ratio of indices of refraction for the two media. Using this fact, and the knowledge that the light is moving at its maximum speed in the air, the students should be able to infer that the higher speed is in the medium with the least amount of bending.

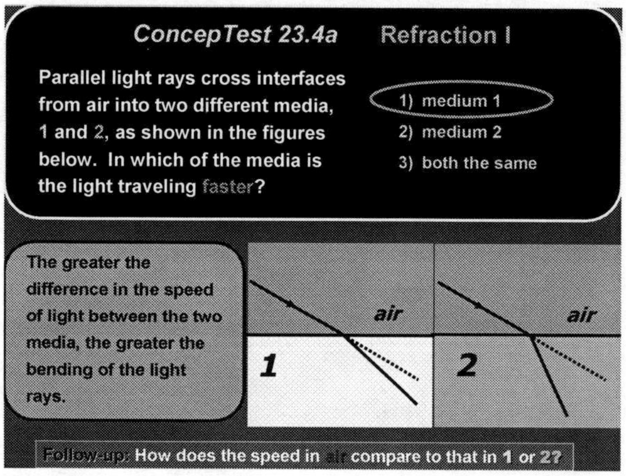

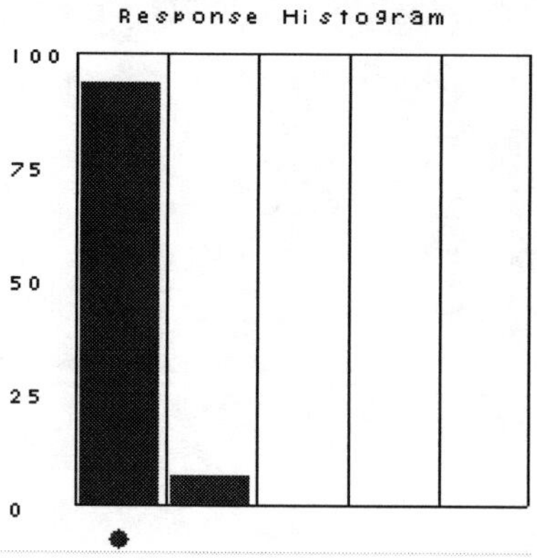

ConcepTest 23.4b: Refraction II
Difficulty Level: 3

The students have to use what they learned in the previous question to infer the ranking of the indices of refraction based on the bending of the light rays. They also need to remember that the ray will bend toward the normal when entering a medium of higher index of refraction (and vice versa).

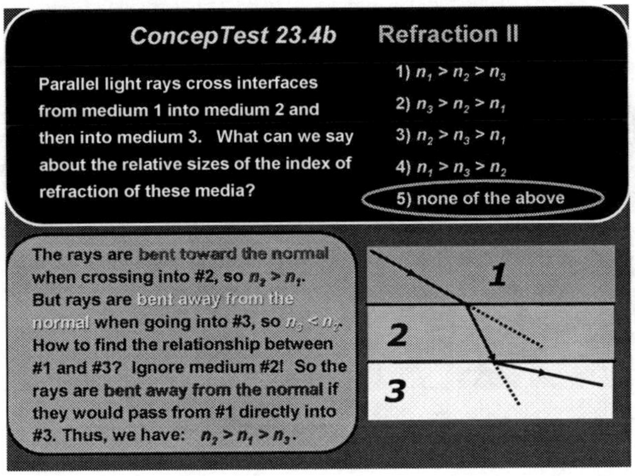

Conceptual Module 23.5 (two parts)

ConcepTest 23.5a: Gone Fishin' I
Difficulty Level: 4

This is a difficult question for students. They must first trace the path of a light ray *from* the object back *toward* the observer in order to establish the image position (which is higher, in this case). Then they must realize that a bullet will not refract and so must be aimed lower than the image to account for the refraction of the light ray emanating from the object.

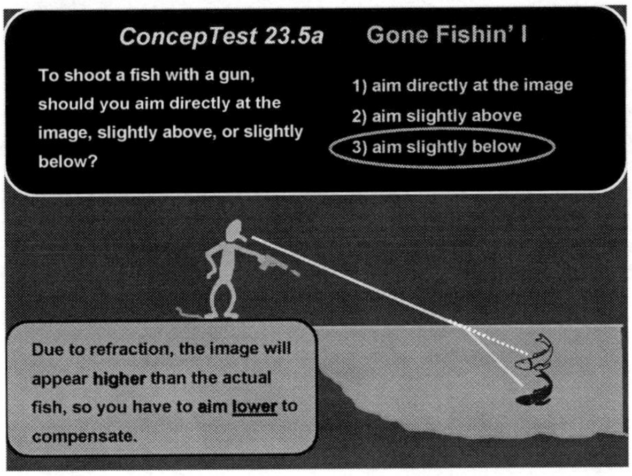

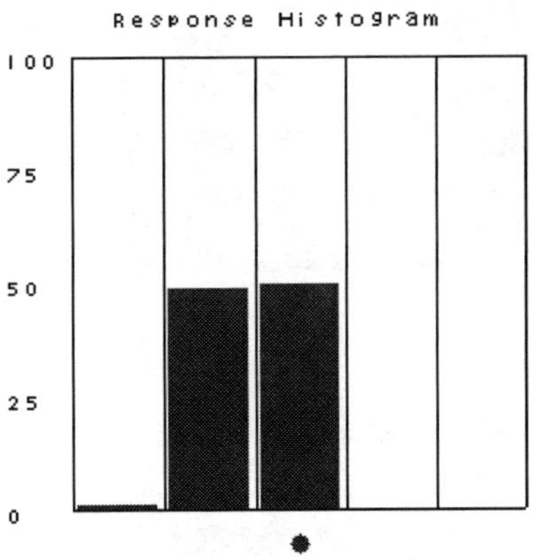

ConcepTest 23.5b: Gone Fishin' II
Difficulty Level: 2

This follow-up question should actually be easier. Students should realize that the laser beam will be affected in exactly the same way as the light ray from the object, so it should be fine to simply aim directly at the image and let the refraction process work for both rays.

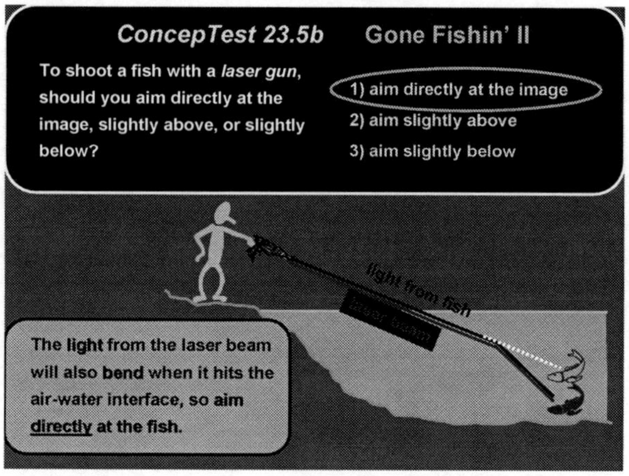

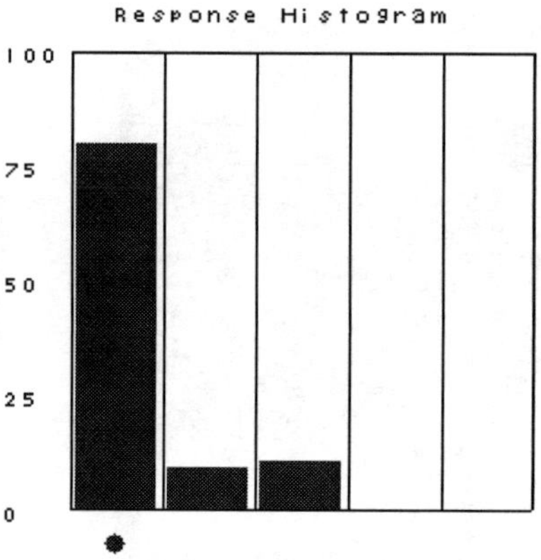

Conceptual Module 23.6

ConcepTest 23.6: Parallel lines
Difficulty Level: 4
Suggested follow-up conceptual question:

- **What happens when the top glass moves toward the bottom glass?**

The refraction of the two parallel rays occurs identically for both rays, so in the end, the angled piece of glass accomplishes nothing more than a lateral shift in the position of the lines on the paper. The shift occurs because of the angle imposed on the rays when they refract into the glass slab.

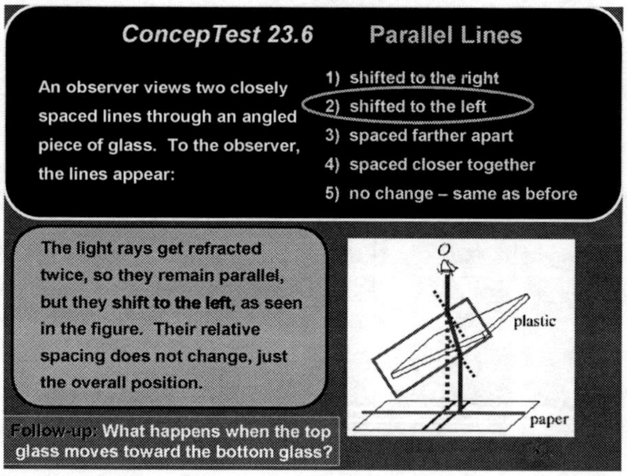

Conceptual Module 24.1

ConcepTest 24.1: Superposition
Difficulty Level: 2

Students have encountered superposition in terms of vectors, but the general notion of superposition as "adding" of quantities may not have been driven home. The superposition of two waves is simply the addition of the amplitudes at each point along the wave. This is easily illustrated in this question, where the wave forms involve simple square pulses.

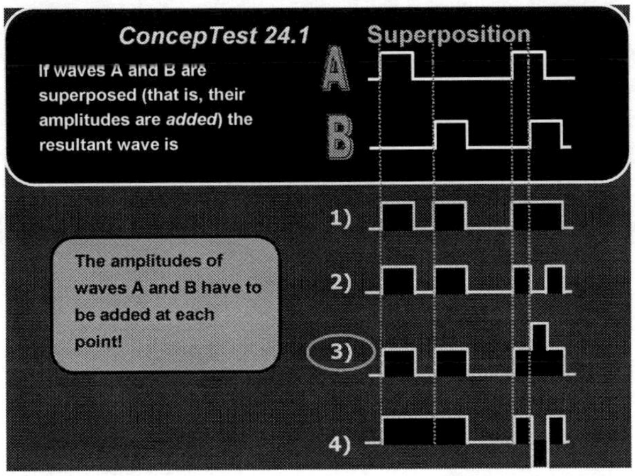

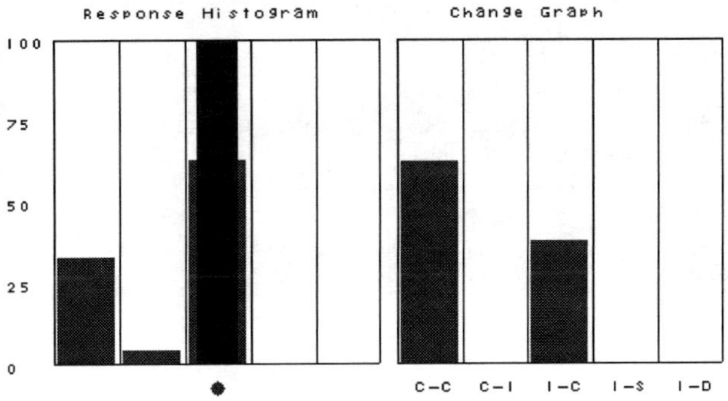

Conceptual Module 24.2 (two parts)

ConcepTest 24.2a: Phase Difference I
Difficulty Level: 2
Suggested follow-up conceptual question:

- **What would the wave look like for no. 4 to be correct?**

This question examines the phase difference between two waves. An understanding of phase is needed for the upcoming work on interference. The notion that a half-wavelength corresponds to a phase difference of 180° and a quarter-wavelength is 90° is important for the students to get straight.

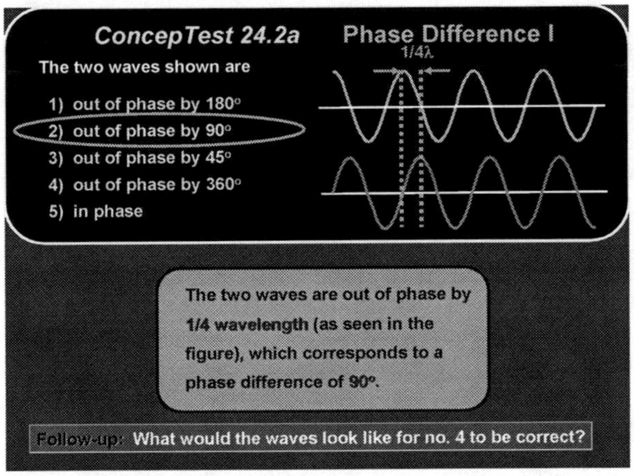

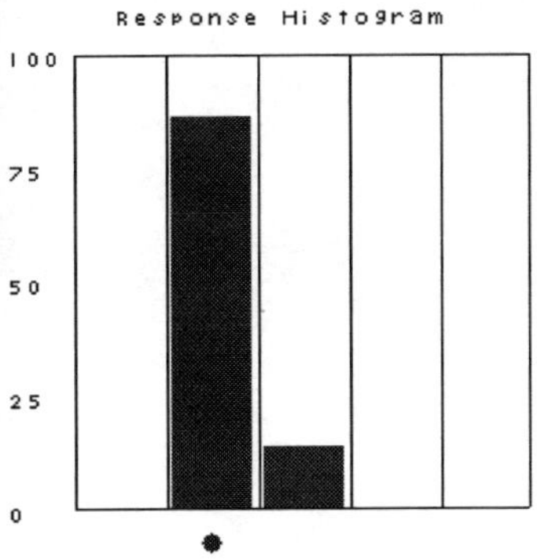

ConcepTest 24.2b: Phase Difference II
Difficulty Level: 2

This is a simple interference problem where the students must realize that waves displaced by an integral number of wavelengths will still be in phase. The periodic nature of the waves guarantees that each successive wavelength is identical to its predecessors, so that such a shift alters nothing.

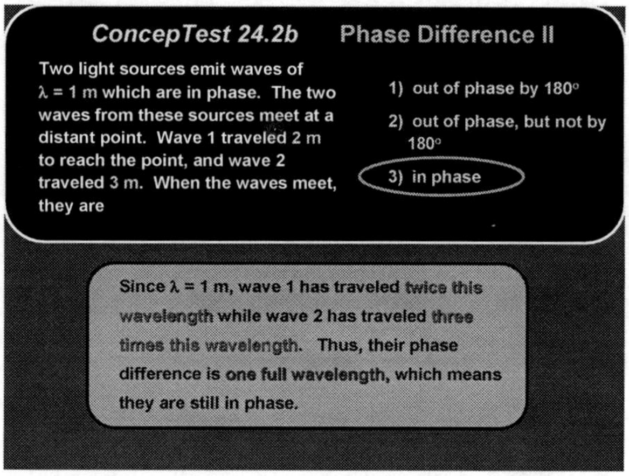

Conceptual Module 24.3 (two parts)

ConcepTest 24.3a: Double Slits I
Difficulty Level: 3

This question examines the basic formula governing double-slit interference. The general idea is to give the students a feeling for the interplay of the slit spacing, the wavelength and the position of the fringes on the screen (as determined by the angle θ and the distance of the screen to the slits). In this case, with the slit spacing fixed, the pattern spreads out (θ increases) as the wavelength increases.

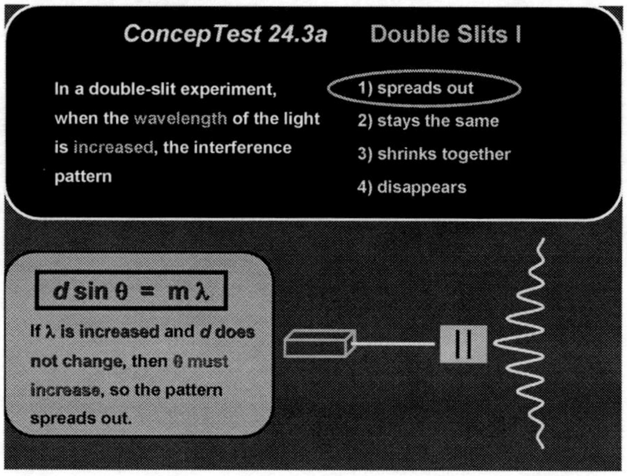

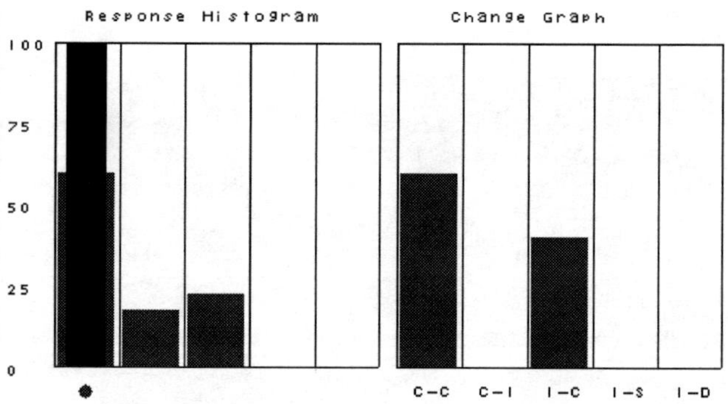

ConcepTest 24.3b: Double Slits II
Difficulty Level: 3
Suggested follow-up conceptual question:

- **When would the interference pattern disappear?**

This follow-up question continues to explore the double-slit interference relation. In this case, with the wavelength fixed, the pattern gets tighter (θ decreases) as the slit spacing increases.

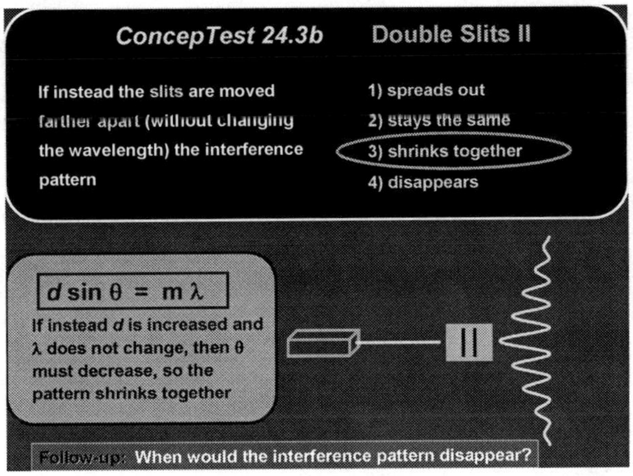

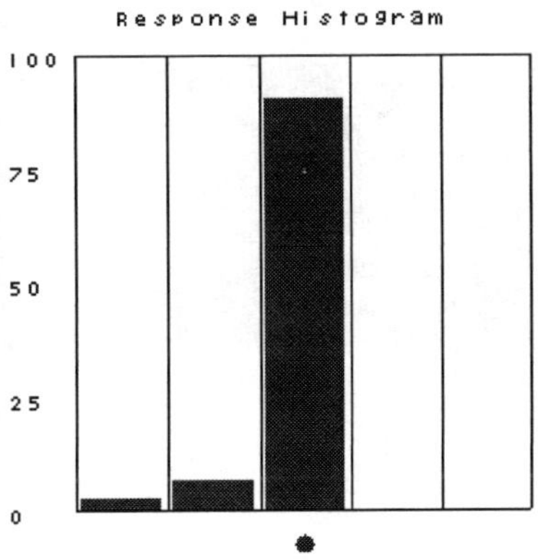

Conceptual Module 24.4

ConcepTest 24.4: Path Difference
Difficulty Level: 2

One of the central ideas of interference relates to the path difference between two different waves. This is essentially what determines whether the waves will be in phase or out of phase when they arrive at the observer. In the double-slit interference pattern, each successive fringe corresponds to an additional phase shift of a full wavelength. It is important for students to understand this relationship.

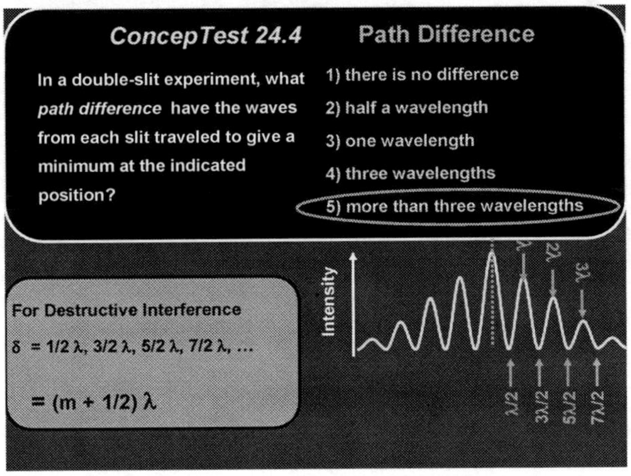

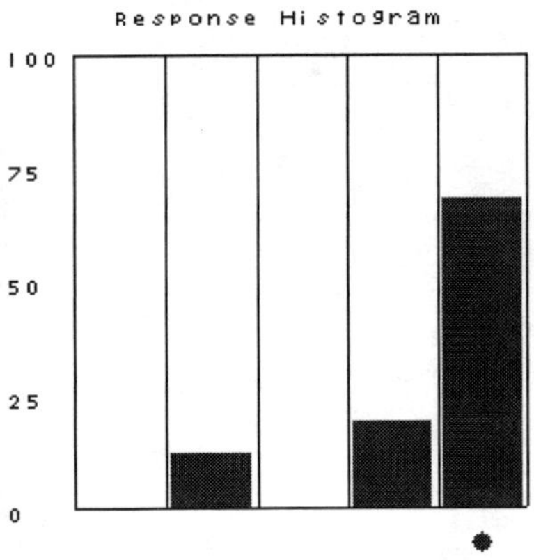

Conceptual Module 24.5
ConcepTest 24.5: Interference Pattern
Difficulty Level: 4
Suggested follow-up conceptual question:

- **What happens when the phase difference is 90°?**

This question is asking students to examine the phase relationship between interfering waves more carefully. While they have learned that the phase difference at the observer depends largely on the path-length difference, this question purposely alters the *initial* phase relationship of the waves to see if students can realize that this will simply "flip" the pattern on the screen.

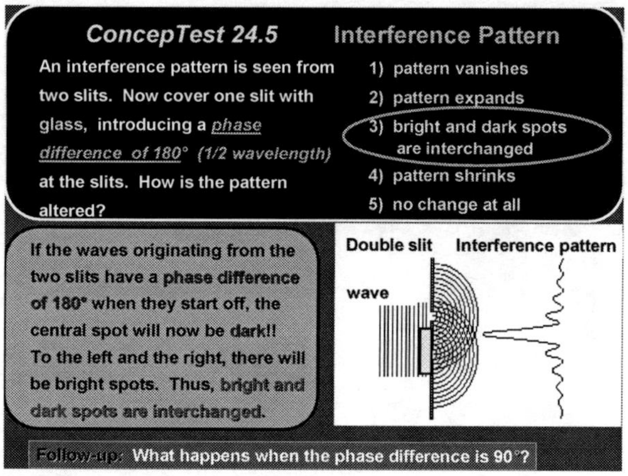

Conceptual Module 24.5 (two parts)
ConcepTest 24.5a: Diffraction I
Difficulty Level: 3

This question examines the expression for diffraction maxima and minima, which relates the wavelength and slit width to the position of interference fringes on the screen. Based on this relation, the wider the slit, the narrower the diffraction pattern.

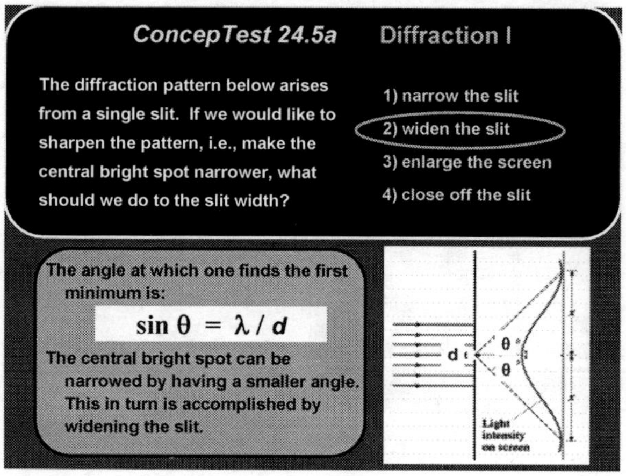

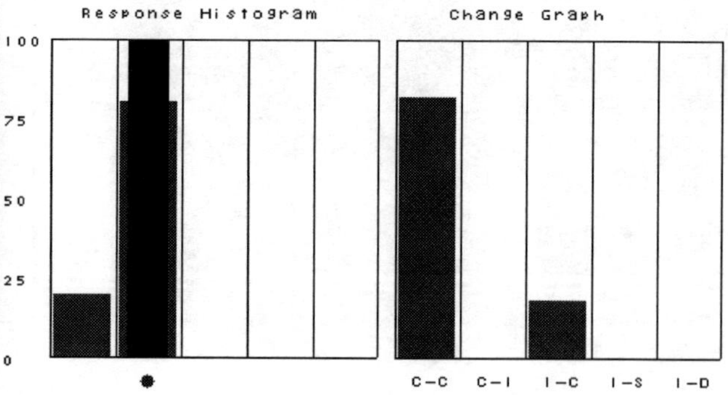

ConcepTest 24.5b: Diffraction II
Difficulty Level: 4

This two-step question requires students to recognize the shift in the diffraction pattern that will occur when the wavelength is changed, and then to deduce what must be done to another parameter (slit width) in order to recover the original pattern. Both of these steps must be performed with the same diffraction formula utilized in the previous question.

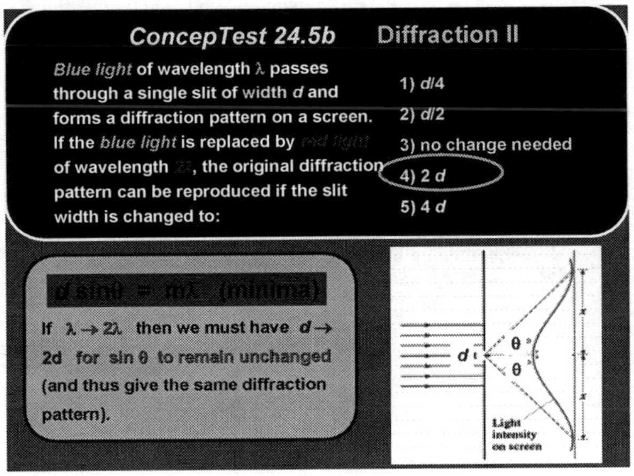

Conceptual Module 24.6

ConcepTest 24.6: Diffraction Disk
Difficulty Level: 2
Suggested follow-up conceptual question:

- **What if the disk is oval and not circular?**

This question emphasizes the relevance of the path-length difference. At the center of the shadow, the waves emanating from the edge of the disk have all traveled the same distance, therefore they will all be in phase and should give a maximum intensity at that point.

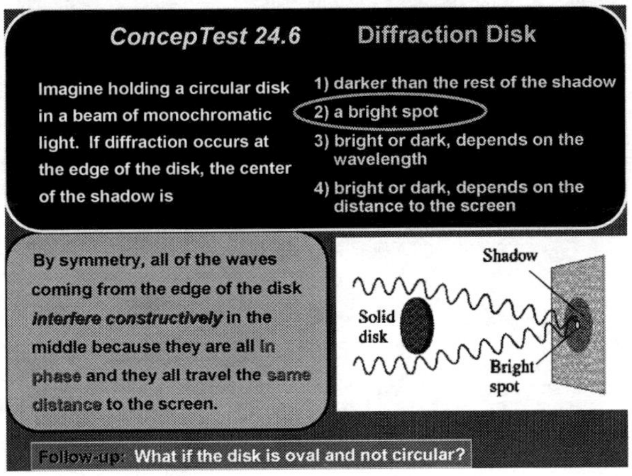

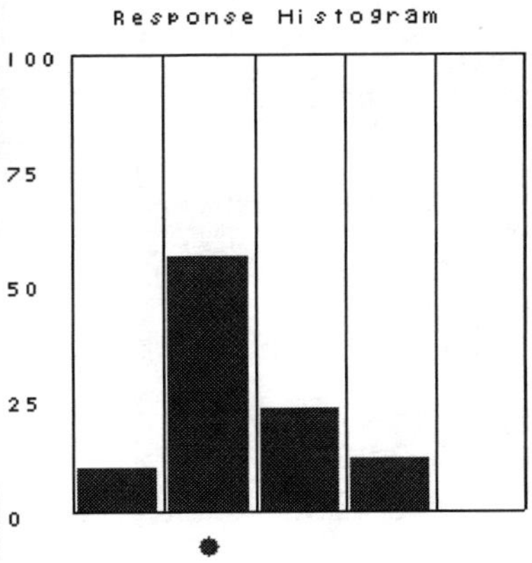

Conceptual Module 24.7 (four parts)
ConcepTest 24.7a: Parallel Slides I
Difficulty Level: 2

This question emphasizes the path-length difference in interference. For a given slide position, the interference is uniformly dark or light across the entire slide. As the slide positions change, the path-length difference between interfering waves is constantly changing, so the pattern changes as well.

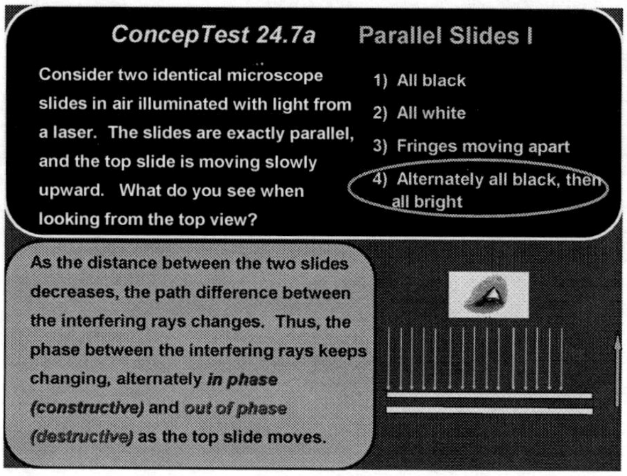

ConcepTest 24.7b: Parallel Slides II
Difficulty Level: 1

In this case, the slides are fixed, but the path-length difference arises from the angle between the slides, leading to a variable separation across the slide. The interference at point A might be expected to be constructive, due to the a path-length difference of zero, but due to the additional phase shift that occurs at the interface, point A in fact indicates destructive interference. This is the same issue that gives a dark spot in the center of Newton's rings.

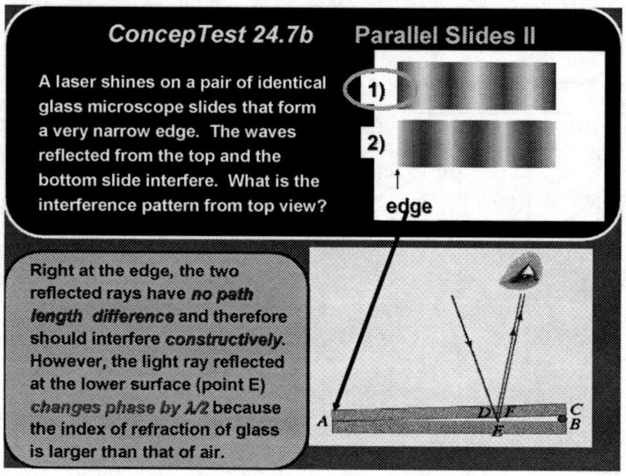

ConcepTest 24.7c: Parallel Slides III
Difficulty Level: 4

Extending the previous question, the angled slides are now brought closer together. This reduces the path-length difference and causes the interference pattern to spread out (assuming fixed wavelength). That is, the ith fringe (which corresponds to a specific path-length difference) will now move to a new position farther out on the slide in order to maintain that same path-length difference (since the slides have now moved closer).

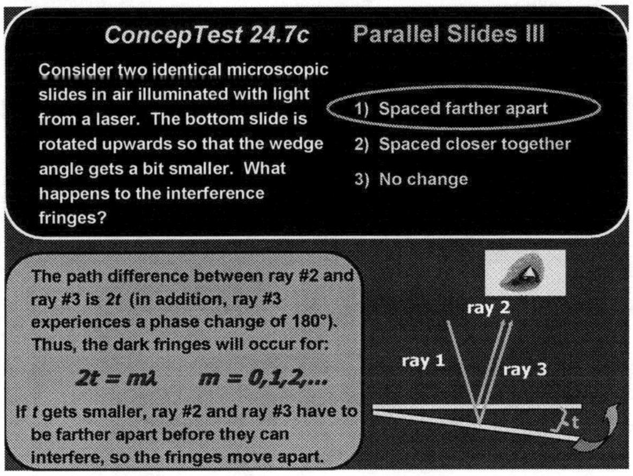

ConcepTest 24.7d: Parallel Slides IV
Difficulty Level: 4

This variation of the question alters the wavelength of the waves by having them travel in water. The decrease in wavelength that occurs in water (where $n > 1$) leads to a tighter interference pattern.

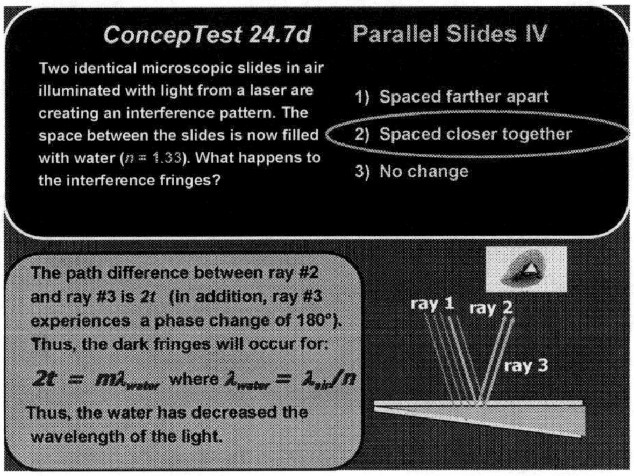

ConcepTest 24.8: Polarization
Difficulty Level: 4

Transmission of light through polarizers depends on the angle between adjacent elements, as given by Malus's Law. In cases #1 and #3, there are two adjacent perpendicular elements, which will cancel any transmitted light. However, in case #2, the intermediate 45° element will rotate the plane of polarization such that it can get through the last polarizer.

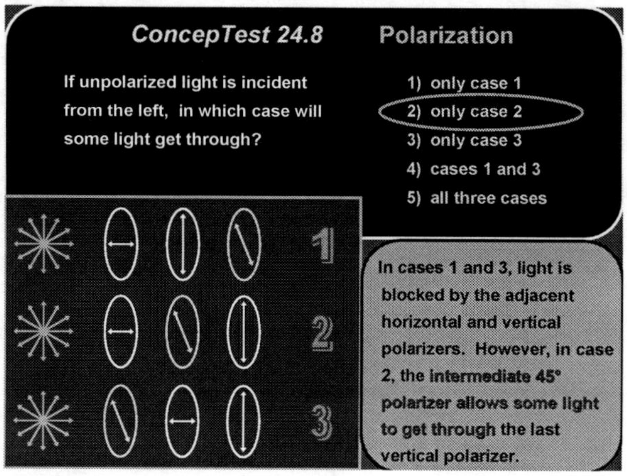

Conceptual Module 26.1

ConcepTest 26.1: Playing Ball on the Train

Difficulty Level: 2

Suggested follow-up conceptual question:

- **What velocity does the ball have, as measured by an observer at rest on the station platform?**

A basic concept to understand before embarking on special relativity is the notion of reference frames. Students have a lot of experience with reference frames (sitting in class vs. driving in a car) but have probably not given this much consideration. This question tries to get them to relate their experience to the physical quantities in the respective frames.

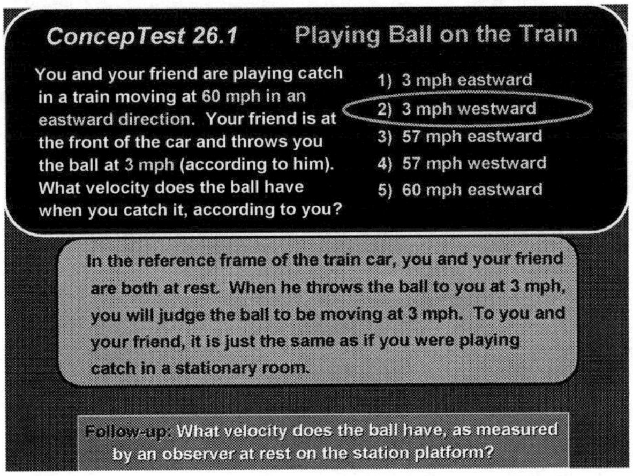

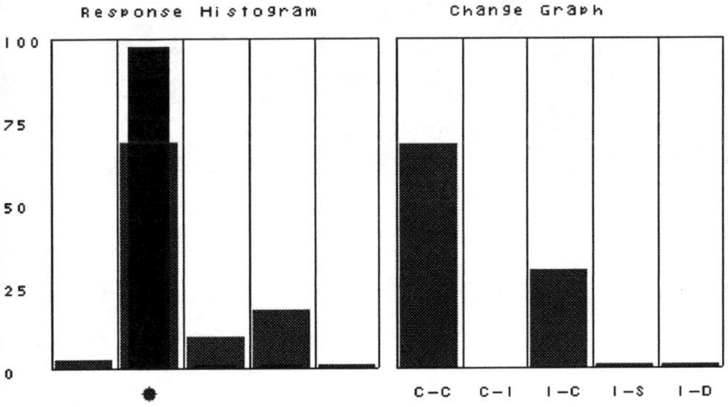

Conceptual Module 26.2

ConcepTest 26.2: Running with an Electron

Difficulty Level: 2

The students know that magnetic fields are generated by moving charges. However, that motion is relative, and it can in fact be the observer who is moving relative to a fixed charge. The idea that the presence of a magnetic field is frame-dependent will probably come as a surprise to students.

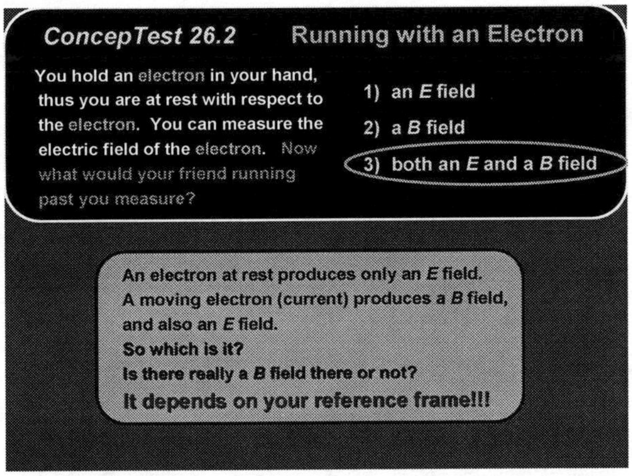

Conceptual Module 26.3
ConcepTest 26.3: Inertial Reference Frames
Difficulty Level: 1

The definition of an inertial reference frame is one that is not accelerating. This question asks students to make this distinction for several choices, keeping in mind that uniform circular motion does indeed involve an acceleration.

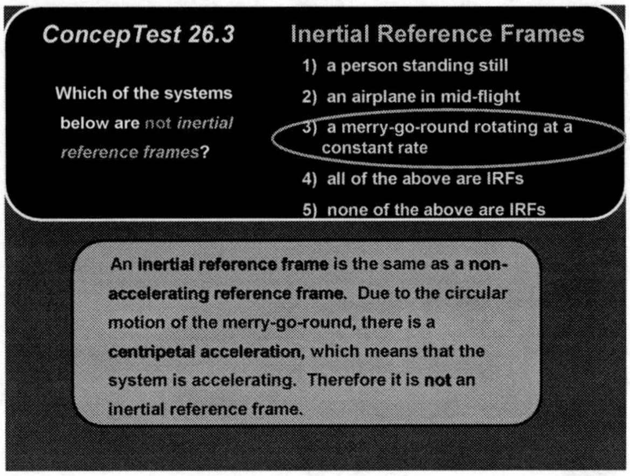

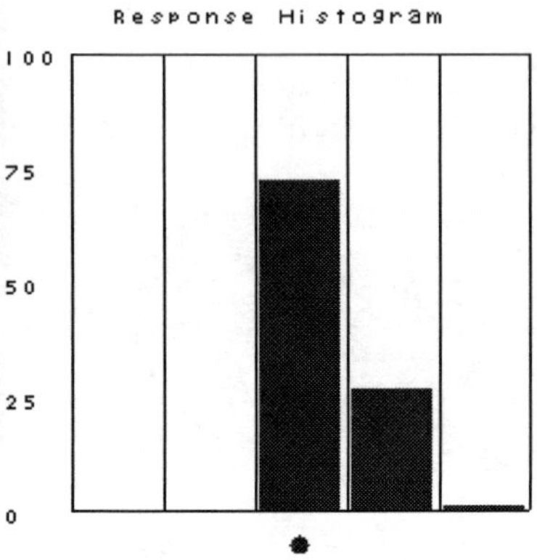

Conceptual Module 26.4 (two parts)
ConcepTest 26.4a: Changing Reference Frames I
Difficulty Level: 2

Certain quantities are frame-dependent but other quantities are invariant. The students are asked to examine several kinematic quantities to deduce whether they depend on the reference frame or not. For example, velocity is obviously dependent on the frame, as evidenced by riding in a car.

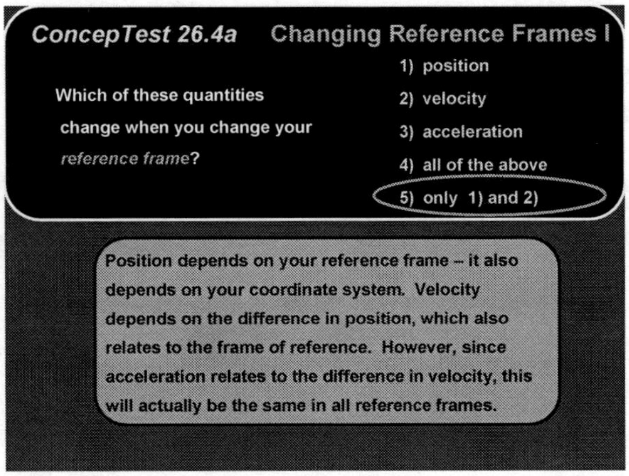

ConcepTest 26.4b: Changing Reference Frames II
Difficulty Level: 2

This question continues to examine the frame dependence of various physical quantities.

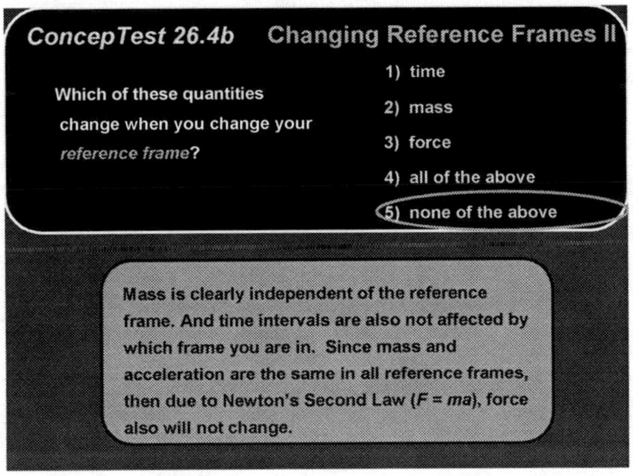

Conceptual Module 26.5
ConcepTest 26.5: Windowless Spaceship
Difficulty Level: 2

Since the physics of inertial reference frames is the same in all such frames, it is therefore impossible to distinguish a frame at rest from a frame moving at constant velocity. Any kinematics experiment that would be performed in either frame would yield the same result.

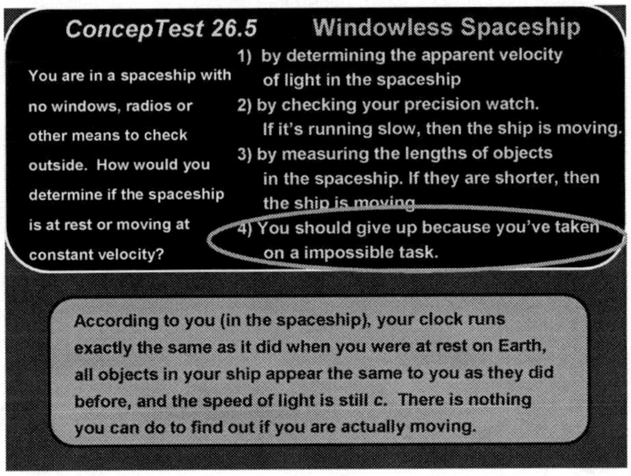

Conceptual Module 26.6 (three parts)

ConcepTest 26.6a: Borg Ship I
Difficulty Level: 2

This initial question ignores special relativity (for the moment) and asks students to identify the relative speed of two approaching objects. This is essentially an exercise in the "classical" addition of velocities, mainly to establish the meaning of the concept of relative velocity.

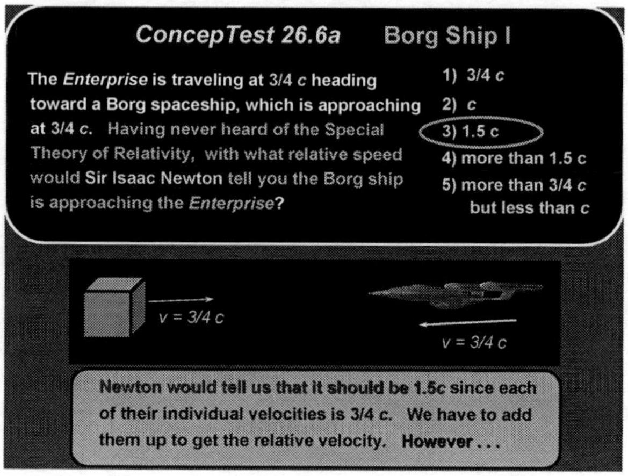

ConcepTest 26.6b: Borg Ship II
Difficulty Level: 2

Now taking special relativity into account, the students must assert that the speed of light is the ultimate speed. In this case, while the relative velocity will certainly be greater than $0.75c$, it cannot exceed c, which is the speed of light.

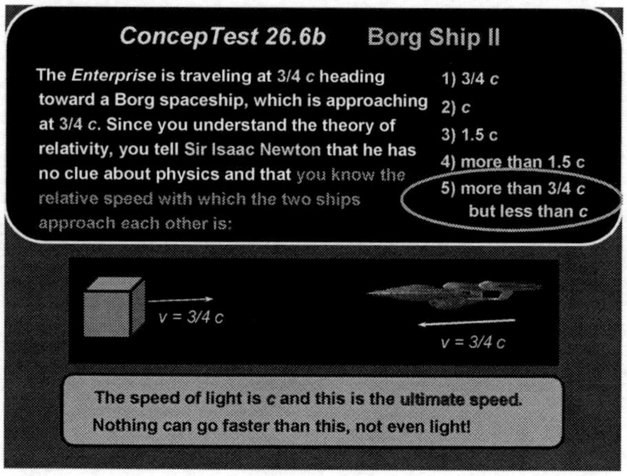

ConcepTest 26.6c: Borg Ship III
Difficulty Level: 1

Here, again, the fact that the speed of light is the ultimate speed is re-emphasized. Even though the speed of the source is non-zero, the speed of light does not change, for neither the source nor the observer.

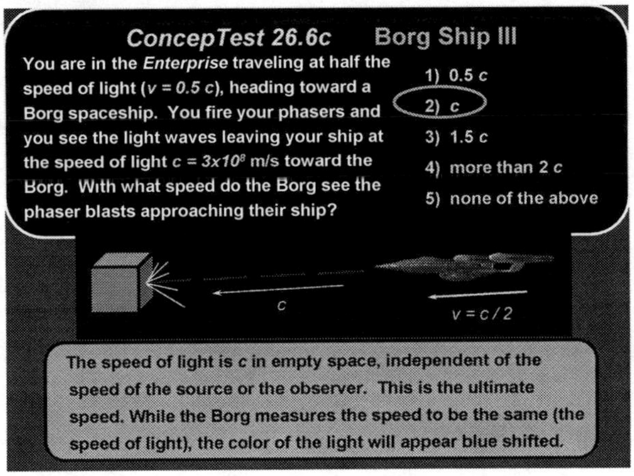

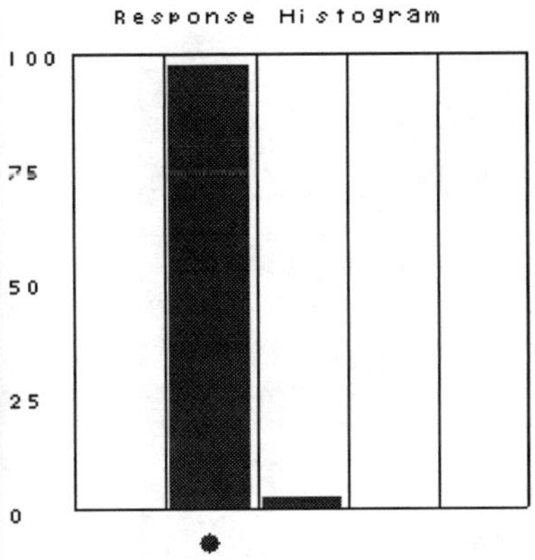

Conceptual Module 26.7 (two parts)

ConcepTest 26.7a: Speed of Light I
Difficulty Level: 2

This classic statement illustrates the postulate that the speed of light *c* is the same in all inertial reference frames. Thus, even though the observer is moving rapidly, the speed of light in that observer's frame is still *c*.

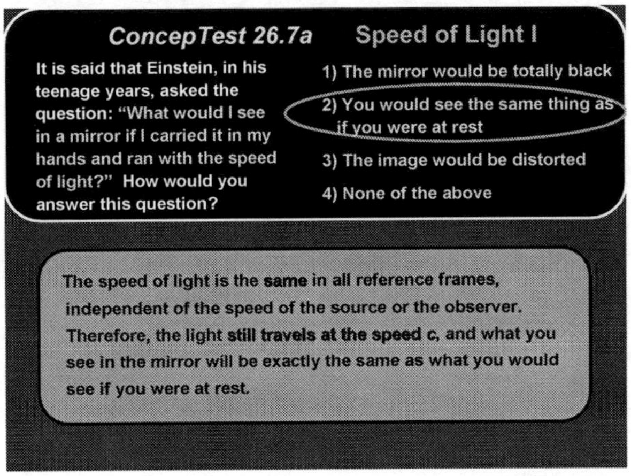

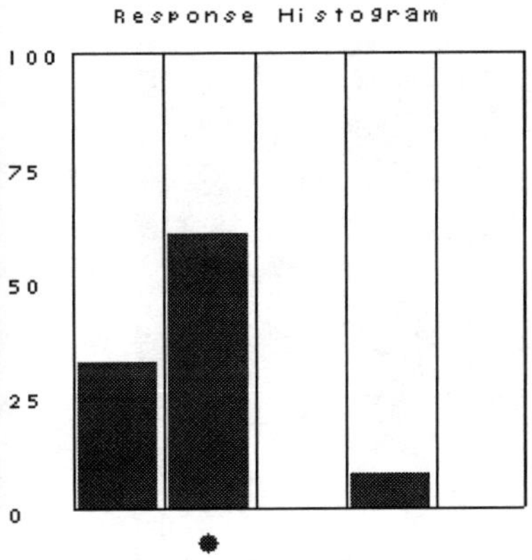

ConcepTest 26.7b: Speed of Light II
Difficulty Level: 2

Since the speed of light is reduced in a medium with index of refraction $n > 1$, it is indeed possible for a highly relativistic particle to move faster than light (in that medium!). This leads to a shock wave called Cherenkov radiation. However, it must be understood that no particle can travel faster than c (where this is the speed of light in vacuum).

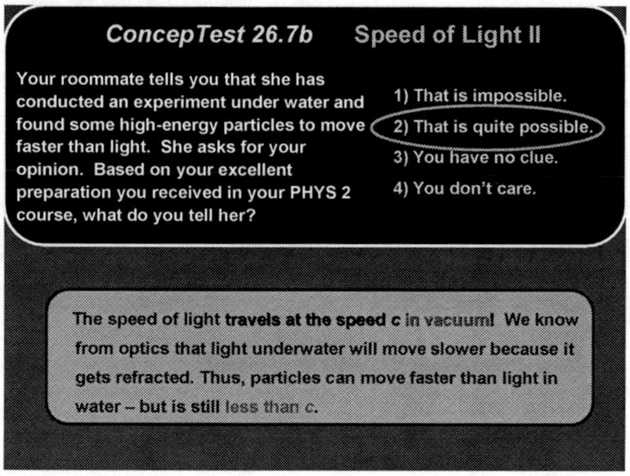

Conceptual Module 26.8

ConcepTest 26.8: Foghorns
Difficulty Level: 2

This is a "warm-up" to the discussion of simultaneity. In this case, while the observer hears the sounds at the same time, the fact that one is softer than the other allows the observer to extrapolate that this sound must have originated farther away and thus must have sounded first.

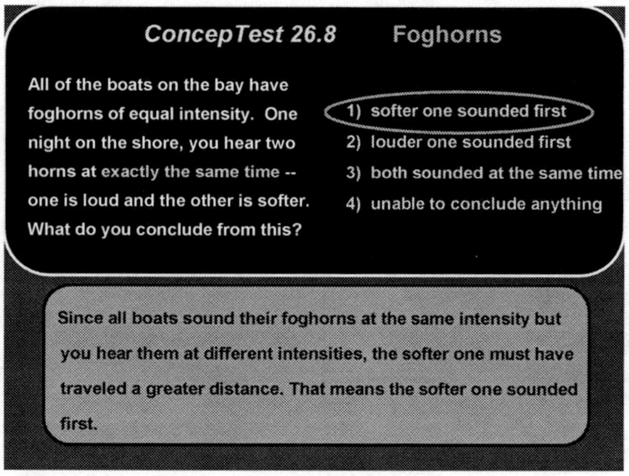

Conceptual Module 26.9

ConcepTest 26.9: Balls in Boxcar
Difficulty Level: 3

In this case, the classical understanding of relative velocity applies. According to the ground observer, the green ball has a higher initial speed, but the far end of the car is receding from it. The blue ball has a lower initial speed, but the near end of the car is approaching it. Thus, the travel times in both cases will end up being the same, and the balls strike the respective walls *simultaneously*.

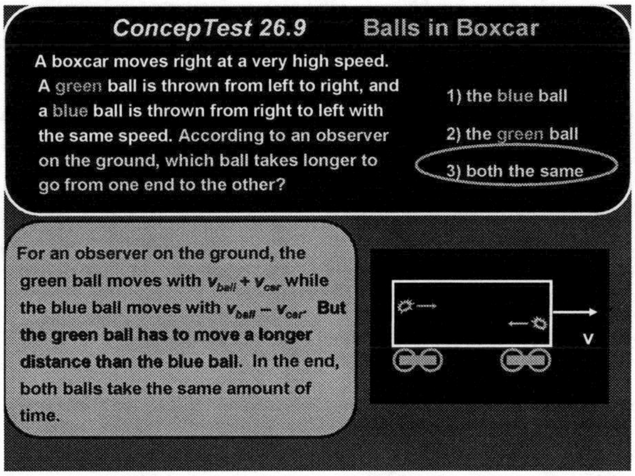

Conceptual Module 26.10 (two parts)

ConcepTest 26.10a: Light Flashes in Boxcar I
Difficulty Level: 2

In the reference frame of the boxcar, the light flashes traverse the same distance at the same velocity, so the travel times in both cases will be the same, and the flashes will arrive at the respective walls *simultaneously*.

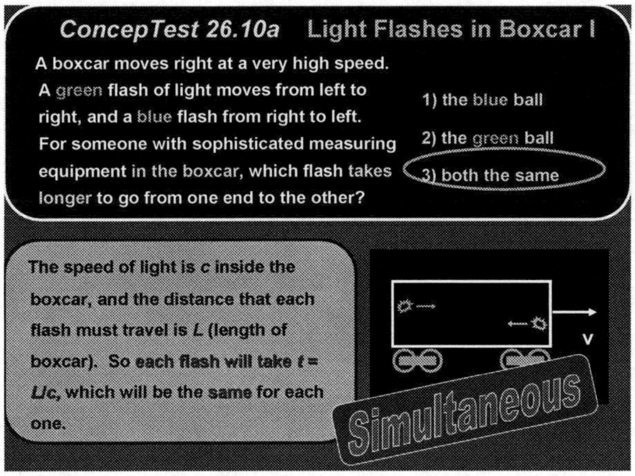

ConcepTest 26.10b: Light Flashes in Boxcar II
Difficulty Level: 4

This question considers the same situation from the perspective of the ground observer, but the result is different from ConcepTest 26.9 above. The key point here is that the light flashes *still travel at the speed of light*, independent of the motion of the boxcar (which is quite different from the case of the two balls). Since the distance traveled is different as the boxcar moves, but the speed of the flashes is the same, therefore the hits *will not be simultaneous*, according to the ground observer.

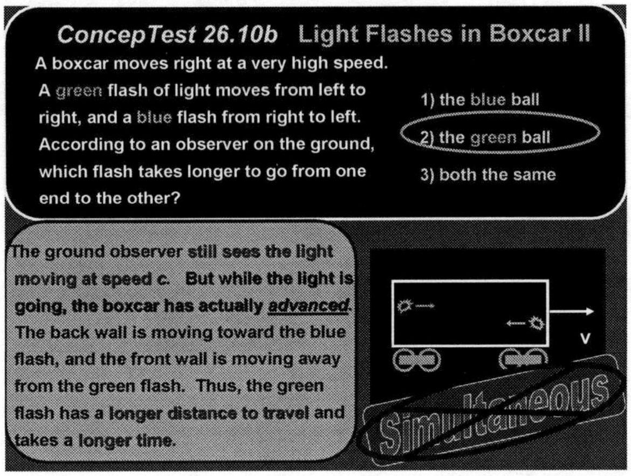

Conceptual Module 26.11

ConcepTest 26.11: Causality
Difficulty Level: 2

Seeing the result of the previous ConcepTest, students realize that as the boxcar speeds up, the blue flash will hit earlier and earlier before the green flash, and the time interval between the emission of the blue flash and its contact with the wall decreases. However, the contact can never actually *precede* the emission of the flash, since that would require the boxcar to move faster than c!

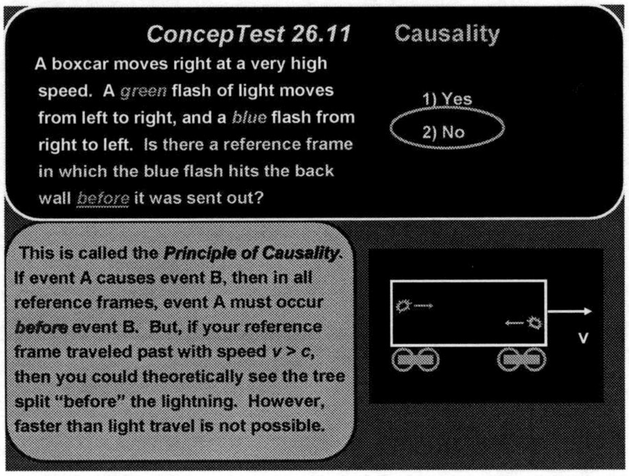

Conceptual Module 26.12 (two parts)
ConcepTest 26.12a: Boxcar I
Difficulty Level: 3

In this case, the classical understanding of relative velocity applies. According to the ground observer, the soccer ball has a higher initial speed, but the far end of the car is receding from it. Since the difference in each case is simply due to the speed of the boxcar, the higher speed and longer distance end up giving the same travel time for the soccer ball.

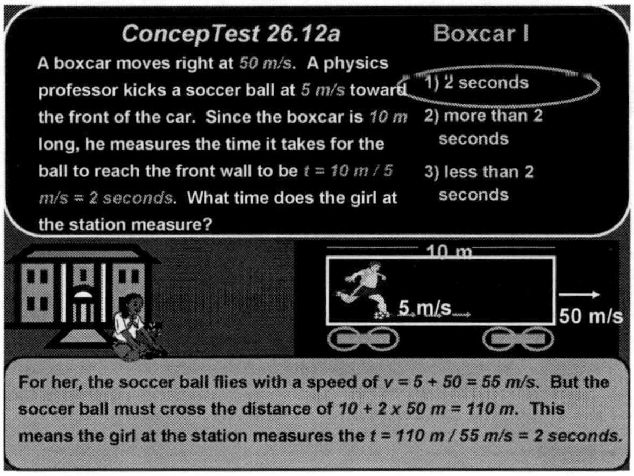

ConcepTest 26.12b: Boxcar II
Difficulty Level: 3

This case is similar to above, except that a light flash is involved (instead of a soccer ball). Since both the ground and boxcar observers see the light move at the same speed, the ground observer will measure a longer transit time because the far end of the car is receding from the flash. This is the first glimpse of the effect of time dilation for the students.

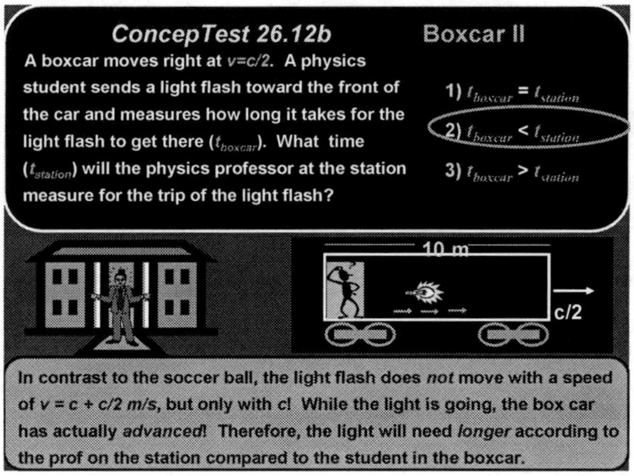

Conceptual Module 26.13 (three parts)
ConcepTest 26.13a: Time Dilation I
Difficulty Level: 3

The time dilation effect ("moving clocks run slow") implies that an observer who is *not* in the frame of the event will measure a longer time interval. Thus, the astronaut's pulse rate would appear to be slower, as determined by the Earth-bound observer. It can be very confusing for students to identify which frame is the "rest" frame and which frame is the "moving" frame. This is especially difficult since the concept of motion is all relative!

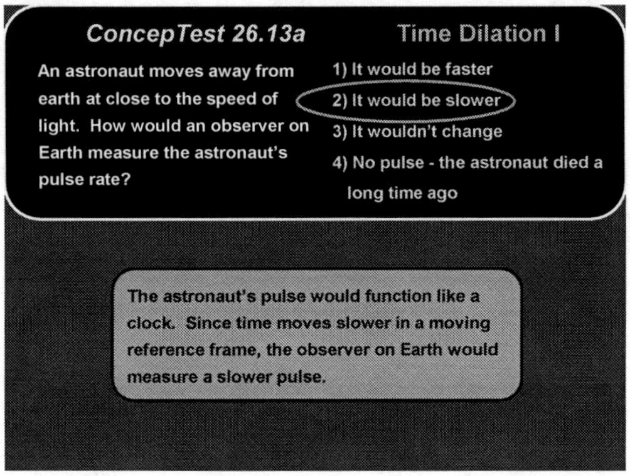

ConcepTest 26.13b: Time Dilation II
Difficulty Level: 3
Suggested follow-up conceptual question:

- **What would the astronaut in the spaceship measure?**

This follow-up question on time dilation is very similar to the previous one. Here again, since the observer is *not* in the rest frame of the pendulum, that observer will measure the period to be slower than the proper period (measured when an observer is at rest with respect to the pendulum).

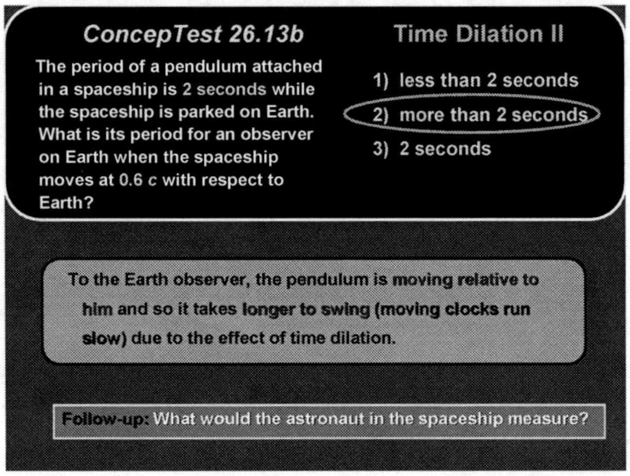

ConcepTest 26.13c: Time Dilation III
Difficulty Level: 2

This question is intended to dispel any notion that the time dilation effect is related to the direction of motion. By the logic posed in this question, if moving away from a clock means that time runs slower, then moving toward the clock will make time speed up (which is untrue!).

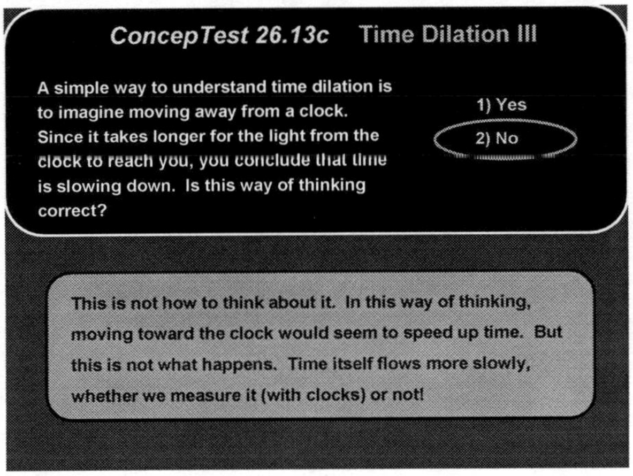

Conceptual Module 26.14

ConcepTest 26.14: Length Contraction
Difficulty Level: 3
Suggested follow-up conceptual question:

- **What would the astronaut measure about his spaceship?**

In length contraction, an observer *not* in the rest frame of the object will see it to be shorter than its proper length. In this case, the Earth-bound observer is clearly not in the rest frame of the moving spaceship, so the length will be contracted. Since the length contraction effect is related to the speed of the moving object, the contraction will become more pronounced as the speed increases.

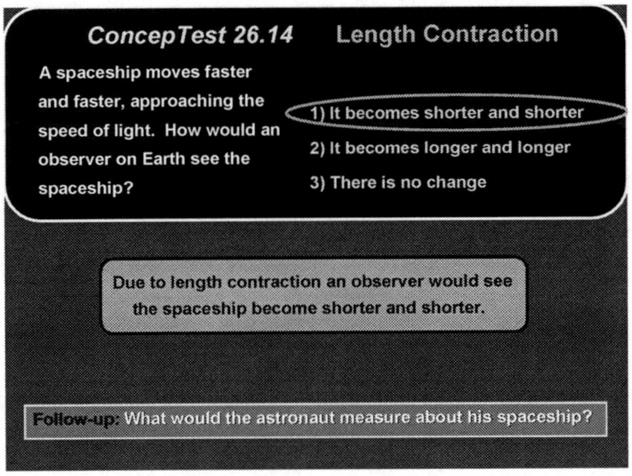

Conceptual Module 26.15

ConcepTest 26.15: Pancake or Cigar?

Difficulty Level: 3

Students must realize that the length contraction affects the dimension *along the direction of motion*, and not the other dimensions that are perpendicular to the motion. Thus, the spherical spaceship in this case will appear contracted in the horizontal direction of motion, but unaffected vertically – this will lead to a pancake-like shape.

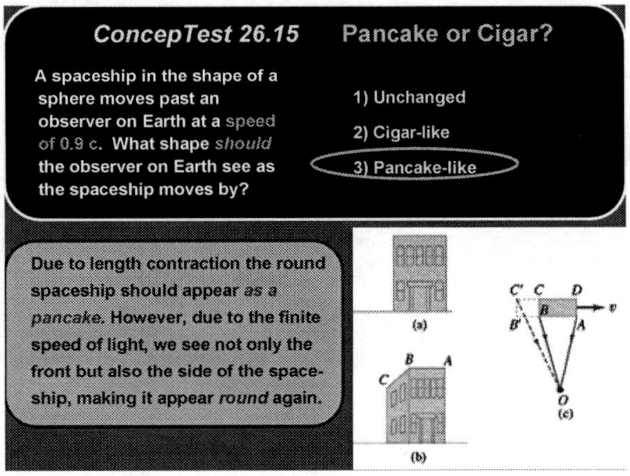

Conceptual Module 26.16 (two parts)

ConcepTest 26.16a: The Tunnel I

Difficulty Level: 4

This is a classic paradox that the students are asked to resolve. In this first part, the spacecraft is readily seen to be length contracted according to the lady (ground observer), so it will fit inside the tunnel when both doors are closed simultaneously (in the lady's frame!) Generally, there is little difficulty with this part of the two-part series of questions.

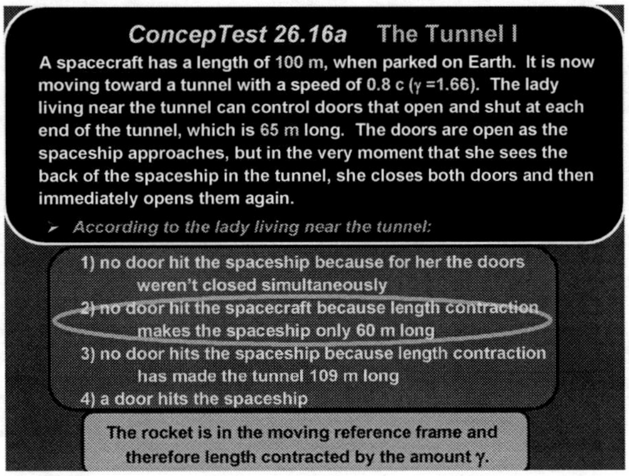

ConcepTest 26.16b: The Tunnel II
Difficulty Level: 5

This second part often causes conceptual difficulties. According to the spacecraft observers, the tunnel is now length contracted, so the situation is even worse for them! However, it must be agreed that if the spacecraft is not destroyed according to the lady (previous question), then that result must also hold as viewed in the moving frame. The resolution to the paradox lies in the fact that the people in the spacecraft *do not see the doors closed simultaneously*, and this is how they explain their safe passage.

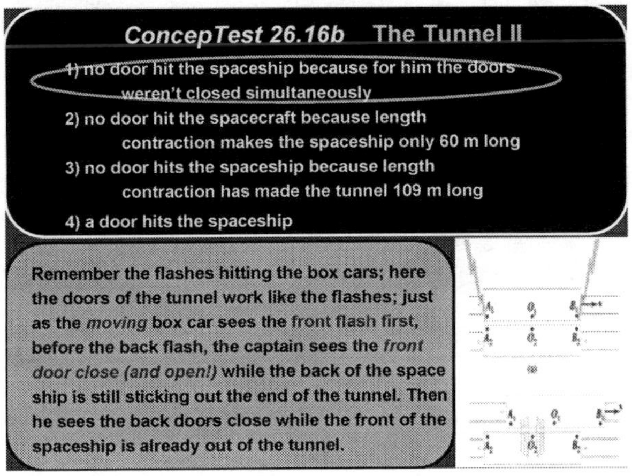

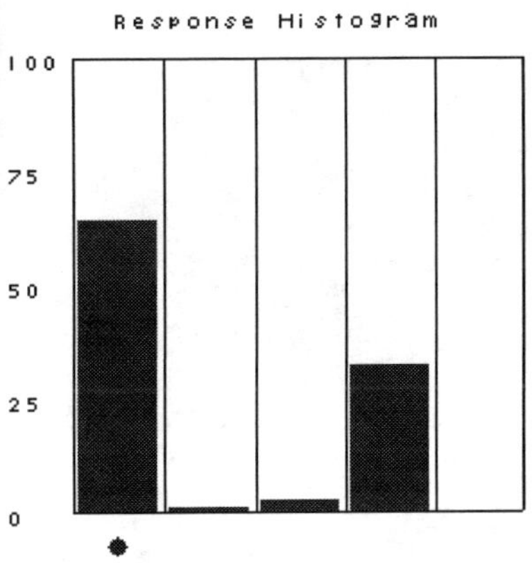

Conceptual Module 26.17

ConcepTest 26.17: Relativistic Mass
Difficulty Level: 3

This is a straightforward two-step question. The students must use the length information to infer the relativistic factor (sometimes called γ), which must then be applied to obtain the relativistic mass. In this case, the factor is 2.0, which leads to a doubling of the mass. Note that some choices can be eliminated immediately, since the mass can never decrease.

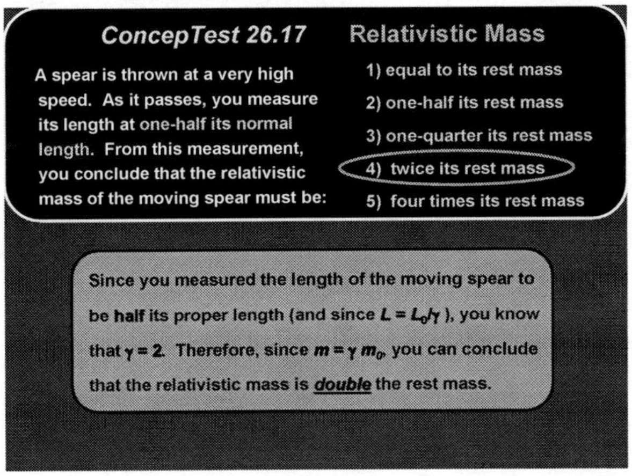

Conceptual Module 26.18

ConcepTest 26.18: Muon Decay
Difficulty Level: 4

It is important for the students to realize that the outcome of two events must be agreed upon by observers in different frames. In this case, the muons reach Earth in both cases. From the frame of reference of an Earth-bound observer, we say that time dilation made the muon lifetime exceed the proper lifetime, which enabled the muons to travel farther and therefore reach Earth's surface. However, from the muon's frame, the story is different (even though the result is the same!). The muons say that the distance from the upper atmosphere to the surface was length contracted, and that is how they could reach the surface during their proper lifetime.

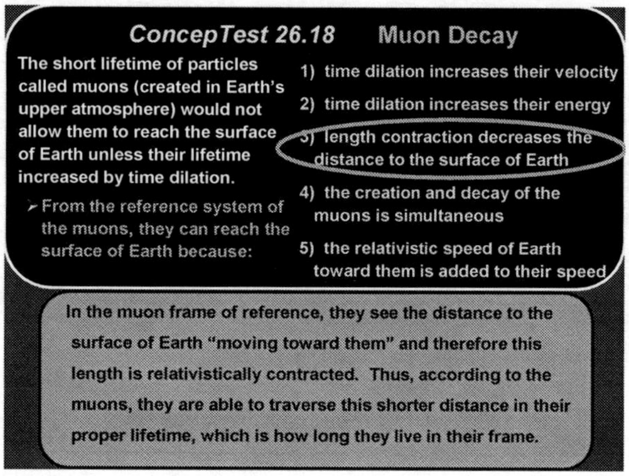

Conceptual Module 27.1
ConcepTest 27.1: Photons
Difficulty Level: 2

It is useful for the students to make a close association between the Planck relation and the visible spectrum. Generally, the visible spectrum is characterized by wavelength and color, whereas the students learn the Planck relation as $E = hf$. This question allows the students to connect color to wavelength, and then wavelength to energy.

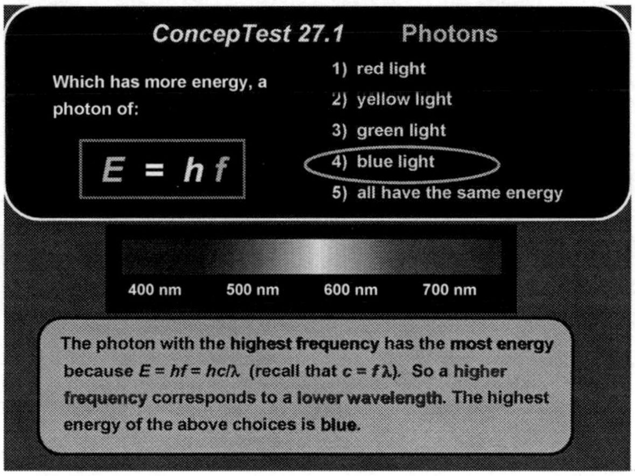

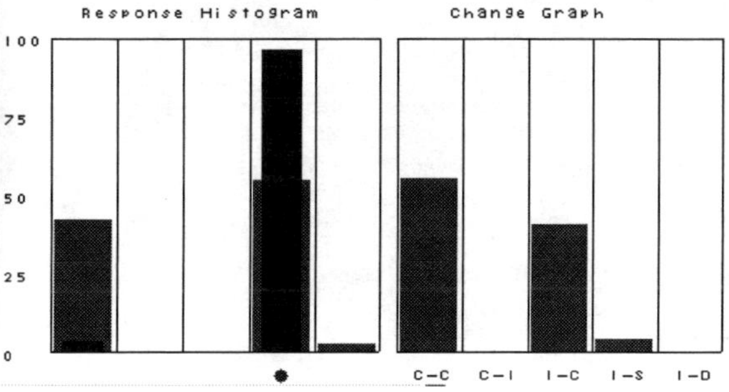

Conceptual Module 27.2 (five parts)
ConcepTest 27.2a: Photoelectric Effect I
Difficulty Level: 3
Suggested follow-up conceptual question:

- **What do you expect to happen to the work function of a metal if the metal was heated up?**

The work function is what gives rise to a particular cutoff in energy for the photoelectric effect. This characteristic energy can be represented by a frequency, which is a direct result of the Planck relation ($E = hf$). This question emphasizes (again) this connection between energy and frequency.

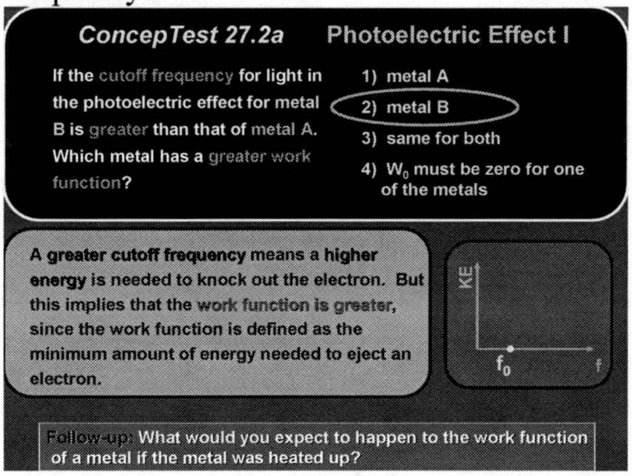

ConcepTest 27.2b: Photoelectric Effect II
Difficulty Level: 4

This question comes back to the relation between color, wavelength and energy. In this case, red light has a longer wavelength, which means it has lower energy. In fact, as an additional twist, the energy is actually *below* the cutoff energy, so no electrons can be emitted. Another way to look at this is that the wavelength of red light is *longer than the cutoff wavelength*, which again implies a lower energy.

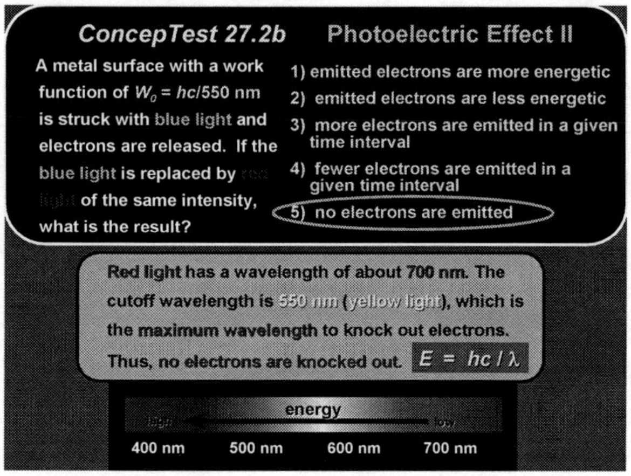

ConcepTest 27.2c: Photoelectric Effect III
Difficulty Level: 2

In this case, the wavelength is decreased, which means that the energy of the incident light has increased. The issue here is what is the effect of the increased energy. This does not affect the number of electrons emitted, but does increase the individual energies of the electrons emitted.

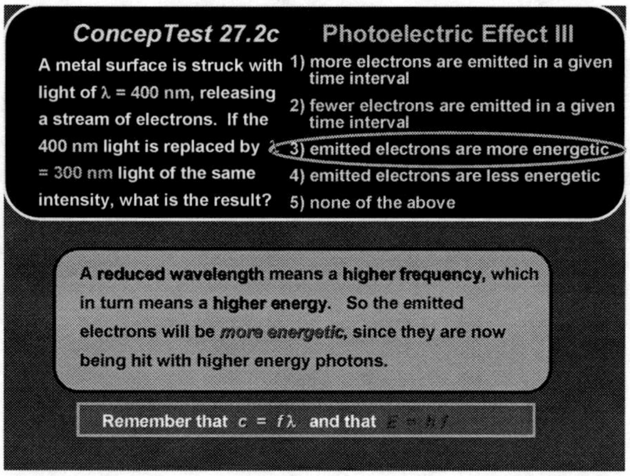

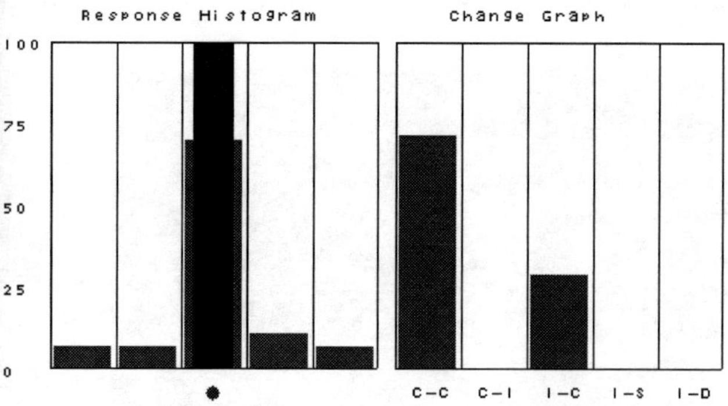

ConcepTest 27.2d: Photoelectric Effect IV
Difficulty Level: 2

In this follow-up question, the wavelength is held fixed, but the intensity of the light has increased. This means that more photons are impinging on the metal surface, which results in the release of more electrons. However, the electron energies do not change, since the energy of the photons has not changed (there are just more photons). This emphasizes the difference between the energy and the intensity of the incident light.

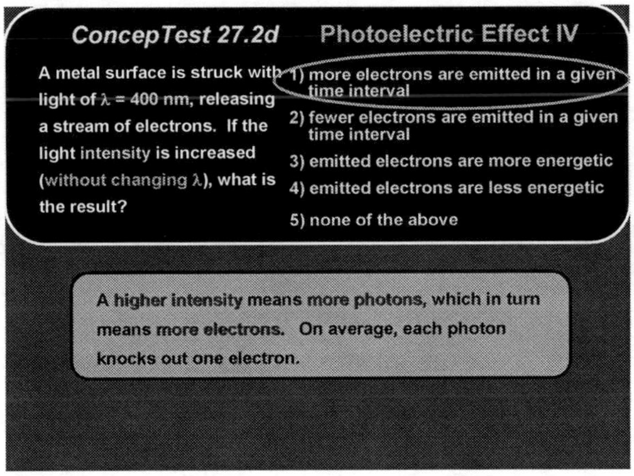

ConcepTest 27.2e: Photoelectric Effect V
Difficulty Level: 2

The measured current is related to the amount of charge flowing, which relates directly to the number of electrons (and not their energies). The students have just seen that this quantity is proportional to the intensity of the light (and not its energy, which is related to frequency or wavelength).

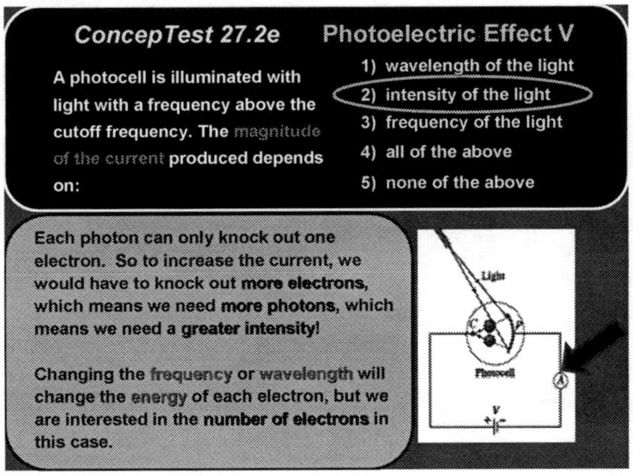

Conceptual Module 27.3 (four parts)
ConcepTest 27.3a: Wave-Particle Duality I
Difficulty Level: 2

The wave-particle duality of nature specifies that both photons and matter can have wave-like and particle-like properties. The wavelength of a matter wave is inversely proportional to its momentum, so the proton with the lower momentum (in this case, the lower speed, because both have the same mass) will have the longer wavelength.

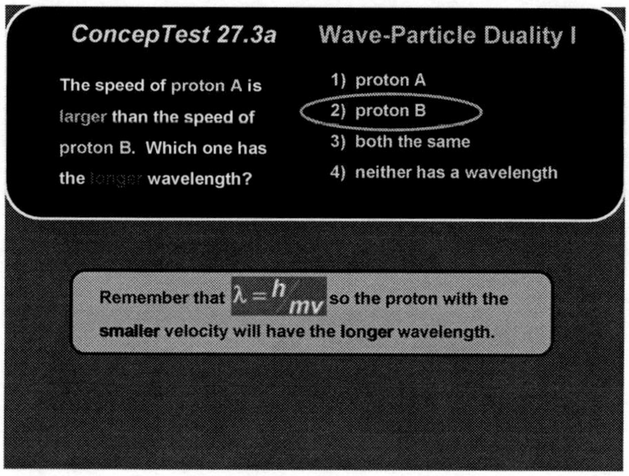

ConcepTest 27.3b: Wave-Particle Duality II
Difficulty Level: 2

The wavelength of a matter wave is inversely proportional to its momentum, so the particle with the lower momentum (in this case, the smaller mass, because both have the same speed) will have the longer wavelength. Here, it is the electron with the longer wavelength (rather than the proton).

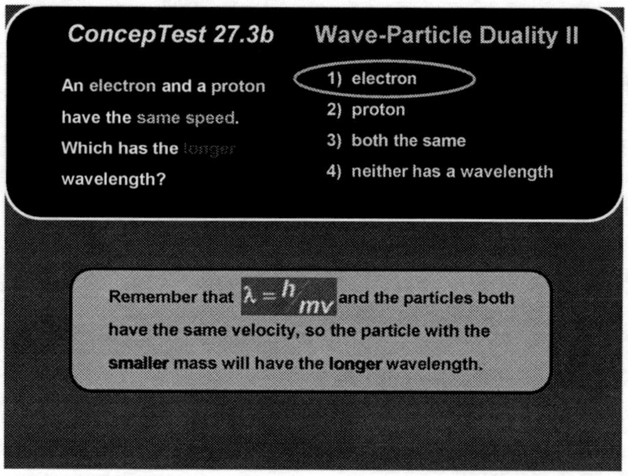

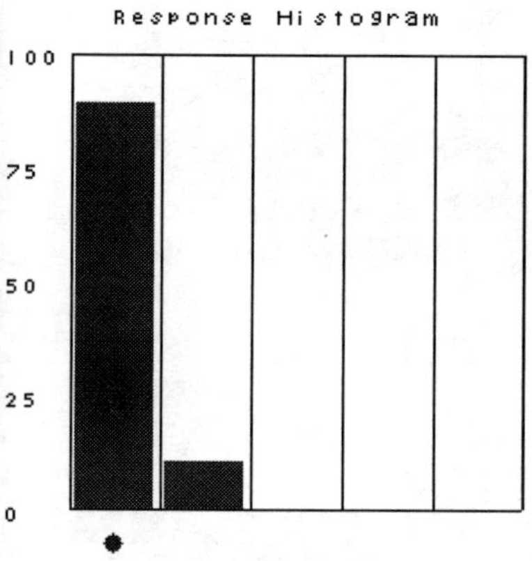

ConcepTest 27.3c: Wave-Particle Duality III
Difficulty Level: 4

This question sets up the same type of problem, but where the proton and electron have the same kinetic energy. This must be related to the momentum of the particle, in order to draw a conclusion about the wavelength. Rewriting the kinetic energy as KE = $p^2/2m$ indicates that the lighter particle (electron) will also have the smaller momentum (and hence, the longer wavelength).

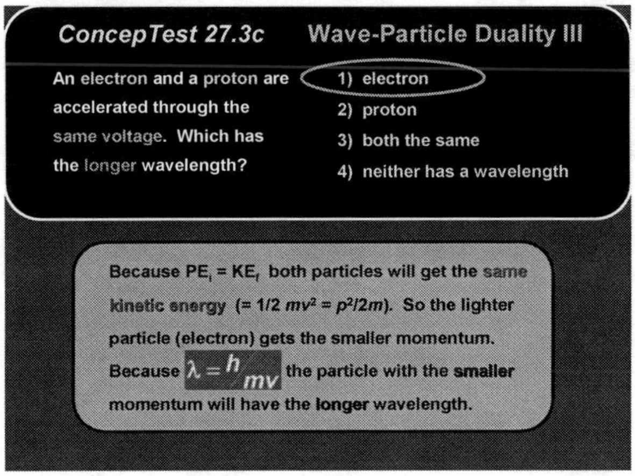

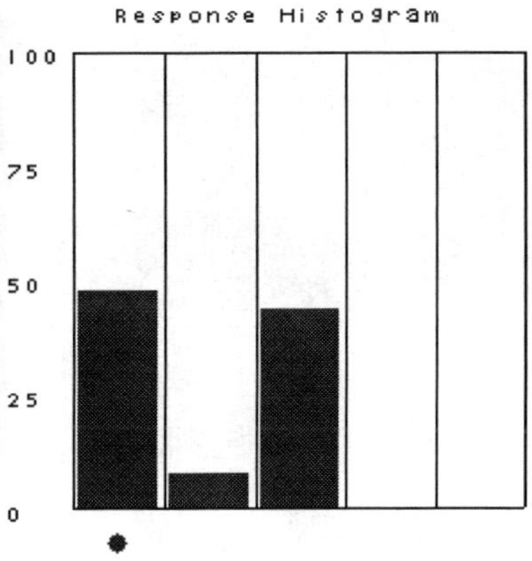

ConcepTest 27.3d: Wave-Particle Duality IV
Difficulty Level: 1

This question directly addresses the relation between the wavelength of a matter wave and the momentum of the particle.

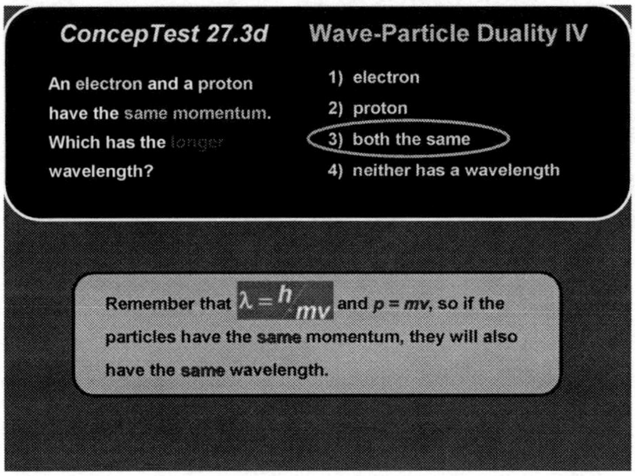

Conceptual Module 27.4

ConcepTest 27.4: Ionization

Difficulty Level: 1

Suggested follow-up conceptual question: How much energy does it take to change a He^+ ion into a He^{++} ion? Keep in mind that $Z = 2$ for helium.

This is a quick check to make sure that the students know the ground-state energy of the hydrogen atom and also that they understand what is meant by ionization. Discrete transitions to higher bound levels is something they will also deal with, but they need to realize that transitions to the continuum (unbound) states are possible as well (i.e., ionization).

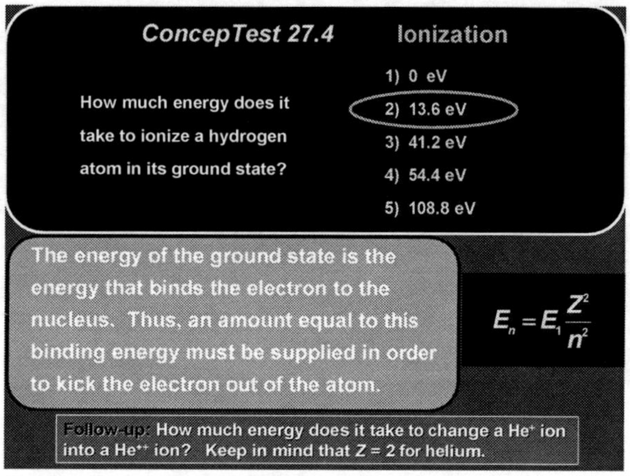

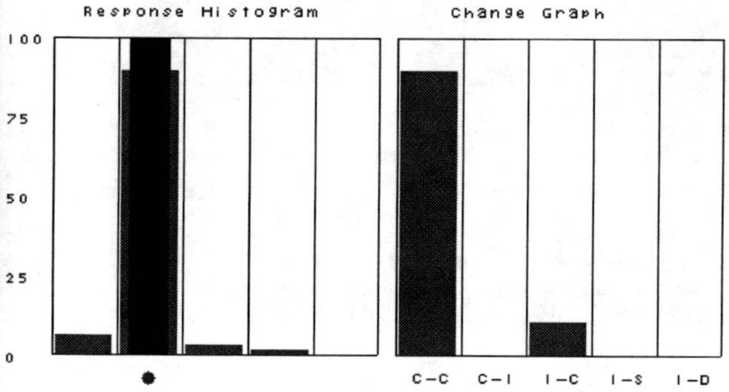

Conceptual Module 27.5 (two parts)

ConcepTest 27.5a: Atomic Transitions I

Difficulty Level: 3

Suggested follow-up conceptual question:

- **Which transition will emit the shortest wavelength photon?**

Students must get accustomed to the energy level diagram, in terms of the spacing of the energy levels. They must also understand that electrons will gain energy when making upward transitions or lose energy when making downward transitions.

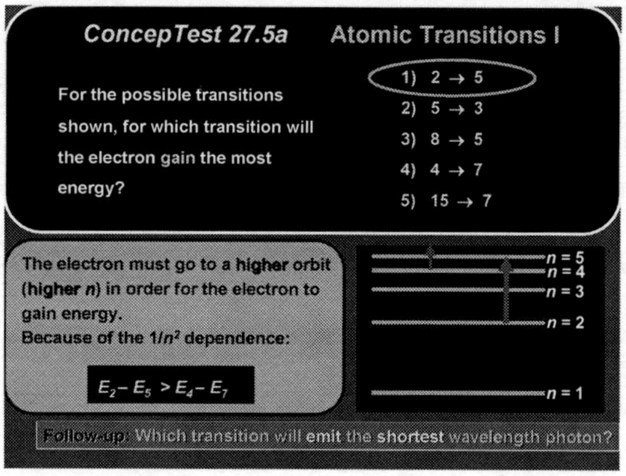

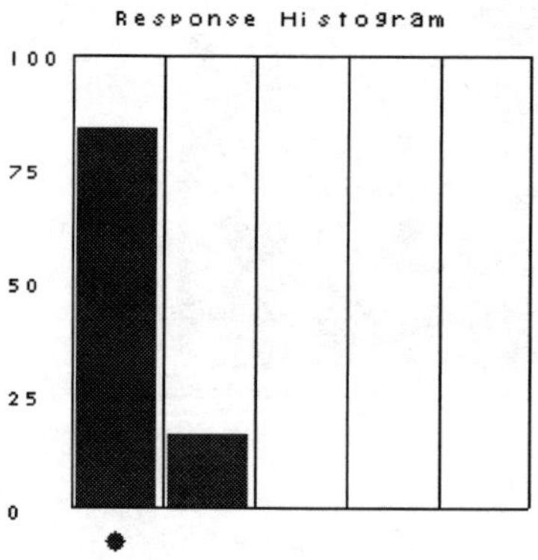

ConcepTest 27.5b: Atomic Transitions II
Difficulty Level: 3
Suggested follow-up conceptual question:

- **Which transition leads to the shortest wavelength photon?**

Students should understand that the energy of an emitted photon is the same as the energy difference between the two levels in a transition. In this case, the "reddest" photon corresponds to a lower-energy transition, which would pertain to a transition between the most closely spaced levels.

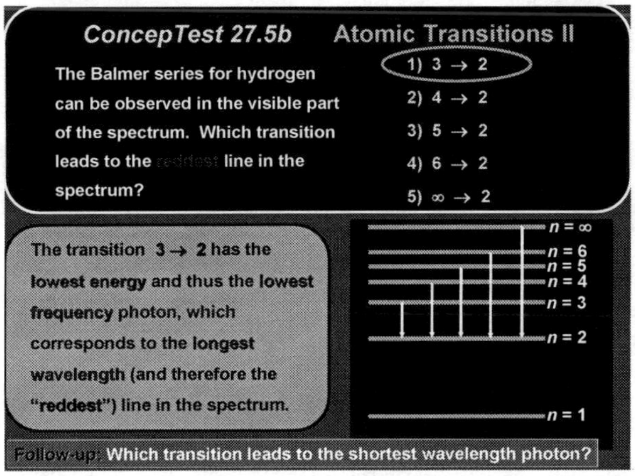

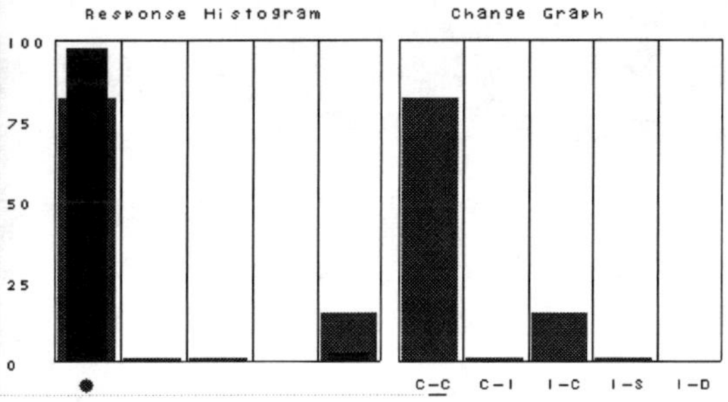

Conceptual Module 27.3

ConcepTest 27.6: Balmer Series
Difficulty Level: 2
Suggested follow-up conceptual question:

- **From the diagram at right, where in the EM spectrum is the Lyman series located?**

This is more of a question of common sense. The Balmer series is the only series that involves visible light, so none of the other transitions that are occurring can be observed in the visible range.

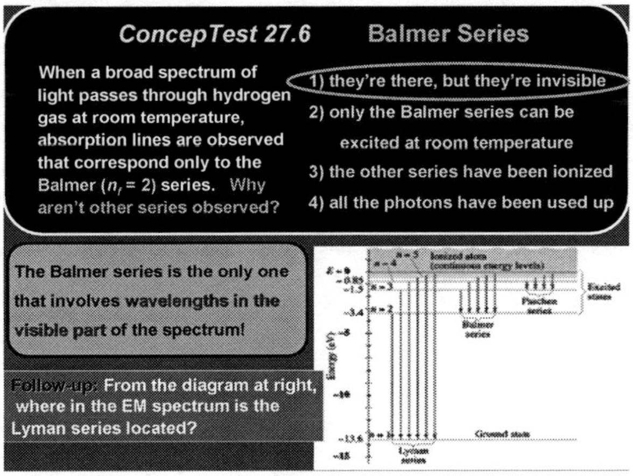

Conceptual Module 27.7 (two parts)
ConcepTest 27.7a: Energy Levels I
Difficulty Level: 2

This is an exercise in counting transitions. The students must systematically consider each initial state and add up the number of transitions possible, and then add up the total.

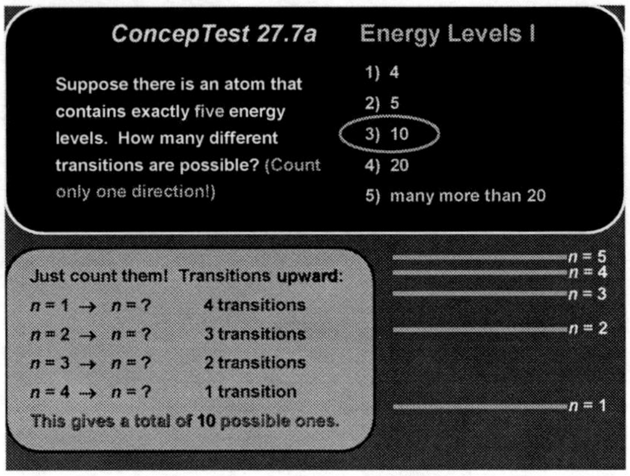

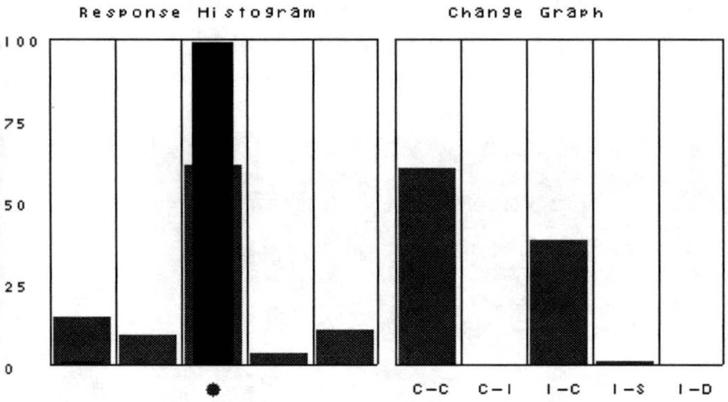

ConcepTest 27.7b: Energy Levels II
Difficulty Level: 4

This is a more challenging question, in which students must infer 4 high-energy transitions (from the blue and violet lines) and 2 lower-energy transitions (from the yellow and orange lines). Then they must search the available energy level diagrams to match this pattern (which is satisfied by #2).

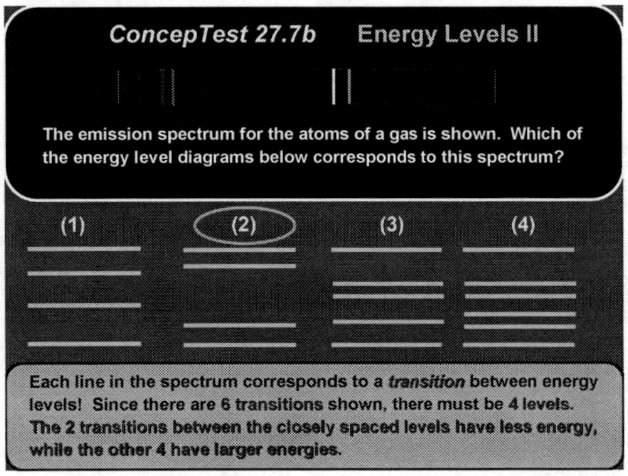

Conceptual Module 27.8
ConcepTest 27.8: Rutherford Model
Difficulty Level: 2

The Rutherford atom was characterized by a continuum of levels (not discrete levels, as in the Bohr atom). Since this would allow any energy transition to be possible, then all energies of light incident on the atom could potentially be absorbed and the spectrum would appear totally dark.

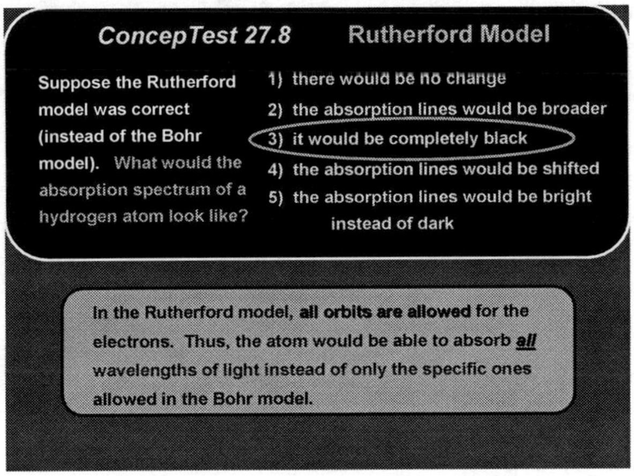

Conceptual Module 29.1
ConcepTest 29.1: Ionic Bonding
Difficulty Level: 2

The likely partners in an ionic bond are atoms that can either readily give up a valence electron (to leave behind a closed shell) or readily accept one in a vacant valence orbital (to fill a shell).

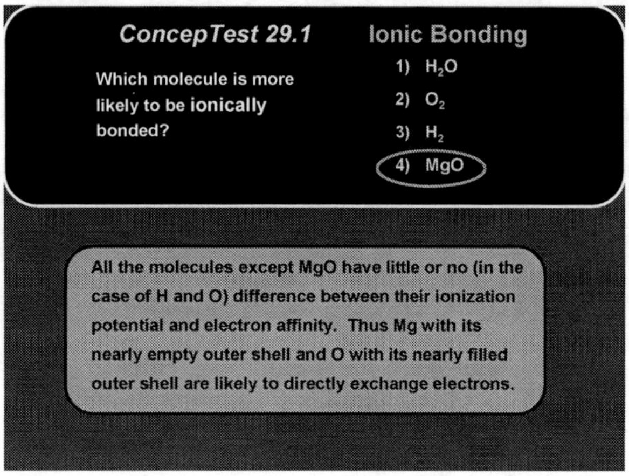

Conceptual Module 29.2
ConcepTest 29.2a: Electron Spin I
Difficulty Level: 1

This is a simple question about the Pauli Exclusion Principle. Students need to remember the definition of an antisymmetric wave function.

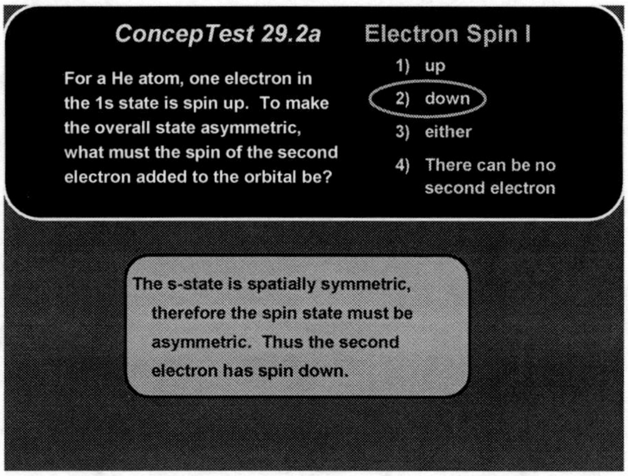

Conceptual Module 29.2
ConcepTest 29.2b: Electron Spin II
Difficulty Level: 2

Students need to recognize how the Pauli Principle works in the combination of several quantum numbers. The above answer in the previous ConcepTest does not apply since the electron now enters a *p*-orbital.

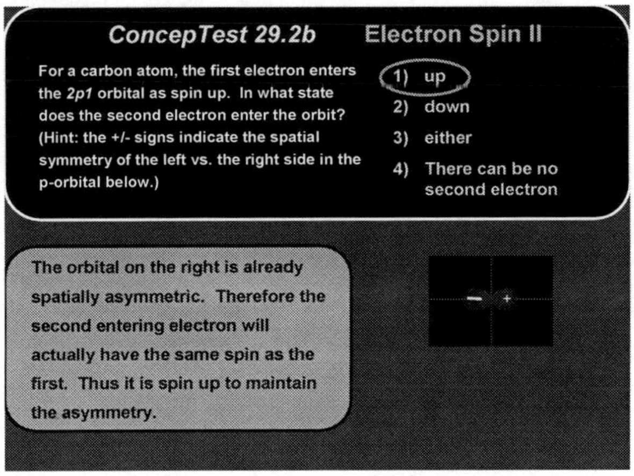

Conceptual Module 29.3
ConcepTest 29.3: Vibrational Levels
Difficulty Level: 2

Students need to remember the angular momentum selection rule for vibrational levels.

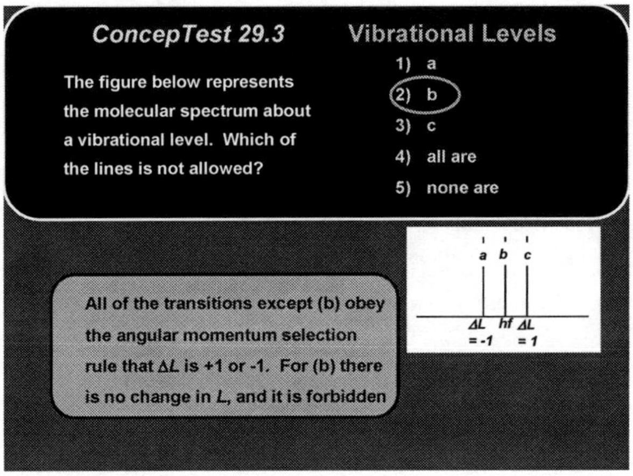

Conceptual Module 29.4
ConcepTest 29.4: More Bonding
Difficulty Level: 1

Students need to remember the definition of ionic bonding.

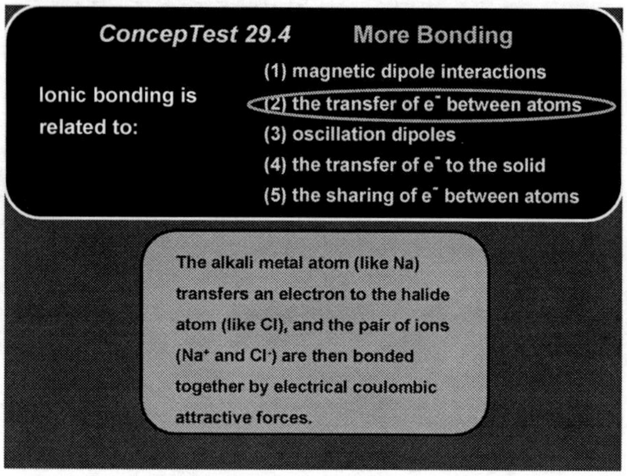

Conceptual Module 29.5
ConcepTest 29.5: Solid H and He
Difficulty Level: 2

This is a simple question where students learn how to translate their knowledge from atomic levels to band structure in solids.

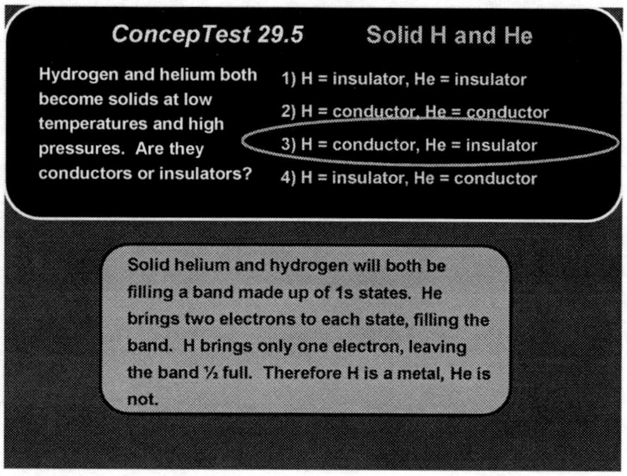

Conceptual Module 29.6
ConcepTest 29.6: Metallic Conduction
Difficulty Level: 2

This question tests students' understanding of an energy gap in the band structure of solids and its relationship to conduction.

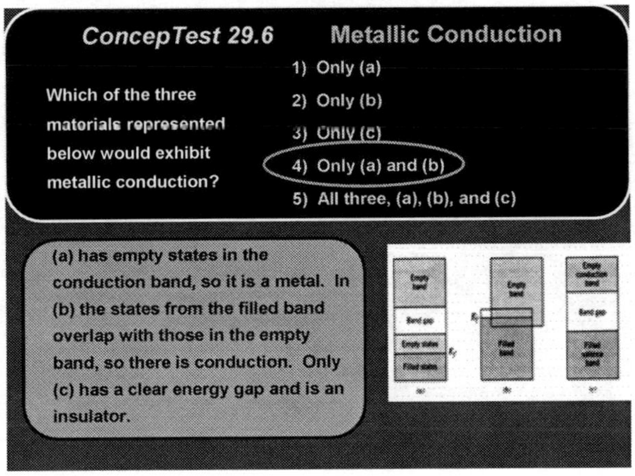

Conceptual Module 30.1
ConcepTest 30.1: The Nucleus
Difficulty Level: 1

Regarding the four fundamental forces of nature, the nuclear interaction is the strongest. It is much stronger than the Coulomb force that causes like charges to repel. This introductory question is intended to highlight this point.

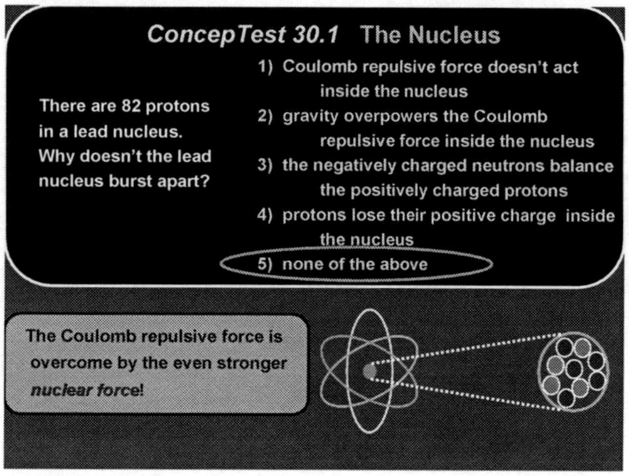

Conceptual Module 30.2 (three parts)
ConcepTest 30.2a: Binding Energy I
Difficulty Level: 2

The concept of binding energy is often confusing for students. The notion that there is more energy in a free proton and electron than in a bound hydrogen atom (due to the binding energy!) means that the former combination weighs more than the latter. The equivalency of mass and energy (by $E = mc^2$) allows us to talk about mass and energy interchangeably. This example of an atomic system is meant to serve as a gentle introduction to the concept of binding energy, since the students should be quite familiar with the hydrogen atom.

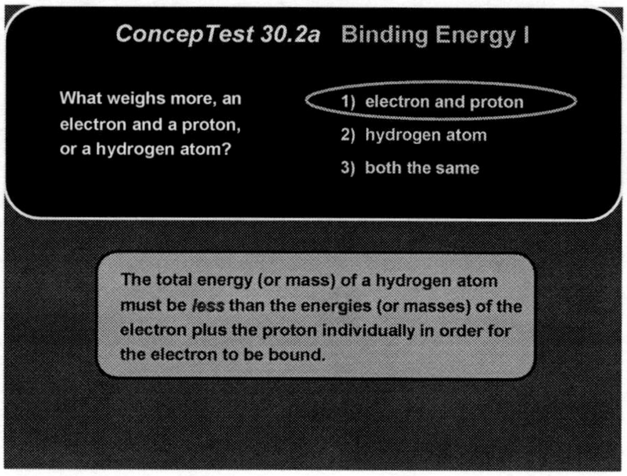

ConcepTest 30.2b: Binding Energy II
Difficulty Level: 2

This follow-up question emphasizes the mass difference alluded to in the previous question. Since the binding energy is negative for a bound system, the mass (or energy) of a hydrogen atom is actually less than that of a free proton and electron by 13.6 eV.

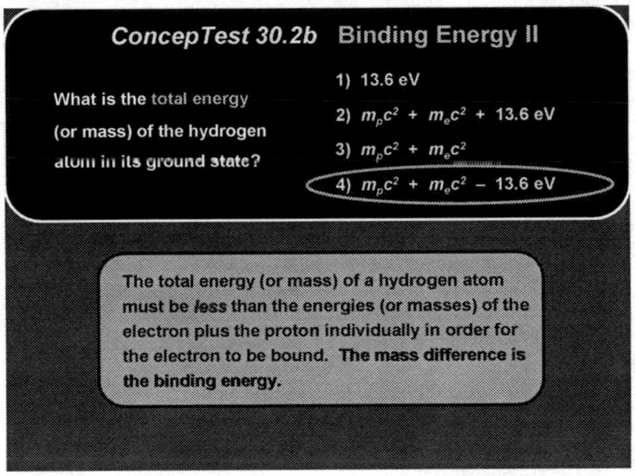

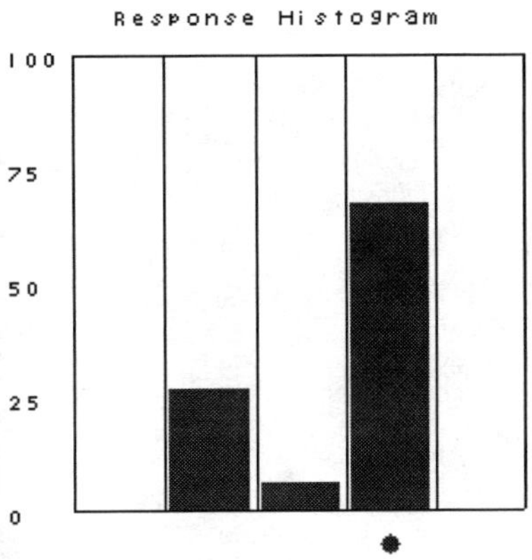

ConcepTest 30.2c: Binding Energy III
Difficulty Level: 2

Now we switch gears to a nuclear system, but the concept of binding energy remains the same. In this case, the bound tritium nucleus has a lower total energy (weighs less) than the separate constituents. The difference in energy is simply the binding energy, which in this case is 8.5 MeV.

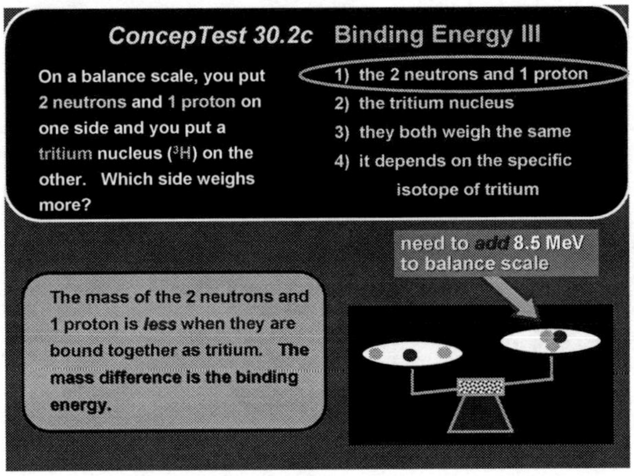

Conceptual Module 30.3

ConcepTest 30.3: Separation Energy
Difficulty Level: 2

The proton will tend to be repelled by the other protons in the nucleus, so in fact the removal of a proton is aided by the Coulomb repulsion. The uncharged neutron does not have this advantage, and so it will take more energy to remove a neutron (15.7 MeV) than to remove a proton (12.1 MeV).

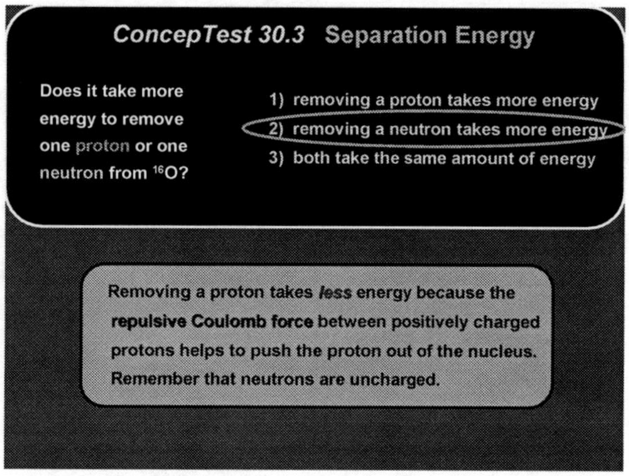

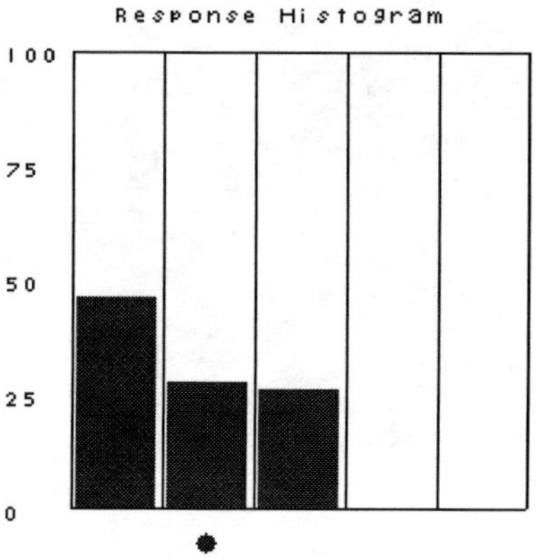

Conceptual Module 30.4 (two parts)

ConcepTest 30.4a: Particle Emission I
Difficulty Level: 2

Alpha particles are positively charged. The students will have to remember the right-hand rule in order to determine the direction of the magnetic force on a moving positively charged particle.

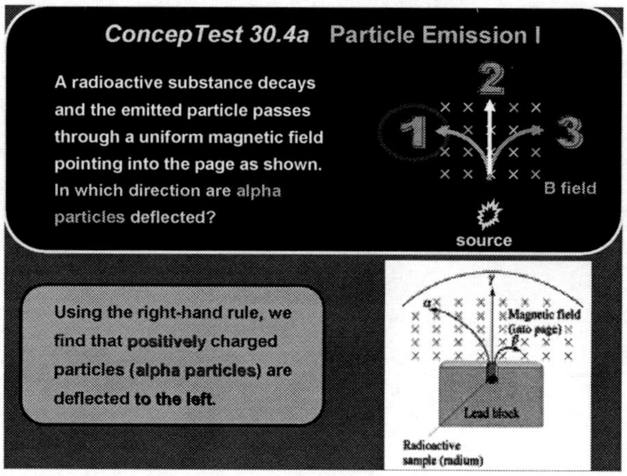

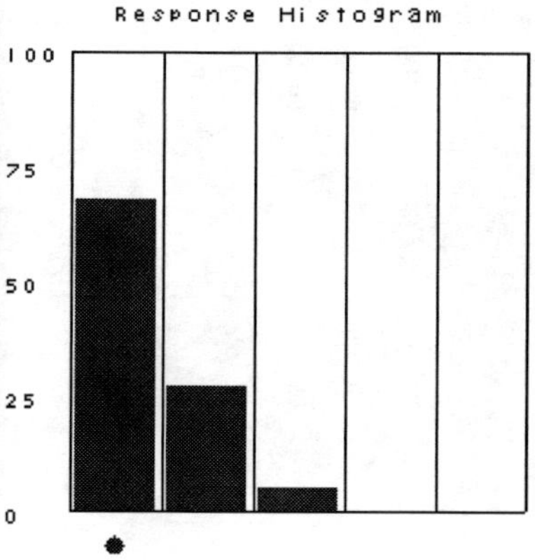

ConcepTest 30.4b: Particle Emission II
Difficulty Level: 2
Suggested follow-up conceptual question:

- **What particles are bent to the right?**

Gamma rays are uncharged – they are essentially photons – and uncharged particles are not deflected in a magnetic field.

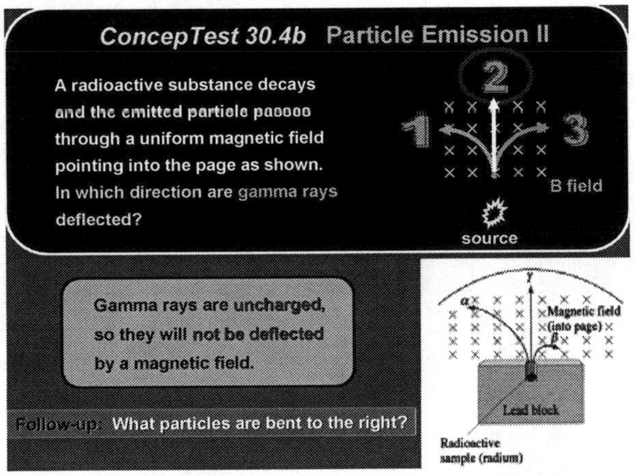

Conceptual Module 30.5

ConcepTest 30.5: Radioactive Decay Energy
Difficulty Level: 2
Suggested follow-up conceptual question:

- **What process could release a photon with billions of eV?**

The students have already seen that energies of a few eV pertain to atomic systems. Binding energy of nuclear systems has already been discussed, and so the students should be familiar with the fact that in nuclear systems, we are talking about energies of several MeV (millions of eV). The energy scale is determined largely by the strength of the interaction (nuclear force vs. Coulomb force).

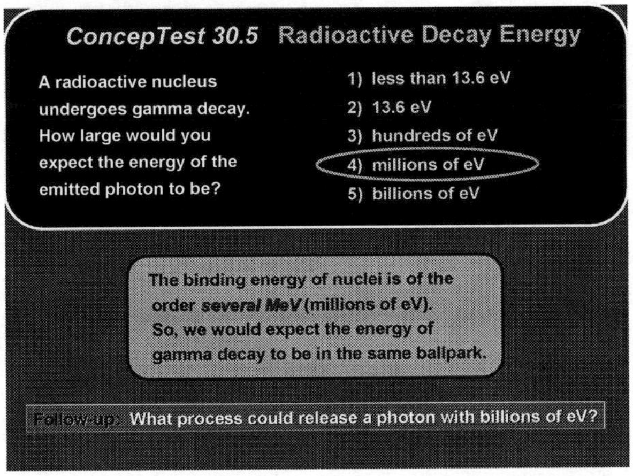

Conceptual Module 30.6 (three parts)

ConcepTest 30.6a: Alpha Decay I
Difficulty Level: 2
Suggested follow-up conceptual question:

- **In what directions are the two products emitted?**

This question relates back to the conservation of momentum. Since the initial system has zero momentum, then momentum conservation dictates that the final system will have zero momentum, so that the momenta of the reaction products must be equal and opposite.

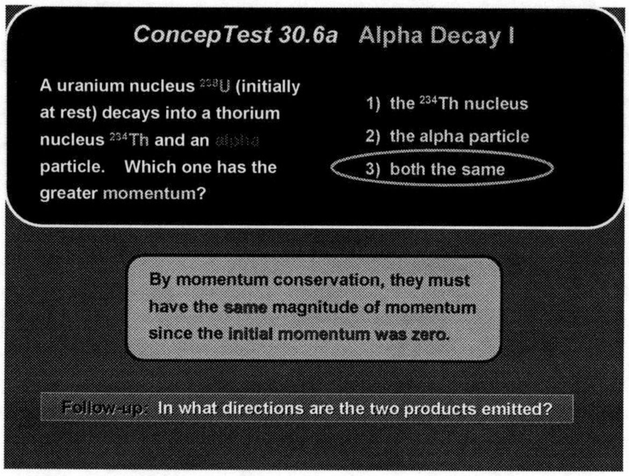

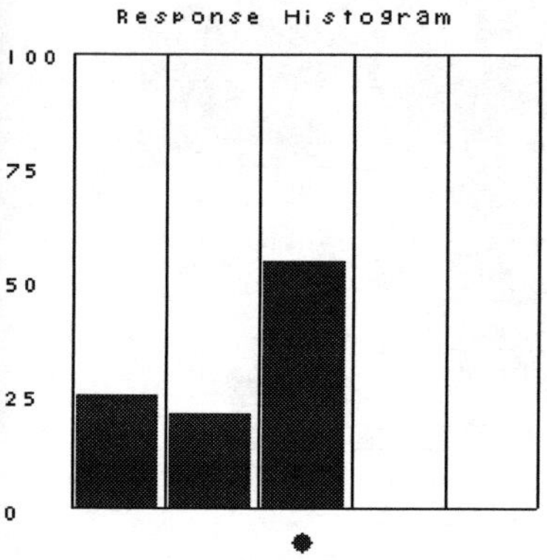

ConcepTest 30.6b: Alpha Decay II
Difficulty Level: 3

Since the momenta have the same magnitude, then the particle with the smaller mass must have the greater velocity. In this case, the alpha particle is moving faster than the ^{234}Th nucleus.

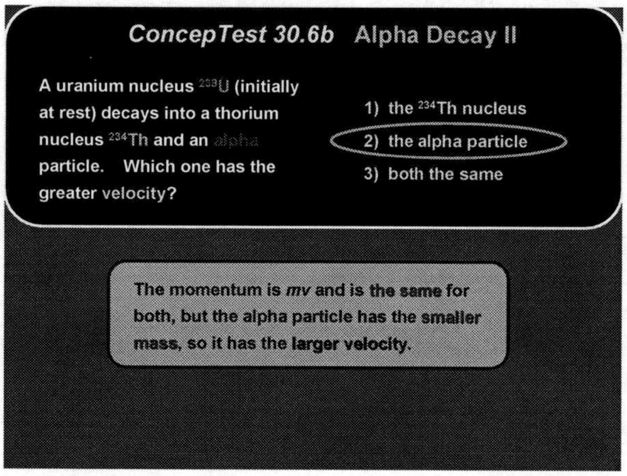

ConcepTest 30.6c: Alpha Decay III
Difficulty Level: 4

For this question, the students must recall that kinetic energy can be rewritten as $KE = p^2/2m$. With the momenta of the two particles being equal, then the lighter mass particle (the alpha particle) will have the greater kinetic energy.

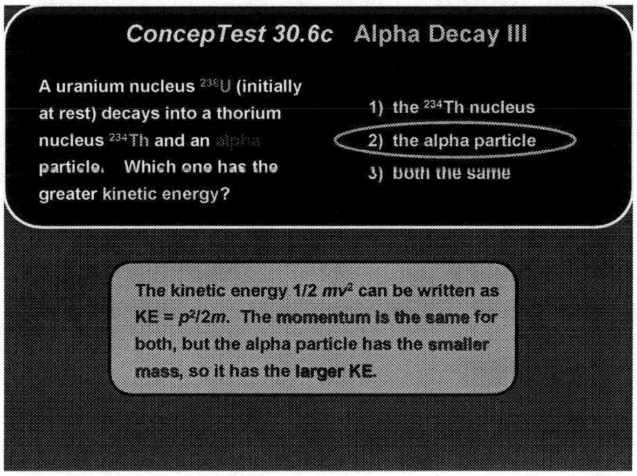

Conceptual Module 30.7

ConcepTest 30.7: Beta Decay
Difficulty Level: 3
Suggested follow-up conceptual question:

- **How would you turn ^{14}C into ^{15}N?**

In beta decay, a neutron turns into a proton and emits a beta particle (and a neutrino). In this type of reaction, the number of nucleons *(A)* does not change, but the number of protons *(Z)* does change. Thus, a new element is formed *(Z+1)* with the same atomic mass *(A)* as the original nucleus.

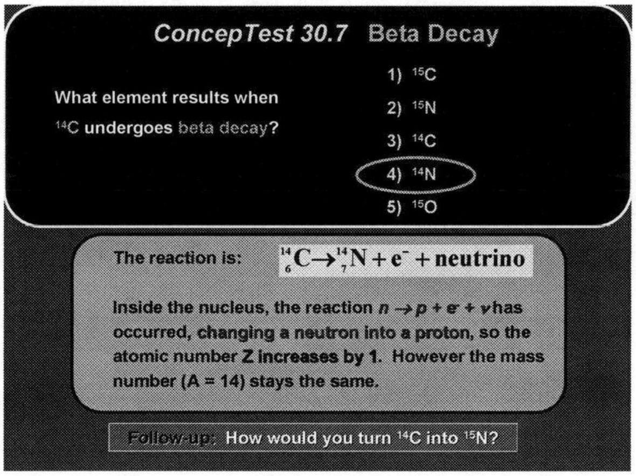

Conceptual Module 30.8 (four parts)

ConcepTest 30.8a: Radioactive Decay Law I
Difficulty Level: 3
Suggested follow-up conceptual question:

- **When will the sample be reduced to nothing?**

In this question, the students must simply count the number of half-lives that have passed, with a reduction by a factor of 2 in mass for each passing half-life. In this case, three half-lives have gone by, leading to a reduction factor of $2^3 = 8$ in mass.

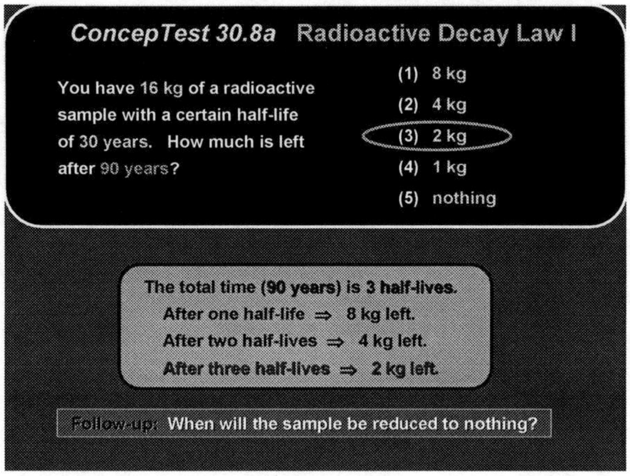

ConcepTest 30.8b: Radioactive Decay Law II
Difficulty Level: 3
Suggested follow-up conceptual question:

- **How much of the sample is left after another 10 years?**

In this case, the initial and final masses are given, and the students are asked to infer the half-life from the total time (10 years) that has passed. In this case, the mass is reduced by a factor of $2^2 = 4$, which means that two half-lives have gone by. Thus, each half-life must be 5 years.

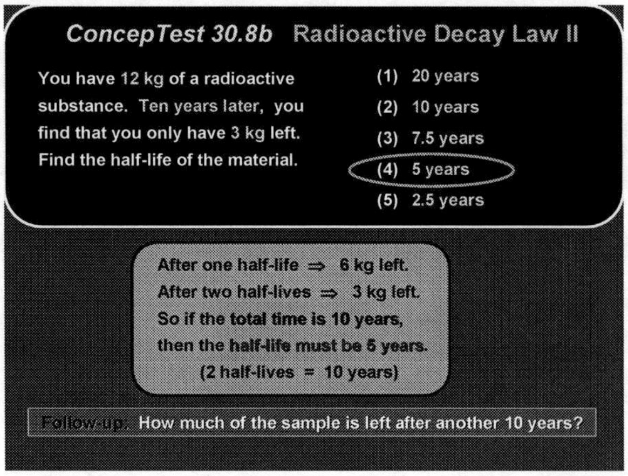

ConcepTest 30.8c: Radioactive Decay Law III
Difficulty Level: 4

Radioactive decay is not a linear process – it follows an exponential decay law. The time involved in this problem is not an integral number of half-lives. The students will probably realize that 40 years (two half-lives) would leave 100 g and that 60 years (three half-lives) would leave 50 g. But the fact that 50 years is in between these choices *does not mean* that 75 g will be left over! In fact, there will be less than 75 g, due to the rate of decay governed by the exponential law.

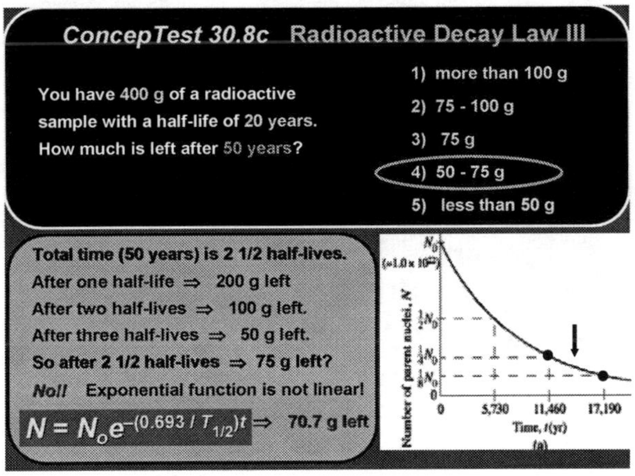

ConcepTest 30.8d: Radioactive Decay Law IV
Difficulty Level: 4
Suggested follow-up conceptual question

- **When will the samples again have equal amounts?**

This is a multi-step problem in which the students must relate what they know about one sample and the decay time given to the other sample. First, the final amount of the known sample must be found, and then, working backwards from there, the initial amount of the unknown can be determined.

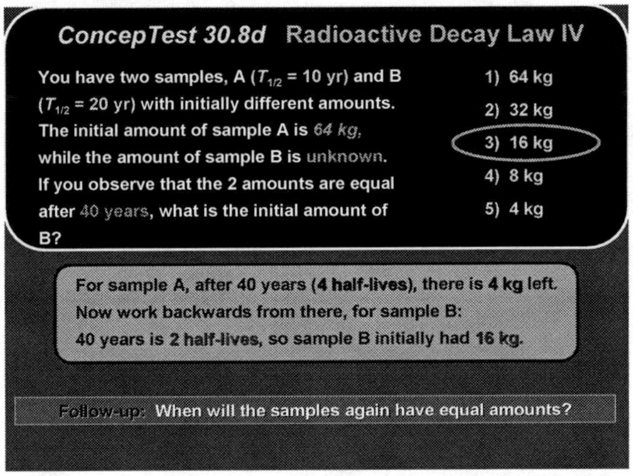

Conceptual Module 30.9 (two parts)

ConcepTest 30.9a: Activity and Half-Life I
Difficulty Level: 1
Suggested follow-up conceptual question

- **What is the ratio of activities for the two samples?**

The activity is inversely related to the half-life (for samples of equal size). That is, a sample with a high activity will decay quickly and therefore have a shorter half-life.

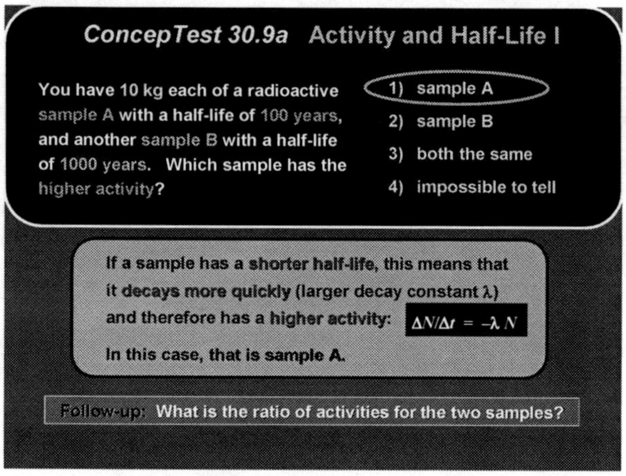

ConcepTest 30.9b: Activity and Half-Life II
Difficulty Level: 2

This question is quite similar to the previous one, except for the quantitative feature added. In this case, the factor of 5 in activity (for equal size samples) must be appropriately applied to the half-life.

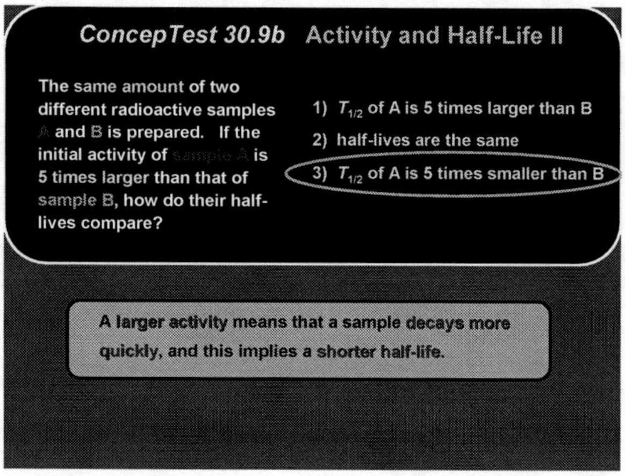

Conceptual Module 31.1

ConcepTest 31.1: Nuclear Reactions
Difficulty Level: 2
Suggested follow-up conceptual question:

- **Is radioactive decay an endothermic or exothermic reaction?**

Some reactions give off energy when they occur (exothermic) and therefore can occur spontaneously, whereas other reactions require some input energy in order to occur (endothermic). Radioactive decay occurs spontaneously with a corresponding release of energy, so the reaction is exothermic and therefore the *Q*-value must be positive.

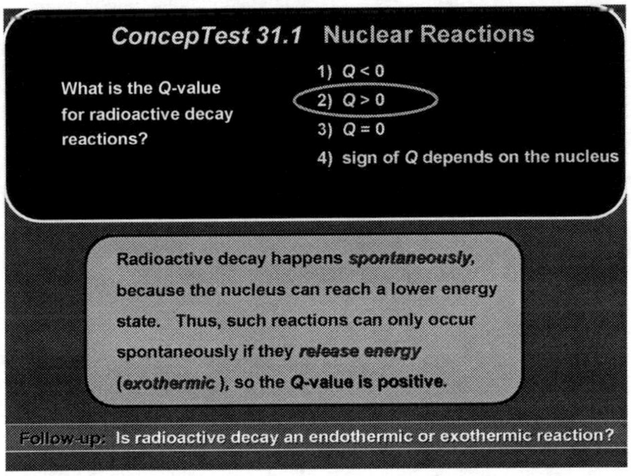

Conceptual Module 31.2

ConcepTest 31.2: Nuclear Reaction Products
Difficulty Level: 3
Suggested follow-up conceptual question:

- **What would you get if you started with p + ^{16}O instead?**

For this question, it is a matter of counting the neutrons and protons on both sides and then requiring a balance. In the nuclear reaction, constituents are exchanged, but the total number of nucleons (neutrons and protons) is conserved.

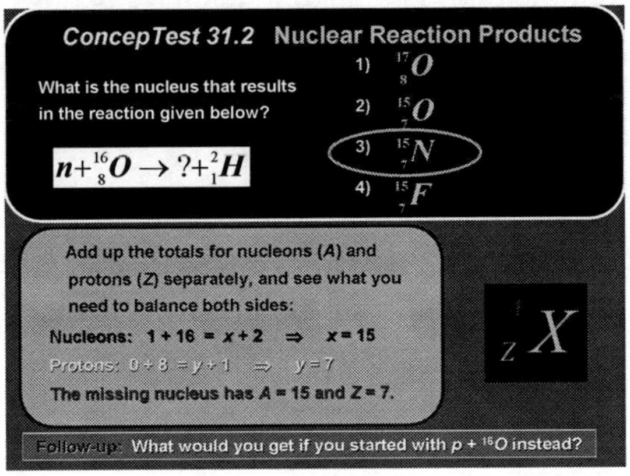

Conceptual Module 31.3

ConcepTest 31.3: Beta Decay Products

Difficulty Level: 3

Suggested follow-up conceptual question:

- **How would you turn ^{14}C into ^{15}N?**

In beta decay, a neutron turns into a proton and emits a beta particle (and a neutrino). In this type of reaction, the number of nucleons *(A)* does not change, but the number of protons *(Z)* does change. Thus, a new element is formed *(Z+1)* with the same atomic mass *(A)* as the original nucleus.

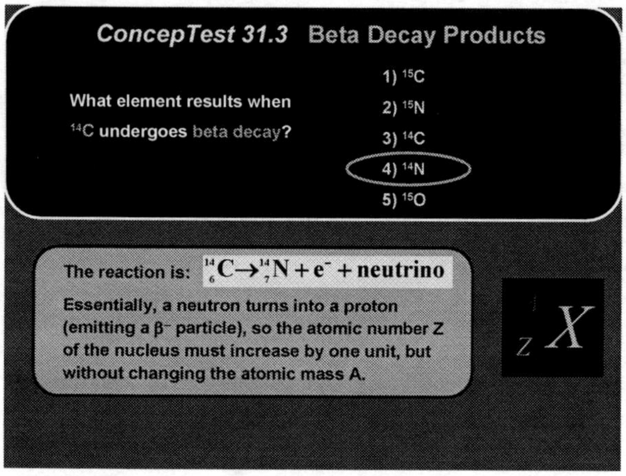

Conceptual Module 31.4

ConcepTest 31.4: Nuclear Fission

Difficulty Level: 2

Suggested follow-up conceptual question:

- **Where are the fission fragments located relative to the original nucleus on the curve of binding energy per nucleon?**

This is essentially a binding-energy question. The nucleus undergoes fission because it can reach a lower energy (mass) state by splitting apart. Thus, the fragments must have less mass than the original nucleus when it was intact. This is the opposite of what was seen for light nuclei (such as tritium), but it is all related to the position of the components on the curve of binding energy per nucleon.

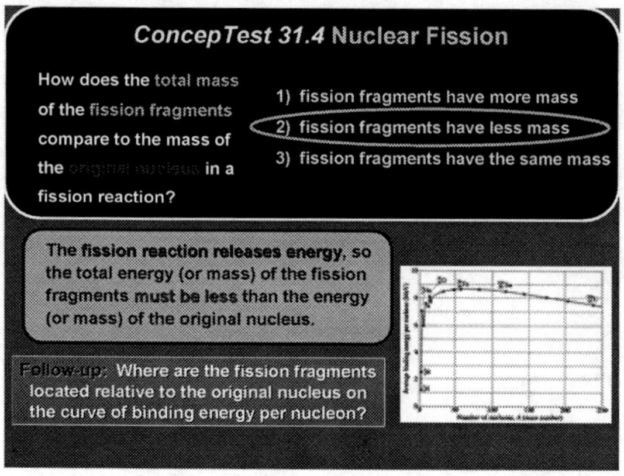

Conceptual Module 31.5

ConcepTest 31.5: Nuclear Fusion

Difficulty Level: 2

Suggested follow-up conceptual question:

- **Which weighs more: the fusion products or the pieces?**

This question gets to the heart of why light nuclei prefer to undergo fusion and heavy nuclei prefer to undergo fission. Increasing the binding energy per nucleon makes the result more stable.

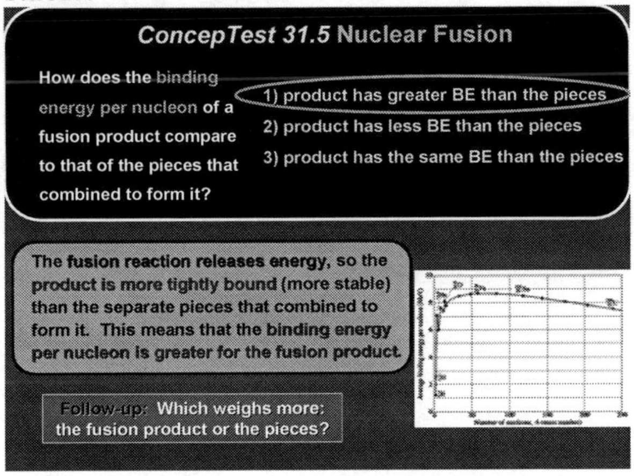

Conceptual Module 31.6

ConcepTest 31.6: Radiation Shielding

Difficulty Level: 2

Gamma rays are the least interacting of the three types of radioactive decay because they are not charged. Alpha particles are doubly charged, so they lose the most energy. Betas are lighter and singly charged, so they lose less energy, but gamma rays are the hardest to stop.

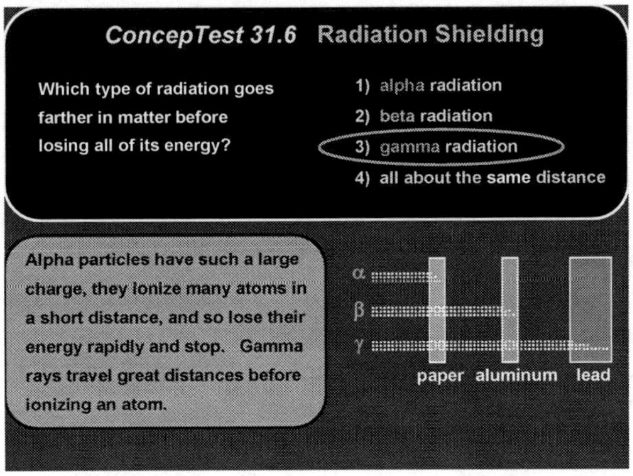

Conceptual Module 31.7 (two parts)

ConcepTest 31.7a: Radiation Exposure I
Difficulty Level: 3

The intensity of radiation follows an inverse square law, so that the intensity drops as the square of the distance. Thus, an increase in distance by a factor of 2 will result in a drop of intensity by a factor of 4.

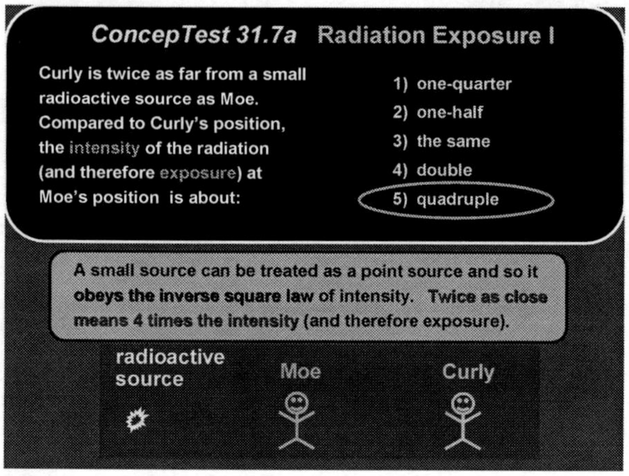

ConcepTest 31.7b: Radiation Exposure II
Difficulty Level: 3

In this question, the reduction factor is specified and the students must determine the safe distance. An estimate is required in this case because the required factor of 10 is not a perfect square. Thus, a factor of 3 in distance (15 m) is *not* enough and the next choice (20 m) must be selected.

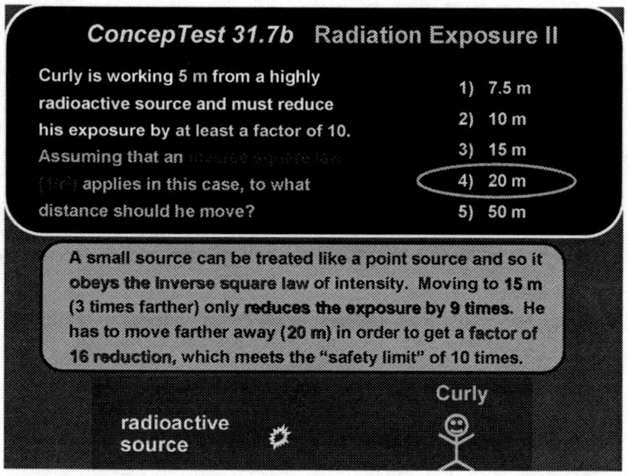

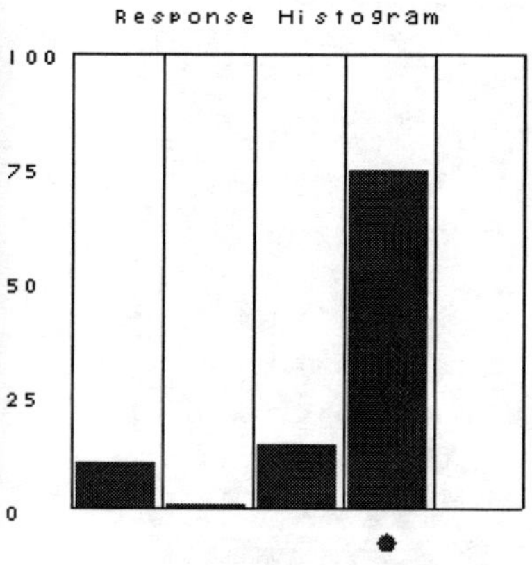

Conceptual Module 31.8

ConcepTest 31.8: Radiation Damage
Difficulty Level: 1
Suggested follow-up conceptual question:

- **What type of radiation will tend to do the most damage?**

This question addresses the issue of the mechanism of radiation damage to materials (biological or inorganic). The damage occurs due to ionization of the materials.

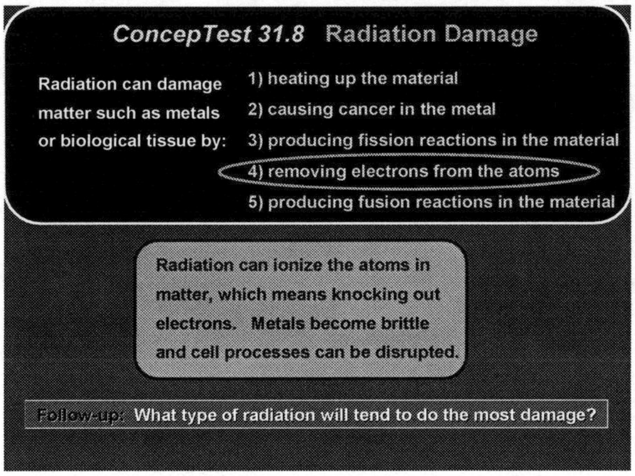